Lecture Notes in Computer Science　　7020

Commenced Publication in 1973
Founding and Former Series Editors:
Gerhard Goos, Juris Hartmanis, and Jan van Leeuwen

Werner Kuich   George Rahonis (Eds.)

Algebraic Foundations in Computer Science

Essays Dedicated to Symeon Bozapalidis
on the Occasion of His Retirement

 Springer

Volume Editors

Werner Kuich
Technische Universität Wien
Institut für Diskrete Mathematik und Geometrie
1040 Wien, Austria
E-mail: kuich@tuwien.ac.at

George Rahonis
Aristotle University of Thessaloniki
Department of Mathematics
54124 Thessaloniki, Greece
E-mail: grahonis@math.auth.gr

The illustration appearing on the cover of this book is the work of Daniel Rozenberg (DADARA).

ISSN 0302-9743   e-ISSN 1611-3349
ISBN 978-3-642-24896-2   ISBN 978-3-642-24897-9 (eBook)
DOI 10.1007/978-3-642-24897-9
Springer Heidelberg Dordrecht London New York

Library of Congress Control Number: 2011938838

CR Subject Classification (1998): F.4, D.3, F.3.2, I.1, F.4.1, D.2.4

LNCS Sublibrary: SL 1 – Theoretical Computer Science and General Issues

Typesetting: Camera-ready by author, data conversion by Scientific Publishing Services, Chennai, India

Printed on acid-free paper

Springer is part of Springer Science+Business Media (www.springer.com)

Symeon Bozapalidis

Preface

This Festschrift is published in honor of Symeon Bozapalidis (born in Thessaloniki, Greece on April 5, 1948) on the occasion of his retirement. Symeon completed his undergraduate studies at the Department of Mathematics of the Aristotle University of Thessaloniki in 1970, his doctorate at the Department of Mathematics of the University of Ioannina in 1973, and his second doctorate, in 1976, at Université Paris 7, France. He became a full professor of algebra at the age of 32 at the University of Ioannina, and moved to the Aristotle University of Thessaloniki in 1982.

His initial research was focused on algebra and especially on categories theory. In the 1980s he turned to theoretical computer science. His main interests are closely connected with algebraic foundations in computer science, which is why we chose this title for this volume. In particular, he contributed to the development of the theory of tree languages and series, the axiomatization of graphs, a deep consideration of picture theory, and fuzzy languages. He never stopped working and producing seminal results despite his serious sight problems. He was awarded twice, in 2000 and 2005, for his research by the most prestigious organization in Greece, the *Academy of Athens.*

The contribution of Symeon to algebraic informatics is not limited to his publications. He organized the Third International Conference of Developments in Language Theory (DLT 1997), the Workshop on Current Trends and Developments in Fuzzy Logic in 1998, and he initiated the International Conference on Algebraic Informatics (CAI) series in 2005.

Symeon *is* (because after his official retirement in 2010 his students are still being taught by him) also an excellent teacher. He taught for more than 35 years several subjects such as algebra, linear algebra, mathematical logic, number theory, and in addition, after 1982, automata theory, tree languages and series, algebraic semantics, and fuzzy languages. Seven PhD theses were completed under his supervision.

This Festschrift volume contains 15 invited papers which are all connected to Symeon's research topics. They are written by colleagues, friends, and students of Symeon. All the papers were refereed according to the usual journal standards. We are deeply grateful to all the authors for their kind contribution; in fact the whole project turned out to be a nice and friendly procedure. Most of the papers were presented at the Workshop of Algebraic Foundations in Computer Science, held in Thessaloniki, during November 7–8, 2011. Last but not least we would like to express our gratitude to the team at Springer for their excellent, as usual, cooperation.

August 2011

Werner Kuich
George Rahonis

Workshop on Algebraic Foundations in Computer Science

Organizers

Werner Kuich, Vienna
Dimitrios Poulakis, Thessaloniki
George Rahonis, Thessaloniki

Local Organizing Committee

Eleni Mandrali
Irini Eleftheria Mens
Dimitrios Poulakis (Co-chair)
George Rahonis (Chair)
Eleni Maria Vretta

Table of Contents

Selected Decision Problems for Square-Refinement Collage Grammars*

Frank Drewes

Department of Computing Science,
Umeå University, Sweden
drewes@cs.umu.se

Abstract. We consider collage grammars whose rules subdivide the unit square into smaller and smaller rectangles. The decidability status of selected decision problems for this type of grammars is surveyed: the membership problem, the emptiness and finiteness problems, connectedness and disconnectedness of the generated pictures, and the question whether a generated collage contains a rectangle whose lower-left corner is a point on the diagonal.

1 Introduction

Picture generation by means of grammars in the tradition of Chomsky grammars (in a wide sense) is a field that has a long history. Many different kinds of pictures and grammatical formalisms to generate them have been studied. Array grammars of different descriptions belong to the oldest class of models of this kind. They were proposed in various forms in the late 1960s and early 1970s by Rosenfeld and others (see [24] and the more recent [25]). The underlying notion of pictures is that of a two-dimensional array of symbols, thus generalizing strings from one to two dimensions. Although many different kinds of rules have been proposed, the basic idea is usually to replace a rectangular sub-array by another array.[1] A problem with this kind of replacement is that a context-free notion of replacement is not easily conceived, because it would replace a single cell (a pixel) by an array, an operation that tends to destroy the rectangular nature of arrays.

A type of picture-generating grammars that avoids this problem is the collage grammar by Habel and Kreowski [21,22], which is inspired by hyperedge-replacement graph grammars (see, e.g., [2,20,19,12]). Here, pictures are sets of geometric objects in $\mathbb{R}^d$ (squares, triangles, circles, ... in the case $d = 2$), so-called parts. Nonterminals are labelled hyperedges, entities that are attached to finitely many points (typically $d+1$ points) in $\mathbb{R}^d$, and a right-hand side of a rule replacing such a nonterminal has as many distinguished points, so-called external points. Replacing a hyperedge e attached to points $p_1, \ldots, p_k$ by a right-hand

---

* Dedicated to Symeon Bozapalidis on the occasion of his retirement.
[1] Another one is to generate arrays column-wise, by considering the individual columns as strings.

W. Kuich and G. Rahonis (Eds.): Bozapalidis Festschrift, LNCS 7020, pp. 1–29, 2011.

side C with external points $q_1, \ldots, q_k$ means to remove e and add $\tau(C)$, where τ is an affine transformation such that $\tau(q_i) = p_i$ for $i \in [k]$.[2] In this way, a truly context-free picture generation mechanism is obtained that, moreover, is not confined to the generation of rectangular arrays of pixels of equal size.

The formalization of collage grammars by means of hyperedges is sometimes not very convenient. Therefore, equivalent definitions of collage grammars have been developed. Most notably, a "tree-based" definition in terms of tree grammars avoids the use of hyperedges. Instead, operations composed of affine transformations are applied to collages which are now simply sets of geometric parts. Similar tree-based definitions can be given for many other picture-generating formalisms as well, which has been exploited in [8] to present the area in a coherent way.

In this article, a type of two-dimensional collages is considered that shows certain similarities with arrays: all parts are non-overlapping rectangular subsets of the unit square, with sides parallel to the axes. In this case, (labelled) rectangles can be used not only as terminals, but also as nonterminals (corresponding to hyperedges attached to the four corners of the rectangle). Replacing such a nonterminal N by a collage C now means to remove N and add $\tau(C)$, where τ is the (non-uniform) scaling and translation that mapps the unit square onto N. Thus, τ makes C fit into the space previously occupied by N. We call these collage grammars square-refinement collage grammars, as they successively refine the unit square by replacing rectangles by smaller ones.

We note here that, with a starting point in array grammars rather than in collage grammars, Bozapalidis proposed a somewhat similar notion of picture grammars called picture-refinement grammars [3]. These grammars are based on elegant algebraic notions of arrays and their deformation. Intuitively, deformation corresponds to the transformation τ that fits a right-hand side C into a given rectangle (i.e., a cell of the array). Consequently, pictures consist of rectangles that are not necessarily the same size any more, similarly to the case of square-refinement collage grammars. The algebraic definition of picture-refinement grammars complements the geometric one of collage grammars in a nice way, providing access to different toolboxes of proof methods.

In the literature, some undecidability, decidability, and complexity results for square-refinement collage grammars and some of their special cases can be found. Not all of them have originally been formulated and proved for this type of grammars, but in most cases it is fairly easy to transfer them to this setting. Readers who would like to get general introductions to the theory of collage grammars should consult [14,8]. In the following sections, we define square-refinement collage grammars and the special cases that are of interest in order to formulate these results, and we present the decision algorithms and undecidability proofs, including a few new or strengthened results. For complexity considerations, we assume that the algorithms are implemented on a random-access machine

---

[2] Usually, it is required that τ is uniquely determined by the points given, which is the case if $d + 1$ points are considered, such that none of these points lies in the sub-space spanned by the remaining ones.

having indirect addressing and instructions that correspond to the basic arithmetic operations on integers.

The remainder of this article is structured as follows. In Sect. 2, we define square-refinement collage grammars and the special cases that are of interest for this paper: partial-array collage grammars, array collage grammars, and grid collage grammars. We also prove a useful normal-form result for square-refinement collage grammars, similar to the removal of useless symbols, chain rules and epsilon rules from a context-free grammar. Sect. 3 discusses the decidability and complexity of different flavours of membership, emptiness, and finiteness problems. Sect. 4 shows that it is undecidable whether all collages generated by a partial-array collage grammar are connected, and whether all of them are disconnected. A decidability result for the rather special case of so-called framed square-refinement collage grammars is also included. Sect. 5 shows the undecidability of the question whether a grid collage grammar generates a (collage containing a) part whose lower-left corner lies on the diagonal of the unit square. Finally, Sect. 6 concludes the paper.

The examples shown throughout the rest of this paper generate coloured pictures. These colours are visible only in the electronic version of the paper, whereas they are converted to shades of grey in the printed version. Thus, wherever particular colours are mentioned in examples, this refers to the electronic version. The colours are, however, not central for the readability of the paper.

2 Square-Refinement Collage Grammars

Let us first compile some basic notation and conventions. The sets of all natural numbers (including zero), rational numbers, and real numbers are denoted by $\mathbb{N}$, $\mathbb{Q}$, and $\mathbb{R}$, resp. For $n \in \mathbb{N}$, $[n] = \{i \in \mathbb{N} \mid 1 \le i \le n\}$. The set of all strings over an alphabet Σ is denoted by Σ^*. It includes, in particular, the empty string ϵ.

Given a function $f\colon A \to B$, the canonical extension of f to a function from the power set of A to the power set of B is denoted by f as well, i.e., $f(A') = \{f(a) \mid a \in A\}$ for all $A' \subseteq A$. Given a binary relation $\Rightarrow \subseteq A^2$, $\Rightarrow^n$ denotes the n-fold composition of $\Rightarrow$ with itself (for $n \in \mathbb{N}$), where $\Rightarrow^0$ is the identity on A. Moreover, $\Rightarrow^+ = \bigcup_{n \ge 1} \Rightarrow^n$ and $\Rightarrow^* = \bigcup_{n \in \mathbb{N}} \Rightarrow^n$ denote the transitive and the reflexive and transitive closures of $\Rightarrow$, respectively.

Square-refinement collage grammars are two-dimensional collage grammars whose terminal and nonterminal parts are labelled rectangular subsets of the unit square. Starting with a nonterminal that occupies the unit square, rules replace nonterminal rectangles by smaller ones (which may be nonterminal or terminal).

More formally, we work in the unit square $U = [0,1]^2$, where $[0,1]$ denotes the closed interval of real numbers between 0 and 1. A rectangle ρ (in U) is given by two points $p = (x,y), p' = (x',y') \in U$ such that $x < x'$ and $y < y'$, where $\rho = \{(a,b) \in U \mid x \le a \le x' \text{ and } y \le b \le y'\}$. Thus, p and p' are the lower-left and upper-right corners of ρ. Such a rectangle is also denoted by $[p:p']$. To avoid representation and computability issues encountered when dealing with arbitrary real numbers, we generally assume that all rectangles $[p:p']$ considered in this paper are given by rational coordinates, i.e., $p, p' \in U \cap \mathbb{Q}^2$. Two rectangles *overlap* if their intersection is a rectangle, and they *touch* if their intersection is nonempty. Every rectangle ρ defines a unique transformation τ consisting of a translation and scaling, such that $\tau(U) = \rho$. In the following, this transformation will be denoted by τ_ρ.

Throughout this paper, let Λ be a countably infinite set of labels. A *part* is a pair $P = (\lambda, \rho)$ consisting of a label $\lambda \in \Lambda$ and a rectangle ρ. If geometric and set-theoretic terminology is used in connection with parts, it refers to their second components. Thus, for example, P overlaps with (touches) another part (λ', ρ') if ρ and ρ' overlap (touch, respectively). Given a transformation $\tau \colon \mathbb{R}^2 \to \mathbb{R}^2$, the transformation of parts by τ is defined in the obvious way: $\tau((\lambda, \rho)) = (\lambda, \tau(\rho))$.[3] A *collage* (over Λ) is a finite set of parts that are pairwise non-overlapping. For $T \subseteq \Lambda$, the set of collages with labels exclusively taken from T is denoted by $\mathcal{C}_T$. By abuse of notation, we shall identify a singleton collage $\{P\}$ with its unique part P.

We note here that most of the literature on collage grammars deals with black-and-white collages, i.e., the terminal parts do not carry labels or colours. The extension to labelled terminals is, however, small and entirely uncritical for most of the theory of collage grammars (and, in particular, for the results discussed in this paper) as long as only finitely many colours are considered. In [8, Chap. 7], a much more general way do introduce and generate "real" colours is studied, based on operations that continuously affect the colour of parts by manipulating red-green-blue components.

Definition 1 (square-refinement collage grammar). *A* square-refinement collage grammar *is a quadruple* $G = (X, T, R, S)$ *consisting of*

- *finite disjoint sets* $X, T \subseteq \Lambda$ *of nonterminal and terminal labels,*
- *a finite set* R *of rules* $A ::= K$, *where* $A \in X$ *and* $K \in \mathcal{C}_{X \cup T}$, *and*
- *an initial nonterminal label* $S \in X$.

A part whose label is terminal (nonterminal) is a terminal *(nonterminal, resp.). If R contains a rule $r = (A ::= K)$ and C is a collage that contains a nonterminal $N = (A, \rho)$, then C derives $C' = C[K/N] = C \setminus \{N\} \cup \tau_\rho(K)$. Such a derivation step is denoted by $C \Rightarrow_r C'$ or $C \Rightarrow_R C'$. The language generated by G is*

$$L(G) = \{C \in \mathcal{C}_T \mid (S, U) \Rightarrow_R^* C\},$$

and is called a square-refinement language.

In the following, we will write the derivation relation as $\Rightarrow$, omitting the subscript, whenever the set R of rules in question is obvious from the context.

Example 2 (square-refinement collage grammar). An example is given in Fig. 1, together with four of the collages it generates. Here, the conventions are as

---

[3] We will only use transformations that preserve the property of being a rectangle.

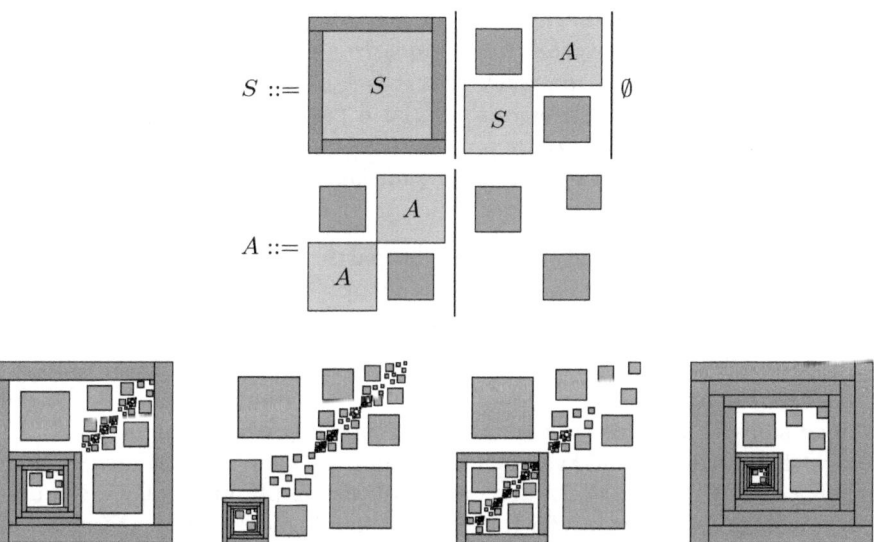

Fig. 1. A square-refinement collage grammar (top) and four of its generated collages (bottom); the generated collages are magnified relative to the size in which the rules are depicted

follows. Nonterminals are drawn in grey with a thin black border and the label inside. The labels of terminals are interpreted as colours, in this case light blue and orange. The left-hand side of the first rule is the initial nonterminal. Rules sharing the same left-hand side are drawn by separating their right-hand sides with vertical bars in a BNF-like manner, instead of drawing the individual rules separately.

In the introduction, array grammars and Bozapalidis' picture-refinement grammars were briefly mentioned. As the example above illustrates, the right-hand sides of square-refinement collage grammars are not very array like, as they do not necessarily consist of rows and columns. We shall now define special cases that have this property. Here, the parts are cells in a grid given by a set of horizontal and vertical grid lines (though not necessarily evenly spaced). We shall consider three types of such grammars. In the most general one, the right-hand sides of rules are required to be partial arrays. In a partial array, the rectangles are determined by a grid as described above. A slightly more restricted case is the array collage grammar, in which right-hand sides are required to be total arrays, in the sense that they fill all of U with rectangles, leaving no empty spaces in between. This type of square-refinement grammars is the one among those studied in this paper that is closest to Bozapalidis' picture-refinement grammars. The third special case of square-refinement collage grammars we consider is the grid collage grammar, where all right-hand sides are required to be partial arrays over the same grid.

Definition 3 (grid and array). *Let $m, n \in \mathbb{N} \setminus \{0\}$. An $m \times n$-grid or simply grid is a pair $\Gamma = (v_0 \cdots v_m, h_0 \cdots h_n)$ consisting of two finite sequences of real numbers $v_0, \ldots, v_m, h_1, \ldots, h_n$ such that $0 = v_0 < v_1 < \cdots < v_m = 1$ and $0 = h_0 < h_1 < \cdots < h_n = 1$. A collage C is a partial Γ-array, if every part in C is of the form $(\lambda, [p : p'])$, where $p = (v_{i-1}, h_{j-1})$ and $p' = (v_i, h_j)$ for suitable $(i, j) \in [m] \times [n]$. C is a Γ-array if it, in addition, contains such a part for every $(i, j) \in [m] \times [n]$.*

We simply speak of (partial) arrays if the particular grid Γ is understood or irrelevant. Thus, partial arrays may contain empty spaces, whereas arrays fill all of U with their parts. As mentioned above, we are going to consider three types of collage grammars whose right-hand sides are (partial) arrays:

- A *partial-array collage grammar* is a collage grammar such that the right-hand sides of its rules are partial arrays (possibly over different grids). If all right-hand sides are arrays, then the grammar is an *array collage grammar*.
- For a grid Γ, a *Γ-grid collage grammar* (or simply grid collage grammar) is a collage grammar in which each right-hand side is either a terminal unit square or a partial Γ-array.

Grid collage grammars were originally introduced and studied in [5]. In that original definition, the grid was required to be an evenly spaced $n \times n$ grid. Here, we drop this restriction, using arbitrary grids instead, as was first discussed in [8, Sect. 5.5]. Another difference between the original definition and the one employed here is that the former one allowed nonterminal squares to be rotated by multiples of 90 degrees and possibly also reflected. Clearly, this additional degree of freedom does not affect the generating power of grid collage grammars, because we can always implement the finitely many possible rotations and reflections by additional nonterminals whose right-hand sides are accordingly rotated and reflected versions of those the original rules (cf. Example 4 below, which illustrates this in the case of array collage grammars). Hence, as long as we do not limit the number of nonterminals a grammar may have, the restriction does not really make a difference.

Example 4 (array collage grammar). Fig. 2 shows an example of an array collage grammar whose terminals are blue, orange, and white. The difference between the nonterminals S and A on the one hand, and S' and A' on the other hand is that the latter generate collages that are rotated by 90 degrees with respect to those generated by the former. A partial-array collage grammar (that is not an array collage grammar) would be obtained from this one by leaving out some of the parts, e.g., by replacing the terminating rules by $S ::= \emptyset$ and $S' ::= \emptyset$. Conversely, every partial-array collage grammar can be turned into an array collage grammar by filling empty spaces with suitable white (i.e., invisible) parts without a visible difference (provided that we omit the frames around parts). Hence, from the point of view of pattern generation, array collage grammars are as powerful as partial-array collage grammars. Nevertheless, the latter are more difficult to handle in some algorithmic respects, namely if the problems to be solved require us to distinguish between collages that yield the same pattern.

Fig. 2. An array collage grammar and three of the collages it generates

Example 5 (grid collage grammar). An example of a grid collage grammar over the evenly spaced 2×2-grid is shown in Fig. 3.

The types of collages that can be derived by (partial) array collage grammars or Γ-grid collage grammars may be called nested (partial) array collages and nested Γ-grid collages, resp. It should be clear that the generative power of the three types of collage grammars is not affected by allowing, in their right-hand sides, nested (partial) arrays and nested Γ-grid collages, resp. This is because, in the usual way, we may always introduce new nonterminal labels to be able to decompose such a rule into a finite set of rules of the non-nested kind.

A standard result that carries over from the string case in a straightforward manner makes it possible to remove chain rules and epsilon rules from a square-refinement collage grammar $G = (X, T, R, S)$. Here, a chain rule is a rule whose right-hand side is a nonterminal unit square, i.e., a rule of the form $A ::= (B, U)$ with $B \in X$. An epsilon rule is a rule whose right-hand side is $\emptyset$. Unfortunately, the removal of epsilon rules may lead to an exponential size increase, which is highly undesirable from the point of view of efficiency. Therefore, we will use

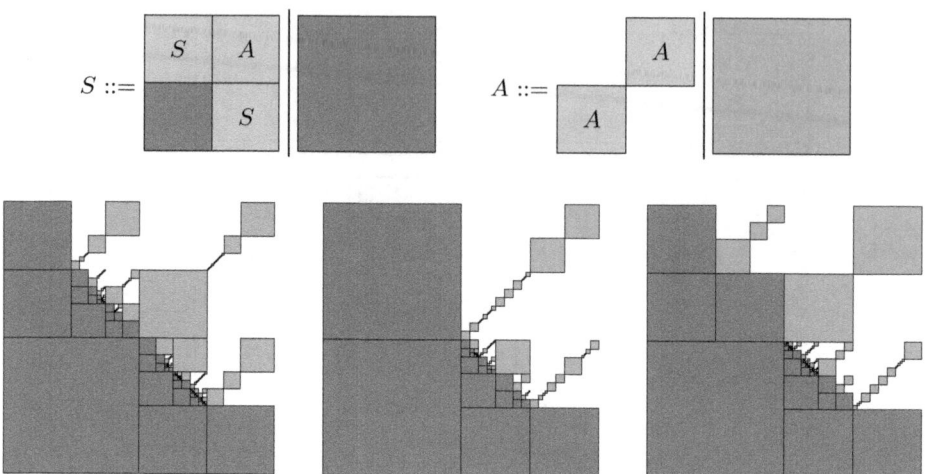

Fig. 3. A grid collage grammar and three of the collages it generates

a slightly relaxed condition. We say that a derivation $(A, U) \Rightarrow^n C$ is a (non-trivial) *epsilon derivation* if $n > 1$ and $C = \emptyset$.

Another standard normal-form result that is often useful concerns the removal of useless nonterminal labels. As usual, call $A \in X$ useful if there is a derivation $(S, U) \Rightarrow^* C \Rightarrow^* C'$ such that $(A, \rho) \in C$ for some rectangle ρ, and $C' \in \mathcal{C}_T$. Clearly, because of the context-free nature of collage grammars, all nonterminal labels are useful if and only if every $A \in X$ is reachable (i.e., there is a derivation $(S, U) \Rightarrow^* C$ with $(A, \rho) \in C$ for some rectangle ρ) and productive (i.e., $(A, U) \Rightarrow^* C$ for a collage $C \in \mathcal{C}_T$). For complexity considerations, it is useful to show that both normal forms can simultaneously be achieved in time $O(n^4)$.

Lemma 6. *For every square-refinement collage grammar G such that $L(G) \neq \emptyset$, there is a square-refinement collage grammar G' containing only useful nonterminals and neither chain rules nor epsilon derivations, such that $L(G') = L(G)$. The computation of G' from G can be performed in time $O(n^4)$ by an algorithm that also detects whether $L(G) = \emptyset$.*

Proof. Let $G = (X, T, R, S)$, and consider first the removal of useless nonterminal labels. As in the string case, we can use iterative procedures (that run in quadratic time) to compute the set of all productive nonterminal labels and restrict X and R accordingly. Afterwards, we do the same with respect to reachability. The resulting grammar contains only useful nonterminal labels.

Next, we remove epsilon derivations from G (where X and R now stand for the possibly smaller sets left after the first phase of the algorithm). By the standard iterative procedure, we first determine the set E of all $A \in X$ such that $(A, U) \Rightarrow^* \emptyset$. This can be done in quadratic time. Now, let $X' = X \cup \{A' \mid A \in X\}$, where A' is a fresh nonterminal label for each $A \in X$. The idea is to use a guess-and-verify strategy to make A' generate the same collages as A, except $\emptyset$. For every rule $A ::= K$ in R, let R' contain

- the rules $A ::= K'$ and $A' ::= K'$, for all K' that can be obtained from K by replacing exactly one nonterminal (B, ρ) with (B', ρ),
- the original rule $A ::= K$ if $K \in \mathcal{C}_T$, and
- the rule $A' ::= K$ if $K \in \mathcal{C}_T \setminus \{\emptyset\}$.

The reader should easily be able to check that, for all $A \in X$,

$$\{C \in \mathcal{C}_T \mid (A', U) \Rightarrow_{R'}^* C\} = \{C \in \mathcal{C}_T \mid (A, U) \Rightarrow_R^* C\} \setminus \{\emptyset\}.$$

From this and the construction of rules in R' with left-hand sides in X, it follows that (X', T, R', S) does not contain epsilon derivations and generates the same language as G. Note that the size of the grammar increases at most quadratically.

Finally, we follow the usual strategy for removing chain rules from a context-free Chomsky grammar. Thus, we determine the set D of all $(A, B) \in X' \times X'$ such that $(A, U) \Rightarrow_{R'}^* (B, U)$. Again, this can be done by an iterative procedure that takes quadratic time (in $|R'|$). Now, replace R' by the set R'' of all rules of the form $A ::= K$, such that there is a rule $B ::= K$ in R' for which $(A, B) \in D$. Clearly, this modification neither affects the language generated, nor does it re-introduce useless nonterminal labels or epsilon derivations. Thus, $G' = (X', T, R'', S)$ is as required, and the procedure runs in time $O(n^4)$ in total, as claimed.   □

3   Membership, Emptiness, and Finiteness

As our first decision problem regarding square-refinement collage grammars, we consider the most basic one for languages of any kind: the membership problem. Thus, we are given a square-refinement collage grammar G and a collage C, and the question to be answered is whether $C \in L(G)$. Recall that there are two flavours of this problem, namely the uniform and the non-uniform one. In the first, G is part of the input (which means that the algorithm has to work for all G), whereas G is arbitrary but fixed in the non-uniform case (meaning that a family of algorithms indexed by G would do).

To the best of the author's knowledge, the (mostly rather straightforward) results of this section have not explicitly been stated in the literature. However, somewhat similar results were proved without complexity analyses in [8, Chap. 5] for the case of grid collage grammars (see also the remark following the proof of Theorem 11).

We first observe that membership is decidable for square-refinement collage grammars in general, by a rather straightforward (and inefficient) approach.

Theorem 7. *The uniform membership problem for square-refinement collage grammars is decidable.*

Proof. By Lemma 6, we may assume that the given square-refinement collage grammar $G = (X, T, R, S)$ contains neither chain nor epsilon rules. The procedure for removing epsilon rules already gives us the answer to the membership problem if $C = \emptyset$, as this is the case if and only if $S \in E$. Otherwise, let α be the area of the largest nonterminal part appearing in the right-hand sides of rules in R. If a derivation tree of a derivation in G (defined in the obvious way) is of depth d, then the generated collage contains a part whose area is α^d or smaller. Thus, if α_0 is the area of the smallest rectangle in C, we only have to enumerate all derivations whose derivation trees are of depth at most $\log_\alpha \alpha_0$ to find out whether $C \in L(G'')$.   □

It may be worth pointing out that the argument above works for all collage grammars (i.e., not just square-refinement collage grammars) whose parts have a nonzero but finite area and whose rules strictly decrease the area of nonterminals.
    A better result can be obtained for array collage grammars.

Theorem 8. *The uniform membership problem for array collage grammars is in P. The non-uniform variant of the problem is solvable in cubic time.*

Proof. By following the well-known construction of the Chomsky normal form for context-free Chomsky grammars, a given array collage grammar G may be transformed into a normal form in which each right-hand side is either

- an array consisting of two nonterminals or
- an array consisting of a single terminal.

For picture-refinement grammars, this normal form was established by Bozapalidis in [3]. The transformation can be done in polynomial time and yields a grammar whose size is polynomial in the size of the original grammar. Now, an adapted version of the Cocke-Younger-Kasami algorithm [27] can be used to check whether a given collage C is in $L(G)$. It builds a table τ whose cells $\tau_p^{p'}$ correspond to rectangles $[p : p']$ such that p and p' are the lower-left and upper-right corners, respectively, of parts in C. Given such a rectangle, let $C_p^{p'}$ be the sub-collage of C consisting of all parts in C which are subsets of $[p : p']$. In a bottom-up procedure similar to the ordinary CYK algorithm, the cells are filled with nonterminal labels in such a way that $\tau_p^{p'}$ contains the label A if and only if $(A, [p : p']) \Rightarrow^* C_p^{p'}$. When $\tau_{(0,0)}^{(1,1)}$ has been filled, we know whether $C \in L(G)$, because this is the case if and only if $\tau_{(0,0)}^{(1,1)}$ contains the initial nonterminal label.
    For a fixed language, the complexity analysis of the algorithm is similar to that of the CYK algorithm in the string case, which yields a cubic time bound.   □

It is unclear whether the membership problem for partial-array collage grammars can also be solved in polynomial time. In that case, the difficulty is that, intuitively, the corners of nonterminals may have to be placed somewhere in the empty space between terminals, for which there may be far to many possibilities. Hence, determining the complexity of the membership problem for partial-array collage grammars is an interesting open problem.

Another interesting question is whether we can also decide the membership problem for the patterns generated. This is because, if we interpret the terminal labels as colours and draw collages by simply drawing rectangles in their respective colours, then two distinct collages may yield identical patterns. If we are only interested in whether or not a certain pattern is generated by a collage grammar, we thus have to solve the *pattern-membership problem*. Apparently, essentially nothing can be found in the literature about the solvability and complexity of this problem.

Formally, for a collage C, define

$$pattern(C) = \{(\lambda, p) \in \Lambda \times U \mid \exists \rho \colon p \in \rho \text{ and } (\lambda, \rho) \in C\}.$$

C is *pattern equivalent* with C', denoted $C \sim C'$, if $pattern(C) = pattern(C')$. Given a collage grammar G and a collage C, the pattern-membership problem asks whether there is a pattern-equivalent collage $C' \in L(G)$. Below, the relatively easy result is proved that this problem is in P for grid collage grammars. The reader is encouraged to have a look at the more general cases and find out whether the problem is solvable and, if so, whether we can solve it efficiently.

In the following theorem, we assume that the input is a nested Γ-grid collage, where Γ is the grid underlying the grammar in question. If more general collages are allowed as input, the problem is likely to become more complex, because the algorithm must then also convert the input into a Γ-grid collage.

Theorem 9. *The uniform pattern-membership problem for grid collage grammars can be solved in quadratic time. The non-uniform variant of the problem is solvable in linear time.*

Proof. Let C, C' be collages consisting of Γ-parts. Let us say that C' *is coarser than* C if C and C' are pattern equivalent and, for every part $(\lambda, \rho) \in C$, there is a part $(\lambda, \rho') \in C'$ such that $\rho \subseteq \rho'$. Clearly, the set of all collages that are pattern equivalent with C has a unique coarsest element $\widehat{C}$ that can be computed in linear time by a depth-first strategy. The algorithm described below turns the given grid collage grammar into one whose language is closed under taking $\widehat{C}$ (i.e., for each member C of the generated language, $\widehat{C}$ is a member as well), and checks whether the grammar generates $\widehat{C}$, where C is the input collage.

Let $G = (X, T, R, S)$ be a Γ-grid collage grammar and C the input collage. In quadratic time, using the standard iterative technique, we can compute the set $fill(A)$ for every $A \in X$, where

$$fill(A) = \{\lambda \in T \mid (A, U) \Rightarrow^* C \text{ for a collage } C \sim (\lambda, U)\}.$$

Now, add to G all rules $A ::= (\lambda, U)$ with $A \in X$ and $\lambda \in fill(A)$. Then the resulting grammar G' satisfies $L(G') = \widehat{L(G)}$. Hence, to answer the membership question for G and C, it suffices to check whether $\widehat{C} \in L(G')$. This can be done with the help of Theorem 8, but we actually do not need the full CYK for this purpose. Since the grid Γ is fixed, every nested Γ-grid collage has a unique structure (similar to a quadtree), and checking whether the collage is generated by the grammar corresponds to running a bottom-up finite-state tree automaton on this tree, which takes linear time.

Altogether, this yields a quadratic time bound for the uniform case and a linear one for the non-uniform one (because, in the non-uniform case, G' is fixed and need not be computed). □

Besides membership, two of the most fundamental questions regarding languages are those that concern their emptiness and finiteness. In the remainder of this section, these will briefly be discussed.

An algorithm for deciding emptiness for square-refinement languages is easily established, using the corresponding algorithm for context-free string languages as a blueprint. In fact, one may rightfully claim that these two algorithms are *the same*. This is because emptiness is a property independent of what sort of objects is generated. Consider a square-refinement collage grammar $G = (X, T, R, S)$ and a rule $A ::= K$. If K contains the (pairwise distinct) nonterminals $(A_1, \rho_1), \ldots, (A_k, \rho_k)$, then we can regard this rule as a context-free Chomsky rule $A ::= A_1 \cdots A_k$ instead. The context-free Chomsky grammar G' obtained in this way generates a string (namely the empty string) if and only if $L(G) \neq \emptyset$. Thus, the following theorem just repeats well-known results from the theory of context-free Chomsky grammars "in disguise".

Theorem 10. *The emptiness problem for square-refinement languages is*

1. *P-complete and*
2. *solvable in quadratic time.*

For the finiteness problem, the situation is somewhat similar, although not entirely as trivial, as we have to make use of Lemma 6 to establish it.

Theorem 11. *The finiteness problem for square-refinement languages is solvable in time $O(n^4)$.*

Proof. By Lemma 6, we have to show that the problem is solvable in linear time for square-refinement collage grammars that have no useless symbols and contain neither chain rules nor epsilon derivations. However, for such grammars G, $L(G)$ is infinite if and only if there is a nonterminal label A that admits a derivation $(A, U) \Rightarrow^+ C$ such that C contains a nonterminal $N = (A, \rho)$ for some rectangle ρ. This is because, if (A, U) derives a (non-empty) terminal collage C' (which it, by assumption, does), then (A, ρ) derives $\tau_\rho(C')$. The latter is necessarily distinct from C', because $\rho \neq U$, which means that the smallest part in $\tau_\rho(C')$ is smaller than the smallest one in C'. As we can derive a collage containing A from (S, U), by iteration, we generate more and more collages.

The question whether the required derivation $(A, U) \Rightarrow^+ C$ exists is equivalent to checking whether a graph contains a cycle, a problem which can be decided in linear time. (The nodes of the graph are the nonterminal labels, and there is an edge from A to B if there is a rule $A ::= K$ such that K contains a part labelled with B.) □

It may be worth noting that Theorem 11 (without the polynomial time bound) can, in fact, be extended to square-refinement collage grammars that employ more powerful derivation modes than the purely context-free one considered in this paper. This makes use of a general result from [9] stating that the finiteness problem is decidable for the images of regular tree languages under arbitrary compositions of macro tree transductions. In the tree-based approach to picture generation, a collage grammar is a pair consisting of a so-called collage algebra and a tree-generating device. A collage algebra is a set of operations on collages. Intuitively, every collage K (such as the right-hand side of a rule) with n non-terminals $N_1, \ldots, N_n$ yields an n-ary operation that, when applied to collages $C_1, \ldots, C_n$, returns $K[C_1/N_1] \cdots [C_n/N_n]$. Then, every device that generates trees (i.e., formal expressions) over such operations defines a collage language, namely the set of all collages obtained by evaluating the trees generated. Collage languages in the sense of this paper are obtained if the tree-generating device is a regular tree grammar. By replacing regular tree grammars with more powerful devices, such as compositions of regular tree grammars with arbitrary chains of macro tree transducers, more sophisticated languages can be generated. Using a generalization of Lemma 6 together with the main result of [9], it is possible to show that the finiteness problem is decidable for this class of collage grammars as well. For grid collage grammars, this was shown in [8, Theorem 5.3.11 and Sect. 5.5.2].

While the preceding theorem may be useful in some cases, it is conceivable that one will usually be interested in knowing whether the set of patterns generated by G is finite, i.e., whether the quotient $L(G)/\sim$ is finite. (For emptiness, the cases of collages and patterns do, of course, coincide. Thus, we only need to discuss finiteness.) To which extent this problem can be solved is an interesting open question that, apparently, has not been studied in the literature on collage grammars before. Here, we only consider the easiest case, namely that of grid collage languages. This case can be solved by more or less the same algorithm as in the proof of Theorem 11.

Theorem 12. *For grid collage grammars G, finiteness of $L(G)/\sim$ is decidable in polynomial time.*

Proof. Let $G = (X, T, R, S)$. Similar to the definition of epsilon derivations, say that a (non-trivial) *filler derivation* is a derivation $(A, U) \Rightarrow^n C$ such that $n > 1$ and $C \sim (\lambda, U)$ for some $\lambda \in T$ (cf. the definition of *fill(A)* in the proof of Theorem 9). Using a similar construction as for the removal of epsilon derivations, we can extend the proof of Lemma 6 to ensure that the resulting grammar does not contain filler derivations. Now, two generated collages are pattern equivalent if and only if they are identical. Hence, the theorem follows from Theorem 11.

Note that, as the removal of filler derivations is indeed entirely similar to the removal of epsilon derivations, it comes at the expense of increasing the degree of the polynomial time bound by another factor of 2. Hence, the time bound becomes $O(n^8)$. This might be avoidable by a more clever construction

that removes both epsilon and filler derivations simultaneously, but we will not
pursue this direction of thought here.   □

As mentioned above, it is unclear to what extent the preceding theorem can
be improved to cover more general types of square-refinement collage grammars.
The problem seems likely to be solvable for array languages, but to find a rigorous
proof is bound to be more difficult than in the case of Theorem 12: as the right-
hand sides of different rules may be based on different grids, the major argument
in its proof does not work any more. To see this, have a look at the (in all other
respects completely uninteresting) example in Fig. 4.Clearly, the infinitely many
collages generated by this grammar are all pattern equivalent, even though the
grammar does not contain filler derivations.

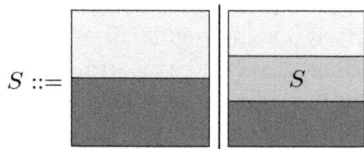

Fig. 4. An array collage grammar without U-derivations whose generated language is
infinite, yet it generates only one pattern

4   Connectedness

We shall now consider the question whether the language generated by a partial-
array collage grammar contains a connected collage, and whether it contains
a disconnected collage. For general collage grammars (as opposed to square-
refinement collage grammars) the disconnectedness question was shown to be
undecidable in [13]. Here, we use basically the same reduction, but strengthen
the result in two respects:

1. We show that the proof idea can be used to show the undecidability of both
   the connectedness and the disconnectedness question.
2. Rather than using general collage grammars, we show the result for partial-
   array collage grammars.

At the end of this section, we include a small decidability result regarding both
questions, for the case of so-called framed square-refinement collage grammars,
a result that has also been taken from [13].

   We employ the obvious definition of connectedness: a collage C is said to be
connected if, for all parts $P = (\lambda, \rho)$ and $P' = (\lambda', \rho')$ in C, there is a $n \geq 1$
and there are parts $P_1 = (\lambda_1, \rho_1), \ldots, P_n = (\lambda_n, \rho_n) \in C$ such that $P = P_1$,
$P' = P_n$, and $\rho_i \cap \rho_{i+1} \neq \emptyset$ for all $i \in [n-1]$. We shall show that this problem
is undecidable, by a reduction from a modified version of Post's correspondence

problem over an alphabet Σ. Let us first recall this problem, which we denote by MPCP. An instance of MPCP consists of a finite sequence Π of pairs of nonempty strings over Σ, say $\pi_1 = (\pi_1^1, \pi_1^2), \ldots, \pi_k = (\pi_k^1, \pi_k^2)$. The question to be answered is whether there exists an index sequence $i_1 \ldots i_l \in [k-1]^*$ such that $\pi_{i_1}^1 \cdots \pi_{i_l}^1 \pi_k^1 = \pi_{i_1}^2 \cdots \pi_{i_l}^2 \pi_k^2$. Hence, the pair π_k is special, and is appended at the end, while it is prepended at the beginning in the usual definitions of MPCP. Clearly, this does not make a difference with respect to decidability, because we only need to reverse all strings in Π. We call an index sequence $i_1 \ldots i_l k$ that satisfies this requirement a witness. An example with $\Sigma = [3]$ and $k = 3$ is $\Pi = (11, 112), (211, 1), (3, 13)$. This instance is a *yes* instance, a witness being 123, as $11\,211\,3 = 112\,1\,13$.

It is well known that MPCP is undecidable; see, e.g., [23]. Given an instance Π of MPCP as above, we shall construct a partial-array collage grammar G_{PCP} such that $L(G_{\text{PCP}})$ contains a connected collage if and only if Π has a solution. In the following, we shall sketch the construction of this grammar. As in the example above, we assume without loss of generality that $\Sigma = [m]$ for some $m \geq k$. Furthermore, we assume that the last symbol in the two strings in π_k does not occur anywhere else in the instance. Clearly, if this is not the case, it is easy to achieve by choosing a new symbol and appending it to both π_k^1 and π_k^2.

Since connectedness is independent of the labelling (or colouring) of parts, we use only one terminal label interpreted as black, say β. Therefore, we mostly leave out the label of terminal parts and regard them as pure rectangles. The basic building blocks of the reduction are partial arrays that encode the numbers in $q \in [m]$ in two different ways: Let $e = 1/m$ and $0 < d < e$, and define

$$\boxed{q} = [((q-1)e, 0) : ((q-1)e + d, 1)] \text{ and}$$
$$\underline{q} = \{[((q-1)e, d) : ((q-1)e + d, 1)], [(0, 0) : (1, d)]\}.$$

Thus, as illustrated in Fig. 5, $\boxed{q}$ is a vertical bar of height 1 and thickness d whose position within U encodes q in such a way that the encodings of different $q \in [m]$ do not intersect. The encoding $\underline{q}$ is similar, but of the form $\perp$, i.e., with a horizontal base.

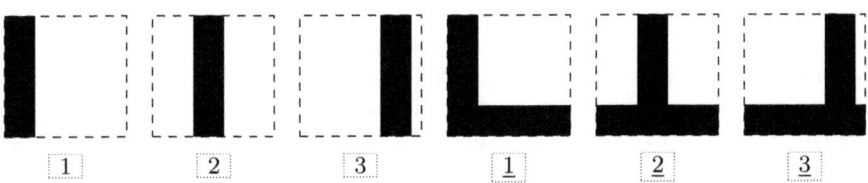

Fig. 5. The basic building blocks of the reduction, illustrated for the case where $m = 3$ and $d = 1/4$

Using these building blocks, we can assemble larger ones. Let σ_1 be the transformation that scales its argument horizontally by $1/2$ (but does not affect its vertical extension), and let σ_2 be σ_1 followed by a translation by $(1/2, 0)$. Now, for a string $s \in [m]^*$ and a collage C, we let

$$
\boxed{s \longrightarrow \boxed{C}} = \begin{cases} C & \text{if } s = \epsilon \\ \boxed{q} \cup \boxed{s' \longrightarrow \boxed{C}} & \text{if } s = qs' \text{ for a } q \in [m] \end{cases}
$$

$$
\boxed{\underline{s} \longrightarrow \boxed{C}} = \begin{cases} C & \text{if } s = \epsilon \\ \underline{\boxed{q}} \cup \boxed{\underline{s'} \longrightarrow \boxed{C}} & \text{if } s = qs' \text{ for a } q \in [m] \end{cases}
$$

In the case where C consists of a single part (λ, U), we shall denote $\boxed{s \longrightarrow \boxed{C}}$ and $\boxed{\underline{s} \longrightarrow \boxed{C}}$ by $\boxed{s \longrightarrow \lambda}$ and $\boxed{\underline{s} \longrightarrow \lambda}$, resp.

The collages that will be generated consist of three rectangular regions that contain suitably rotated and/or mirrored copies of blocks of the types defined above. The idea is that the three regions must fit together in a certain way for the collage to be connected. The lower-left region encodes a guessed index sequence $i_1 \cdots i_u k$ at its top edge and the corresponding string $\pi_{i_1}^1 \cdots \pi_{i_u}^1 \pi_k^1$ at its right edge. The lower-right region encodes a (possibly different) guessed index sequence $j_1 \cdots j_v k$ at its top edge and the string $\pi_{j_1}^2 \cdots \pi_{j_v}^2 \pi_k^2$ at its left edge. The idea is that the right edge of the lower-left region will fit the left edge of the lower-right region if and only if $\pi_{i_1}^1 \cdots \pi_{i_u}^1 \pi_k^1 = \pi_{j_1}^2 \cdots \pi_{j_v}^2 \pi_k^2$. The upper region encodes two copies of a third, also guessed, index sequence, say $z_1, \ldots, z_w k$. One copy extends from the left edge to the middle, while the other one is mirrored, extending from the right edge to the middle. The purpose of this region is to check that the left copy coincides with $i_1 \cdots i_u k$ and the right copy coincides with $j_1 \cdots j_v k$, thus establishing that the three guessed index sequences are, in fact, identical.

The collage that would be generated corresponding to the index sequence 123 of the instance mentioned above is shown on the left-hand side of Fig. 6, where the three regions are indicated by different background colours. On the right-hand side of the figure, a generated collage based on inconsistent guesses is shown, that moreover, yields sequences that do not coincide. In that collage, the top and lower-left regions both correspond to the index sequence 13. As a consequence, these regions fit together at their common edge. However, the guess in the lower-right square has been 123, so that the vertical bar encoding the 3 in the top region does not fit the one encoding the 2 at the top edge of the lower-right region. Moreover, the index sequence 13 gives rise to the (encoding of the) string 113 at the right edge of the lower-left region, while the sequence 123 gives rise to 112113, resulting in a misfit where the 3 on the left meets the 2 on the right.

Let us now define the partial-array collage grammar G_{PCP} obtained from an instance $\Pi = \pi_1, \ldots, \pi_k$. For convenience, we shall use right-hand sides that are nested partial arrays. As mentioned earlier, this does not add power, because such rules can be broken down into ordinary ones using a larger set of nonterminals. With this we need, besides the initial nonterminal label S, three additional

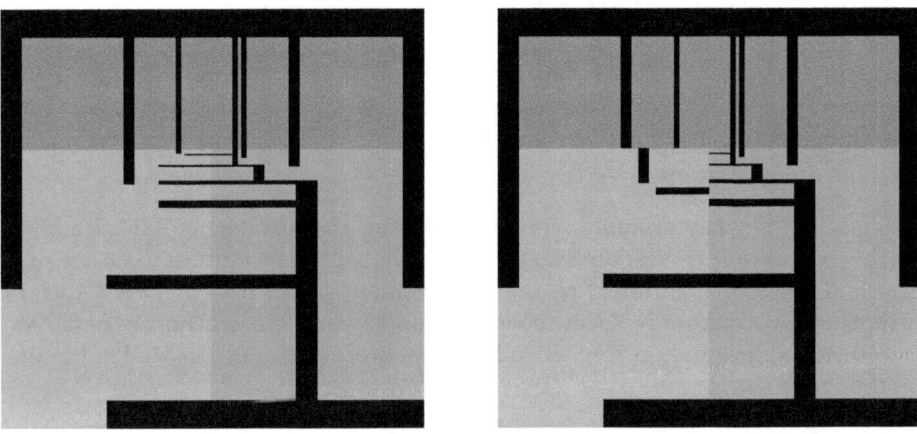

Fig. 6. A generated collage corresponding to a witness (left) and one that does not (right)

nonterminal labels: I, A, and B. They will be responsible for generating the upper, lower-left, and lower-right regions, resp. Thus, the only rule for with the left-hand side S is

$$S ::= \begin{array}{|c c|} \hline \multicolumn{2}{|c|}{I} \\ \hline A & B \\ \hline \end{array}$$

The rules with left-hand sides I, A, and B are the following, where h ranges over $[k-1]$:

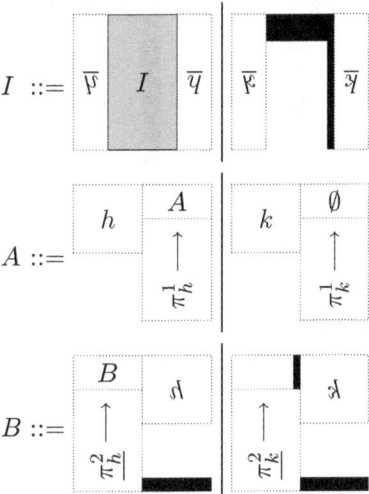

For a nonterminal label $Z \in \{A, B, I\}$, let $r_{Z,1}, \ldots, r_{Z,k}$ denote the k rules whose left-hand side is Z, i.e., $r_{Z,1}, \ldots, r_{Z,k-1}$ are obtained by letting h above run from 1 to $k - 1$. Clearly, every derivation that generates a terminal collage corresponds to three index sequences $i_1, \ldots, i_u$, $j_1, \ldots, j_v$, $z_1, \ldots, z_w$, such that the rules applied to A, B, and I are $r_{A,i_1}, \ldots, r_{A,i_u}$, $r_{B,j_1}, \ldots, r_{B,j_v}$, and $r_{I,z_1}, \ldots, r_{I,z_w}$, resp. Moreover, $i_u = j_v = z_w = k$.

Example 13. For the grammar constructed from the instance of MPCP above, the derivation that corresponds to the index sequence 123 applies the rules $r_{Z,1}$, $r_{Z,2}$, and $r_{Z,3}$ (in this order) to the nonterminals labelled with $Z \in \{A, B, I\}$ (after the first step that replaces the initial nonterminal). This sequence of derivation steps is shown in Fig. 7, where all three nonterminals are replaced in parallel in each step.

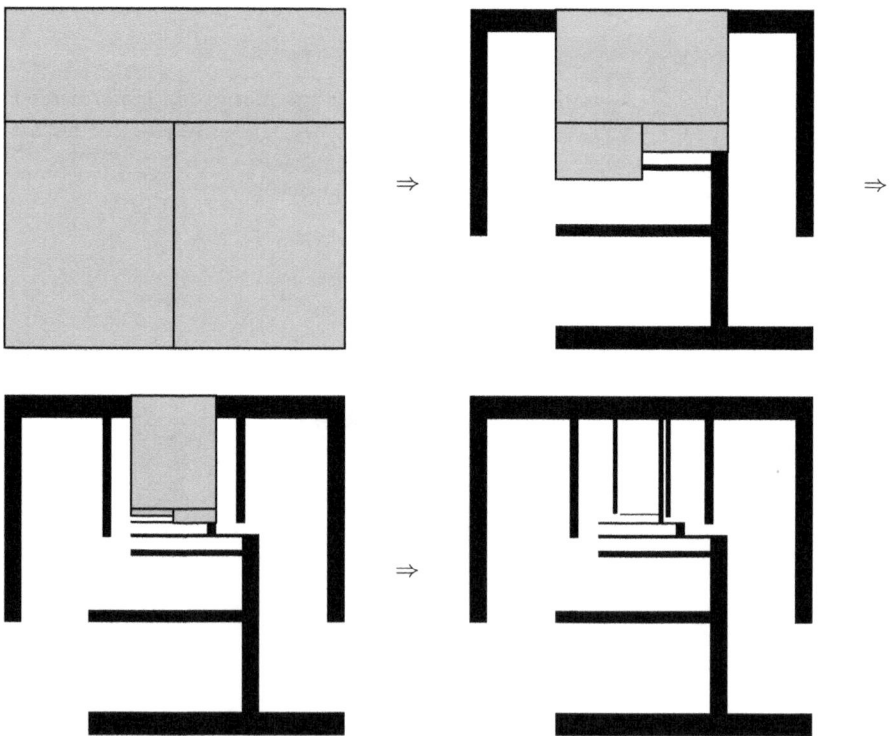

Fig. 7. The steps corresponding to the witness 123 of our sample instance of MPCP

Let us say that a derivation is a witness if the three index sequences coincide and, moreover, $\pi_{i_1}^1 \cdots \pi_{i_u}^1 = \pi_{i_1}^2 \cdots \pi_{i_u}^2$. The following lemma shows that the reduction works correctly.

Lemma 14. *Let G_{PCP} be constructed from an MPCP instance as described above. A derivation generates a connected collage if and only if it is a witness.*

Proof. The reader should easily be able to check that a witness generates a connected collage, because the encodings of identical symbols are placed on opposite sides of the edges at which the regions generated by A, B, and I intersect. What is perhaps less obvious is that every derivation that is not a witness generates a disconnected collage. Thus, let us consider such a derivation, and let $i_1, \ldots, i_u$, $j_1, \ldots, j_v$, and $z_1, \ldots, z_w$ be the guessed index sequences. There are two cases:

Case 1: The index sequences differ
Assume that $i_1 \cdots i_u \neq z_1 \cdots z_w$, as the other case is similar. Since $i_u = k = z_w$, whereas $k \notin \{i_1, \ldots, i_{u-1}, z_1, \ldots, z_{w-1}\}$, there necessarily has to be an $l \in [\min(u,w)]$ such that $i_l \neq z_l$. Consequently, the vertical bar inserted by the application of r_{A,i_l} is not connected to any other part in the generated collage, as its horizontal position does not coincide with the horizontal position of the vertical bar inserted by the application of r_{I,z_l}. (Note that the copies of $\boxed{i_l}$ and $\boxed{z_l}$ inserted at these steps are horizontally aligned with each other.) Hence, the generated collage is not connected.

Case 2: The index sequences are identical, but $\pi^1_{i_1} \cdots \pi^1_{i_u} \neq \pi^2_{i_1} \cdots \pi^2_{i_u}$
The arguments are similar to those used in the first case. Let $\pi^1_{i_1} \cdots \pi^1_{i_u} = s_1 \cdots s_n$ and $\pi^2_{i_1} \cdots \pi^2_{i_u} = t_1 \cdots t_{n'}$. Again, the copy of $\boxed{s_l}$ encoding the lth symbol in $s_1 \cdots s_n$ is aligned with the copy of $\boxed{t_l}$ encoding the lth symbol in $t_1, \ldots, t_{n'}$ (though this time along the vertical edge), for all $l \in [\min(n, n')]$. Since i_u is the only index among $i_1, \ldots, i_u$ that is equal to k, and both π^1_k and π^2_k end in a symbol that does not occur anywhere else, it follows that there is an $l \in [\min(n, n')]$, such that $s_l \neq t_l$. This means that the horizontal bar in the copy of $\boxed{s_l}$ is not connected to any other part in the generated collage, which proves that the collage is not connected.   □

Using Lemma 14, we obtain the desired theorem, saying that it is undecidable whether a partial-array collage grammar generates a connected collage. By a slight variation of the construction, we can also show that the same holds for disconnectedness.[4]

Theorem 15. *For partial-array collage grammars G, the following questions are undecidable:*

- *Does $L(G)$ contain a connected collage?*
- *Does $L(G)$ contain a disconnected collage?*

Proof. By Lemma 14, the first question is undecidable. The second can be shown to be undecidable by modifying the reduction in two steps, as follows. In the first step, the (patterns of the) collages generated by G_{PCP} are inverted, i.e.,

---

[4] This is, up to the adjustments necessary to adhere to the definition of partial-array collage grammars, the original construction from [13].

we modify the grammar in such a way that the parts generated cover exactly the areas of U that the grammar above leaves uncovered. In addition, we place a black frame around the whole collage (which means that we have to reduce the size of the interior a bit, to fit everything into U). Thus, after this step, the witness in Fig. 6 would look as shown in Fig. 8.

Fig. 8. Inverting the collages generated by G_{PCP}

In the second step, we place a "skeleton" (constructed from slim parts) on the inside of the white area in such a way that, if the collage is a witness, the skeleton is separated from the rest of the collage by the white area. For the witness above, the resulting collage is shown at the top of Fig. 9.[5] A collage that does not correspond to a witness is displayed at the bottom of the figure. In such collages, the (torn apart) skeleton becomes connected to the rest of the collage, thus yielding a connected collage altogether. Hence, the grammar generates a disconnected collage if and only if the given MPCP instance has a witness.   □

As we can turn every partial-array collage grammar into an array collage grammar by filling the empty space with appropriate terminals carrying a new label, we obtain the following corollary.

Corollary 16. *For array collage grammars and terminal labels* λ, *the following questions are undecidable:*

- *Does* $L(G)$ *contain a collage* C *such that* $\{(\mu, \rho) \in C \mid \mu = \lambda\}$ *is connected?*
- *Does* $L(G)$ *contain a collage* C *such that* $\{(\mu, \rho) \in C \mid \mu = \lambda\}$ *is disconnected?*

---

[5] The reader may have to zoom into the pictures or look at a high-resolution printout in order to be able to see all relevant details of the figure appropriately.

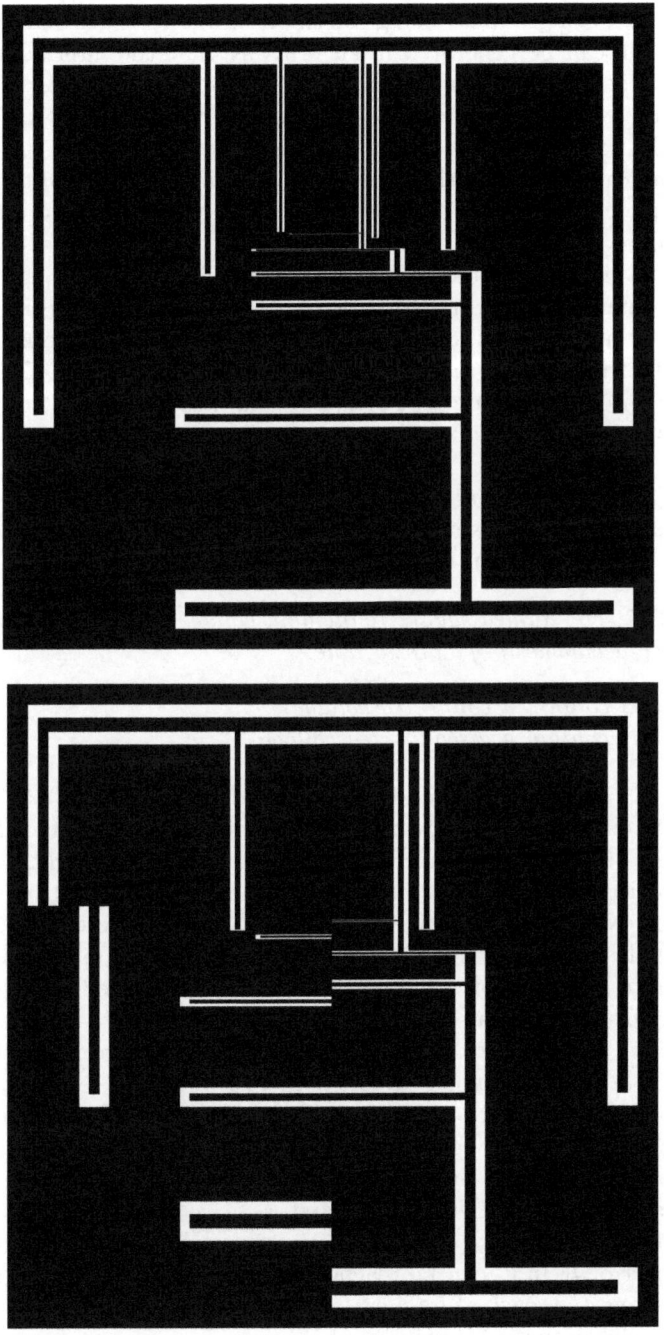

Fig. 9. Proving that the existence of a disconnected generated collage is undecidable

It may be interesting to note that the reductions construct grammars of finite index, the index of a grammar being the smallest bound b such that every collage in its language can be generated in such a way that the intermediate collages contain at most b nonterminals. Thus, the undecidability results of this section hold even for (partial-)array collage grammars of finite index. In fact, looking closely at the construction of G_{PCP} and re-writing the rules a bit to allow for more economical derivations (in the sense of minimizing the need to use simultaneously active nonterminals) one easily checks that this holds even for the fixed index $b = 3$.

It seems to be mostly unknown for which special cases of square-refinement collage grammars connectedness or disconnectedness can be decided. Even for grid collage grammars, the decidability status of these questions seems to be unknown. To the author's knowledge, only one positive result of this type has been shown for collage grammars in the literature, using a very strong restriction that makes the result become almost immediate.[6] Let us say that a square-refinement collage grammar is *framed* if, for every right-hand side K in that grammar, $\bigcup K$ contains every point that belongs to the boundary of U. Then, we can easily prove the desired decidability result, which is [13, Theorem 4.10] stated for square-refinement collage grammars.

Theorem 17. *For framed square-refinement collage grammars G, the following questions are decidable in quadratic time:*

- *Does $L(G)$ contain a connected collage?*
- *Does $L(G)$ contain a disconnected collage?*

Proof. Let $G = (X, T, R, S)$, and consider a derivation $(S, U) \Rightarrow^n C$. Let us say that a part is a *boundary part* if it intersects the boundary of U. We show by induction on n that C is connected if and only if the right-hand sides of all rules applied in that derivation are connected. For $n = 0$, this statement is trivially true. Now, assume that $(S, U) \Rightarrow^{n-1} C' \Rightarrow C$, where $C = C'[K/N]$ for a nonterminal $N \in C'$ and a right-hand side K of an appropriate rule in R.

"$\Rightarrow$"   Since K is nonempty, it follows that C' is connected. By the induction hypothesis, this means that only rules with connected right-hand sides are applied to derive C'. Now, to derive a contradiction, assume that the (framed) right-hand side K is disconnected. Then there is a part $P \in K$ that cannot be reached from any boundary part $P' \in K$, because each of the boundary parts can be reached from any other boundary part. It follows that, in C, the image of P cannot be reached from the image of P' either, contradicting the assumption that C is connected. Thus, K is connected.

"$\Leftarrow$"   By the induction hypothesis, C' is connected. Thus, the image of every boundary part of K in C can be reached from every part in $C' \setminus \{N\}$. Further, since K is connected, each of its parts can be reached from each of its boundary parts, which shows that C is connected.

---

[6] However, see Sect. 6 for an interesting positive result regarding iterated function systems.

Now, we can decide whether $L(G)$ contains a disconnected collage by applying Lemma 6 to remove useless nonterminal labels. If the resulting grammar contains a rule whose right-hand side is disconnected, G generates a disconnected collage. Since we do not need the full Lemma 6 (for the problem at hand, there is no use in removing chain rules and epsilon derivations), this can be implemented to run in quadratic time.

To decide whether G generates a connected collage, we delete all rules whose right-hand sides are disconnected and apply Theorem 10 afterwards (which, again, takes quadratic time).   □

5   Parts Resting on the Diagonal

Let us say that a part $(\lambda, [p : p'])$ *rests* on its lower-left corner p, and that it *rests on the diagonal* if $p = (x, x)$ for some x. We shall now show that it is undecidable whether a grid collage grammar generates a part that rests on the diagonal. The proof makes use of a proof idea by Dube [15,16]. In his original proof, Dube showed that it is undecidable whether the attractor of an iterated function system (IFS[7]) contains a point on the diagonal. This was strengthened in [8] to show that the problem is undecidable even for the special case of Γ-grid IFSs, a grid IFS being an IFS obeying a restriction similar to that used to define Γ-grid collage grammars. Below, we modify the construction to establish the undecidability of the question whether a grid collage grammar generates a part that rests on the diagonal.

For the undecidability proof, we use another variant of Post's correspondence problem. Recall that, in the ordinary PCP, an instance consists of a finite sequence Π of pairs $\pi_1 = (\pi_1^1, \pi_1^2), \ldots, \pi_k = (\pi_k^1, \pi_k^2)$ of nonempty strings over Σ, and the question to be answered is whether there exists an index sequence $i_1 \ldots i_l \in [k]^+$ such that $\pi_{i_1}^1 \cdots \pi_{i_l}^1 = \pi_{i_1}^2 \cdots \pi_{i_l}^2$. For our purposes, we need a restricted version, namely the prefix-free PCP. Here, it is additionally required that, for all distinct $u, v \in [k]$, neither of $\pi_{i_u}^1$ and $\pi_{i_v}^1$ is a prefix of the other and, similarly, neither of $\pi_{i_u}^2$ and $\pi_{i_v}^2$ is a prefix of the other. The undecidability of PCP is usually proved by constructing an instance in which a witness simulates to an accepting run of a Turing machine. By using reversible Turing machines instead, i.e., Turing machines in which every configuration has exactly one predecessor, Ruohonen [26] showed that even the prefix-free PCP is undecidable. (In fact, the original result by Ruohonen concerns the version of PCP which is both prefix- and suffix-free, but we shall not need suffix-freeness here.)

Lemma 18. *The prefix-free PCP is undecidable.*

Let us now assume that $\pi_1 = (\pi_1^1, \pi_1^2), \ldots, \pi_k = (\pi_k^1, \pi_k^2)$ is such a prefix-free instance of PCP. Without loss of generality, we may assume that $\Sigma = [k-1]$. This enables us to interpret a string $w = d_1 \cdots d_l \in \Sigma^*$ as a number in the

---

[7] See, e.g., [1] for an introduction to the mathematical theory of iterated function systems.

unit interval $[0, 1]$, namely $num(w) = 0.d_1 \ldots d_l$, written in base-$k$ notation. In other words, $num(w) = \sum_{i \in [l]} d_i k^{-i}$. Thus, a pair (w, w') of strings over Σ can be encoded as a point in U, namely $(num(w), num(w'))$. Note that num is injective, because $0 \notin \Sigma$. Hence, $(num(w), num(w'))$ lies on the diagonal if and only if $w = w'$. We will also need to encode the lengths of w and w'. This is done by representing (w, w') as the rectangle $rec[w, w'] = [p : p']$, where $p = (num(w), num(w'))$ and $p' = (num(w) + k^{-|w|}, num(w') + k^{-|w'|})$.

The following lemma is the reason why we have to assume prefix-freeness.

Lemma 19. *Let $w, w' \in \Sigma^*$ be such that neither is a prefix of the other. If $num(w) < num(w')$, then $num(w) + k^{-|w|} \leq num(w')$.*

Proof. Let $w = d_1 \cdots d_l$ and $w' = d'_1 \cdots d'_m$. Since w is not a prefix of w', there is a $j \in [\min(l, m)]$ such that $d_1 = d'_1, \ldots, d_{j-1} = d'_{j-1}$, and $d_j < d'_j$. Hence,

$$\begin{aligned}
num(w) + k^{-|w|} &= \sum_{i \in [l]} d_i k^{-i} + k^{-l} \\
&= \sum_{i \in [j-1]} d_i k^{-i} + (d_j + 1)k^{-j} \\
&\leq \sum_{i \in [j]} d'_i k^{-i} \\
&\leq num(w').
\end{aligned}$$

$\square$

Now, consider the collage grammar

$$G'_{\text{PCP}} = (\{S\}, \{\beta\}, r_1, \ldots, r_k, r'_1, \ldots, r'_k, S),$$

where $r_i = (S ::= (S, rec[\pi_i^1, \pi_i^2]))$, and $r'_i = (S ::= rec[\pi_i^1, \pi_i^2])$ for all $i \in [k]$.[8] By Lemma 19, this collage grammar is a grid collage grammar, where the grid is determined by $num(\pi_1^1), num(\pi_1^1) + k^{-|\pi_1^1|}, \ldots, num(\pi_k^1), num(\pi_k^1) + k^{-|\pi_k^1|}$ and $num(\pi_1^2), num(\pi_1^2) + k^{-|\pi_1^2|}, \ldots, num(\pi_k^2), num(\pi_k^2) + k^{-|\pi_k^2|}$ on the horizontal and the vertical axis, respectively. The following lemma shows that derivations by these rules correspond to concatenation of the strings encoded by the rectangles $rec[\pi_i^1, \pi_i^2]$.

Lemma 20. *For every derivation $(S, U) = C_0 \Rightarrow_{r_{i_1}} C_1 \Rightarrow_{r_{i_2}} \cdots \Rightarrow_{r_{i_l}} C_l$ in G'_{PCP}, it holds that $C_l = (S, rec[\pi_{i_1}^1 \cdots \pi_{i_l}^1, \pi_{i_1}^2 \cdots \pi_{i_l}^2])$.*

Proof. By induction on l. The statement obviously holds for the case $l = 0$. For the inductive step, it suffices to establish that, if $C = \{N\}$ with $N = (S, rec[u, u'])$, and $C' = (S, rec[w, w'])$, then $C[C'/N] = (S, rec[uw, u'w'])$. By the definition of replacement and the construction of $rec[u, u']$, $C[C'/N] = (S, \rho)$, where ρ is obtained from rec by

1. scaling it horizontally and vertically by the factors $k^{-|u|}$ and $k^{-|u'|}$, resp., and
2. translating it horizontally and vertically by $num(u)$ and $num(v)$, resp.

---

[8] Again, since there is only one terminal label, we identify a terminal with its second component.

Thus, if $u = d_1 \ldots d_l$ and $w = d_1' \ldots d_m'$, it follows that the left edge of ρ has the coordinate

$$num(u) + num(w)k^{-|u|} = \sum_{i \in [l]} d_i k^{-i} + \sum_{i \in [m]} d_i' k^{-(l+i)} = num(uw)$$

and the horizontal size of the rectangle is $k^{-(l+m)}$, as required. Clearly, the argument is entirely similar for the vertical direction.   □

Now, we can easily establish the correctness of the reduction.

Lemma 21. *Let G'_{PCP} be constructed from an instance Π of the prefix-free PCP as described above. Then Π has a witness if and only if G'_{PCP} generates a terminal that rests on the diagonal.*

Proof. By Lemma 20, G'_{PCP} generates exactly the set of all terminals of the form $rec[w, w']$, such that there is a non-empty index sequence $i_1 \cdots i_l$ with $w = \pi_{i_1}^1 \cdots \pi_{i_l}^1$ and $w' = \pi_{i_1}^2 \cdots \pi_{i_l}^2$. If $w = w'$, then $rec[w, w']$, by its very definition, rests on the diagonal. Otherwise, by the injectivity of num, we have $num(w) \neq num(w')$, which means that $rec[w, w']$ does not rest on the diagonal.   □

We have thus established the undecidability of the question whether a linear grid collage grammar generates a part that rests on the diagonal. (Recall that a grammar is called linear if every right-hand side contains at most one non-terminal.) By putting all the nonterminals into one right-hand side and all the terminals into another, we obtain the same result for grid collage grammars with only two rules.

Theorem 22. *For grid collage grammars G, it is undecidable whether $L(G)$ generates a (collage containing a) part that rests on the diagonal. This remains true if G is required to be linear or, alternatively, required to have at most two rules.*

Proof. For linear grid collage grammars, this is immediate from Lemma 21, because the reduction constructs a linear grammar. Alternatively, we may consider $G''_{\mathrm{PCP}} = (\{S\}, \{\beta\}, r, r', S)$, where $r = (S ::= \{(S, rec[\pi_i^1, \pi_i^2]) \mid i \in [k]\})$, and $r' = (S ::= \{rec[\pi_i^1, \pi_i^2] \mid i \in [k]\})$. Clearly, $\bigcup L(G''_{\mathrm{PCP}}) = \bigcup L(G'_{\mathrm{PCP}})$, which proves the assertion for grammars with two rules.   □

6   Concluding Remarks

We have defined square-refinement collage grammars and surveyed decidability and complexity results regarding these grammars or their special cases, namely (partial-) array collage grammars, which are closely related to Boza-palidis' picture-refinement grammars, and grid collage grammars.

A result that has been excluded from the presentation in this paper, because its proof can be formulated much more conveniently in the tree-based setting, concerns raster images of grid collage grammars over an evenly spaced grid. These have been studied in [11]. Let us call such grid collage grammars uniform. A raster is simply another evenly spaced grid, say an $m \times n$-grid Γ. Let $Pixels_\Gamma$ be the Γ-array consisting of black parts only. (As before, we shall consider $Pixels_\Gamma$ to be a set of rectangles, disregarding their label.) The intuition is that $Pixels_\Gamma$ corresponds to the set of pixels of a screen or another rectangular displaying device. Now, given a collage C, we can *rasterize* it by selecting those pixels in $Pixels_\Gamma$ that intersect with or are covered by parts in C. In this way, we get the upper and lower raster image of C, respectively (in black and white). Formally,

$$upper_\Gamma(C) = \{pix \in Pixels_\Gamma \mid pix \cap \bigcup C \neq \emptyset\} \text{ and}$$
$$lower_\Gamma(C) = \{pix \in Pixels_\Gamma \mid pix \subseteq \bigcup C\}.$$

Given a collage grammar G, the *upper and lower galleries generated by G* are $\mathcal{G}_\Gamma^u(G) = \{upper_\Gamma(C) \mid C \in L(G)\}$ and $\mathcal{G}_\Gamma^l(G) = \{lower_\Gamma(C) \mid C \in L(G)\}$, respectively. Note that both these galleries are finite, because $Pixels_\Gamma$ is finite. Now, the main result of [11] (see also [8, Theorem 5.2.2]) states that $\mathcal{G}_\Gamma^u(G)$ and $\mathcal{G}_\Gamma^l(G)$ are computable:

Theorem 23. *There is an algorithm that, given as input a uniform grid collage grammar G and a raster Γ, computes $\mathcal{G}_\Gamma^u(G)$ and $\mathcal{G}_\Gamma^l(G)$.*

It seems likely that the theorem can be extended to general grid collage grammars (thus dropping the restriction to evenly spaced grids) and to at least certain types of partial-array collage grammars, but this question has not been studied yet. As a corollary to the theorem, we get the following.

Corollary 24. *The following questions are decidable:*

1. *Given a uniform grid collage grammar G, a raster Γ, and a set $img \subseteq Pixels_\Gamma$ as input, is img in $\mathcal{G}_\Gamma^u(G)$? Is img in $\mathcal{G}_\Gamma^l(G)$?*
2. *Given two uniform grid collage grammars G, G' (possibly over different grids) and a raster Γ as input, are $\mathcal{G}_\Gamma^u(G)$ and $\mathcal{G}_\Gamma^u(G')$ equal? Are $\mathcal{G}_\Gamma^l(G)$ and $\mathcal{G}_\Gamma^l(G')$ equal?*

Restrictions that are similar to those applied to collage grammars in this paper can also be used to restrict other picture-generating mechanisms, to make them refine the unit square in a stepwise fashion. In particular, this includes iterated function systems (IFS, that were already mentioned in Sect. 5) and their extensions. Intuitively, these are deterministic square-refinement collage grammars that generate a single, usually infinitely detailed picture called a fractal, which is the limit of an infinite derivation. In the same way, but dropping the requirement of determinism, languages of fractals can be generated and studied. Devices of these kinds have been studied in [5,6,7,8] in the collage grammar setting and in [3] in the algebraic setting of picture-refinement grammars.

In view of the undecidability results of Sect. 4, it may be of interest to mention that connectedness of the fractals generated by grid IFS was shown to be a decidable problem in [5] (see also [8, Sect. 5.4.4]). The reader may perhaps wonder why the question is undecidable for collage grammars but decidable for IFS, even though the restriction to grids is imposed on both. There are two major differences between the two situations. An IFS is deterministic, making it impossible employ "guessing" as in the grammar G_{PCP} used in Sect. 4, and it has only one nonterminal, whereas the idea underlying the construction of G_{PCP} requires to use more than one in order to create the different regions that, together, represent a witness. Thus, there are several interesting open problems on either side left to be studied: Can the decidability result for grid IFS be extended to so-called networked IFS (which can make use of any number of nonterminals) or to nondeterministic IFS? Similarly, is connectedness decidable for collage grammars with only one nonterminal? Can we restrict the nondeterminism in a reasonable way for connectedness to become decidable?

Most of the positive results shown here, especially those in Sect. 3, remain correct for certain types of grammars with more powerful derivation modes, such as ET0L, branching synchronization [10], or even the collage languages whose underlying derivation trees are in $MT^*(REGT)$, the closure of the class of regular tree languages under macro tree transductions. For details, the interested reader should consult [8].

An interesting extension of partial-array collage grammars that has been studied in the literature is the random-context picture grammar invented by Ewert and van der Walt [18] (see [17] for further references). Random context is a form of regulated rewriting [4], where every rule is equipped with two sets of so-called permitting and forbidding nonterminal labels. By definition, a rule is applicable to an occurrence of its left-hand side in the sentential form if all permitting nonterminal labels occur elsewhere in the sentential form and none of the forbidding ones does. Thus, the term "random" indicates that the context may be arbitrarily distributed in the sentential form. As a consequence, this form of context dependency is well suited not only for string grammars, but also for other types of grammars based on the replacement of nonterminal items. While there are several structural results known for random-context picture grammars, in particular pumping and shrinking lemmas, not much seems to be known about decidability questions (except for the undecidability results that follow trivially from results such as those surveyed in this paper).

Acknowledgment. I thank George Rahonis for proofreading the paper and pointing out a number of typos.

References

1. Barnsley, M.: Fractals Everywhere. Academic Press, Boston (1988)
2. Bauderon, M., Courcelle, B.: Graph expressions and graph rewriting. Mathematical Systems Theory 20, 83–127 (1987)

3. Bozapalidis, S.: Picture deformation. Acta Informatica 45, 1–31 (2008)
4. Dassow, J., Păun, G.: Regulated Rewriting in Formal Language Theory. EATCS Monographs in Theoretical Computer Science. Springer, Berlin (1989)
5. Drewes, F.: Language theoretic and algorithmic properties of d-dimensional collages and patterns in a grid. Journal of Computer and System Sciences 53, 33–60 (1996)
6. Drewes, F.: Tree-based generation of languages of fractals. Report 2/99, Univ. Bremen (1999); revised version appeared in Theoretical Computer Science
7. Drewes, F.: Tree-based picture generation. Theoretical Computer Science 246, 1–51 (2000)
8. Drewes, F.: Grammatical Picture Generation – A Tree-Based Approach. Texts in Theoretical Computer Science. An EATCS Series. Springer, Heidelberg (2006)
9. Drewes, F., Engelfriet, J.: Decidability of the finiteness of ranges of tree transductions. Information and Computation 145, 1–50 (1998)
10. Drewes, F., Engelfriet, J.: Branching synchronization grammars with nested tables. Journal of Computer and System Sciences 68, 611–656 (2004)
11. Drewes, F., Ewert, S., Klempien-Hinrichs, R., Kreowski, H.-J.: Computing raster images from grid picture grammars. Journal of Automata, Languages and Combinatorics 8, 499–519 (2003)
12. Drewes, F., Habel, A., Kreowski, H.-J.: Hyperedge replacement graph grammars. In: Rozenberg, G. (ed.) Handbook of Graph Grammars and Computing by Graph Transformation. Foundations, vol. 1, ch. 2, pp. 95–162. World Scientific, Singapore (1997)
13. Drewes, F., Kreowski, H.-J.: (Un-)decidability of geometric properties of pictures generated by collage grammars. Fundamenta Informaticae 25, 295–325 (1996)
14. Drewes, F., Kreowski, H.-J.: Picture generation by collage grammars. In: Ehrig, H., Engels, G., Kreowski, H.-J., Rozenberg, G. (eds.) Handbook of Graph Grammars and Computing by Graph Transformation. Applications, Languages, and Tools, vol. II, ch. 11, pp. 397–457. World Scientific, Singapore (1999)
15. Dube, S.: Undecidable problems in fractal geometry. Complex Systems 7, 423–444 (1993)
16. Dube, S.: Fractal geometry, Turing machines, and divide-and-conquer recurrences. RAIRO Theoretical Informatics and Applications 28, 405–423 (1994)
17. Ewert, S.: Random context picture grammars: The state of the art. In: Drewes, F., Habel, A., Hoffmann, B., Plump, D. (eds.) Manipulation of Graphs, Algebras and Pictures. Essays Dedicated to Hans-Jörg Kreowski on the Occasion of His 60th Birthday, pp. 135–147 (2009)
18. Ewert, S., v.d. Walt, A.P.: Random context picture grammars. Publicationes Mathematicae (Debrecen) 54 (Supp.), 763–786 (1999)
19. Habel, A.: Hyperedge Replacement: Grammars and Languages. LNCS, vol. 643. Springer, Heidelberg (1992)
20. Habel, A., Kreowski, H.-J.: May we introduce to you: Hyperedge replacement. In: Ehrig, H., Nagl, M., Rosenfeld, A., Rozenberg, G. (eds.) Graph Grammars 1986. LNCS, vol. 291, pp. 15–26. Springer, Heidelberg (1987)
21. Habel, A., Kreowski, H.-J.: Pretty patterns produced by hyperedge replacement. In: Göttler, H., Schneider, H.-J. (eds.) WG 1987. LNCS, vol. 314, pp. 32–45. Springer, Heidelberg (1988)
22. Habel, A., Kreowski, H.-J.: Collage grammars. In: Ehrig, H., Kreowski, H.-J., Rozenberg, G. (eds.) Graph Grammars 1990. LNCS, vol. 532, pp. 411–429. Springer, Heidelberg (1991)

23. Hopcroft, J.E., Ullman, J.D.: Introduction to Automata Theory, Languages and Computation. Addison-Wesley, Reading (1979)
24. Rosenfeld, A.: Picture Languages: Formal Models for Picture Recognition. Academic Press, New York (1979)
25. Rosenfeld, A., Siromoney, R.: Picture languages – a survey. Languages of Design 1, 229–245 (1993)
26. Ruohonen, K.: Reversible machines and Post's correspondence problem for biprefix morphisms. EIK – Journal on Information Processing and Cybernetics 12, 579–595 (1985)
27. Younger, D.H.: Recognition and parsing of context-free languages in time n^3. Information and Control 10, 189–208 (1967)

Weighted Tree Automata over Valuation Monoids and Their Characterization by Weighted Logics

Manfred Droste[1], Doreen Götze[1], Steffen Märcker[2], and Ingmar Meinecke[1]

[1] Institut für Informatik,
Universität Leipzig, D-04109 Leipzig, Germany
{droste,goetze,meinecke}@informatik.uni-leipzig.de
[2] Institut für Angewandte Informatik e.V.,
Universität Leipzig, D-04109 Leipzig, Germany
maercker@infai.org

Abstract. Quantitative aspects of systems can be modeled by weighted automata. Here, we deal with such automata running on finite trees. Usually, transitions are weighted with elements of a semiring and the behavior of the automaton is obtained by multiplying the weights along a run. We turn to a more general cost model: the weight of a run is now determined by a global valuation function. An example of such a valuation function is the average of the weights. We establish a characterization of the behaviors of these weighted finite tree automata by fragments of weighted monadic second-order logic. For bi-locally finite bimonoids, we show that weighted tree automata capture the expressive power of several semantics of full weighted MSO logic. Decision procedures follow as consequences.

1 Introduction

Trees or terms are one of the most fundamental concepts both in mathematics and in computer science. J. R. Büchi [19] wrote: "It is very easy to tell what these terms are and why they merit an investigation. They are those famous (some will say infamous) formulas that distinguish mathematical texts from others. [...] And of course, these very same terms make up the bulk, if not all, of modern programming languages." The early papers by Thatcher and Wright [54] and by Doner [25] on regular tree languages established the connection between finite tree automata and monadic second-order logic, thus extending the corresponding results on string languages by Büchi [18] and Elgot [37]. Within the last decades, the theory of tree languages developed rapidly, see [24, 42] for surveys. Not surprisingly, also quantitative aspects gained attention since the beginning of the 1980s [2]. Many authors dealt with different aspects of weighted tree automata and their behavior [5, 31, 39, 46, 47, 38, 53]. Bozapalidis and his co-workers contributed a huge amount of concepts and results to the theory of formal tree series, especially concerning an algebraic treatment of the topic [1, 6–11, 17, 12–16]. For an overview of recent results on weighted automata over different structures including trees see [29].

W. Kuich and G. Rahonis (Eds.): Bozapalidis Festschrift, LNCS 7020, pp. 30–55, 2011.

In order to obtain a counterpart in terms of logic for weighted word and tree automata, Droste and Gastin [26, 27] developed a weighted MSO logic with weights stemming from a semiring and proved a Büchi-like characterization for formal power series on words. The generalization of this result to tree series followed quickly by Droste and Vogler [33]. Moreover, similar results have been obtained for unranked trees [35], for infinite trees [52], and for trees over multi-operator monoids [40].

Here, we present another concept of weighted tree automata and their characterization by weighted logics. The conceptual difference is the following: the weight of a run of the automaton is not any longer computed in a local way by a binary operation (e.g., semiring multiplication), but in a global way, i.e., given a run of the automaton we apply a valuation function to all the weights appearing along the run. A natural example is the arithmetic mean of the weights; it models, e.g., average consumption of resources. The concept of a valuation function traces back to a study by Chatterjee, Doyen, and Henzinger [20] where they considered such functions for string automata with real numbers as weights. The expressiveness and various decision problems for these automata were explored in several papers [21–23]. The model of Chatterjee et al. was generalized in [30] to a more general weight structure, called valuation monoid, and the class of behaviors of these weighted automata was characterized by weighted MSO logic both for finite and infinite words. In [44], this approach was extended to finite traces.

For words, an evaluation function maps sequences of weights to a single weight. For trees, we adapt the concept such that a tree valuation function maps trees which are labeled with weights to a single weight. An additive monoid equipped with a tree valuation function forms a tree valuation monoid. These structures subsume semirings but are much more comprehensive. Concerning automata, the weight of a run is computed by the valuation function and then the weights of several runs on the same tree are combined by addition. We note that the weighted tree automata over absorptive multioperator monoids recently investigated in [39, 40] can also be seen as weighted tree automata over particular valuation monoids.

With respect to weighted MSO logic, more structure is necessary to define the semantics. Here, we follow the approach in [30]. Negation is pushed to the atomic level. Disjunction and existential quantification are interpreted by addition, and the formula 'false' is mapped to the neutral element of the monoid, the zero element $\mathbb{0}$. First-order universal quantification makes a statement about all positions of a tree. Therefore, we will define the semantics of this quantor by the application of the valuation function. But what about conjunction, what about the simple formula 'true'? For this, we will introduce an additional binary operation, called product, and a unit for this product, the element $\mathbb{1}$. Now the semantics of 'true' will be $\mathbb{1}$ and the semantics of conjunction will be multiplication by the new product. Tree valuation monoids with these additional features are called product tree valuation monoids. It is of interest that, first, the semantics of universal quantification and of conjunction, respectively, are not defined by the same operation anymore and, second, the interpretation of conjunction in general needs neither to be commutative nor associative anymore. Also, distributivity of the product over addition is not necessarily required. We would like to stress that in our setting the loss of those properties is not substituted by local finiteness conditions. Such non-standard properties of logics appeared also already in lattice valued fuzzy logics [36], in multi-valued logics [45, 49], and in quantum logics [3].

Our main result is as follows. We characterize the behavior of weighted tree automata by three different fragments of weighted MSO logics, cf. Theorem 5.5. Which fragment can be used depends on the properties of the underlying product tree valuation monoid. The restrictions on the fragments are purely syntactic. The use of universal quantification is restricted in the same way as it was already done for semiring-weighted automata [27, 28]. The fragments differ in the way to which extent the use of conjunction is confined. Our result generalizes both the result of [33], also cf. [41], about semiring-weighted tree automata and the result of [30] about weighted automata over finite words and valuation monoids.

Moreover, we consider strong bimonoids as a special case of (tree) valuation monoids. Strong bimonoids can be viewed as particular semirings missing distributivity and, in this context, as product tree valuation monoids by using multiplication both for the product and the valuation function. However, for the valuation function we need also an enumeration of the positions of a tree to define in which order the weights of a run are multiplied. We show that different enumerations may yield different classes of recognizable tree series for the same underlying strong bimonoid, cf. Proposition 3.4. The characterization by weighted MSO logic follows as a corollary of our main result. However, we do not have to restrict the logic if certain local finiteness assumptions are made. In this case, all weighted MSO formulas have a recognizable semantics, cf. Theorem 6.3. In fact, for bi-locally finite bimonoids every tree series is of a simple form: it is a so-called recognizable step function, see Theorem 6.2. In the more general setting of tree valuation monoids, the concept of local finiteness still has to be explored.

2   Trees and Tree Valuation Monoids

Let $\mathbb{N} = \{1, 2, \ldots\}$ be the set of all natural numbers and $\mathbb{N}_0 = \mathbb{N} \cup \{0\}$. A *ranked alphabet* is a pair $(\Sigma, \mathrm{rk}_\Sigma)$ consisting of a finite alphabet Σ and a mapping $\mathrm{rk}_\Sigma : \Sigma \to \mathbb{N}_0$ which assigns to each symbol of Σ its rank. We write $\Sigma^{(m)}$ for the set of all symbols with rank $m \in \mathbb{N}_0$ and $a^{(m)}$ to denote that $a \in \Sigma^{(m)}$. If rk_Σ is known from the context, we just write Σ for $(\Sigma, \mathrm{rk}_\Sigma)$. Let $\max_\Sigma = \max\{\mathrm{rk}_\Sigma(a) \mid a \in \Sigma\}$, the maximum rank of Σ. Subsequently, we always assume that $\Sigma^{(0)} \neq \emptyset$.

Let $\mathbb{N}^*$ be the set of all finite words over $\mathbb{N}$. A *tree domain* $\mathcal{B}$ is a finite, non-empty subset of $\mathbb{N}^*$ such that for all $u \in \mathbb{N}^*$ and $i \in \mathbb{N}$, $u.i \in \mathcal{B}$ implies $u, u.1, \ldots, u.(i-1) \in \mathcal{B}$. Note that, since $u \in \mathcal{B}$ whenever $u.i \in \mathcal{B}$, the tree domain $\mathcal{B}$ is prefix-closed. A *tree over a set M (of labels)* is a mapping $t : \mathcal{B} \to M$

such that $\text{dom}(t) = \mathcal{B}$ is a tree domain. The elements of $\text{dom}(t)$ are called *positions* of t and $t(u)$ is called *label* of t at $u \in \text{dom}(t)$. The set of all labels of t is defined by the image of t, thus by $\text{im}(t) = \{t(u) \mid u \in \text{dom}(t)\}$. We denote the set of all trees over M by T_M. A *tree language* is a subset of T_M. A *tree over a ranked alphabet* $(\Sigma, \text{rk}_\Sigma)$ is a tree over the set Σ such that for all $u \in \text{dom}(t)$, $|\{u.i \mid i \geq 1, u.i \in \text{dom}(t)\}| = k$ whenever $t(u) \in \Sigma^{(k)}$.

Analogously to Droste and Meinecke in [30], we define a structure which we use to describe the behavior of weighted tree automata. By abuse of notation let $d \in T_D$ denote the tree only consisting of the position ε which is labeled with $d \in D$.

Definition 2.1. *A* tree valuation monoid *(tv-monoid for short) is a quadruple* $(D, +, \text{Val}, \mathbb{0})$ *such that* $(D, +, \mathbb{0})$ *is a commutative monoid and* $\text{Val} : T_D \to D$ *is a function with* $\text{Val}(d) = d$ *for every tree* $d \in T_D$ *and* $\text{Val}(t) = \mathbb{0}$ *whenever* $\mathbb{0} \in \text{im}(t)$ *for* $t \in T_D$.

Val is called a *(tree) valuation function*. We will use $+$ to deal with the potential non-determinism of weighted tree automata and the valuation function to combine the weights assigned to the positions of a run, see Section 3.

Next we give examples of tv-monoids.

Example 2.2. $\mathbb{Q}_{\max} = (\mathbb{Q} \cup \{-\infty\}, \max, \text{avg}, -\infty)$ with

$$\text{avg}(t) = \frac{\sum_{u \in \text{dom}(t)} t(u)}{|\text{dom}(t)|}$$

for all $t \in T_{\mathbb{Q} \cup \{-\infty\}}$ is a tv-monoid. The valuation function of this tv-monoid calculates the average of all weights of a tree. The value $-\infty$ is the zero of the tv-monoid $\mathbb{Q}_{\max}$.

Similarly, $\mathbb{Q}_{\min} = (\mathbb{Q} \cup \{\infty\}, \min, \text{avg}, \infty)$ is a tv-monoid with ∞ as zero.

Example 2.3. Let $L(t)$ be a longest path from the root to a leaf of a tree t. If there is more than one such path with maximal length, we choose the leftmost one. Then $(\mathbb{N}_0, +, \text{Val}_l, 0)$ with

$$\text{Val}_l(t) = \begin{cases} \prod_{u \in L(t)} t(u) & \text{if } 0 \notin \text{im}(t), \\ 0 & \text{otherwise} \end{cases}$$

for all $t \in T_{\mathbb{N}_0}$ is a tv-monoid. Its valuation function multiplies all values labeling the positions of the leftmost longest path of t.

Example 2.4. The tv-monoid $(\mathbb{Q}_0^+, +, \text{W}_{\text{root}}, 0)$ with

$$\text{W}_{\text{root}}(t) = \begin{cases} t(\varepsilon) & \text{if } 0 \notin \text{im}(t) \\ 0 & \text{otherwise} \end{cases}$$

for all $t \in T_{\mathbb{Q}_0^+}$ is equipped with a valuation function W_{root} which returns the weight of the root position of a tree.

Example 2.5. We consider $\mathbb{R}_{disc} = (\mathbb{R} \cup \{-\infty\}, \max, \mathrm{disc}_\Lambda, -\infty)$ where the tree valuation function disc_Λ models a discounting on trees, cf. [50].

Here, $\Lambda = (\lambda_i)_{i \in \mathbb{N}}$ with $\lambda_i > 0$ for $i \in \mathbb{N}$. Now the weight of some $t \in T_{\mathbb{R}_{disc}}$ at position $u = k_1 k_2 \dots k_m \in \mathbb{N}^*$ is defined as

$$\mathrm{wgt}_u(t) = \begin{cases} t(u) & \text{if } u = \varepsilon, \\ (\prod_{i=1}^m \lambda_{k_i}) \cdot t(u) & \text{otherwise.} \end{cases}$$

Then we put

$$\mathrm{disc}_\Lambda(t) = \sum_{u \in \mathrm{dom}(t)} \mathrm{wgt}_u(t).$$

Here, the discounting depends on the distance of the node from the root. Moreover, we allow different discount factors for different directions within the tree. In [50], a more general setting of discounting on trees is explored.

3   Weighted Tree Automata

Now we introduce weighted automata running on finite trees with weights from a tv-monoid. Let Σ be a ranked alphabet and $(D, +, \mathrm{Val}, \mathbb{0})$ a tv-monoid.

Definition 3.1. *A weighted bottom-up tree automaton (wta for short) over a tv-monoid D is a quadruple $\mathcal{M} = (Q, \Sigma, \mu, F)$ where Q is a non-empty finite set of states, Σ is a ranked alphabet, $\mu = (\mu_m)_{0 \le m \le \max_\Sigma}$ is a family of transition mappings $\mu_m : \Sigma^{(m)} \to D^{Q^m \times Q}$, and $F \subseteq Q$ is a set of final states.*

We define the behavior of a weighted tree automaton $\mathcal{M}$ by a run semantics. A *run* r of $\mathcal{M}$ on a tree $t \in T_\Sigma$ is a mapping $r : \mathrm{dom}(t) \to Q$. For all positions $u \in \mathrm{dom}(t)$ labeled with $t(u) \in \Sigma^{(m)}$, we call $\mu_m(t(u))_{r(u.1)\dots r(u.m).r(u)}$ the *weight of r on t at u*. Note that $\mu_m(t(u))_{r(u.1)\dots r(u.m).r(u)}$ is an abbreviation for $\mu_m(t(u))(r(u.1), \dots, r(u.m), r(u))$ inspired by the matrix notation. Since the domain of a run is a tree domain, each run r on t defines a tree $\mu(t, r) \in T_D$ where $\mathrm{dom}(\mu(t,r)) = \mathrm{dom}(t)$ and $\mu(t,r)(u) = \mu_m(t(u)^{(m)})_{r(u.1)\dots r(u.m).r(u)}$ for all $u \in \mathrm{dom}(t)$. We call r on t *valid* if $\mathbb{0} \notin \mathrm{im}(\mu(t,r))$ and *successful* if $r(\varepsilon) \in F$. Furthermore, $\mathrm{succ}(\mathcal{M}, t)$ denotes the set of all successful runs of $\mathcal{M}$ on t. We call $\mathrm{Val}(\mu(t,r))$ the *weight of r on t*. We set $\mathrm{Val}(\mu(t,r)) = \mathbb{0}$ if r is not valid. The *behavior* of a wta $\mathcal{M}$ is the function $\|\mathcal{M}\| : T_\Sigma \to D$ defined by

$$\|\mathcal{M}\|(t) = \sum (\mathrm{Val}(\mu(t,r)) \mid r \in \mathrm{succ}(\mathcal{M}, t))$$

for all $t \in T_\Sigma$. If no successful run on t exists, we put $\|\mathcal{M}\|(t) = \mathbb{0}$.

A *(formal) tree series* is a mapping $S : T_\Sigma \to D$. A tree series S is called *recognizable* if $S = \|\mathcal{M}\|$ for some wta $\mathcal{M}$. Then we say that $\mathcal{M}$ *recognizes* S.

Example 3.2. Let $\mathbb{Q}_{\max}$ be the tv-monoid of Example 2.2. We consider the wta $\mathcal{M} = (\{q\}, \Sigma, \mu, \{q\})$ with $\mu_0(a).q = 1$ for all $a \in \Sigma^{(0)}$ and $\mu_m(b)_{q\dots q.q} = 0$ for all $b \in \Sigma^{(m)}$ and $m \geq 1$. For every tree t there is a unique run r on t that assigns to each position the state q. This run is valid and successful. Thus, we get

$$\|\mathcal{M}\|(t) = \text{avg}(\mu(t,r)) = \frac{\sum_{u \in \text{dom}(t)} \mu(t,r)(u)}{|\text{dom}(t)|} = \frac{\text{``number of leaves of } t\text{''}}{\text{``size of } t\text{''}}$$

where the size of a tree is the number of all nodes of the tree. So $\mathcal{M}$ calculates the leaves-to-size ratio of every tree t.

With another wta $\mathcal{M}'$ over $\mathbb{Q}_{max}$ we can also calculate the height-to-size ratio of trees. The height of a tree is the number of nodes of one of the longest paths decremented by one. To calculate the height, a run of the wta chooses a path from a leaf to the root non-deterministically and weights its positions with 1 except for the leaf. All other positions of the tree are weighted with 0. To achieve this, the wta possesses two states, say, the state p to signalize that a position belongs to the chosen path and the state n to signalize that a position does not belong to the chosen path. Since the non-determinism of the wta is resolved by maximum, $\mathcal{M}'$ takes the run which chooses one of the longest paths and applies the valuation function to this run. Formally, $\mathcal{M}' = (\{p,n\}, \Sigma, \mu', \{p\})$ with $\mu'_0(a).p = \mu'_0(a).n = 0$ for all $a \in \Sigma^{(0)}$ and

$$\mu'_m(b)_{q_1 \ldots q_m \cdot q} = \begin{cases} 1 & \text{if } \exists i : q = q_i = p \wedge \forall j \neq i : q_j = n, \\ 0 & \text{otherwise} \end{cases}$$

for all $b \in \Sigma^{(m)}$ with $m \geq 1$. For $t \in T_\Sigma$, let $L(t)$ be a longest path in t. Then

$$\|\mathcal{M}'\|(t) = \max_{r \in \text{succ}(\mathcal{M}',t)} \left(\frac{\sum_{u \in \text{dom}(t)} \mu(t,r)(u)}{|\text{dom}(t)|} \right) = \frac{\sum_{u \in L(t)} 1}{|\text{dom}(t)|} = \frac{\text{``height of } t\text{''}}{\text{``size of } t\text{''}}.$$

Remark 3.3. Due to Alexandrakis and Bozapalidis [1], also cf. [41, Def. 3.2], weighted tree automata over a semiring $(S, +, \cdot, \mathbb{0}, \mathbb{1})$ are defined as quadruple $\mathcal{A} = (Q, \Sigma, \mu, \nu)$ where Q is a non-empty finite set of states, Σ a ranked alphabet, $\mu = (\mu_m)_{0 \leq m \leq \max_\Sigma}$ a family of transition mappings $\mu_m : \Sigma^{(m)} \to S^{Q^m \times Q}$, and $\nu \in S^Q$ is a final weight vector. The behavior $\|\mathcal{A}\|$ of such an automaton $\mathcal{A}$ can either be defined by an initial algebra semantics or by a run semantics (using depth first search, see below) which turn out to be the same [41, p. 324]. Due to the distributivity of semiring multiplication over addition, weighted tree automata can be normalized, i.e., the final weights can be replaced by final states, cf. [41, Thm. 3.6]. Thus, it is no restriction for the class of tree series recognized by these automata to assume that $\nu \in \{\mathbb{0}, \mathbb{1}\}^Q$.

Now we can construct a tv-monoid $S' = (S, +, \text{Val}, \mathbb{0})$ by setting

$$\text{Val}(t) = \prod_{u \in \text{dom}(t)} t(u)$$

for all $t \in T_S$ where the weights are multiplied in the order induced by depth first search (for commutative semirings the order does not matter). Then we can consider $\mathcal{A}$ also as a weighted tree automaton over the tree valuation monoid S' with $\|\mathcal{A}\|_S = \|\mathcal{A}\|_{S'}$. Thus, wta over tree valuation monoids generalize those over semirings.

We have defined our automata with final states instead of final weights. Whereas this makes no difference for semirings, this is not clear for tv-momoids because the valuation function evaluates only weights assigned to positions of the tree and final weights are thus out of the scope of a valuation function.

Next we show that wta over tv-monoids also subsume wta over strong bimonoids investigated recently, for words, in [32]. A *bimonoid* $(K, +, \circ, \mathbb{0}, \mathbb{1})$ is an algebraic structure where $(K, +, \mathbb{0})$ and $(K, \circ, \mathbb{1})$ are monoids. A *strong bimonoid* is a bimonoid such that $+$ is commutative and $\mathbb{0}$ is a zero for multiplication. Hence, a semiring is a strong bimonoid in which multiplication distributes over addition. In general, multiplication in a bimonoid is not required to be commutative. Consequently, if we calculate the weight of a tree $t \in T_K$ as a product of the weights at the single positions, the value depends on the order of these weights. On words this order is given naturally, but for trees we have to provide well defined orders. Therefore, we introduce *enumerations of tree positions* as mappings en $: T_K \to [\mathbb{N} \dashrightarrow \mathbb{N}^*]$ such that en(t) is a bijective mapping between $[1, |\mathrm{dom}(t)|]$ and dom(t) for each $t \in T_K$. As before, a weighted tree automaton over a strong bimonoid K is a quadruple $\mathcal{M} = (Q, \Sigma, \mu, F)$. Using an enumeration en of positions, we define the behavior of $\mathcal{M}$ as the function $\|\mathcal{M}\|_{\mathrm{en}} : T_\Sigma \to K$ given by

$$\|\mathcal{M}\|_{\mathrm{en}}(t) = \sum_{r \in \mathrm{succ}(\mathcal{M}, t)} \prod_{i=1}^{|\mathrm{dom}(t)|} \mu(t, r)\Big(\mathrm{en}(\mu(t, r))(i)\Big)$$

for all $t \in T_\Sigma$. We say that a wta $\mathcal{M}$ over a bimonoid K *recognizes a tree series* S *with* en if $S = \|\mathcal{M}\|_{\mathrm{en}}$. If K is commutative, this semantics is independent of the chosen enumeration en.

Weighted tree automata over tv-monoids are an extension of those over bimonoids. In order to derive a tv-monoid $K' = (K, +, \circ_{\mathrm{en}}, \mathbb{0})$ from a given strong bimonoid $K = (K, +, \circ, \mathbb{0}, \mathbb{1})$, we define the valuation function $\circ_{\mathrm{en}}$ by

$$\circ_{\mathrm{en}}(t) = \prod_{i=1}^{|\mathrm{dom}(t)|} t(\mathrm{en}(t)(i))$$

for all $t \in T_K$. Then $\|\mathcal{M}\|_{\mathrm{en}} = \|\mathcal{M}\|_{K'}$.

Now we investigate two special classes of enumerations. First, we give the linear order $\prec_{\mathrm{df}}$ on $\mathbb{N}^*$ as an example derived from depth first search. For all positions $u, v \in \mathbb{N}^*$, inductively let

$$u \prec_{\mathrm{df}} v \iff \exists x \in \mathbb{N} \; \exists u' \in \mathbb{N}^* : u = x.u' \wedge$$
$$\forall y \in \mathbb{N} \; \forall v' \in \mathbb{N}^* : [v = y.v' \to (x < y \vee (x = y \wedge u' \prec_{\mathrm{df}} v'))].$$

The corresponding enumeration is denoted by DF and is used to define the run semantics of bottom-up tree automata including the semantics of wta over general (non-commutative) semirings, cf. Remark 3.3.

The second class of enumerations is based on a linear order of the weight structure K. Weights appearing in a run are collected and then multiplied following their order in K. Let $<$ be a strict linear order on the underlying set of a tv-monoid K. Then we obtain a linear order $\prec_{\mathrm{wo}}$ (the *weight ordering*) on $\mathrm{dom}(t)$ for $t \in T_K$ by requiring

$$t(u) < t(v) \Rightarrow u \prec_{\mathrm{wo}} v$$

for all $u, v \in \mathrm{dom}(t)$; note that $\prec_{\mathrm{wo}}$ may not be unique because a weight d can appear at several positions. But this does not effect the valuation function. We denote the corresponding enumeration by $\mathrm{WO}_<$.

Next we show that the classes of recognizable tree series for two distinct enumerations may be different. Thus, each enumeration induces a specific class of recognizable tree series. Henceforth, for the sake of brevity, we identify a semiring and a bimonoid, respectively, with its derived tv-monoid. For any set M, we let $\wp(M)$ be the power set of M.

Proposition 3.4. *There exist a ranked alphabet Σ, a semiring K, a strict linear order $<'$ on K, and two tree series $S, S' : T_\Sigma \to K$ such that*

1. *S is DF-recognizable but not $\mathrm{WO}_<$-recognizable for any strict linear order $<$ on K and*
2. *S' is $\mathrm{WO}_{<'}$-recognizable but not DF-recognizable.*

Proof. We consider the semiring $K = (\wp(\{1,2\}^*), \cup, \cdot, \emptyset, \{\varepsilon\})$ of languages of finite words over the alphabet $\{1,2\}$ with set union as addition and concatenation of languages as multiplication. Let $\Sigma = \{a^{(0)}, b^{(1)}, c^{(2)}\}$. Note that $\max_\Sigma > 1$.

1. For a word $u = u_1 \ldots u_n \in \{1,2\}^*$ let $\overleftarrow{u} = u_n \ldots u_1$ be the reverse of u. Now let the tree series $S : T_\Sigma \to K$ be defined for all $t \in T_\Sigma$ by

$$S(t) = \{\overleftarrow{u} \mid u \in \mathrm{dom}(t)\},$$

i.e., S collects the reversed positions of the tree t.

We indicate that we can build a wta recognizing S with DF. The basic idea for this wta is that each successful run marks a path to some single node. Now the weight function of the wta maps each marked node at position $u \in \{1,2\}^*$ to $\{i\}$ for $i \in \{1,2\}$ if $u.i$ is the next marked node. In contrast, in a successful run each unmarked node is mapped to $\{\varepsilon\}$. Thus, the evaluation of a successful run with enumeration DF computes the position of a single node in reverse order.

Next, we make some observations on the free monoid $\{1,2\}^*$ and eventually apply them to K. Let $\pi \in \{1,2\}^\omega$ be aperiodic, i.e., there is no factorization such that $\pi = \pi_1 \pi_2^\omega$ for $\pi_1, \pi_2 \in \{1,2\}^*$. Furthermore, let $\pi_1, \ldots, \pi_m$ with $m \in \mathbb{N}$ be an arbitrary sequence of words from $\{1,2\}^*$. Then the set

$$P_m = \left\{ \pi_1^{k_1} \ldots \pi_m^{k_m} \mid k_1, \ldots, k_m \in \mathbb{N}_0 \right\}$$

(where $\pi_i^0 = \varepsilon$) contains only a finite number of prefixes of π. This can be shown by an induction on m. We say, the sequence $\pi_1, \ldots, \pi_m$ forms a *finite number of prefixes of π*.

Now we assume there is a linear order $<$ on K and let $\mathcal{M} = (Q, \Sigma, \mu, F)$ be a wta. Let $W_1 < \ldots < W_m$ with $W_i \subseteq \{1,2\}^*$ for $i \in \{1, \ldots, m\}$ be the sequence of all weights from $\mathrm{im}(\mu)$. Then we have at most $|W_1| \cdot \ldots \cdot |W_m|$ many sequences $\mathsf{w} = \pi_1, \ldots, \pi_m$, such that $\pi_i \in W_i$ for all $i \in \{1, \ldots, m\}$. Next we choose some aperiodic word $\pi \in \{1,2\}^\omega$. Due to the statement above, each sequence w forms only a finite number of prefixes of π. Let k be the length of the longest prefix of π which a sequence w can form. Consequently, no value $W \in \mathrm{im}(\|\mathcal{M}\|)$ contains a word from $\{1,2\}^*$ that is a prefix of π of length greater than k. Let $u \in \{1,2\}^*$ be a prefix of π of length at least $k+1$ and let $t \in T_\Sigma$ be a tree with $\overleftarrow{u} \in \mathrm{dom}(t)$. Then $u \in S(t)$ but $u \notin \|\mathcal{M}\|(t)$. Hence, $\|\mathcal{M}\| \neq S$.

2. We fix a linear order $<'$ on K such that $\{1\} <' \{2\}$. Now we define the tree series S' for all $t \in T_\Sigma$ by

$$
S'(t) = \begin{cases}
\{1^n 2^n\} & \text{if } t \in T_{\{a^{(0)}, b^{(1)}\}} \wedge \exists n \in \mathbb{N}_0 : |\mathrm{dom}(t)| = 2n, \\
\{1^{n+1} 2^n\} & \text{if } t \in T_{\{a^{(0)}, b^{(1)}\}} \wedge \exists n \in \mathbb{N}_0 : |\mathrm{dom}(t)| = 2n + 1, \\
\emptyset & \text{otherwise.}
\end{cases}
$$

A straightforward automaton construction shows that S' is actually recognizable with WO$_{<'}$.

Suppose there is a wta $\mathcal{M} = (Q, \Sigma, \mu, F)$ recognizing the series S' with DF. Let $m = |Q|$ and $\ell_{\max} = \max\{|w| \mid w \in W \text{ and } W \in \mathrm{im}(\mu)\}$, the length of the longest word appearing in a weight of $\mathrm{im}(\mu)$. Subsequently, we apply a pumping argument. Therefore, we choose $x = (m+1) \cdot \ell_{\max}$ and consider the tree

$$
t = a - \overbrace{b - \ldots - b}^{x} - \overbrace{b - \ldots - b}^{x}.
$$

Then $S'(t) = \{1^{x+1} 2^x\}$. Since addition of K is set union, the weight of every valid and successful run of $\mathcal{M}$ on t is $\{1^{x+1} 2^x\}$. Let r be such a run. The run r assigns states to all $2x + 1$ positions of t. The set Q contains m states, thus at least two of the $m+1$ first positions of t (counted from the leaf) are mapped by r to the same state. Hence, r runs through at least one cycle C within its first $m+1$ positions. Now recall that $\mathcal{M}$ uses DF as enumeration and $x = (m+1) \cdot \ell_{\max}$. Therefore, the weight W_C of cycle C contains only words consisting of the letter 1, i.e., $W_C \subseteq \{1\}^*$.

If $W_C = \{\varepsilon\}$, we consider the run r' constructed from r by cutting cycle C. Then r' is a successful run on a tree $t' \in T_{\{a^{(0)}, b^{(1)}\}}$ of size less than t. But since $W_C = \{\varepsilon\}$, the weight of r' is still $\{1^{x+1} 2^x\}$. Hence, $\{1^{x+1} 2^x\} \subseteq \|\mathcal{M}\|(t')$ and thus $\|\mathcal{M}\| \neq S'$.

Now we assume that W_C contains at least one non-empty word. Then we build a new run r' of $\mathcal{M}$ which runs through cycle C twice, but apart from that performs like r. Let t' be the tree on which r' runs. Since $W_C \subseteq \{1\}^*$, every word in the weight of r' has the form $1^y 2^x$ with $y > x + 1$, a contradiction to the definition of $S'(t')$. Hence, $\|\mathcal{M}\| \neq S'$ also in this case.

Altogether, S' is not recognizable with DF. $\square$

Remark 3.5. Next we show that wta over tv-monoids also subsume wta over (absorptive) multioperator monoids investigated recently in [39, 40]. A *multi-operator monoid* (for short: a M-monoid) is a quadruple $A = (A, +, \mathbb{0}, \Omega)$ such that $(A, +, \mathbb{0})$ is a commutative monoid and $\Omega = (\Omega^{(k)})_{k \in \mathbb{N}_0}$ is a collection of operations on A of different arity k. Especially, Ω contains for every arity k the operations $\mathbb{0}^{(k)}$ mapping each k-tuple to $\mathbb{0}$. An M-monoid is called *absorptive* if $\mathbb{0}$ is absorbing for every $\omega \in \Omega$.

Now a weighted tree automaton over a ranked alphabet Σ and an M-monoid A (for short: wmta) is a triple $\mathcal{M} = (Q, \mu, F)$ where Q is a finite non-empty set of states, $\mu = (\mu_k)_{k \in \mathbb{N}_0}$ is a family of mappings $\mu_k : Q^k \times \Sigma^{(k)} \times Q \to \Omega^{(k)}$, and $F \subseteq Q$ is the set of final states. Runs and successful runs are defined in the same way as we do. A tree t and a run r on t determine a tree $\omega(r, t) : \mathrm{dom}(t) \to \Omega$ which is defined canonically by μ. The weight of the run r on t can be computed by evaluating $\omega(r, t)$ within the M-monoid A (apply the operation of a node to the weights already computed for the children of this node). Finally, the weights of all successful runs of $\mathcal{M}$ on t are summed up using the addition of A.

Then we can simulate a wmta $\mathcal{M} = (Q, \mu, F)$ over an absorptive M-monoid $(A, +, \mathbb{0}, \Omega)$ and a ranked alphabet Σ by a wta $\mathcal{M}' = (Q, \Sigma, \mu, F)$ over a tv-monoid A' in our setting. For this, we define the tree valuation monoid $A' = (\Omega, \oplus, \mathrm{Val}, \mathbb{0}^{(0)})$ such that the addition $\oplus$ simulates on $\Omega^{(0)}$ (which we identify with A) the addition $+$ of A and is otherwise defined in a way such that $(\Omega, \oplus, \mathbb{0}^{(0)})$ is a commutative monoid (which can be done easily). Now the valuation function Val maps trees $t \in T_\Omega$ to $\Omega^{(0)}$ by evaluating t within the M-monoid A as described above. Then it is easy to see that $\mathcal{M}$ and $\mathcal{M}'$ compute the same tree series.

4   Weighted MSO Logic for Trees

Now we introduce our weighted MSO logic and its semantics over tv-monoids. We follow [27] incorporating an idea of [4].

Let $\mathcal{V}_1$ and $\mathcal{V}_2$ be a countable, infinite set of first order and second order variables, respectively. Furthermore, let $\mathcal{V} = \mathcal{V}_1 \dot{\cup} \mathcal{V}_2$. Lower-case letters like $x, y, \ldots$ denote variables of $\mathcal{V}_1$ and capital letters like $X, Y, \ldots$ denote variables of $\mathcal{V}_2$. Furthermore, let D be a tv-monoid. The syntax of the weighted MSO logic over D is defined by the grammar:

$$\beta ::= \mathrm{label}_a(x) \mid \mathrm{edge}_i(x, y) \mid x \in X \mid \neg\beta \mid \beta \wedge \beta \mid \forall x\beta \mid \forall X\beta$$
$$\varphi ::= d \mid \beta \mid \varphi \vee \varphi \mid \varphi \wedge \varphi \mid \exists x\varphi \mid \forall x\varphi \mid \exists X\varphi$$

where $d \in D$, $a \in \Sigma$, $1 \leq i \leq \max_\Sigma$, $x, y \in \mathcal{V}_1$, and $X \in \mathcal{V}_2$. We call the formulas β *boolean formulas* and the formulas φ *weighted MSO formulas* (or *wMSO formulas*).

Next we like to define the semantics of the weighted MSO logic. Like a wta, the semantics of a formula shall valuate trees by elements of D. The semantics of boolean formulas will turn out as the usual boolean one. We will use $\mathbb{0}$ to define the semantics of the truth value 'false'. However, it is not clear how the semantics of 'true' should be defined. For this, an additional constant of the tv-monoid has to be provided. Note that negation is only applied to boolean formulas. Hence, we do not have to provide a quantitative semantics for negation. Now consider weighted formulas φ. Since disjunction and existential quantification provide an opportunity for non-determinism, we use the monoid operation $+$ to define their semantics. Since first order universal quantification makes a statement about all positions of a tree, we will use the valuation function to define the semantics of this quantifier. But it is not clear how to define the semantics of conjunction so far. Thus, we have to extend the valuation monoids by an additional binary operation.

Definition 4.1. *A* product tree valuation monoid *(for short: a ptv-monoid)* $(D, +, \mathrm{Val}, \diamond, \mathbb{0}, \mathbb{1})$ *consists of a valuation monoid* $(D, +, \mathrm{Val}, \mathbb{0})$, *a constant* $\mathbb{1} \in D$ *with* $\mathrm{Val}(t) = \mathbb{1}$ *whenever* $\mathrm{im}(t) = \{\mathbb{1}\}$ *for* $t \in T_D$, *and an operation* $\diamond \colon D^2 \to D$ *with* $\mathbb{0} \diamond d = d \diamond \mathbb{0} = \mathbb{0}$ *and* $\mathbb{1} \diamond d = d \diamond \mathbb{1} = d$ *for all* $d \in D$.

The operation $\diamond$ has to be neither commutative nor associative. But note that the restriction of $\diamond$ to $\{\mathbb{0}, \mathbb{1}\}$ has both properties. We use $\diamond$ to define the semantics of the conjunction. The unit element $\mathbb{1}$ will represent the truth value 'true'.

To define the semantics of the weighted MSO logic, we follow the common approach for MSO logics using assignments and extended alphabets to deal with free variables, cf. [55]. The set free(φ) of free variables occurring in φ is defined as usual. A formula without free variables is called a *sentence*. Let φ be a wMSO formula, $\mathcal{V}$ a finite set of variables with free(φ) $\subseteq \mathcal{V}$, and $t \in T_\Sigma$. A $(\mathcal{V}, t)$-*assignment* is a mapping $\sigma : \mathcal{V} \to \mathrm{dom}(t) \cup \wp(\mathrm{dom}(t))$ with $\sigma(x) \in \mathrm{dom}(t)$ and $\sigma(X) \subseteq \mathrm{dom}(t)$. As usual we encode $(\mathcal{V}, t)$-assignments by an extended alphabet. An extended ranked alphabet $\Sigma_\mathcal{V}$ is defined by

$$\Sigma_\mathcal{V} = (\Sigma \times \{0, 1\}^\mathcal{V}, \mathrm{rk})$$

with $\mathrm{rk}((a, f)) = \mathrm{rk}(a)$ for all $a \in \Sigma$, $f \in \{0, 1\}^\mathcal{V}$. Now $(a, f)_1 = a$ denotes the first component of (a, f) and $(a, f)_2 = f$ the second. We call a tree $s \in T_{\Sigma_\mathcal{V}}$ *valid* if for all first order variables $x \in \mathcal{V}$ the equation $s(u)_2(x) = 1$ holds true for exactly one position $u \in \mathrm{dom}(t)$. A tree $s \in T_{\Sigma_\mathcal{V}}$ that is not valid is called *invalid*. Let t be a Σ-tree, σ a $(\mathcal{V}, t)$-assignment, and $s \in T_{\Sigma_\mathcal{V}}$. The pair (t, σ) and s correspond to each other if $\mathrm{dom}(t) = \mathrm{dom}(s)$ and $s(u) = (t(u), f_u)$ for every $u \in \mathrm{dom}(t)$ where $f_u \in \{0, 1\}^\mathcal{V}$ is defined by

$$\forall x \in \mathcal{V}_1, X \in \mathcal{V}_2 : (f_u(x) = 1 \Leftrightarrow u = \sigma(x)) \wedge (f_u(X) = 1 \Leftrightarrow u \in \sigma(X)).$$

From now on we identify s and (t, σ) if they correspond to each other. The update $s[x \to u] \in T_{\Sigma_{\mathcal{V} \cup \{x\}}}$ for position $u \in \mathrm{dom}(t)$ is defined by $s[x \to u] = (t, \sigma[x \to u]) = (t, \sigma')$ where $\sigma'|_{\mathcal{V} \setminus \{x\}} = \sigma|_{\mathcal{V} \setminus \{x\}}$ and $\sigma'(x) = u$. The update $s[X \to I] \in T_{\Sigma_{\mathcal{V} \cup \{X\}}}$ for $I \subseteq \mathrm{dom}(t)$ is defined similarly.

The *semantics of a wMSO formula* φ over a ptv-monoid $(D, +, \mathrm{Val}, \diamond, \mathbb{0}, \mathbb{1})$ and a ranked alphabet Σ is the tree series $[\![\varphi]\!]_\mathcal{V} : T_{\Sigma_\mathcal{V}} \to D$ which equals $\mathbb{0}$ for invalid

Table 1. The semantics of wMSO formulas

$$[\![label_a(x)]\!]_{\mathcal{V}}(s) = \begin{cases} 1 & \text{if } t(\sigma(x)) = a, \\ 0 & \text{otherwise} \end{cases} \qquad [\![d]\!]_{\mathcal{V}}(s) = d$$

$$[\![\varphi \vee \psi]\!]_{\mathcal{V}}(s) = [\![\varphi]\!]_{\mathcal{V}}(s) + [\![\psi]\!]_{\mathcal{V}}(s)$$

$$[\![edge_i(x,y)]\!]_{\mathcal{V}}(s) = \begin{cases} 1 & \text{if } \sigma(y) = \sigma(x).i, \\ 0 & \text{otherwise} \end{cases} \qquad [\![\varphi \wedge \psi]\!]_{\mathcal{V}}(s) = [\![\varphi]\!]_{\mathcal{V}}(s) \diamond [\![\psi]\!]_{\mathcal{V}}(s)$$

$$[\![x \in X]\!]_{\mathcal{V}}(s) = \begin{cases} 1 & \text{if } \sigma(x) \in \sigma(X), \\ 0 & \text{otherwise} \end{cases} \qquad [\![\exists x\, \varphi]\!]_{\mathcal{V}}(s) = \sum_{u \in \mathrm{dom}(s)} [\![\varphi]\!]_{\mathcal{V} \cup \{x\}}(s[x \rightarrow u])$$

$$[\![\neg\beta]\!]_{\mathcal{V}}(s) = \begin{cases} 1 & \text{if } [\![\beta]\!]_{\mathcal{V}}(s) = 0, \\ 0 & \text{otherwise} \end{cases} \qquad [\![\exists X\, \varphi]\!]_{\mathcal{V}}(s) = \sum_{I \subseteq \mathrm{dom}(s)} [\![\varphi]\!]_{\mathcal{V} \cup \{X\}}(s[X \rightarrow I])$$

$$[\![\forall X\, \beta]\!]_{\mathcal{V}}(s) = \begin{cases} 1 & \text{if } [\![\beta]\!]_{\mathcal{V} \cup \{X\}}(s[X \rightarrow I]) = 1 \text{ for all } I \subseteq \mathrm{dom}(s), \\ 0 & \text{otherwise} \end{cases}$$

$$[\![\forall x\, \varphi]\!]_{\mathcal{V}}(s) = \mathrm{Val}(s_D) \text{ for } s_D \in T_D \text{ with } \mathrm{dom}(s_D) = \mathrm{dom}(s) \text{ and}$$
$$s_D(u) = [\![\varphi]\!]_{\mathcal{V} \cup \{x\}}(s[x \rightarrow u]) \text{ for all } u \in \mathrm{dom}(s)$$

trees and which is defined inductively for valid trees as shown in Table 1. Note that universal second order quantification is only applied to boolean formulas and its semantics is defined reflecting the unweighted case.

We write $[\![\varphi]\!]$ for $[\![\varphi]\!]_{\mathrm{free}(\varphi)}$. Let β be a boolean wMSO formula. Then β can be viewed as a classical MSO formula which defines the recognizable language $L_{\mathcal{V}}(\beta)$ and we can easily show that $[\![\beta]\!]_{\mathcal{V}} = \mathbb{1}_{L_{\mathcal{V}}(\beta)}$. Furthermore, we can prove by induction that $[\![\varphi]\!]_{\mathcal{V}}(t, \sigma) = [\![\varphi]\!](t, \sigma|_{\mathrm{free}(\varphi)})$ for every wMSO formula φ, $(t, \sigma) \in T_{\Sigma_{\mathcal{V}}}$, and set of variables $\mathcal{V}$ with $\mathrm{free}(\varphi) \subseteq \mathcal{V}$.

Example 4.2. Let $\mathbb{Q}_{\max}$ be the tv-monoid from Example 2.2. We extend this tv-monoid to the ptv-monoid $(\mathbb{Q} \cup \{\infty, -\infty\}, \max, \mathrm{avg}, \min, -\infty, \infty)$ and consider the boolean formula[1] $\mathrm{leaf}(x) = \bigvee_{a \in \Sigma^{(0)}} \mathrm{label}_a(x)$. For $t \in T_\Sigma$ and an assignment σ we obtain:

$$[\![\mathrm{leaf}(x)]\!](t, \sigma) = \max_{a \in \Sigma^{(0)}} ([\![\mathrm{label}_a(x)]\!](t, \sigma)) = \begin{cases} \infty & \text{if } \sigma(x) \text{ is a leaf,} \\ -\infty & \text{otherwise.} \end{cases}$$

For the formula $\varphi = \forall x((\mathrm{leaf}(x) \wedge 1) \vee (\neg\,\mathrm{leaf}(x) \wedge 0))$ we get $[\![\varphi]\!](t, \sigma) = \mathrm{avg}(t')$ where $t' \in T_{\mathbb{Q} \cup \{\infty, -\infty\}}$ with $\mathrm{dom}(t') = \mathrm{dom}(t)$ and

$$t'(u) = [\![(\mathrm{leaf}(x) \wedge 1) \vee (\neg\,\mathrm{leaf}(x) \wedge 0)]\!](t, \sigma[x \rightarrow u]) = \begin{cases} 1 & \text{if } u \text{ is a leaf,} \\ 0 & \text{otherwise} \end{cases}$$

for all $u \in \mathrm{dom}(t')$. Thus, the semantics of this formula equals for every tree t the leaves-to-size ratio which was previously computed by the first wta of Example 3.2.

---

[1] For boolean formulas, we use $\beta_1 \underline{\vee} \beta_2$ as an abbreviation for $\neg(\neg\beta_1 \wedge \neg\beta_2)$.

There is also a formula whose semantics calculates the same tree series as the second wta of Example 3.2. It is easy to show that there is a boolean formula $\mathrm{path}(X)$ whose semantics is ∞ if $\sigma(X)$ is a path of a tree (t, σ) (without the leaf) and $-\infty$ if $\sigma(X)$ is not such a path. Then

$$\exists X \Big(\mathrm{path}(X) \wedge \forall x \big((x \in X \wedge 1) \vee (x \notin X \wedge 0) \big) \Big)$$

is a formula defining the height-to-size ratio for every tree t.

Similarly to [30] we introduce some interesting fragments of the weighted MSO logic. An *almost boolean formula* is a wMSO formula consisting of finitely many conjunctions and disjunctions of boolean formulas and elements of D. A wMSO formula φ is $\forall$-*restricted* if ψ is almost boolean for each sub-formula $\forall x \psi$ occurring in φ. Let $\mathrm{const}(\varphi)$ be the set of all $d \in D$ occurring in φ. Two subsets $D_1, D_2 \subseteq D$ *commute* if $d_1 \diamond d_2 = d_2 \diamond d_1$ for all $d_1 \in D_1$, $d_2 \in D_2$. We call φ

- *strongly* $\wedge$-*restricted* if whenever φ contains a sub-formula $\varphi_1 \wedge \varphi_2$, then either both φ_1 and φ_2 are almost boolean or φ_1 or φ_2 is boolean,
- $\wedge$-*restricted* if whenever φ contains a sub-formula $\varphi_1 \wedge \varphi_2$, then φ_1 is almost boolean or φ_2 is boolean and
- *commutatively* $\wedge$-*restricted* if whenever φ contains a sub-formula $\varphi_1 \wedge \varphi_2$, then φ_1 is almost boolean or $\mathrm{const}(\varphi_1)$ and $\mathrm{const}(\varphi_2)$ commute.

Obviously, each strongly $\wedge$-restricted wMSO formula is $\wedge$-restricted. If a sub-formula φ_2 of $\varphi_1 \wedge \varphi_2$ is boolean, then $\mathrm{const}(\varphi_2) = \emptyset$, so $\mathrm{const}(\varphi_1)$ and $\mathrm{const}(\varphi_2)$ commute. Hence, each $\wedge$-restricted wMSO formula is commutatively $\wedge$-restricted.

5   Weighted Tree Automata and Weighted MSO Logic

In this section, we characterize the relationship between wta over ptv-monoids and the fragments of the weighted MSO logic which we introduced above. As we will see later, the larger the particular fragment gets, the more restrictions on the underlying ptv-monoid we need. So, similarly to [30] we define properties of ptv-monoids which we will use for this purpose.

First, a ptv-monoid D is called *regular* if for all $d \in D$ and all ranked alphabets Σ a wta $\mathcal{M}_d$ exists with $\|\mathcal{M}_d\|(t) = d$ for each $t \in T_\Sigma$. For every strong bimonoid $(K, +, \circ, \mathbb{0}, \mathbb{1})$ and its associated ptv-monoid $(K, +, \circ_{\mathrm{en}}, \circ, \mathbb{0}, \mathbb{1})$ we can build for all $d \in K$ such a wta $\mathcal{M}_d$ by weighting a deterministic tree automaton with dead-end final states recognizing T_Σ so that the transitions to a final state are weighted with d and every other transition is weighted with $\mathbb{1}$. Thus, we obtain the following lemma.

Lemma 5.1. *Let* $(K, +, \circ, \mathbb{0}, \mathbb{1})$ *be a strong bimonoid and* en *an enumeration of positions. Then* $(K, +, \circ_{\mathrm{en}}, \circ, \mathbb{0}, \mathbb{1})$ *is regular.*

For the second class of ptv-monoids, we have to define several properties. A ptv-monoid D is *left-$\diamond$-distributive* if $d \diamond (d_1 + d_2) = d \diamond d_1 + d \diamond d_2$ for all $d, d_1, d_2 \in D$. We call D *left-multiplicative* if $d \diamond \mathrm{Val}(t) = \mathrm{Val}(t')$ for all $d \in D$, $t, t' \in T_D$ with $\mathrm{dom}(t) = \mathrm{dom}(t')$, $t'(\varepsilon) = d \diamond t(\varepsilon)$, and $t'(u) = t(u)$ for every $u \in \mathrm{dom}(t) \setminus \{\varepsilon\}$. Furthermore, D is *left-Val-distributive* if $d \diamond \mathrm{Val}(t) = \mathrm{Val}(t')$ for all $d \in D$, $t, t' \in T_D$ with $\mathrm{dom}(t) = \mathrm{dom}(t')$ and $t'(u) = d \diamond t(u)$ for every $u \in \mathrm{dom}(t)$. It is easy to see that each left-multiplicative or left-Val-distributive ptv-monoid is regular. Now, a ptv-monoid D is called *left-distributive* if it is left-multiplicative or left-Val-distributive and, moreover, left-$\diamond$-distributive. An example of a left-distributive ptv-monoid is given in Example 5.7.

The third class of ptv-monoids is in fact a class of certain semirings. A ptv-monoid is *right-$\diamond$-distributive* if $(d_1 + d_2) \diamond d = d_1 \diamond d + d_2 \diamond d$ for all $d_1, d_2, d \in D$. Moreover, D is *$\diamond$-distributive* if it is both left- and right-$\diamond$-distributive, and it is *associative* if $\diamond$ is associative. Note that, if D is associative, then $(D, +, \diamond, 0, \mathbb{1})$ is a strong bimonoid. If D is moreover $\diamond$-distributive, then $(D, +, \diamond, 0, \mathbb{1})$ is a semiring. In this case, we call $(D, +, \mathrm{Val}, \diamond, 0, \mathbb{1})$ a *tree valuation semiring* *(tv-semiring)*.

Recall that $D_1 \subseteq D$ and $D_2 \subseteq D$ *commute* if $d_1 \diamond d_2 = d_2 \diamond d_1$ for all $d_1 \in D_1$, $d_2 \in D_2$. We call D *conditionally commutative* if $\mathrm{Val}(t_1) \diamond \mathrm{Val}(t_2) = \mathrm{Val}(t)$ for all $t_1, t_2, t \in T_D$ with $\mathrm{dom}(t_1) = \mathrm{dom}(t_2) = \mathrm{dom}(t)$, $\mathrm{im}(t_1)$ and $\mathrm{im}(t_2)$ commute and $t(u) = t_1(u) \diamond t_2(u)$ for all $u \in \mathrm{dom}(t)$. The definition of conditionally commutative differs for trees and words. In [30] it is only necessary for two sequences $(d_1, \ldots, d_n)$ and $(d'_1, \ldots, d'_n)$ that d_i and d'_j commute for $1 \leq i < j \leq n$. But such a restriction would be reasonable in the case of trees only for particular enumerations.

A *cctv-semiring* is a conditionally commutative tv-semiring which is, moreover, left-multiplicative or left-Val-distributive. Obviously, each cctv-semiring is left-distributive.

In [30] some examples for product valuation monoids are given. We apply those examples to our setting.

Example 5.2. Let $(S, +, \cdot, 0, 1)$ be a commutative semiring. Then $(S, +, \mathrm{Val}, \cdot, 0, 1)$ with $\mathrm{Val}(t) = \prod_{u \in \mathrm{dom}(t)} t(u)$ for all $t \in T_S$ is a left-multiplicative cctv-semiring.

The choice of the product $\diamond$ and the associated $\mathbb{1}$ may influence the properties of the ptv-monoids significantly as the following example shows.

Example 5.3. Let us consider $(\mathbb{Q} \cup \{\infty, -\infty\}, \max, \mathrm{avg}, \min, -\infty, \infty)$ with avg defined as in Example 2.2. This ptv-monoid is a regular tv-semiring, but neither left-multiplicative nor left-Val-distributive, nor conditionally commutative.

Next we choose as the product $\diamond$ the average of two numbers. The resulting ptv-monoid is $(\mathbb{Q} \cup \{\infty, -\infty\}, \max, \mathrm{avg}, \mathrm{avg}', -\infty, \infty)$ where ∞ acts as $\mathbb{1}$ and thus $\mathrm{avg}'(d, \infty) = \mathrm{avg}'(\infty, d) = d$ for every $d \in \mathbb{Q} \cup \{\infty, -\infty\}$. This ptv-monoid is left-$\diamond$-distributive and conditionally commutative, but neither left-multiplicative nor left-Val-distributive, and the product $\diamond = \mathrm{avg}'$ is not associative.

Finally, the ptv-monoid $(\mathbb{Q} \cup \{\infty, -\infty\}, \max, \mathrm{avg}, +, -\infty, 0)$ is a left-Val-distributive cctv-semiring.

Example 5.4. The structure $\mathbb{R}_{disc} = (\mathbb{R} \cup \{-\infty\}, \max, \text{disc}_\wedge, -\infty)$ from Example 2.5 together with $\mathbb{1} = 0$ and $\diamond = \dashv$ is a left-multiplicative cctv-semiring.

By means of these three classes of ptv-monoids, we can characterize the relationship between wta over ptv-monoids and our fragments of weighted MSO logic. As before, let D be a ptv-monoid and Σ a ranked alphabet. The following theorem extends the main result of [30] concerning series on finite words to the setting of trees.

Theorem 5.5. *Let $S : T_\Sigma \to D$ be a tree series.*

1. *If D is regular, then S is recognizable iff $S = \llbracket \varphi \rrbracket$ for a $\forall$-restricted and strongly $\wedge$-restricted wMSO sentence φ.*
2. *If D is left-distributive, then S is recognizable iff $S = \llbracket \varphi \rrbracket$ for a $\forall$-restricted and $\wedge$-restricted wMSO sentence φ.*
3. *If D is a cctv-semiring, then S is recognizable iff $S = \llbracket \varphi \rrbracket$ for a $\forall$-restricted and commutatively $\wedge$-restricted wMSO sentence φ.*

Obviously, we need D to be regular. Otherwise, there is at least one $d \in D$ without a wta recognizing d, hence, the semantics of the $\forall$-restricted and strongly $\wedge$-restricted sentence d is not recognizable.

Before we prove Theorem 5.5, we will give two examples taken from [43] showing that it is not possible to dispose of any constraint in the above theorem.

Example 5.6. Let $\max_\Sigma \geq 1$ and $(\mathbb{N}_0, +, \text{Val}_l, 0)$ be the tv-monoid from Example 2.3. Then $(\mathbb{N}_0, +, \text{Val}_l, \hat{\ }, 0, 1)$ with $\hat{\ }(a, b) = a^b$ for all $a, b \in \mathbb{N}_0 \setminus \{0, 1\}$ and $a^0 = 0^a = 0$, $a^1 = 1^a = a$ for all $a \in \mathbb{N}_0$ is a ptv-monoid which is regular, but not left-distributive (indeed, it has none of the other properties defined above). Due to Theorem 5.5(1), $\llbracket \exists x (\bigvee_{a \in \Sigma} \text{label}_a(x)) \rrbracket$ and $\llbracket 2 \rrbracket$ are recognizable. Furthermore, $\llbracket 2 \rrbracket(t) = 2$ and

$$\llbracket \exists x (\bigvee_{a \in \Sigma} \text{label}_a(x)) \rrbracket(t) = \sum_{u \in \text{dom}(t)} \sum_{a \in \Sigma} \llbracket \text{label}_a(x) \rrbracket_{\{x\}}(t[x \to u]) = |\text{dom}(t)|$$

for all $t \in T_\Sigma$. Let us consider the $\forall$-restricted and $\wedge$-restricted, but not strongly $\wedge$-restricted wMSO-formula $\varphi = 2 \wedge \exists x (\bigvee_{a \in \Sigma} \text{label}_a(x))$. We show that $\llbracket \varphi \rrbracket$ is recognizable by $\mathcal{M} = (\{q_1, q_2\}, \Sigma, \mu, \{q_1, q_2\})$ with $\mu_m : \Sigma^{(m)} \to \{1\}^{\{q_1, q_2\}^m \times \{q_1, q_2\}}$ for all $m \geq 0$. For each $t \in T_\Sigma$ there are $2^{|\text{dom}(t)|}$ valid and successful runs of $\mathcal{M}$ on t weighted with 1. Thus,

$$\|\mathcal{M}\|(t) = \sum_{r \in \text{succ}(\mathcal{M}, t)} 1 = 2^{|\text{dom}(t)|} = \llbracket 2 \rrbracket(t)^{\llbracket \exists x (\bigvee_{a \in \Sigma} label_a(x)) \rrbracket(t)} = \llbracket \varphi \rrbracket(t) .$$

However, we can show that the semantics of the $\forall$-restricted and $\wedge$-restricted, but not strongly $\wedge$-restricted wMSO-formula $\varphi' = 2 \wedge (2 \wedge \exists x (\bigvee_{a \in \Sigma} \text{label}_a(x)))$ is not recognizable. Observe that $\llbracket \varphi' \rrbracket = 2^{2^{|\text{dom } t|}}$. Let $\mathcal{M}' = (Q', \Sigma, \mu', F')$ be any wta.

Since Q' and Σ are finite, $\mathcal{M}'$ exhibits only finitely many weights. Hence, there is a $k \in \mathbb{N}$ that is greater than all weights of $\mathcal{M}'$ and

$$\|\mathcal{M}'\|(t) = \sum_{r \in \mathrm{succ}(\mathcal{M}', t)} \mathrm{Val}_l(\mu'(t, r)) \leq \sum_{r \in \mathrm{succ}(\mathcal{M}', t)} \prod_{u \in \mathrm{dom}(t)} (\mu'(t, r))(u)$$

$$\leq \sum_{r \in \mathrm{succ}(\mathcal{M}', t)} k^{|\mathrm{dom}(t)|} = k^{|\mathrm{dom}(t)|} \left(\sum_{r \in \mathrm{succ}(\mathcal{M}', t)} 1 \right) = (k \cdot |Q'|)^{|\mathrm{dom}(t)|}.$$

But obviously, $2^{2^i} > (k \cdot |Q'|)^i$ for i sufficiently large. Thus, $\mathcal{M}'$ does not recognize $[\![\varphi']\!]$.

By the last example, we have seen that for non-left distributive ptv-monoids the semantics of $\wedge$-restricted wMSO sentences are in general not recognizable anymore. Next we show that it is not sufficient to assume the ptv-monoid to be left-distributive to guarantee the recognizability of the semantics of $\forall$-restricted and commutatively $\wedge$-restricted wMSO sentences.

Example 5.7. Let $\max_\Sigma \geq 2$ and $(\mathbb{Q}_0^+, +, \mathrm{W}_{\mathrm{root}}, 0)$ be the tv-monoid from Example 2.4. We define $\mathrm{div} : \mathbb{Q}_0^+ \times \mathbb{Q}_0^+ \to \mathbb{Q}_0^+$ by $\mathrm{div}(a, b) = b \div a$ for all $a, b \in \mathbb{Q}_0^+ \setminus \{0, 1\}$ and $\mathrm{div}(a, 0) = \mathrm{div}(0, a) = 0$, $\mathrm{div}(a, 1) = \mathrm{div}(1, a) = a$ for all $a \in \mathbb{Q}_0^+$. The ptv-monoid $(\mathbb{Q}_0^+, +, \mathrm{W}_{\mathrm{root}}, \mathrm{div}, 0, 1)$ is left-$\diamond$-distributive, left-multiplicative, left-Val-distributive and thus left-distributive. In addition, it is conditionally commutative but neither associative nor right-$\diamond$-distributive and thus no cctv-semiring. Let $\varphi = \exists x (\bigvee_{a \in \Sigma^{(0)}} \mathrm{label}_a(x)) \wedge 2$, a $\forall$-restricted and commutatively $\wedge$-restricted wMSO sentence. Then

$$[\![\exists x (\bigvee_{a \in \Sigma^{(0)}} \mathrm{label}_a(x)) \wedge 2]\!](t) = \mathrm{div}\left(\sum_{\substack{u \in \mathrm{dom}(t) \\ a \in \Sigma^{(0)}}} [\![\mathrm{label}_a(x)]\!]_{\{x\}}(t[x \to u]), [\![2]\!](t) \right)$$

$$= \frac{2}{\text{``number of leaves of } t\text{''}}$$

for all $t \in T_\Sigma$. Let $\mathcal{M}$ be any wta over $(\mathbb{Q}_0^+, +, \mathrm{W}_{\mathrm{root}}, \mathrm{div}, 0, 1)$. Since the set of weights of $\mathcal{M}$ is finite, there is a rational number $q = 2^{-n} > 0$ smaller than all weights from $\mathrm{im}(\mu) \setminus \{0\}$. Let t be a Σ-tree with $2^{n+1} = \frac{2}{q}$ leaves. Hence, either $\|\mathcal{M}\|(t) = 0$ or $\|\mathcal{M}\|(t) > q$ because $\mathcal{M}$ sums up the positive rational weights of the roots of all valid and successful runs of $\mathcal{M}$ on t. But $[\![\varphi]\!](t) = q$, so $[\![\varphi]\!] \neq \|\mathcal{M}\|$. Thus, $[\![\varphi]\!]$ is not recognizable in $(\mathbb{Q}_0^+, +, \mathrm{W}_{\mathrm{root}}, \mathrm{div}, 0, 1)$.

It remains to prove Theorem 5.5. For this, we first consider recognizable step functions in Subsection 5.1. In Subsection 5.2, we derive properties of recognizable tree series and, eventually, in Subsection 5.3, we give the proof of Theorem 5.5.

5.1   Recognizable Step Functions

Let D be a ptv-monoid, S_1, S_2 two tree series, and $d \in D$. We define the *scalar product* $d \diamond S_1$, the *sum* $S_1 + S_2$ and the *product* $S_1 \diamond S_2$ pointwise by $(d \diamond S_1)(t) =$

$d \diamond S_1(t)$, $(S_1 + S_2)(t) = S_1(t) + S_2(t)$ and $(S_1 \diamond S_2)(t) = S_1(t) \diamond S_2(t)$ for all $t \in T_\Sigma$. For every tree language $L \subseteq T_\Sigma$ we call $\mathbb{1}_L$ with $\mathbb{1}_L(t) = \mathbb{1}$ for all $t \in L$ and $\mathbb{1}_L(t) = \mathbb{0}$ for all $t \in T_\Sigma \setminus L$ the *characteristic function of L*.

Definition 5.8. *A tree series S is a* recognizable step function *if there are recognizable tree languages $(L_i)_{1 \leq i \leq k}$ forming a partition of T_Σ and values $d_1, \ldots, d_k \in D$ such that $S = \sum_{i=1}^k d_i \diamond \mathbb{1}_{L_i}$.*

Clearly, a tree series S is a recognizable step function iff $\mathrm{im}(S)$ is finite and $S^{-1}(d)$ is a recognizable tree language for each $d \in D$. Subsequently, we will write $\sum_{i=1}^k d_i \mathbb{1}_{L_i}$ instead of $\sum_{i=1}^k d_i \diamond \mathbb{1}_{L_i}$.

Lemma 5.9. *The class of recognizable step functions over a ranked alphabet Σ and a ptv-monoid $(D, +, \mathrm{Val}, \diamond, \mathbb{0}, \mathbb{1})$ is closed under the operations $+$ and $\diamond$.*

Proof. Let $S_1 = \sum_{i=1}^n d_i \mathbb{1}_{L_i}$ and $S_2 = \sum_{j=1}^m d'_j \mathbb{1}_{L'_j}$ be two recognizable step functions. Since

$$S_1 + S_2 = \sum_{i=1}^n \sum_{j=1}^m (d_i + d'_j) \mathbb{1}_{L_i \cap L'_j} \quad \text{and} \quad S_1 \diamond S_2 = \sum_{i=1}^n \sum_{j=1}^m (d_i \diamond d'_j) \mathbb{1}_{L_i \cap L'_j}$$

and recognizable tree languages are closed under the boolean operations, the claim follows.   □

Recognizable step functions are indeed recognizable tree series provided the underlying ptv-monoid is regular.

Theorem 5.10. *Let D be a regular ptv-monoid. Each recognizable step function S over D is a recognizable tree series.*

Proof (sketch). Let $S = \sum_{i=1}^n d_i \mathbb{1}_{L_i}$. Since D is regular, for each $i \in \{1, \ldots, n\}$ there is a wta $\mathcal{M}_i$ with $\|\mathcal{M}_i\|(t) = d_i$ for all $t \in T_\Sigma$. Furthermore, there is a deterministic tree automaton recognizing L_i which we transform in the usual way into a wta $\mathcal{M}'_i$ recognizing $\mathbb{1}_{L_i}$. Now by a product of $\mathcal{M}_i$ and $\mathcal{M}'_i$, we build automata $\mathcal{P}_i$ with $\|\mathcal{P}_i\|(t) = d_i$ iff $t \in L_i$ for all $t \in T_\Sigma$. Eventually, the disjoint union of the wta $\mathcal{P}_i$ ($i \in \{1, \ldots, n\}$) yields a wta recognizing S.   □

Clearly, $\llbracket \varphi \rrbracket$ is a recognizable step function iff $\llbracket \varphi \rrbracket_\mathcal{V}$ is a recognizable step function for every wMSO formula φ and any finite set of variables $\mathcal{V}$ with $\mathrm{free}(\varphi) \subseteq \mathcal{V}$. Next we show:

Lemma 5.11. *If $S : T_\Sigma \to D$ is a recognizable step function, then $S = \llbracket \varphi \rrbracket$ for some almost boolean sentence φ. Conversely, if φ is an almost boolean formula, then $\llbracket \varphi \rrbracket$ is a recognizable step function.*

Proof. ($\Leftarrow$) If φ is a boolean formula, then $\llbracket \varphi \rrbracket = \mathbb{1}_{L(\varphi)}$ is a recognizable step function. Trivially, $\llbracket d \rrbracket$ is also a recognizable step function for any $d \in D$. Let φ and ψ be two almost boolean formulas and $\mathcal{V} = \mathrm{free}(\varphi) \cup \mathrm{free}(\psi)$. By induction and due to Lemma 5.9, $[\![\varphi \vee \psi]\!] = [\![\varphi]\!]_{\mathcal{V}} + [\![\psi]\!]_{\mathcal{V}}$ and $[\![\varphi \wedge \psi]\!] = [\![\varphi]\!]_{\mathcal{V}} \diamond [\![\psi]\!]_{\mathcal{V}}$ are recognizable step functions.

$(\Rightarrow)$ Let $S = \sum_{i=1}^{n} d_i \mathbb{1}_{L_i}$. Due to Thatcher and Wright [54], there are MSO sentences φ_i with $L(\varphi_i) = L_i$ for all $i \in [1, n]$. Let $\varphi = (d_1 \wedge \varphi_1) \vee \ldots \vee (d_n \wedge \varphi_n)$. Then φ is almost boolean and

$$[\![\varphi]\!] = \sum_{i=1}^{n} d_i \mathbb{1}_{L(\varphi_i)} = \sum_{i=1}^{n} d_i \mathbb{1}_{L_i} = S. \qquad \square$$

5.2 Closure Properties of Recognizable Tree Series

In this subsection, we show that under suitable conditions the operations sum, product, and scalar product preserve the recognizability of tree series.

Theorem 5.12. *1. Let D be a ptv-monoid.*
  (a) The class of recognizable tree series is closed under sum.
  (b) Let L be a recognizable tree language and S a recognizable tree series. Then $\mathbb{1}_L \diamond S$ and $S \diamond \mathbb{1}_L$ are also recognizable.
2. Let D be left-distributive.
  (a) The class of recognizable tree series is closed under scalar product.
  (b) Let S_1 be a recognizable step function and S_2 a recognizable tree series. Then $S_1 \diamond S_2$ is also recognizable.
3. Let D be a cctv-semiring. Furthermore, let $\mathcal{M}_1 = (Q_1, \Sigma, \mu_1, F_1)$ and $\mathcal{M}_2 = (Q_2, \Sigma, \mu_2, F_2)$ be two wta such that $\mathrm{im}(\mu_1)$ and $\mathrm{im}(\mu_2)$ commute. Then $\|\mathcal{M}_1\| \diamond \|\mathcal{M}_2\|$ is a recognizable tree series.

Proof. (sketch) *1.(a)* This can be shown as usual by a disjoint union construction.

(b) We weight a classical deterministic bottom-up tree automaton recognizing L with the values $\mathbb{0}$ (no transition) and $\mathbb{1}$ (transition exists). Then we build a product automaton of this weighted tree automaton and a wta recognizing S.

2.(a) In case D is left-multiplicative, we construct a wta with dead-end final states (i.e., every transition leaving a final state is weighted with $\mathbb{0}$) recognizing a tree series S and multiply the weights of every transition ending in a final state with $d \in D$ from the left. In case D is left-Val-distributive, we multiply the weight of every transition of a wta recognizing S by d from the left. In both cases, the resulting wta recognizes $d \diamond S$.

(b) Let $S_1 = \sum_{i=1}^{n} d_i \mathbb{1}_{L_i}$ for some partition $(L_i)_{1 \leq i \leq n}$ of T_Σ. Thus, $S_1 \diamond S_2 = \sum_{i=1}^{n} \mathbb{1}_{L_i} \diamond (d_i \diamond S_2)$. By parts 1 and 2(a), this series is recognizable.

3. We build a synchronized product $\mathcal{M}$ of $\mathcal{M}_1$ and $\mathcal{M}_2$. Since D is conditionally commutative and the weights of $\mathcal{M}_1$ and $\mathcal{M}_2$ commute, the weight of a successful run of $\mathcal{M}$ on a tree t is the product of the respective runs of $\mathcal{M}_1$ and $\mathcal{M}_2$ on t. Because of the $\diamond$-distributivity, $\|\mathcal{M}\| = \|\mathcal{M}_1\| \diamond \|\mathcal{M}_2\|$. $\qquad \square$

Next, we consider the closure under relabeling, similarly to [33, 30]. Let Σ and Γ be two ranked alphabets and $h : \Sigma \to \wp(\Gamma)$ be a mapping such that $h(a) \subseteq \Gamma^{(m)}$ for all $a \in \Sigma^{(m)}$. Then h can be extended inductively to a mapping $h' : T_\Sigma \to \wp(T_\Gamma)$ via

$$h'(a(t_1, \ldots, t_m)) = \{b(t'_1, \ldots, t'_m) \mid b \in h(a) \wedge \forall i \in [1, m] : t'_i \in h'(t_i)\}$$

for all $m \geq 0$, $a \in \Sigma^{(m)}$ and $t_1, \ldots, t_m \in T_\Sigma$. Next, we define for every tree series S over D and Σ the tree series $h''(S)$ over D and Γ by

$$h''(S)(t) = \sum_{s \in T_\Sigma \wedge t \in h'(s)} S(s)$$

for all $t \in T_\Gamma$. Subsequently, we denote h' and h'' also by h which is called a *relabeling*. The proof for the following lemma works by an automaton construction already applied in [34, 30]. Surprisingly, we do not need any distributivity of the underlying ptv-monoid.

Lemma 5.13. *Recognizable tree series are closed under relabeling.*

5.3   From Weighted Logics to Weighted Tree Automata and Reverse

The following result will be very useful. Using Theorem 5.12(1)(b), it can be proved as the corresponding result in [33].

Proposition 5.14. *Let φ be a wMSO formula and $\mathcal{V}$ a finite set of variables with free$(\varphi) \subseteq \mathcal{V}$. Then $[\![\varphi]\!]$ is recognizable iff $[\![\varphi]\!]_\mathcal{V}$ is recognizable.*

Now we show that our logical operators preserve the recognizability of the semantics of wMSO formulas.

Proposition 5.15. *Let φ and ψ be two wMSO formulas over Σ and a ptv-monoid D. If $[\![\varphi]\!]$ and $[\![\psi]\!]$ are recognizable, then $[\![\varphi \vee \psi]\!]$, $[\![\exists x \varphi]\!]$, and $[\![\exists X \varphi]\!]$ are recognizable.*

Proof (sketch). Let $\mathcal{V} = \text{free}(\varphi \vee \psi) = \text{free}(\varphi) \cup \text{free}(\psi)$. By Proposition 5.14 and Theorem 5.12(1)(a), $[\![\varphi \vee \psi]\!] = [\![\varphi]\!]_\mathcal{V} + [\![\psi]\!]_\mathcal{V}$ is recognizable.

Let $h : \Sigma_{\text{free}(\varphi)} \to \Sigma_{\text{free}(\exists x \varphi)}$ be the relabeling defined by erasing the x-row in $\Sigma_{\text{free}(\varphi)}$ (if existing). Then, for all $s \in T_{\text{free}(\exists x \varphi)}$:

$$[\![\exists x \varphi]\!](s) = \sum ([\![\varphi]\!]_{\text{free}(\varphi) \cup \{x\}}(s[x \to u]) \mid u \in \text{dom}(s))$$

$$= \sum ([\![\varphi]\!](s[x \to u]) \mid u \in \text{dom}(s)) = \sum ([\![\varphi]\!](s') \mid h(s') = s)$$

$$= h([\![\varphi]\!])(s).$$

Since $[\![\varphi]\!]$ is recognizable, by Lemma 5.13 $[\![\exists x \varphi]\!] = h([\![\varphi]\!])$ is recognizable.

Similarly, it follows that $[\![\exists X \varphi]\!]$ is recognizable.   □

Proposition 5.16. *Let φ be an almost boolean formula over D and Σ. Then $[\![\forall x \varphi]\!]$ is recognizable.*

Proof (sketch). We proceed as in [30, 33] which rest on [26, 27]. Now, let $\mathcal{W} = \text{free}(\varphi) \cup \{x\}$ and $\mathcal{V} = \text{free}(\forall x \varphi) = \mathcal{W} \setminus \{x\}$. Since φ is almost boolean, $[\![\varphi]\!]_\mathcal{W}$ is a recognizable step function (see Lemma 5.11). Thus, $[\![\varphi]\!]_\mathcal{W} = \sum_{i=1}^n d_i \mathbb{1}_{L_i}$. We can assume that all L_i only consists of valid trees, since $[\![\varphi]\!]_\mathcal{W}(s) = \mathbb{0}$ for any invalid tree s. We introduce the ranked alphabet $\tilde{\Sigma} = \Sigma \times \{1, \ldots, n\}$ where the rank is determined by the component from Σ. A tree $(t, \nu, \sigma) \in T_{\tilde{\Sigma}_\mathcal{V}}$ consists of a tree $(t, \sigma) \in T_{\Sigma_\mathcal{V}}$ and a mapping $\nu : \text{dom}(t) \to \{1, \ldots, n\}$. Next, let $\tilde{L} \subseteq T_{\tilde{\Sigma}_\mathcal{V}}$ be the tree language of all (t, ν, σ) such that (t, σ) is valid and

$$\nu(u) = i \Rightarrow (t, \sigma[x \to u]) \in L_i$$

for all $i \in \{1, \ldots, n\}$ and $u \in \text{dom}(t)$. Since $(L_i)_{1 \leq i \leq n}$ is a partition, for each (t, σ) there is exactly one mapping ν satisfying the implication above, that means, ν encodes to which L_i the update of (t, σ) and x belongs. Similarly to [41] we can show that $\tilde{L}$ is recognizable. Let $\mathcal{A} = (Q, \tilde{\Sigma}_\mathcal{V}, \delta, F)$ be a deterministic bottom-up tree automaton recognizing $\tilde{L}$. Furthermore, let $\mathcal{M} = (Q, \tilde{\Sigma}_\mathcal{V}, \mu, F)$ with

$$\mu_m((a, i, f))_{q_1 \ldots q_m, q} = \begin{cases} d_i & \text{if } \delta_{(a,i,f)}(q_1, \ldots, q_m) = q \\ \mathbb{0} & \text{otherwise} \end{cases}$$

for all $(a, i, f) \in \tilde{\Sigma}_\mathcal{V}^{(m)}$, $q_1, \ldots, q_m, q \in Q$ and $m \geq 0$. We can show that $\|\mathcal{M}\|((t, \nu, \sigma)) = [\![\forall x \varphi]\!]((t, \sigma))$ for all trees $(t, \nu, \sigma) \in T_{\tilde{\Sigma}_\mathcal{V}}$. Now, let the relabeling $h : \tilde{\Sigma}_\mathcal{V} \to \Sigma_\mathcal{V}$ be defined by $h((a, i, f)) = (a, f)$. Then

$$h(\|\mathcal{M}\|)((t, \sigma)) = \sum \Big(\|\mathcal{M}\|((t, \nu, \sigma)) \mid (t, \nu, \sigma) \in T_{\tilde{\Sigma}_\mathcal{V}}, (t, \nu, \sigma) \in h((t, \sigma)) \Big)$$
$$= \|\mathcal{M}\|(t, \nu', \sigma) = [\![\forall x \varphi]\!]((t, \sigma))$$

for all valid trees (t, σ). Hence, $[\![\forall x \varphi]\!]$ is recognizable by Lemma 5.13.   □

Proposition 5.17. *Let φ and ψ be two wMSO formulas over a ptv-monoid D and a ranked alphabet Σ. Then $[\![\varphi \wedge \psi]\!]$ is recognizable if the following is satisfied: (a) $[\![\varphi]\!]$ is recognizable and ψ is boolean (then $[\![\psi \wedge \varphi]\!]$ is recognizable, too) or (b) D is left-distributive, φ is almost boolean, and $[\![\psi]\!]$ is recognizable.*

Proof. Let $\mathcal{V} = \text{free}(\varphi \wedge \psi)$.

(a) Since ψ is boolean, $[\![\psi]\!]_\mathcal{V} = \mathbb{1}_{L_\mathcal{V}(\psi)}$ and $L_\mathcal{V}(\psi)$ is recognizable. Moreover, $[\![\varphi]\!]_\mathcal{V}$ is recognizable. By Theorem 5.12(1)(b), $[\![\varphi \wedge \psi]\!]$ and $[\![\psi \wedge \varphi]\!]$ are recognizable.

(b) Since φ is almost boolean, $[\![\varphi]\!]$ and thus $[\![\varphi]\!]_\mathcal{V}$ are recognizable step functions. Moreover, $[\![\psi]\!]_\mathcal{V}$ is recognizable. Hence, due to Theorem 5.12(2)(b), $[\![\varphi \wedge \psi]\!]$ is recognizable.   □

With the propositions above, we prove our main result Theorem 5.5.

Proof of Theorem 5.5. The recognizability of the semantics of almost boolean formulas over a regular ptv-monoid D is guaranteed by Lemma 5.11 and Theorem 5.10. Now, in cases of (1) and (2) the tree series $[\![\varphi]\!]$ for a formula φ from the respective fragment is recognizable by Propositions 5.15–5.17. For (3), we show by induction on the structure of φ that there is a wta recognizing $[\![\varphi]\!]$ whose weights are in the subsemiring $\langle \text{const}(\varphi) \cup \{\mathbb{0}, \mathbb{1}\}, +, \diamond \rangle$ generated by $\text{const}(\varphi) \cup \{\mathbb{0}, \mathbb{1}\}$. In the almost boolean case, the statement is obvious. In case of disjunction and quantification we proceed as in the proofs of Propositions 5.15 and 5.16 which are constructive. It is traceable that all constructions retain the set of weights. To deal with $\varphi \wedge \psi$, we can show that the sets of weights of the wta recognizing $[\![\varphi]\!]$ and $[\![\psi]\!]$, respectively, commute with each other, since D is a cctv-semiring. By Theorem 5.12(3), $[\![\varphi \wedge \psi]\!]$ is recognizable (by a wta with weights in $\langle \mathrm{const}(\varphi) \cup \mathrm{const}(\psi) \cup \{\mathbb{0}, \mathbb{1}\}, +, \diamond \rangle$).

For the proof of the opposite direction, we build a $\forall$-restricted and strongly $\wedge$-restricted wMSO sentence describing the run semantics of the weighted tree automaton recognizing S, cf. [33, 41]. □

Remark 5.18. Theorem 5.5 generalizes the main result of [30] on finite words. Since we deal with wta over valuation monoids, the usual translation (cf., e.g., [41]) does not seem to apply, but we can proceed as follows. We translate the alphabet Σ into a ranked alphabet $\Sigma_r = \{a^{(0)} \mid a \in \Sigma\} \cup \{a^{(1)} \mid a \in \Sigma\}$ and define a bijection tree : $\Sigma^+ \to T_{\Sigma_r}$ inductively by $\mathrm{tree}(a) = a^{(0)}$ and $\mathrm{tree}(wa) = a^{(1)}(\mathrm{tree}(w))$ for all $a \in \Sigma$ and $w \in \Sigma^+$. Then Theorem 5.5 can be translated to the respective result on words (note, however, that the present definition of being conditionally commutative is for trees a bit more restrictive than the one of [30] for words).

Theorem 5.5 also generalizes the respective result on wta over semirings [33].

6   Strong Bimonoids and Weighted Logic

Here, we consider weighted logic and weighted tree automata over strong bimonoids. Syntactically, we define the logic exactly like the weighted MSO logic for trees over ptv-monoids. The definition of the semantics differs from the one of the wMSO logic over ptv-monoids with regard to the universal quantification. Here, we use the multiplication of the strong bimonoid instead of the valuation function. Thus, we have to additionally provide an enumeration of tree positions to fix the order of the multiplication. Hence, we write $[\![\varphi]\!]^{\mathrm{en}}$ instead of $[\![\varphi]\!]$. Since every ptv-monoid resulting from a strong bimonoid is regular (see, Lemma 5.1), Theorem 5.5(1) implies the following corollary.

Corollary 6.1. *Let Σ be a ranked alphabet, $(K, +, \circ, \mathbb{0}, \mathbb{1})$ a strong bimonoid, S a tree series over K and Σ, and en an enumeration of tree positions. Then S is recognizable with en iff there is an $\forall$-restricted and strongly $\wedge$-restricted wMSO sentence φ with $[\![\varphi]\!]^{\mathrm{en}} = S$.*

Next we show that for strong bimonoids with local finiteness properties we do not need to restrict the weighted MSO logic to characterize the class of recognizable tree series.

A bimonoid $(K, +, \circ, \mathbb{0}, \mathbb{1})$ is called *additively locally finite* and *multiplicatively locally finite*, respectively, if for every finite subset $K' \subseteq K$ the submonoids $\langle K' \cup \{\mathbb{0}\}, + \rangle$ and $\langle K' \cup \{\mathbb{1}\}, \circ \rangle$, respectively, generated by K' are finite. If K is both additively and multiplicatively locally finite, we call K *bi-locally finite*. Clearly, commutative and idempotent monoids $(K, +, \mathbb{0})$ (i.e., $k + k = k$ for all

$k \in K$) are locally finite. Hence, strong bimonoids $(K, +, \circ, 0, 1)$ with a commutative multiplication where both addition and multiplication are idempotent are always bi-locally finite. A monoid $(K, \circ, 1)$ is called *periodic* if every element $k \in K$ generates a finite submonoid $\langle \{k, 1\}, \circ \rangle$. We call a bimonoid *bi-periodic* if it is additively and multiplicatively periodic. It is easy to prove that commutative, periodic monoids are locally finite. But it is well known that there are non-commutative periodic monoids which are not locally finite, compare [48, Section 3.3]. Thus, the class of bi-periodic bimonoids properly contains the class of bi-locally finite bimonoids.

Theorem 6.2. *Let K be a strong bimonoid and S a tree series over K and Σ.*
(a) Let K be bi-locally finite. Then S is a recognizable step function iff S is recognizable with DF.
(b) Let K be bi-periodic. Then S is a recognizable step function iff S is recognizable with WO$_<$ *for any linear order $<$ on K.*

Proof (sketch). The result of (a) was stated implicitly in [34, p. 162]. We indicate how to use the idea of Droste, Stüber and Vogler [32] also for the proof of (b).

By Lemma 5.1 and Theorem 5.10, if S is a recognizable step function, then S is recognizable with DF and WO$_<$, respectively.

For the opposite direction of (a) and (b), let $\mathcal{M} = (Q, \Sigma, \mu, F)$ be a wta recognizing S (either with DF or with WO$_<$). If K is bi-locally finite, the product closure $Y = \{b_1 \cdot \ldots \cdot b_n \mid n \in \mathbb{N} \wedge b_1, \ldots, b_n \in \text{im}(\mu)\}$ of all weights of $\mathcal{M}$ is finite. Furthermore, if K is bi-periodic, the ordered product closure $Y' = \{b_1 \cdot \ldots \cdot b_n \mid n \in \mathbb{N} \wedge b_1, \ldots, b_n \in \text{im}(\mu), b_1 \leq \cdots \leq b_n\}$ of all weights of $\mathcal{M}$ is finite. We construct for every $\alpha \in Y$ and for every $\alpha \in Y'$, respectively, a wta $\mathcal{M}_\alpha$ over $(\mathbb{N}_0, +, \cdot, 0, 1)$ such that for every $t \in T_\Sigma$ the value $\|\mathcal{M}_\alpha\|(t)$ is the number of all runs r of $\mathcal{M}$ on t with $\text{Val}(\mu(t, r)) = \alpha$. The construction of these wta differs depending on the choice of the enumeration. In case of DF, we recall from [32] that we can use $Q \times Y$ as states to propagate the accumulated product of the original wta $\mathcal{M}$ to the root. Then a state is marked final if its second component is α. In case of WO$_<$, we use multisets to count how often a weight is used along a run in $\mathcal{M}$. Since Y' is finite, the size of the multisets can be bounded and thus the resulting state set is finite. Again, a state is final if the product of all attached coefficients is α. For further details see [51, Theorem 3.2.17].

Eventually, we define tree series S_α by $S_\alpha(t) = \sum_{i=1}^m \alpha$ where $m = \|\mathcal{M}_\alpha\|(t)$. Since addition in K is locally finite and due to a result about recognizability of certain pre-images of recognizable series over $(\mathbb{N}_0, +, \cdot, 0, 1)$ (cf. [33, Lemma 6.3]), we can show that the tree series S_α are recognizable step functions. By Lemma 5.9, $S = \|\mathcal{M}\| = \sum_{\alpha \in Y} S_\alpha$ and $S = \sum_{\alpha \in Y'} S_\alpha$, respectively, is a recognizable step function.   □

By Theorem 5.10, each recognizable step function is recognizable with any enumeration en. But currently, we do not know for which enumerations besides the mentioned ones the reverse holds true. While a recognizable tree series S over a bi-locally finite, strong bimonoid always has a finite image, certain enumerations may exist such that S is not based on recognizable tree languages.

Theorem 6.3. *Let $(K, +, \circ, \mathbb{0}, \mathbb{1})$ be a strong bimonoid and S a tree series over K and Σ.*
(a) Let K be bi-locally finite. Then S is recognizable with DF iff $S = \llbracket \varphi \rrbracket^{\mathrm{DF}}$ for a wMSO sentence φ.
(b) Let K be bi-periodic and $<$ a linear order on K. Then S is recognizable with $\mathrm{WO}_<$ iff $S = \llbracket \varphi \rrbracket^{\mathrm{WO}_<}$ for a wMSO sentence φ.

Proof. The result of (a) was stated (without explicit proof) in [34, p. 162], where it was proved for the word case. We will use the same arguments for a simultaneous proof of (a) and (b).

Subsequently, we use that the results for ptv-monoids carry over to strong bimonoids using the correspondence between the strong bimonoid $(K, +, \circ, \mathbb{0}, \mathbb{1})$ with enumeration en and the associated ptv-monoid $(K, +, \circ_{\mathrm{en}}, \circ, \mathbb{0}, \mathbb{1})$. Now let en be either DF or $\mathrm{WO}_<$.

($\Rightarrow$) Immediate by Theorem 6.2 and Lemma 5.11.

($\Leftarrow$) Let φ be a wMSO sentence. We claim that $\llbracket \varphi \rrbracket^{\mathrm{en}}$ is a recognizable step function (and hence recognizable). We proceed by induction over the structure of φ. Clearly, our claim holds for all boolean formulas β and constants k. For disjunction and conjunction we use Lemma 5.1; for existential and universal quantifications we apply Proposition 5.15, Proposition 5.16, and Theorem 6.2.   □

As a consequence, the class of definable tree series is the same for DF and $\mathrm{WO}_<$.

Corollary 6.4. *Let $(K, +, \circ, \mathbb{0}, \mathbb{1})$ be a bi-locally finite strong bimonoid and $<$ a linear order on D. A tree series S over K and Σ is definable in wMSO with DF iff S is definable in wMSO with $\mathrm{WO}_<$.*

All our constructions are effective. Thus, for en $\in \{\mathrm{DF}, \mathrm{WO}_<\}$ and a wMSO formula φ we can construct classical tree automata $\mathcal{A}_1, \dots, \mathcal{A}_n$ and $d_1, \dots, d_n \in K$ such that $\llbracket \varphi \rrbracket^{\mathrm{en}} = \sum_{i=1}^{n} d_i \diamond \mathbb{1}_{L(\mathcal{A}_i)}$. Since the equivalence of classical tree automata is decidable, we obtain the following corollary.

Corollary 6.5. *Let K be an effectively given, strong bimonoid.*
(a) Let K be bi-locally finite. Then for any two wMSO formulas φ and ψ it is decidable whether $\llbracket \varphi \rrbracket^{\mathrm{DF}} = \llbracket \psi \rrbracket^{\mathrm{DF}}$.
(b) Let K be bi-periodic and $<$ a linear order on K. Then for any two wMSO formulas φ and ψ it is decidable whether $\llbracket \varphi \rrbracket^{\mathrm{WO}_<} = \llbracket \psi \rrbracket^{\mathrm{WO}_<}$.

7 Conclusion

We have presented several Büchi-like characterizations of recognizable tree series by weighted MSO logics in the very general setting of tree valuation monoids allowing for such new aspects like the average cost of a run. Our focus was the equivalence of different formalisms, here automata and logics, even if the weight structure is very general and does not allow for rich properties. Another point is the effective computation of the valuation function. Certainly, only valuation functions that can be computed effectively, e.g. the average cost, are of interest.

A topic we have not addressed so far for valuation monoids is the decidability of typical questions like emptiness, universality, inclusion, and equivalence. All these problems have quantitative counterparts which should be considered for concrete valuation monoids like those with average. This has been done for finite and infinite words [20–23] and should also be explored for trees.

Some questions remain open concerning local finiteness. For bi-locally finite strong bimonoids and bi-periodic strong bimonoids, respectively, we could show for two kinds of enumerations that recognizable tree series are always recognizable step functions. This implies the equivalence between automata and the whole weighted MSO-logic. However, it is open whether this is true for every enumeration. If not, we would like to characterize a class of enumerations for which this equivalence is true. Moreover, we wonder if we can define a notion of local finiteness for general tree valuation monoids such that weighted tree automata and weighted MSO-logic define the same class of tree series.

Valuation functions are even more interesting for infinite words [20] where one can consider the limit superior, the limit average, or discounting as valuation functions. In [30], a similar characterization like the one shown here was given for infinite words. We wonder: what about infinite trees? There, even a definition comprising meaningful examples is not straightforward. However, results concerning discounting were developed in [50].

References

1. Alexandrakis, A., Bozapalidis, S.: Weighted grammars and Kleene's theorem. Information Processing Letters 24(1), 1–4 (1987)
2. Berstel, J., Reutenauer, C.: Recognizable formal power series on trees. Theoretical Computer Science 18(2), 115–148 (1982)
3. Birkhoff, G., Neumann, J.v.: The logic of quantum mechanics. The Annals of Mathematics 37(4), 823–843 (1936)
4. Bollig, B., Gastin, P.: Weighted versus probabilistic logics. In: Diekert, V., Nowotka, D. (eds.) DLT 2009. LNCS, vol. 5583, pp. 18–38. Springer, Heidelberg (2009)
5. Borchardt, B.: The theory of recognizable tree series. Ph.D. thesis, TU Dresden (2004)
6. Bozapalidis, S.: Effective construction of the syntactic algebra of a recognizable series on trees. Acta Informatica 28(4), 351–363 (1991)
7. Bozapalidis, S.: Constructions effectives sur les séries formelles d'arbres. Theoretical Computer Science 77(3), 237–247 (1990)
8. Bozapalidis, S.: Representable tree series. Fundamenta Informaticae 21, 367–389 (1994)
9. Bozapalidis, S.: Convex algebras, convex modules and formal power series on trees. Journal of Automata, Languages and Combinatorics 1(3), 165–180 (1996)
10. Bozapalidis, S.: Positive tree representations and applications to tree automata. Information and Computation 139(2), 130–153 (1997)
11. Bozapalidis, S.: Context-free series on trees. Information and Computation 169(2), 186–229 (2001)

12. Bozapalidis, S., Alexandrakis, A.: Représentations matricielles des séries d'arbre reconnaissables. Informatique Théorique et Applications 23(4), 449–459 (1989)
13. Bozapalidis, S., Ioulidis, S.: Varieties of formal series on trees and Eilenberg's theorem. Information Processing Letters 29(4), 171–175 (1988)
14. Bozapalidis, S., Louscou-Bozapalidou, O.: The rank of a formal tree power series. Theoretical Computer Science 27, 211–215 (1983)
15. Bozapalidis, S., Louscou-Bozapalidou, O.: Finitely presentable tree series. Acta Cybernetica 17(3) (2006)
16. Bozapalidis, S., Louscou-Bozapalidou, O.: Fuzzy tree language recognizability. Fuzzy Sets and Systems 161(5), 716–734 (2010)
17. Bozapalidis, S., Rahonis, G.: On the closure of recognizable tree series under tree homomorphisms. Journal of Automata, Languages and Combinatorics 10(2/3), 185–202 (2005)
18. Büchi, J.R.: Weak second-order arithmetic and finite automata. Zeitschrift für mathematische Logik und Grundlagen der Mathematik 6, 66–92 (1960)
19. Büchi, J.: Finite Automata, Their Algebras and Grammars. Springer, Heidelberg (1989); Siefkes, D. (ed.)
20. Chatterjee, K., Doyen, L., Henzinger, T.: Quantitative languages. In: Kaminski, M., Martini, S. (eds.) CSL 2008. LNCS, vol. 5213, pp. 385–400. Springer, Heidelberg (2008)
21. Chatterjee, K., Doyen, L., Henzinger, T.: Alternating weighted automata. In: Kutyłowski, M., Charatonik, W., Gębala, M. (eds.) FCT 2009. LNCS, vol. 5699, pp. 3–13. Springer, Heidelberg (2009)
22. Chatterjee, K., Doyen, L., Henzinger, T.: Probabilistic weighted automata. In: Bravetti, M., Zavattaro, G. (eds.) CONCUR 2009. LNCS, vol. 5710, pp. 244–258. Springer, Heidelberg (2009)
23. Chatterjee, K., Doyen, L., Henzinger, T.A.: Expressiveness and closure properties for quantitative languages. In: Proceedings of LICS 2009, pp. 199–208. IEEE Computer Society, Los Alamitos (2009)
24. Comon, H., Dauchet, M., Gilleron, R., Löding, C., Jacquemard, F., Lugiez, D., Tison, S., Tommasi, M.: Tree automata techniques and applications (2007), http://www.grappa.univ-lille3.fr/tata
25. Doner, J.: Tree acceptors and some of their applications. Journal of Computer and System Sciences 4(5), 406–451 (1970)
26. Droste, M., Gastin, P.: Weighted automata and weighted logics. In: Caires, L., Italiano, G.F., Monteiro, L., Palamidessi, C., Yung, M. (eds.) ICALP 2005. LNCS, vol. 3580, pp. 513–525. Springer, Heidelberg (2005)
27. Droste, M., Gastin, P.: Weighted automata and weighted logics. Theoretical Computer Science 380, 69–86 (2007)
28. Droste, M., Gastin, P.: Weighted automata and weighted logics. In: Droste, et al. (eds.) [29], ch. 5 (2009)
29. Droste, M., Kuich, W., Vogler, H. (eds.): Handbook of Weighted Automata. EATCS Monographs on Theoretical Computer Science. Springer, Heidelberg (2009)
30. Droste, M., Meinecke, I.: Describing average- and longtime-behavior by weighted MSO logics. In: Hliněný, P., Kučera, A. (eds.) MFCS 2010. LNCS, vol. 6281, pp. 537–548. Springer, Heidelberg (2010)
31. Droste, M., Pech, C., Vogler, H.: A Kleene theorem for weighted tree automata. Theory of Computing Systems 38(1), 1–38 (2005)
32. Droste, M., Stüber, T., Vogler, H.: Weighted finite automata over strong bimonoids. Information Sciences 180(1), 156–166 (2010); special Issue on Collective Intelligence
33. Droste, M., Vogler, H.: Weighted tree automata and weighted logics. Theoretical Computer Science 366(3), 228–247 (2006)

34. Droste, M., Vogler, H.: Kleene and Büchi theorems for weighted automata and multi-valued logics over arbitrary bounded lattices. In: Gao, Y., Lu, H., Seki, S., Yu, S. (eds.) DLT 2010. LNCS, vol. 6224, pp. 160–172. Springer, Heidelberg (2010)
35. Droste, M., Vogler, H.: Weighted logics for unranked tree automata. Theory of Computing Systems 48(1), 23–47 (2011)
36. Edmonds, E.: Lattice fuzzy logics. International Journal of Man-Machine Studies 13(4), 455–465 (1980)
37. Elgot, C.C.: Decision problems of finite automata design and related arithmetics. Transactions of the American Mathematical Society 98, 21–52 (1961)
38. Ésik, Z., Kuich, W.: Formal tree series. Journal of Automata Languages, Combinatorics 8(2), 219–285 (2003)
39. Fülöp, Z., Maletti, A., Vogler, H.: A Kleene theorem for weighted tree automata over distributive multioperator monoids. Theory of Computing Systems 44(3), 455–499 (2009)
40. Fülöp, Z., Stüber, T., Vogler, H.: A Büchi-like theorem for weighted tree automata over multioperator monoids. Theory of Computing Systems, 1–38 (2010)
41. Fülöp, Z., Vogler, H.: Weighted Tree Automata and Tree Transducers. In: Droste, et al. (eds.) [29], ch. 9 (2009)
42. Gécseg, F., Steinby, M.: Tree languages. In: Handbook of Formal Languages, vol. 3, ch. 1, pp. 1–68. Springer, Heidelberg (1997)
43. Götze, D.: Gewichtete Logik und Baumautomaten über Baumbewertungsmonoiden. Master's thesis, Universität Leipzig (2011), http://lips.informatik.uni-leipzig.de/pub/2011-3
44. Huschenbett, M.: Models for quantitative distributed systems and multi-valued logics. In: Dediu, A.-H., Inenaga, S., Martín-Vide, C. (eds.) LATA 2011. LNCS, vol. 6638, pp. 310–322. Springer, Heidelberg (2011)
45. Kreinovich, V.: Towards more realistic (e.g., non-associative) "and"- and "or"-operations in fuzzy logic. Soft Computing 8(4), 274–280 (2004)
46. Kuich, W.: Formal power series over trees. In: Bozapalidis, S. (ed.) DLT 1997, pp. 61–101. Aristotle University of Thessaloniki Press (1997)
47. Kuich, W.: Tree transducers and formal tree series. Acta Cybernetica 14(1), 135–149 (1999)
48. de Luca, A., Varricchio, S.: Finiteness and Regularity in Semigroups and Formal Languages. EATCS Monographs on Theoretical Computer Science. Springer, Heidelberg (1999)
49. Mallya, A.: Deductive multi-valued model checking. In: Gabbrielli, M., Gupta, G. (eds.) ICLP 2005. LNCS, vol. 3668, pp. 297–310. Springer, Heidelberg (2005)
50. Mandrali, E., Rahonis, G.: Recognizable tree series with discounting. Acta Cybernetica 19(2), 411–439 (2009)
51. Märcker, S.: Charakterisierung erkennbarer Baumreihen über starken Bimonoiden durch gewichtete MSO-Logik. Master's thesis, Universität Leipzig (2010), http://lips.informatik.uni-leipzig.de/pub/2010-25
52. Rahonis, G.: Weighted Muller tree automata and weighted logics. Journal of Automata, Languages and Combinatorics 12(4), 455–483 (2007)
53. Seidl, H.: Finite tree automata with cost functions. Theoretical Computer Science 126(1), 113–142 (1994)
54. Thatcher, J.W., Wright, J.B.: Generalized finite automata theory with an application to a decision problem of second-order logic. Theory of Computing Systems 2, 57–81 (1968)
55. Thomas, W.: Languages, automata, and logic. In: Rozenberg, G., Salomaa, A. (eds.) Handbook of Formal Languages, vol. A, pp. 389–455. Springer, Heidelberg (1997)

Partial Conway and Iteration
Semiring-Semimodule Pairs

Zoltán Ésik*

Dept. of Computer Science, University of Szeged, Hungary

Abstract. A Conway semiring is a semiring S equipped with a unary operation $^*: S \to S$, called "star", satisfying the sum star and product star identities. A Conway semiring-semimodule pair consists of a Conway semiring S and a left S-semimodule V together with a function $^\omega: S \to V$, called "omega power", subject to the sum omega and product omega identities. A Kleene type theorem holds in all Conway semiring-semimodule pairs that can be instantiated to give the equivalence of Büchi automata and regular languages over ω-words. However, sometimes the star and omega power operations cannot be defined in an appropriate manner on the whole semiring S. To handle this situation, we introduce partial Conway semiring-semimodule pairs and develop their basic theory in connection with automata. We prove a Kleene theorem, applicable to all partial Conway semiring-semimodule pairs.

1 Semirings and Matrix Theories

A *semiring* [7,13,16] $S = (S, +, \cdot, 0, 1)$ consists of a commutative monoid $(S, +, 0)$ and a monoid $(S, \cdot, 1)$ such that product distributes over finite sums, so that

$$a(b + c) = ab + ac$$
$$(b + c)a = ba + ca$$
$$a \cdot 0 = 0$$
$$0 \cdot a = 0$$

hold for all $a, b, c \in S$. Examples of semirings include the *semiring* $\mathbb{N}$ *of natural numbers*, the *Boolean semiring* $\mathbb{B}$ whose sum and product operations are disjunction and conjunction, and the *language semiring* $P(\Sigma^*)$ over an alphabet Σ, where the sum operation is set union and the product operation is concatenation, and where $0 = \emptyset$ and $1 = \{\epsilon\}$. Morphisms of semirings preserve the operations and the constants 0 and 1.

It is well-known that if S is a semiring, then so is the collection $S^{n \times n}$ of all $n \times n$ matrices over S, equipped with the pointwise sum and the usual matrix

* Partially supported by the project TÁMOP-4.2.1/B-09/1/KONV-2010-0005 "Creating the Center of Excellence at the University of Szeged", supported by the European Union and co-financed by the European Regional Fund, and by grant no. K 75249 from the National Foundation of Hungary for Scientific Research.

W. Kuich and G. Rahonis (Eds.): Bozapalidis Festschrift, LNCS 7020, pp. 56–71, 2011.

product as sum and product. The matrix $0_{n,n}$ whose entries are all 0 serves as the neutral element 0, while the multiplicative identity is the usual $n \times n$ unit matrix E_n.

We can associate a category $\mathbf{Mat}_S$ with every semiring S, called a *matrix theory*, see [3,8]. The objects of this category are the natural numbers, and a morphism $n \to p$ is an $n \times p$ matrix over S, i.e., an element of $S^{n \times p}$. Composition is matrix multiplication and the matrices E_n serve as identity morphisms. Each hom-set $\mathbf{Mat}_S(n, n)$ of morphisms $n \to p$ of the theory $\mathbf{Mat}_S$ is also equipped with a commutative monoid structure $(\mathbf{Mat}_S(n, p), +, 0_{n,p})$, where $+$ is defined pointwise. The usual distributivity laws hold:

$$A(B + C) = AB + AC$$
$$(B + C)D = BD + CD$$

for all $A : m \to n$, $B, C : n \to p$ and $D : p \to q$. Moreover,

$$0_{m,n} \cdot A = 0_{m,p}$$
$$A \cdot 0_{p,q} = 0_{n,q}$$

for all $A : n \to p$.

Every morphism $h : S \to S'$ between semirings S and S' may be lifted to a functor $\mathbf{Mat}_S \to \mathbf{Mat}_{S'}$ defined pointwise. Such functors are called *matrix theory morphisms*. The category of semirings is clearly isomorphic to the category of matrix theories. For an abstract treatment of matrix theories, we refer to [3,8].

2   Partial Conway Semirings

Conway semirings are implicit in Conway [6]. They were defined explicitly in [2,3]. The definition of Conway semirings involves two important identities of regular languages. A general Kleene theorem for Conway semirings is proved in [3]. However, the applicability of Conway semirings is limited, due to the fact that the star operation is total, whereas many important semirings only allow for a partially defined star operation. Moreover, it is not true that all semirings can be embedded into a Conway semiring with a totally defined star operation. These facts led to the introduction of partial Conway semirings in [5].

Definition 1. *A* partial **-semiring is a semiring S equipped with a partially defined star operation $* : S \to S$ whose domain is an* ideal *of S. A *-semiring is a partial *-semiring S such that * is defined on the whole semiring S. A morphism $S \to S'$ of (partial) *-semirings is a semiring morphism $h : S \to S'$ such that for all $s \in S$, if s^* is defined then so is $(sh)^*$, and $s^* h = (sh)^*$.*

Thus, in a partial *-semiring S, 0^* is defined, and if a^* and b^* are defined then so is $(a + b)^*$, finally, if a^* or b^* is defined, then so is $(ab)^*$. When S is a partial *-semiring, we let $D(S)$ denote the domain of definition of the star operation.

Definition 2. *A* partial Conway semiring *is a partial * * -semiring S satisfying the following two axioms:*

1. *Sum star identity:*

$$(a + b)^* = a^*(ba^*)^*$$

   for all $a, b \in D(S)$.
2. *Product star identity:*

$$(ab)^* = 1 + a(ba)^*b,$$

   for all $a, b \in S$ such that $a \in D(S)$ or $b \in D(S)$.

A Conway semiring *is a partial Conway semiring S which is a * * -semiring (i.e., $D(S) = S$). A morphism of (partial) Conway semirings is a (partial) * -semiring morphism.*

In any partial Conway semiring S,

$$aa^* + 1 = a^*$$
$$a^*a + 1 = a^*$$
$$0^* = 1$$

for all $a \in D(S)$. Moreover, if $a \in D(S)$ or $b \in D(S)$, then

$$(ab)^*a = a(ba)^*.$$

It follows that also

$$aa^* = a^*a$$
$$(a + b)^* = (a^*b)^*a^*$$

for all $a, b \in D(S)$. When $a \in D(S)$, we will denote $aa^* = a^*a$ by a^+ and call $^+$ the *plus* operation.

Partial Conway semirings include the partial iterative semirings discussed below. Conway semirings include all continuous or complete semirings [7,10] and the inductive *-semirings of [10] and Kozen's Kleene algebras [14].

Conway semirings give rise to Conway matrix theories [3]. In the same way, partial Conway semirings give rise to partial Conway matrix theories. We say that a collection J of matrices in $\mathbf{Mat}_S$ is a *matrix ideal* if there is an ideal $I \subseteq S$ such that a matrix is in J iff its entries are all in I. In this case we also write $J = M(I)$.

Definition 3. *Suppose that S is a semiring and consider the matrix theory $\mathbf{Mat}_S$ equipped with a distinguished matrix ideal $M(I)$. We say that $\mathbf{Mat}_S$ is a* partial Conway matrix theory *if it is equipped with a star operation $A \mapsto A^*$,* *defined on the square matrices $A : n \to n$ in $M(I)$, $n \geq 0$, such that the matrix versions of the sum and product star identities hold:*

$$(A + B)^* = A^*(BA^*)^*,$$

for all $A, B : n \to n \in M(I)$, and

$$(AB)^* = E_n + A(BA)^*B,$$

for all $A : n \to m$, $B : m \to n$ such that A or B is in $M(I)$. When $\mathbf{Mat}_S$ is a partial Conway matrix theory such that star is defined on all square matrices, i.e., $M(I) = \mathbf{Mat}_S$, then we call $\mathbf{Mat}_S$ a Conway matrix theory. A morphism of (partial) matrix theories is a matrix theory morphism which preserves the distinguished ideal and the star operation.

Note the following special cases:

$$A^* = AA^* + E_n$$
$$A^* = A^*A + E_n$$
$$0^*_{nn} = E_n$$

where $A : n \to n$ in $M(I)$, $n \geq 0$. Also, $AA^* = A^*A$ for all $A : n \to n$ in $M(I)$. Below we will denote AA^* by A^+.

If $\mathbf{Mat}_S$ is a (partial) Conway matrix theory, then by identifying a 1×1 matrix (a) in $\mathbf{Mat}_S$ with the element $a \in S$, the semiring S becomes a (partial) Conway semiring. Conversely, any (partial) Conway semiring determines a (partial) Conway matrix theory, as we show below.

Definition 4. *Suppose that S is a partial Conway semiring with $D(S) = I$. We define a partial star operation on the semirings $S^{k \times k}$, $k \geq 0$, whose domain of definition is $I^{k \times k}$, the ideal of those $k \times k$ matrices with entries in I. When $k = 0$, $S^{k \times k}$ is trivial as is the definition of star. When $k = 1$, we use the star operation on S. Assuming that $k > 1$, we write $k = n + 1$. For a matrix $\begin{pmatrix} a & b \\ c & d \end{pmatrix}$ in $I^{k \times k}$, define*

$$\begin{pmatrix} a & b \\ c & d \end{pmatrix}^* = \begin{pmatrix} \alpha & \beta \\ \gamma & \delta \end{pmatrix} \tag{1}$$

where $a \in S^{n \times n}$, $b \in S^{n \times 1}$, $c \in S^{1 \times n}$ and $d \in S^{1 \times 1}$, and where

$$\alpha = (a + bd^*c)^* \qquad \beta = \alpha bd^*$$
$$\gamma = \delta ca^* \qquad \delta = (d + ca^*b)^*.$$

We have thus defined a star operation on the square matrices in $M(I)$. It is known (cf. [3]) that when S is a Conway semiring, then equipped with the above star operation, $\mathbf{Mat}_S$ is a Conway matrix theory. More generally, but with the same proof, we have:

Theorem 1. *Suppose that S is a partial Conway semiring with $D(S) = I$. Then, equipped with the matrix ideal $M(I)$ and the above star operation, $\mathbf{Mat}_S$ is a partial Conway matrix theory, where the star operation is defined on the square matrices in $M(I)$.*

Corollary 1. *If S is a partial Conway semiring with $D(S) = I$, then so is the semiring $S^{n \times n}$ with $D(S^{n \times n}) = I^{n \times n}$ and the above star operation for each n.*

See also [5].

Corollary 2. *The category of (partial) Conway semirings is equivalent to the category of (partial) Conway matrix theories.*

The following result is known to hold for Conway matrix theories, see [3,5].

Theorem 2. *Suppose that $\mathbf{Mat}_S$ is a partial Conway matrix theory with star defined on the square matrices in $M(I)$. Then the following identities hold.*

1. *The matrix star identity (1) for all possible decompositions of a square matrix in $M(I)$ into four blocks such that a and d are square matrices, i.e., where $a : n \to n$, $b : n \to m$, $c : m \to n$ and $d : m \to m$, $n, m \geq 0$.*
2. *The permutation identity*

$$(\pi A \pi^T)^* = \pi A^* \pi^T,$$

   for all $A : n \to n$ in $M(I)$ and any permutation matrix $\pi : n \to n$, where π^T denotes the transpose (inverse) of π.

The proof is the same as for Conway matrix theories, cf. [3]. For later use we note the following. When $\mathbf{Mat}_S$ is a partial Conway matrix theory with star operation defined on the square matrices in $M(I)$, and if $A = \begin{pmatrix} a & b \\ c & d \end{pmatrix}$ is a matrix with entries in $M(I)$, partitioned as above, then

$$A^+ = \begin{pmatrix} (a + bd^*c)^+ & (a + bd^*c)^*bd^* \\ (d + ca^*b)^*ca^* & (d + ca^*b)^+ \end{pmatrix} \tag{2}$$

By (2), $A^+ \in M(I)$.

3   Partial Conway Semiring-Semimodule Pairs

In this section, we consider semiring-semimodule pairs (S, V) consisting of a semiring S and a left S-semimodule V, called a semimodule for short. We equip (S, V) with (partially defined) star and omega power operations, where the omega power operation maps an ideal of S into V. We define partial Conway semiring-semimodule pairs and develop the basic theory of these structures. In particular, we show that partial Conway semiring-semimodule pairs give rise to partial Conway matricial theories.

Definition 5. *A partial* *-*semiring* $^\omega$-*semimodule pair* (S, V) *consists of a semiring* S *and a left* S-*semimodule* V *together with partially defined operations* * : $S \to S$ *and* $^\omega$: $S \to V$, *whose domain of definition is an ideal* $D(S)$ *of* S. *A partial Conway semiring-semimodule pair is a partial* *-*semiring* $^\omega$-*semimodule pair* (S, V) *such that* S *is a partial Conway semiring and the following identities hold:*

1. Sum omega identity:

$$(a + b)^\omega = (a^*b)^\omega + (a^*b)^*a^\omega,$$

for all $a, b \in D(S)$.
2. Product omega identity:

$$(ab)^\omega = a(ba)^\omega,$$

for all $a, b \in S$ *such that* $a \in D(S)$ *or* $b \in D(S)$.

A Conway semiring-semimodule pair is a partial Conway semiring-semimodule pair (S, V) *such that* $D(S) = S$. *A morphism of (partial) Conway semiring-semimodule pairs* $(S, V) \to (S', V')$ *consists of a semiring morphism* $h_S : S \to S'$ *and a monoid morphism* $h_V : V \to V'$ *with* $D(S)h_S \subseteq D(S')$ *that jointly preserve the action and the star and omega power operations.*

Note that $a^\omega = aa^\omega$ and $0^\omega = 0$ hold in all partial Conway semiring-semimodule pairs (S, V), for all $a \in D(S)$.

An important feature of the above definition is that star and omega are defined on the *same* ideal. Partial Conway semiring-semimodule pairs include the partial iterative semiring-semimodule pairs discussed below. Conway semiring-semimodule pairs include the *bi-inductive semiring-semimodule pairs* of [12].

Let (S, V) be a semiring-semimodule pair. Following Elgot [8], we can define the *matricial theory* $\mathbf{Matr}_{S,V}$ that is the category with the nonnegative integers as objects, and as morphisms $m \to n$ all ordered pairs (A, v), where A is an $m \times n$ matrix over S, so that $A : m \to n$ in $\mathbf{Mat}_S$, and v is an n-dimensional column over V, i.e., an element of V^n. We usually identify a morphism $(, v) : n \to 0$ with $v \in V^n$. The action of S on V can naturally be extended to an action

$$S^{m \times n} \times V^n \to V^m, \ m, n \geq 0$$

$(A, v) \mapsto Av$. Using this action, we define composition in $\mathbf{Matr}_{S,V}$ as follows. Given $(A, u) : m \to n$ and $(B, v) : n \to p$, we let

$$(A, u) \cdot (B, v) = (AB, u + Av) : m \to p.$$

For each n, the identity morphism $n \to n$ is the pair $(E_n, 0^n)$, consisting of the $n \times n$ identity matrix over S and the n-dimensional column $0^n \in V^n$ whose components are all 0. The theory $\mathbf{Matr}_{S,V}$ is also equipped with an additive structure, where $(A, u) + (B, v) = (A + B, u + v)$ is defined pointwise, for all

$(A, u), (B, v) : m \to n$. It is clear that each hom-set $\mathbf{Matr}_{S,V}(m, n)$ is a commutative monoid with neutral element the pair $(0_{m,n}, 0^m)$. Note that each $m \times n$ matrix A in $\mathbf{Mat}_S$ can be naturally identified with the pair $(A, 0^m)$. By this identification, $\mathbf{Mat}_S$ becomes a subtheory of $\mathbf{Matr}_{S,V}$. We call $\mathbf{Mat}_S$ the *underlying matrix theory* of $\mathbf{Matr}_{S,V}$.

When $\mathbf{Matr}_{S,V}$ and $\mathbf{Matr}_{S',V'}$ are both matricial theories, a morphism $h : \mathbf{Matr}_{S,V} \to \mathbf{Matr}_{S',V'}$ is determined by a morphism $(h_S, h_V) : (S, V) \to (S', V')$ of semiring-semimodule pairs. We define $(A, u)h = (Ah, uh)$ for all $(A, u) : n \to p$ in $\mathbf{Matr}_{S,V}$, where Ah and uh are obtained by applying h_S and h_V to the entries of A and u, respectively. (There is also a more abstract definition, cf. [3]). The category of matricial theories is thus isomorphic to the category of semiring-semimodule pairs.

Note that

$$((A, u) + (B, v)) \cdot (C, w) = (A, u) \cdot (C, w) + (B, v) \cdot (C, w)$$
$$(0_{m,n}, 0^m) \cdot (A, u) = (0_{m,p}, 0^m)$$

hold for all $(A, u), (B, v) : n \to p$ and $(C, w) : p \to q$. On the other hand, the equations

$$(A, u) \cdot ((B, v) + (C, w)) = (A, u) \cdot (B, v) + (A, u) \cdot (C, w) \qquad (3)$$
$$(A, u) \cdot (0_{p,q}, 0^p) = (0_{n,q}, 0^n)$$

do not hold. However, (3) holds whenever V is idempotent: $v + v = v$ for all $v \in V$.

Definition 6. *Suppose that (S, V) is a semiring-semimodule pair and consider the matricial theory $\mathbf{Matr}_{S,V}$. We say that $\mathbf{Matr}_{S,V}$ is a partial Conway matricial theory if its underlying matrix theory $\mathbf{Mat}_S$ is equipped with a partial star operation, defined on a matrix ideal $M(I)$, such that it is a partial Conway matrix theory, moreover, there is an omega power operation $A \mapsto A^\omega$, mapping $n \times n$ matrices A in $M(I)$ to columns in V^n, subject to the following conditions.*

$$(A + B)^\omega = (A^* B)^\omega + (A^* B)^* A^\omega$$

for all $A, B \in M(I)$, $A, B : n \to n$, and

$$(AB)^\omega = A(BA)^\omega,$$

for all $A : n \to m$, $B : m \to n$ such that A or B is in $M(I)$. When $\mathbf{Matr}_{S,V}$ is a partial Conway matricial theory such that the star and omega power operations are defined on all square matrices, then we call $\mathbf{Matr}_{S,V}$ a Conway matricial theory. A morphism of (partial) Conway matricial theories is a matricial theory morphism which preserves the star and omega power operations.

Note the following special case:

$$AA^\omega = A^\omega,$$

where $A : n \to n$ in $M(I)$, $n \geq 0$. Also, $0_{n,n}^\omega = 0^n$, $n \geq 0$.

An important fact is that the following holds in all partial Conway matricial theories $\mathbf{Matr}_{S,V}$ with distinguished ideal $M(I)$. Let $A = \begin{pmatrix} a & b \\ c & d \end{pmatrix}$ be a square matrix in $M(I)$ partitioned so that a and d are square matrices. Then

$$A^\omega = \begin{pmatrix} (a + bd^*c)^\omega + (a + bd^*c)^* bd^\omega \\ (d + ca^*b)^\omega + (d + ca^*b)^* ca^\omega \end{pmatrix}. \tag{4}$$

Suppose now that (S, V) is a partial Conway semiring-semimodule pair, so that S is a partial Conway semiring. Suppose that the star and omega power operations are defined on the ideal $I = D(S)$. We have already extended the star operation to all square matrices with entries in I. We now show how the omega power operation can be extended. For this reason, let $A \in I^{n \times n}$. We define A^ω by induction on n. When $n = 0$, we define A^ω as the unique element of V^0. When $n = 1$, so that $A = (s)$ for some $s \in I$, we let $A^\omega = (s^\omega)$. Finally, when $n > 1$, let us partition A as above such that d is 1×1, say. Then we define A^ω by the formula (4). The ω-power operation is well-defined since all entries of a, b, c, d are in I.

Theorem 3. *When (S, V) is a partial Conway semiring-semimodule pair with star and omega power defined on the ideal I, then $\mathbf{Matr}_{S,V}$ is a partial Conway matricial theory with distinguished matrix ideal $M(I)$. Conversely, if $\mathbf{Matr}_{S,V}$ is a partial Conway matricial theory where star and omega power are defined on the square matrices in $M(I)$, then (S, V) is a partial Conway semiring-semimodule pair with distinguished ideal I.*

In fact, the category of (partial) Conway semiring-semimodule pairs is isomorphic to the category of (partial) Conway matricial theories.

The proof is similar to the proof of the corresponding fact for Conway semiring-semimodule pairs and Conway matricial theories, cf. [3].

4 Partial Iterative Semiring-Semimodule Pairs

In this section, we define a subclass of partial Conway semiring-semimodule pairs and partial Conway matricial theories. We call a semiring-semimodule pair (S, V) *positive* if V has at least two elements, and for all $a \in S$ and $v, v' \in V$, if $v + v' = 0$ then $v = v' = 0$, and if $av = 0$ then $a = 0$ or $v = 0$. When (S, V) is positive so is S: For all $a, b \in S$, if $a + b = 0$ then $a = b = 0$, and if $ab = 0$ then $a = 0$ or $b = 0$. Indeed, if (S, V) is positive and $a + b = 0$, then $(a + b)v = av + bv = 0$ for all $v \in V$. Thus, $av = bv = 0$ for all $v \in V$ so that $a = b = 0$. And if $ab = 0$ then $a(bv) = (ab)v = 0$ for all v. Assume that $v \neq 0$. Since $a(bv) = 0$, $a = 0$ or $bv = 0$. But if $bv = 0$ then $b = 0$, since $v \neq 0$. We conclude that $a = 0$ or $b = 0$.

Definition 7. *Suppose that (S, V) is a positive semiring-semimodule pair with a distinguished ideal $I \subseteq S$. We call (S, V) a partial iterative semiring-semimodule pair if for all $a \in I$, $b \in S$ and $v \in V$, the equation $x = ax + b$ has a unique* solution in S, moreover, the equation $y = ay + v$ has either 0 as its unique solution in V, when $a = 0$ and $v = 0$, or it has a unique nonzero solution in V. A morphism of partial iterative semiring-semimodule pairs is a morphism of semiring-semimodule pairs which preserves the distinguished ideal.

The notion of partial iterative semiring-semimodule pairs originates from the Matricial Extension Theorem of [3].

We give some examples of partial iterative semiring-semimodule pairs. Suppose that (S, V) is a positive semiring-semimodule pair such that both S and $V - \{0\}$ are complete metric spaces. Moreover, suppose that $I \subseteq S$ is an ideal such that for each $s, s' \in I$ and $r \in S$ with $s' \neq 0$, the functions

$$x \mapsto sx + r, \quad x \in S$$
$$v \mapsto sv, \quad v \in V - \{0\}$$

are proper contractions. Then (S, V), equipped with the distinguished ideal I, is a partial iterative semiring-semimodule pair, since proper contractions of complete metric spaces have unique fixed points.

In particular, let Σ be an alphabet, and define the usual metric d on Σ^ω: For every $u, v \in \Sigma^\omega$, $d(u, v) = 0$ if $u = v$, and $d(u, v) = 2^{-n}$ if $u = wax$ and $v = wby$ for some $w \in \Sigma^*$ of length $n - 1$ and for some $a, b \in \Sigma$ and $x, y \in \Sigma^\omega$ with $a \neq b$. Then Σ^ω is a complete metric space as is the collection $P'_c(\Sigma^\omega)$ of all nonempty closed subsets of Σ^ω equipped with the Hausdorff metric. Let $P_c(\Sigma^\omega) = P'_c(\Sigma^\omega) \cup \{\emptyset\}$ and define the action LV of a language $L \in P(\Sigma^*)$ on $V \in P_c(\Sigma^\omega)$ to be the closure of the set

$$LV = \{xv : x \in L, \ v \in V\}.$$

Then equipped with set union as $+$, $P_c(\Sigma^\omega)$ is a positive $P(\Sigma^*)$-semimodule. The semiring $P(\Sigma^*)$ may also be turned into a complete metric space with distance function

$$d(L, L') = 2^{-\min\{|u|:u \in L - L' \text{ or } u \in L' - L\}}$$

for all $L, L' \subseteq \Sigma^*$ with $L \neq L'$, and $d(L, L') = 0$ if $L = L'$. Let I be the ideal of all languages in $P(\Sigma^*)$ not containing the empty word. Then equipped with the distinguished ideal I, $(P(\Sigma^*), P_c(\Sigma^\omega))$ is a partial iterative semiring-semimodule pair.

Theorem 4. *Suppose that (S, V) is a partial iterative semiring-semimodule pair with distinguished ideal I. Then (S, V) is a partial Conway semiring-semimodule pair with $D(S) = I$. More precisely, there is a unique way to turn (S, V) into a partial Conway semiring-semimodule pair such that the domain of definition of the star and omega power operations is I and a^ω is nonzero whenever $a \in I$ is not zero.*

Proof. For any $a \in I$, define a^* as the unique solution of the equation $x = ax + 1$, and when $a \neq 0$, define a^ω as the unique *nonzero* solution of the equation $x = ax$. When $a = 0$, define $a^\omega = 0$. Note that the definitions are forced.

It was shown in [5] that, equipped with the above partial star operation, S is a partial Conway semiring. So to complete the proof, it suffices to prove that the sum omega and product omega identities hold.

Suppose that $a, b \in S$. If $a \in I$ or $b \in I$, then $ab, ba \in I$. Moreover, $a(ba)^\omega = aba(ba)^\omega$, showing that $a(ba)^\omega$ is a solution of the equation $x = abx$. If a and b are nonzero, then ab, ba are nonzero. Thus $a(ba)^\omega$ is nonzero, so that $(ab)^\omega = a(ba)^\omega$, by uniqueness. If $a = 0$ or $b = 0$ then both $(ab)^\omega$ and $a(ba)^\omega$ are 0.

If $a, b \in I$, then $a + b \in I$. We show that $(a^*b)^\omega + (a^*b)^*a^\omega$ is a solution of the equation $x = (a + b)x$. Indeed,

$$
\begin{aligned}
(a + b)((a^*b)^*a^\omega + (a^*b)^\omega) &= \\
&= (a(a^*b)^*a^\omega + (ba^*)^*ba^\omega) + (a(a^*b)^\omega + (ba^*)^\omega) \\
&= (a + aa^*(ba^*)^*b + (ba^*)^*b)a^\omega + (aa^*(ba^*)^\omega + (ba^*)^\omega) \\
&= (a^\omega + (aa^* + 1)(ba^*)^*ba^\omega) + (aa^* + 1)(ba^*)^\omega \\
&= (1 + a^*(ba^*)^*b)a^\omega + a^*(ba^*)^\omega \\
&= (a^*b)^*a^\omega + (a^*b)^\omega.
\end{aligned}
$$

Now, if $a + b$ is nonzero then a and b are also nonzero, therefore $(a^*b)^*a^\omega + (a^*b)^\omega$ is not zero. It follows that $(a + b)^\omega = (a^*b)^*a^\omega + (a^*b)^\omega$. When $a + b = 0$ then $a = b = 0$ so that $(a + b)^\omega = 0 = (a^*b)^*a^\omega + (a^*b)^\omega$. □

We also note that any morphism $(S, V) \to (S', V')$ between partial iterative semiring-semimodule pairs is a partial Conway semiring-semimodule pair morphism.

Proposition 1. *Suppose that (S, V) is a partial iterative semiring-semimodule pair with distinguished ideal I. Then $s^\omega + s^\omega = s^\omega$ holds for all $s \in I$. Moreover, $s's^\omega = s^\omega$ for all $s \in I$ and nonzero $s' \in S$ with $ss' = s's$.*

Proof. When $s = 0$, $s^\omega = 0$, and our claims are obvious. Suppose now that $s \neq 0$. Since $s(s^\omega + s^\omega) = ss^\omega + ss^\omega = s^\omega + s^\omega$ and $s^\omega + s^\omega$ is not zero, it follows by uniqueness that $s^\omega + s^\omega = s^\omega$. Also, if $s' \neq 0$ and $ss' = s's$, then $s(s's^\omega) = s'(ss^\omega) = s's^\omega$ is not zero, so that $s's^\omega = s^\omega$. □

If (S, V) is a partial iterative semiring-semimodule pair then it is also a partial Conway semiring-semimodule pair, so that $\mathbf{Matr}_{S,V}$ is a partial Conway matricial theory. We end this section with an important feature of these partial Conway matricial theories, see [2].

Theorem 5. *Suppose that (S, V) is a partial iterative semiring-semimodule pair with distinguished ideal I, so that $\mathbf{Matr}_{S,V}$ is a partial Conway matricial theory with distinguished ideal $M(I)$. Then for all $A : n \to n$ in $M(I)$ and $B : n \to p$ in $\mathbf{Mat}_S$, A^*B is the unique solution of the equation $X = AX + B$ in $\mathbf{Mat}_S$. Moreover, for all $A : n \to n$ in $M(I)$ and $v \in V^n$, $u = A^\omega + A^*v$ is the unique "maximal" solution of the equation $x = Ax + v$ in V^n, so that whenever w is another solution, then for all $i \in [n]$, either the ith entry of w agrees with the ith entry of u, or the ith entry of w is 0.*

5 Kleene Theorem

In this section, extending a result from [11], we establish a Kleene theorem for partial Conway semiring-semimodule pairs. To this end, we first generalize the omega power operation on matrices.

Definition 8. *Suppose that* (S, V) *is a partial Conway semiring-semimodule pair and* $A : n \to n$ *in* $D(S)^{n \times n}$. *When* $0 \le k \le n$, *let us partition* A *as*

$$A = \begin{pmatrix} a & b \\ c & d \end{pmatrix},$$

where a *is* $k \times k$, *etc. Then we define*

$$A^{\omega_k} = \begin{pmatrix} (a + bd^*c)^\omega \\ d^*c(a + bd^*c)^\omega \end{pmatrix} \tag{5}$$

This definition makes sense since all entries of d, bd^*c and $a + bd^*c$ are in $D(S)$. Note that $A^{\omega_n} = A^\omega$. For later use we note that

$$\begin{pmatrix} A \\ C & D \end{pmatrix}^{\omega_k} = \begin{pmatrix} A^{\omega_k} \\ D^*CA^{\omega_k} \end{pmatrix} \tag{6}$$

for all $A \in D(S)^{n \times n}$, $C \in D(S)^{n \times m}$ and $D \in D(S)^{m \times m}$ with $k \le n$. Also,

$$AA^{\omega^k} = A^{\omega_k} \tag{7}$$

for all $A \in D(S)^{n \times n}$ and $k \le n$, as the reader can easily verify.

Suppose for the rest of this section that (S, V) is a partial Conway semiring-semimodule pair. Let S_0 be a fixed subsemiring of S and let $\Sigma \subseteq D(S)$. We denote by $S_0\Sigma$ the set of all finite linear combinations over Σ with coefficients in S_0.

Definition 9. *An automaton over* (S_0, Σ) *(of dimension* n*) is a triplet* $\mathbf{A} = (\alpha, A, \beta)$, *where* $\alpha \in S_0^{1 \times n}$, $A \in (S_0\Sigma)^{n \times n}$ *and* $\beta \in S_0^{n \times 1}$. *The behavior of* A *is*

$$|\mathbf{A}| = \alpha A^* \beta \in S.$$

Definition 10. *A Büchi automaton over* (S_0, Σ) *is a system* $\mathbf{A} = (\alpha, A, k)$, *where* $\alpha \in S_0^{1 \times n}$, $A \in (S_0\Sigma)^{n \times n}$ *and* $0 \le k \le n$ *for some* n. *The behavior of* $\mathbf{A}$ *is*

$$|\mathbf{A}| = \alpha A^{\omega_k} \in V.$$

We let $\mathbf{Rec}_{(S,V)}(S_0, \Sigma)$ and $\mathbf{Rec}^\omega_{(S,V)}(S_0, \Sigma)$ respectively denote the set of all behaviors of automata and Büchi automata over (S_0, Σ).

Definition 11. *We say that* $s \in S$ *is* rational over (S_0, Σ) *if it is of the form* $s = x + a$, *where* $x \in S_0$ *and* a *is in the least set* $\mathbf{Rat}'_{(S,V)}(S_0, \Sigma)$ *of elements of* S *that can be constructed from* $\Sigma \cup \{0\}$ *by the operations* $+, \cdot$ *and* $^+$, *and by multiplication on the left or on the right with elements of* S_0.

Moreover, we say that $v \in V$ *is* rational over (S_0, Σ) *if it can be generated from the elements of* $\mathbf{Rat}'_{(S,V)}(S_0, \Sigma)$ *by* $+$ *and* $^\omega$, *and by the action of rational elements of* S *over* (S_0, Σ).

The above definition is correct, since if s is in $\mathbf{Rat}'_{(S,V)}(S_0, \Sigma)$, then $s \in D(S)$. Note that $v \in V$ is rational over (S_0, Σ) iff it can be written as a finite sum $\sum_{i=1}^{n} s_i r_i^\omega$, where each s_i is rational over (S_0, Σ) and each r_i is in $\mathbf{Rat}'_{(S,V)}(S_0, \Sigma)$.

We let $\mathbf{Rat}_{(S,V)}(S_0, \Sigma)$ and $\mathbf{Rat}^\omega_{(S,V)}(S_0, \Sigma)$ respectively denote the collection of all rational elements in S and V over (S_0, Σ). The following Kleene theorem was proved in [5]. For special cases, see also [3,11].

Theorem 6. $\mathbf{Rat}_{(S,V)}(S_0, \Sigma) = \mathbf{Rec}_{(S,V)}(S_0, \Sigma)$.

Remark 1. The proof actually shows that when $s \in \mathbf{Rat}'_{(S,V)}(S_0, \Sigma)$, then there is an automaton $\mathbf{A} = (\alpha, A, \beta)$ over (S_0, Σ) with $\alpha\beta = 0$ and $|\mathbf{A}| = s$.

Our aim is to prove the corresponding fact for recognizable and rational elements in V:

Theorem 7. $\mathbf{Rat}^\omega_{(S,V)}(S_0, \Sigma) = \mathbf{Rec}^\omega_{(S,V)}(S_0, \Sigma)$.

The proof of the inclusion $\mathbf{Rat}_{(S,V)}(S_0, \Sigma) \subseteq \mathbf{Rec}_{(S,V)}(S_0, \Sigma)$ is divided into several lemmas. Suppose that $\mathbf{A} = (\alpha, A, k)$ and $\mathbf{B} = (\beta, B, m)$ are Büchi automata over (S_0, Σ), where

$$A = \begin{pmatrix} a & b \\ c & d \end{pmatrix} \qquad B = \begin{pmatrix} e & f \\ g & h \end{pmatrix}$$

where a is $k \times k$ and e is $m \times m$. Write $\alpha = (\alpha_1, \alpha_2)$ and $\beta = (\beta_1, \beta_2)$ where α_1 is $1 \times k$ and β_1 is $1 \times m$. Then we define the following Büchi automaton:

$$\mathbf{A} + \mathbf{B} = \left((\alpha_1 \ \beta_1 \ \alpha_2 \ \beta_2), \begin{pmatrix} a & b & & \\ & e & f & \\ c & & d & \\ & g & & h \end{pmatrix}, k + m \right)$$

Lemma 1. $|\mathbf{A} + \mathbf{B}| = |\mathbf{A}| + |\mathbf{B}|$

Proof. First note that

$$\left(\begin{pmatrix} a \\ & e \end{pmatrix} + \begin{pmatrix} b \\ & f \end{pmatrix} \begin{pmatrix} d \\ & h \end{pmatrix}^* \begin{pmatrix} c \\ & g \end{pmatrix} \right)^\omega = \begin{pmatrix} (a + bd^*c)^\omega \\ (e + fh^*g)^\omega \end{pmatrix}$$

and thus

$$\begin{pmatrix} a & b & & \\ & e & f & \\ c & & d & \\ & g & & h \end{pmatrix}^{\omega_{k+m}} = \begin{pmatrix} (a + bd^*c)^\omega \\ (e + fh^*g)^\omega \\ d^*c(a + bd^*c)^\omega \\ h^*g(e + fh^*g)^\omega. \end{pmatrix}$$

Using this,

$$|\mathbf{A} + \mathbf{B}| =$$

$$= \begin{pmatrix} \alpha_1 & \beta_1 & \alpha_2 & \beta_2 \end{pmatrix} \begin{pmatrix} (a + bd^*c)^\omega \\ (e + fh^*g)^\omega \\ d^*c(a + bd^*c)^\omega \\ h^*g(e + fh^*g)^\omega \end{pmatrix}$$

$$= \alpha_1(a + bd^*c)^\omega + \alpha_2 d^*c(a + bd^*c)^\omega + \beta_1(e + fh^*g)^\omega + \beta_2 h^*g(e + fh^*g)^\omega$$

$$= |\mathbf{A}| + |\mathbf{B}|.$$

□

Suppose next that $\mathbf{A} = (\alpha, A, k)$ is a Büchi automaton as before and $\mathbf{B} = (\beta, B, \gamma)$. Then we define the Büchi-automaton

$$\mathbf{B} \cdot \mathbf{A} = \left(\begin{pmatrix} 0 & \beta \end{pmatrix}, \begin{pmatrix} A & \\ \gamma\alpha A & B \end{pmatrix}, k \right).$$

Lemma 2. $|\mathbf{B} \cdot \mathbf{A}| = |\mathbf{B}||\mathbf{A}|$

Proof. First note that

$$\begin{pmatrix} A & \\ \gamma\alpha A & B \end{pmatrix}^{\omega_k} = \begin{pmatrix} A^{\omega_k} \\ B^*\gamma\alpha A^{\omega_k} \end{pmatrix},$$

by (6) and (7). Thus,

$$|\mathbf{B} \cdot \mathbf{A}| = \begin{pmatrix} 0 & \beta \end{pmatrix} \begin{pmatrix} A^{\omega_k} \\ B^*\gamma\alpha A^{\omega_k} \end{pmatrix}$$

$$= \beta B^*\gamma\alpha A^{\omega_k}$$

$$= |\mathbf{B}||\mathbf{A}|.$$

□

Finally, we treat the omega power operation. Let $\mathbf{A} = (\alpha, A, \beta)$ be an automaton of dimension n. Then we define a Büchi automaton of dimension $2n$:

$$\mathbf{A}^\omega = \left(\begin{pmatrix} 0 & \alpha \end{pmatrix}, \begin{pmatrix} \beta\alpha A & \beta\alpha A \\ A & A \end{pmatrix}, n \right).$$

Lemma 3. If $\alpha\beta = 0$ then $|\mathbf{A}^\omega| = |\mathbf{A}|^\omega$

Proof. Since $\alpha\beta = 0$, $|\mathbf{A}| = \alpha A^+ \beta$. Since

$$(\beta\alpha A + \beta\alpha A A^* A)^\omega = (\beta\alpha A^+)^\omega$$

$$= \beta(\alpha A^+ \beta)^\omega,$$

we have

$$\begin{pmatrix} \beta\alpha A & \beta\alpha A \\ A & A \end{pmatrix}^{\omega_n} = \begin{pmatrix} \beta(\alpha A^+ \beta)^\omega \\ A^+ \beta(\alpha A^+ \beta)^\omega \end{pmatrix}.$$

Thus,

$$|\mathbf{A}^\omega| = \alpha A^+ \beta (\alpha A^+ \beta)^\omega$$
$$= (\alpha A^+ \beta)^\omega$$
$$= |\mathbf{A}|^\omega.$$

□

Corollary 3. $\mathbf{Rat}^\omega_{(S,V)}(S_0, \Sigma) \subseteq \mathbf{Rec}^\omega_{(S,V)}(S_0, \Sigma)$

Proof. Clear from Theorem 6, Lemmas 1, 2, 3 and Remark 1.   □

Proposition 2. $\mathbf{Rec}^\omega_{(S,V)}(S_0, \Sigma) \subseteq \mathbf{Rat}^\omega_{(S,V)}(S_0, \Sigma)$

Proof. When all entries of an $n \times n$ matrix A are in $\mathbf{Rat}'_{(S,V)}(S_0, \Sigma)$, then the same holds for all entries of A^+ by (2). Using this, it follows from (2), (4) and (5) that all entries of A^{ω_k} are in $\mathbf{Rat}^\omega_{(S,V)}(S_0, \Sigma)$. It follows that $\mathbf{Rec}^\omega_{(S,V)}(S_0, \Sigma) \subseteq \mathbf{Rat}^\omega_{(S,V)}(S_0, \Sigma)$.   □

The set $\mathbf{Rat}_{(S,V)}(S_0, \Sigma)$ may not be closed under star. But there are two important special cases when it is: When $S_0 \subseteq D(S)$ and S_0 is closed under star, so that $D(S) = S$, or if whenever $x + a \in D(S)$ for some $x \in S_0$ and $a \in D(S)$, then it follows that $x = 0$, cf. [5].

Corollary 4. *If either of the above two assumptions holds, then* $\mathbf{Rec}_{(S,V)}(S_0, \Sigma)$ *and* $\mathbf{Rec}^\omega_{(S,V)}(S_0, \Sigma)$ *are the smallest sets of elements of* S *and* V, *respectively, that can be constructed from* $S_0 \cup \Sigma$ *by the rational operations of* $+, \cdot,^*$ *and* $^\omega$.

6   Partial Iteration Semiring-Semimodule Pairs

Most of the natural partial Conway semiring-semimodule pairs and partial Conway matricial theories satisfy several additional identities that do not hold in all partial Conway semiring-semimodule pairs or partial Conway matricial theories. In [6], Conway associated an identity of regular languages with every finite group. A generalization of Conway's group identities meaningful in all Conway theories was introduced in [9].

Let (S, V) be a partial Conway semiring-semimodule pair with distinguished ideal $D(S) \subseteq S$. Suppose that G is a finite group of order n with multiplication operation denoted $\circ$. Without loss if generality we may assume that the elements of G are the integers $1, \ldots, n$. We say that the *star group identity* associated with G holds in (S, V) or in the partial Conway matricial theory $\mathbf{Matr}_{S,V}$ if for all $a_1, \ldots, a_n$ in $D(S)$, the sum of the entries of the first row of M_G^* is equal to $(a_1 + \ldots + a_n)^*$, where $M_G = M_G(a_1, \ldots, a_n)$ is the $n \times n$ matrix over $D(S)$ whose (i,j)th entry is a_k iff $i \circ k = j$, for all $i, j, k = 1, \ldots, n$. Thus, denoting by 1_n the n-dimensional row matrix whose first component is 1 and whose other components are all 0, and by u_n the n-dimensional column matrix whose entries are all 1, the star group identity associated with G takes the form

$$1_n \cdot M_G^* \cdot u_n = (a_1 + \ldots + a_n)^*.$$

Moreover, we say that the *omega group identity* holds in (S, V), or in $\mathbf{Matr}_{S,V}$, if the first component of M_G^ω is equal to $(a_1 + \ldots + a_n)^\omega$, i.e. when

$$1_n \cdot M_G^\omega = (a_1 + \ldots + a_n)^\omega.$$

Since the star and omega permutation identities

$$(\pi \cdot A \cdot \pi^T)^* = \pi \cdot A^* \cdot \pi^T$$
$$(\pi \cdot A \cdot \pi^T)^\omega = \pi \cdot A^\omega$$

hold for all $A \in D(S)^{n \times n}$ and all $n \times n$ permutation matrices π, whether the star and omega group identities associated with G hold do not depends on the enumeration of the group elements.

The group identities can sometimes be simplified. For example, when G is the cyclic group of order 2, then the star and omega group identities take the form

$$(a + ba^*b)^*(1 + ba^*) = (a + b)^* \tag{8}$$
$$(a + ba^*b)^\omega + (a + ba^*b)^* ab^\omega = (a + b)^\omega. \tag{9}$$

where a and b range over $D(S)$. Using the sum star and product star identities, the first is equivalent to

$$a^*(ba^*ba^*)^*(1 + ba^*) = (a + b)^*. \tag{10}$$

Now by substituting 0 for a we obtain

$$(b^2)^*(1 + b) = b^*. \tag{11}$$

Conversely, if this holds for all $b \in D(S)$, then so do (10) and (8) for all $a, b \in D(S)$, since

$$a^*(ba^*b^*a^*)^*(1 + ba^*) = a^*(ba^*)^* = (a + b)^*.$$

Thus, (8) may be simplified to (11). Similarly, in partial Conway semiring-semimodule pairs (S, V) satisfying (11) for all $b \in D(S)$, (9) may be replaced by the simpler

$$(b^2)^\omega = b^\omega,$$

for all $b \in D(S)$.

Proposition 3. *The group identities hold in all partial iterative semiring-semimodule pairs.*

Proof. Suppose that (S, V) is a partial iterative semiring-semimodule pair with $D(S) = I$. Let $A \in I^{n \times n}$, $B \in I^{m \times m}$ and $C \in S^{n \times m}$ with $AC = CB$. Then

$$ACB^* + C = CBB^* + C = C(BB^* + E_m) = CB^*$$

and

$$ACB^\omega = CBB^\omega = CB^\omega.$$

Thus, $A^*C = CB^*$. Moreover, if no component CB^ω is zero, then $A^\omega = CB^\omega$ by uniqueness of maximal solutions.

In particular, let $M_G = M_G(a_1, \ldots, a_n)$, where $a_1, \ldots, a_n \in I$, and consider the column matrix u_n whose components are all 1. Then $M_G u_n = u_n(a_1 + \ldots + a_n)$, so that $M_G^* = u_n(a_1 + \ldots + a_n)^*$. If $a_i = 0$ for all i, then both all components of M_G^ω and $u_n(a_1 + \ldots + a_n)^\omega$ are 0. Otherwise no component of $u_n(a_1 + \ldots + a_n)^\omega$ is 0 and thus $M_G^\omega = u_n(a_1 + \ldots + a_n)^\omega$ by Theorem 5. $\qquad\square$

The group identities have been utilized in several completeness results. See e.g., [4,15].

References

1. Berstel, J., Reutenauer, C.: Noncommutative Rational Series With Applications. In: Encyclopedia of Mathematics and its Applications, vol. 137. Cambridge University Press, Cambridge (2010)
2. Bloom, S.L., Ésik, Z.: Matrix and matricial iteration theories, Parts I and II. J. Comput. Sys. Sci. 46, 381–408, 409–439 (1993)
3. Bloom, S.L., Ésik, Z.: Iteration Theories: The Equational Logic of Iterative Processes. EATCS Monographs on Theoretical Computer Science. Springer, Heidelberg (1993)
4. Bloom, S.L., Ésik, Z.: Axiomatizing rational power series over natural numbers. Information and Computation 207, 793–811 (2009)
5. Bloom, S.L., Ésik, Z., Kuich, W.: Partial Conway and iteration semirings. Fundamenta Informaticae 86, 19–40 (2008)
6. Conway, J.C.: Regular Algebra and Finite Machines. Chapman and Hall, London (1971)
7. Eilenberg, S.: Automata, Languages, and Machines, vol. A. Academic Press, London (1974)
8. Elgot, C.C.: Matricial theories. J. of Algebra 42, 391–421 (1976)
9. Ésik, Z.: Group axioms for iteration. Information and Computation 148, 131–180 (1999)
10. Ésik, Z., Kuich, W.: Inductive *-semirings. Theoret. Comput. Sci. 324, 3–33 (2004)
11. Ésik, Z., Kuich, W.: A semiring-semimodule generalization of ω-regular languages, Parts 1 and 2. J. Automata, Languages, and Combinatorics 10, 203–242, 243–264 (2005)
12. Ésik, Z., Kuich, W.: On iteration semiring-semimodule pairs. Semigroup Forum 75, 129–159 (2007)
13. Golan, J.S.: Semirings and their Applications. Kluwer Academic Publishers, Dordrecht (1999)
14. Kozen, D.: A completeness theorem for Kleene algebras and the algebra of regular events. Inform. and Comput. 110, 366–390 (1994)
15. Krob, D.: Complete systems of B-rational identities. Theoretical Computer Science 89, 207–343 (1991)
16. Kuich, W., Salomaa, A.: Semirings and Formal Power Series. EATCS Monograph Series in Theoretical Computer Science. Springer, Heidelberg (1986)

Kleene Theorem in Partial Conway Theories with Applications

Zoltán Ésik and Tamás Hajgató*

Department of Computer Science,
University of Szeged, Hungary

Abstract. Partial Conway theories are algebraic theories equipped with
a partially defined dagger operation satisfying some natural identities.
We prove a Kleene type theorem for partial Conway theories and discuss
several applications of this result.

1 Introduction

Fixed point operations occur in just about all areas of theoretical computer
science including automata and languages, semantics of programming languages,
process algebra, logical theories of computational systems, programming logics,
recursive types and proof theory, computational complexity, etc. The equational
properties of the fixed point, or dagger operation can be best described in the
context of Lawvere theories of functions over a set equipped with structure, or
more generally, in the context of abstract Lawvere theories (or just theories), or
cartesian or co-cartesian categories, cf. [26,7,13,33].

Iteration theories were introduced in [5], and independently in [15] in order
to describe the equational properties of the dagger operation in iterative and
rational algebraic theories, cf. [13,35]. In an iterative theory, dagger is defined
by unique fixed points, and in rational theories, by least fixed points. In both
types of theories, the dagger operation satisfies the same set of identities. These
identities define iteration theories. In [33], it is argued that any reasonable class
of fixed point models satisfies either exactly the iteration theory identities, or
the identities of iteration theories satisfying one more extra identity.

Iteration theories can be axiomatized by the Conway theory identities and a
group identity associated with each finite (simple) group, cf. [18]. Whereas the
group identities are needed for completeness, several constructions in automata
and language theory and other areas of computer science only require the Conway
identities.

In [7], a general Kleene type theorem was proved for all Conway theories.
However, in many models of interest, the dagger operation is only partially

---

* Both authors were partially supported by the project TÁMOP-4.2.1/B-09/1/KONV-
2010-0005 "Creating the Center of Excellence at the University of Szeged", supported
by the European Union and co-financed by the European Regional Fund. The first
author was also supported by the grant no. K 75249 from the National Foundation
of Hungary for Scientific Research.

W. Kuich and G. Rahonis (Eds.): Bozapalidis Festschrift, LNCS 7020, pp. 72–93, 2011.

defined. In this paper, we define partial Conway theories and provide a Kleene theorem for partial Conway theories. We also discuss several application of this generic result.

2   Basic Definitions

In this section, we review some basic concepts used in this paper. For more details, the reader is referred to [7]. In any category whose objects are the non-negative integers we will denote the composite of the morphisms $f : n \to p$ and $g : p \to q$ in diagrammatic Al order as $f \cdot g$. The identity morphism corresponding to object p will be denoted 1_p. When n is a nonnegative integer, we will denote the set $\{1, 2, \ldots, n\}$ by $[n]$. Thus, $[0]$ is the empty set.

Let us recall from [7] that a *(Lawvere) theory* T is a small category with objects the nonnegative integers such that each nonnegative integer n is the n-fold coproduct of the object 1 with itself. We assume that each theory T comes with distinguished coproduct injections $i_n : 1 \to n$, $i \in [n]$, called *distinguished morphisms*, turning n to an n-fold coproduct of object 1 with itself. By the coproduct property, for each finite sequence of *scalar morphisms* $f_1, \ldots, f_n : 1 \to p$ there is a unique morphism $f : n \to p$ such that $i_n \cdot f = f_i$, for each $i \in [n]$. This unique morphism is denoted $\langle f_1, \ldots, f_n \rangle$. The operation implicitly defined by the coproduct property is called *tupling*. In particular, when $n = 0$, tupling defines a unique morphism $0_p : 0 \to p$, for each $p \geq 0$. Note that $1_n = \langle 1_n, \ldots, n_n \rangle$ for all nonnegative integers n. In addition, we will always assume that $1_1 = 1_1$, so that $\langle f \rangle = f$ for each $f : 1 \to p$. A theory T is termed *trivial* if $1_2 = 2_2$. In a trivial theory, there is at most one morphism $n \to p$, for each $n, p \geq 0$.

Tuplings of distinguished morphisms are called *base morphisms*. For example, 0_n and 1_n are base morphisms. When ρ is a mapping $[n] \to [p]$, there is an associated base morphism $n \to p$, the tupling $\langle (1\rho)_p, \ldots, (n\rho)_p \rangle$ of the distinguished morphisms $(1\rho)_p, \ldots, (n\rho)_p$. A *base permutation* is a base morphism associated with a bijective mapping. Note that in any theory, a base permutation corresponding to a bijection $\pi : [n] \to [n]$ is an isomorphism with inverse the base permutation corresponding to the inverse function of π.

When $f : n \to p$ and $g : m \to p$ in a theory T, we define $\langle f, g \rangle$ to be the morphism $h : n + m \to p$ with $i_{n+m} \cdot h = i_n \cdot f$ and $(n + j)_{n+m} \cdot h = j_m \cdot g$ for all $i \in [n]$ and $j \in [m]$. Moreover, for each $f : n \to p$ and $g : m \to q$, we define $f \oplus g = \langle f \cdot \kappa, g \cdot \lambda \rangle : n + m \to p + q$, where κ is the base morphism corresponding to the inclusion $[p] \hookrightarrow [p + q]$ and λ is the base morphism corresponding to the translated inclusion $[q] \hookrightarrow [p + q]$ mapping j in $[q]$ to $p + j$ in $[p + q]$, for all $j \in [q]$. Note that the *pairing* operation $\langle f, g \rangle$ and the *separated sum* operation $f \oplus g$ are associative. Moreover, $\langle f, 0_p \rangle = f = \langle 0_p, f \rangle$ and $f \oplus 0_0 = f = 0_0 \oplus p$ for all $f : n \to p$. Also, $\langle f, g \rangle \cdot h = \langle f \cdot h, g \cdot h \rangle$ for all $f : n \to p$, $g : m \to p$ and $h : p \to q$, and $(f \oplus g) \cdot \langle h, k \rangle = \langle f \cdot h, g \cdot k \rangle$ for all $f : n \to p$, $g : m \to q$, $h : p \to r$ and $k : q \to r$. Finally, $(f \oplus g) \cdot (h \oplus k) = (f \cdot h) \oplus (g \cdot k)$ for all appropriate morphisms f, g, h, k.

A *morphism* $T \to T'$ between theories T and T' is a functor which preserves the objects and the distinguished morphisms. It follows that any theory morphism preserves the pairing, tupling and separated sum operations. A theory T is a *subtheory* of a theory T' if T is a subcategory of T' and has the same distinguished morphisms as T, so that the inclusion $T \hookrightarrow T'$ is a theory morphism.

Example 1. A basic example of a theory is the *theory of functions over a set A*. In this theory, a morphism $n \to p$ is a function $f : A^p \to A^n$. Note the reversal of the arrow. The composite of morphisms $f : n \to p$ and $g : p \to q$ is their function composition written from right to left, which is a function $A^q \to A^n$. The distinguished morphisms are the projection functions.

Example 2. Let $S = (S, +, \cdot, 0, 1)$ be a semiring [23]. The *theory of matrices* $\mathbf{Mat}_S$ over S has as morphisms $n \to p$ all $n \times p$ matrices in $S^{n \times p}$. Composition is matrix multiplication defined in the usual way. For each $i \in [p]$, $p \geq 0$, the distinguished morphism $i_p : 1 \to p$ is the $1 \times p$ row matrix with a 1 on the ith position and 0's elsewhere. It is known that in each matrix theory, each object n is also the n-fold product of the object 1 with itself. The transposes i_n^T of the distinguished morphisms serve as the projection morphisms $n \to 1$, see [7,14]. The theory $\mathbf{Mat}_S$ comes with a sum operation $+$ defined on each hom-set $\mathbf{Mat}_S(n, p) = S^{n \times p}$. For each $n, p \geq 0$, $(\mathbf{Mat}_S(n, p), +, 0_{n,p})$ is a commutative monoid, where $0_{n,p}$ is the $n \times p$ matrix whose entries are all 0. Moreover, composition distributes over finite sums both on the left and on the right. Thus, for each n, $(\mathbf{Mat}_S(n, n), +, \cdot, 1_n, 0_{n,n})$ is itself a semiring, since the product of two $n \times n$ matrices is an $n \times n$ matrix. In particular, $\mathbf{Mat}_S(1, 1)$ is isomorphic to S. We will usually identify a morphism $1 \to 1$ with the corresponding element of S.

Example 3. Suppose that S is a semiring and $V = (V, +, 0)$ is a (left) S-semimodule, cf. [23]. Then the *matricial theory* $\mathbf{Matr}_{S,V}$ [14,7] over (S, V) has as morphisms $n \to p$ all ordered pairs $(A; v)$ consisting of a matrix $A : n \to p$ in $\mathbf{Mat}_S$ and an n-dimensional column vector $v \in V^n$. When $p = 0$, we usually write $(; v)$ or just v. Composition is defined by the rule

$$(A; v) \cdot (B; w) = (AB; v + Aw)$$

for all $(A; v) : n \to p$ and $(B; w) : p \to q$. For each $i \in [p]$, $p \geq 0$, the distinguished morphism i_p is the ordered pair $(i_p; 0)$, where somewhat ambiguously, i_p also denotes the corresponding distinguished morphism in $\mathbf{Mat}_S$. The theory $\mathbf{Matr}_{S,V}$ comes with the pointwise sum operation and the zero morphisms $0_{n,p} = (0_{n,p}; 0^n)$, where 0^n denotes the n-dimensional column vector of 0's in V^n. Each hom-set $\mathbf{Matr}_{S,V}(n, p) = (\mathbf{Matr}_{S,V}(n, p), +, 0_{n,p})$ is a commutative monoid and composition distributes over finite sums on the right, but usually not on the left. Note that $\mathbf{Mat}_S$ may be identified with the subtheory of $\mathbf{Matr}_{S,V}$ determined by the morphisms of the sort $(A; 0^n) : n \to p$, $n, p \geq 0$. We call $\mathbf{Mat}_S$ the *underlying matrix theory* of $\mathbf{Matr}_{S,V}$.

Example 4. A *ranked alphabet* Σ is a family of pairwise disjoint sets $(\Sigma_n)_n$, where n ranges over the nonnegative integers. We assume that the reader is familiar with the notion of (total) Σ-*trees* over a set $X_p = \{x_1, \ldots, x_p\}$ of variables, defined as usual, see e.g. [7]. Below we will denote the collection of finite and infinite Σ-trees over X_p by $T^\omega_\Sigma(X_p)$ and the collection of just the finite trees by $T_\Sigma(X_p)$. We call a tree *proper* if it is not one of the trees x_i. Σ-trees form a theory ΣTR whose morphisms $n \to p$ are all n-tuples of trees in $T^\omega_\Sigma(X_p)$. Composition is defined by substitution for the variables x_i, and for $i \in [p]$, the tree with a single vertex labeled x_i serves as the ith distinguished morphism $1 \to p$. Thus, if $t : 1 \to n$ and $t'_1, \ldots, t'_n : 1 \to p$ in ΣTR, then $t \cdot \langle t'_1, \ldots, t'_n \rangle : 1 \to p$ is the tree obtained by substituting a copy of t'_i for each leaf of t labeled x_i, for $i \in [n]$. See [7] for details. A tree is called *regular* if up to isomorphism it has a finite number of subtrees. The subtheory of ΣTR containing only the regular Σ-trees is denoted Σtr, and the subtheory containing only the finite Σ-trees is denoted ΣTerm. As usual, we identify each letter σ in Σ_n with the corresponding *atomic tree* $\sigma(x_1, \ldots, x_n)$ in $T_\Sigma(X_n)$ whose root is labeled σ and has n immediate successors labeled $x_1, \ldots, x_n$, respectively.

It is known that ΣTerm is freely generated by Σ in the category of theories. In particular, when Σ is is empty, ΣTerm is an initial theory.

Example 5. The theory Θ has as morphisms $n \to p$ all functions $[n] \to [p]$. Composition is defined by function composition written from left to right. For each n and $i \in [n]$, the distinguished morphism $1 \to n$ is the function $[1] \to [n]$ selecting the integer i. When T is a nontrivial theory, the base morphisms form a subtheory of T isomorphic to Θ. Moreover, when Σ is the empty ranked alphabet, the theory ΣTR (or ΣTerm) is isomorphic to Θ.

3   Partial Conway Theories

Let T be a theory. A nonempty collection of morphisms I is an *ideal* [7] in T if it is closed under tupling, composition with base morphisms on the left, and composition with arbitrary morphisms on the right. Note that every ideal contains the morphisms 0_p, $p \geq 0$.

Definition 1. A partial dagger theory[1] *is a theory* T *equipped with a distinguished ideal* $D(T)$ *and a partially defined dagger operation*

$$^\dagger : T(n, n+p) \to T(n, p), \; n, p \geq 0$$

defined on morphisms $n \to n+p$ *in* $D(T)$. *A dagger theory is a partial dagger theory* T *with* $D(T) = T$. *(An equivalent condition is that* $\mathbf{1}_1$, *or at least one distinguished morphism is in* $D(T)$.)

Let T, T' be (partial) dagger theories. A *(partial) dagger theory morphism* $\varphi : T \to T'$ is a theory morphism $T \to T'$ which preserves the distinguished ideal and the dagger operation, so that $D(T)\varphi \subseteq D(T')$ and $(f^\dagger)\varphi = (f\varphi)^\dagger$ for all $f : n \to n+p$ in $D(T)$ and $n, p \geq 0$. We say that T is a *(partial) dagger subtheory*

---

[1] These theories are called partial preiteration theories in [7].

of T' if T is a subtheory of T', $D(T) = D(T') \cap T$, and the dagger operation of T is the restriction of the dagger operation of T', so that the inclusion $T \hookrightarrow T'$ is a morphism of (partial) dagger theories.

In the sequel, we will consider partial dagger theories satisfying certain identities that we will define now. For the origins of these identities, the reader is referred to [2,1,15,28,29,34,35].

Definition 2. *We say that the partial dagger theory T satisfies:*

1. *the* fixed point identity, *if*

$$f^\dagger = f \cdot \langle f^\dagger, 1_p \rangle$$

   for each $f : n \to n + p$ in $D(T)$,
2. *the* left zero identity, *if*

$$(0_n \oplus f)^\dagger = f$$

   for each $f : n \to p$ in $D(T)$,
3. *the* right zero identity, *if*

$$(f \oplus 0_q)^\dagger = f^\dagger \oplus 0_q$$

   for each $f : n \to n + p$ in $D(T)$,
4. *the (*base*) parameter identity, *if*

$$f^\dagger \cdot g = (f \cdot (1_n \oplus g))^\dagger$$

   for each $f : n \to n + p$ in $D(T)$ and $g : p \to q$ in T (such that g is a base morphism),
5. *the* permutation identity, *if*

$$(\pi \cdot f \cdot (\pi^{-1} \oplus 1_p))^\dagger = \pi \cdot f^\dagger$$

   for each $f : n \to n + p$ in $D(T)$ and base permutation $\pi : n \to n$ with inverse π^{-1},
6. *the* pairing identity (or Bekić identity), *if for all $f : n \to n + m + p$ and $g : m \to n + m + p$ in $D(T)$,*

$$\langle f, g \rangle^\dagger = \langle f^\dagger \cdot \langle h^\dagger, 1_p \rangle, h^\dagger \rangle,$$

   where $h = g \cdot \langle f^\dagger, 1_{m+p} \rangle : m \to m + p$,
7. *the* double dagger identity, *if*

$$(f \cdot (\langle 1_n, 1_n \rangle \oplus 1_p))^\dagger = (f \cdot \langle f^\dagger, 1_{n+p} \rangle)^\dagger$$

   for each $f : n \to n + n + p$ in $D(T)$,
8. *the* composition identity, *if*

$$f \cdot \langle (g \cdot \langle f, 0_m \oplus 1_p \rangle)^\dagger, 1_p \rangle = (f \cdot \langle g, 0_n \oplus 1_p \rangle)^\dagger$$

   for each $f : n \to m + p$ and $g : m \to n + p$ in $D(T)$,

9. the simplified composition identity, *if*

$$(f \cdot g)^\dagger = f \cdot (g \cdot (f \oplus \mathbf{1}_p))^\dagger$$

for each $f : n \to m$ *and* $g : m \to n + p$ *in* $D(T)$.

Remark 1. Note that the composition identity implies the simplified composition identity and the fixed point identity implies the left zero identity. If the fixed point identity holds, then $f^\dagger$ is in $D(T)$ for each $f : n \to n + p$ in $D(T)$, since $f^\dagger = f \cdot \langle f^\dagger, \mathbf{1}_p \rangle$. Then we say that T satisfies

10. the simplified form of the double dagger identity, *if*

$$(f \cdot (\langle \mathbf{1}_n, \mathbf{1}_n \rangle \oplus \mathbf{1}_p))^\dagger = f^{\dagger\dagger}$$

for each $f : n \to n + n + p$ *in* $D(T)$.

If the fixed point identity holds, then the simplified form of the double dagger identity is equivalent to the double dagger identity.

The *scalar versions* of the fixed point, left zero, right zero, (base) parameter and double dagger identities are obtained by taking $n = 1$ in the corresponding identity. The scalar versions of the composition and simplified composition identities are obtained by taking $n = m = 1$ in those identities.

Definition 3. *A (partial) Conway theory is a (partial) dagger theory satisfying the fixed point, right zero, pairing, (simplified) double dagger and permutation identities.*

A morphism of (partial) Conway theories is a (partial) dagger theory morphism. A (partial) Conway theory T is a *(partial) Conway subtheory* of the partial Conway theory T' if T is a (partial) dagger subtheory of T'. The following fact is known from [7].

Proposition 1. *All of the identities listed above hold in all partial Conway theories.*

It is in effect shown in [7] that partial Conway theories may also be axiomatized by the scalar versions of the fixed point, right zero, pairing and double dagger identities and the permutation identity for morphisms $f : 2 \to 2 + p$ in the distinguished ideal and the non-trivial base permutation $\pi : 2 \to 2$.

The axioms of Conway theories may be simplified. The following fact is known, cf. [7], Chapter 6.

Proposition 2. *Suppose that T is a dagger theory. Then the following are equivalent:*

- T *satisfies the left zero, right zero, pairing and permutation identities.*
- T *satisfies the base parameter, composition and double dagger identities.*
- T *satisfies the fixed point, base parameter, simplified composition and double dagger identities.*

- T satisfies the scalar versions of the parameter, composition and double dagger identities, and the pairing identity for $m = 1$.
- T satisfies the scalar versions of the fixed point, parameter, simplified composition and double dagger identities, and the pairing identity for $m = 1$.

We give several examples of (partial) Conway theories.

Example 6. A *partial iterative theory* is a theory T equipped with a distinguished ideal $D(T)$ such that the fixed point equation

$$\xi = f \cdot \langle \xi, \mathbf{1}_p \rangle \tag{1}$$

has a unique solution in T for each $f : n \to n+p$ in $D(T)$. Each partial iterative theory may be turned into a partial Conway theory by defining $f^\dagger$ as the unique solution of the above equation for each $f : n \to n+p$ in $D(T)$. It can be seen that a theory T, equipped with a distinguished ideal $D(T)$, is a partial iterative theory if (1) has a unique solution for each *scalar* $f : 1 \to 1+p$ in $D(T)$, see [11]. Partial iterative theories generalize the notion of Elgot's *iterative theories*, cf. [13].

Example 7. Examples of Conway theories are the theories of continuous or monotone functions over directed complete partial orders (dcpo's) equipped with the *least fixed point operation* as dagger. In a *continuous theory* T, each hom-set $T(n,p)$ is a dcpo so that $f \leq g$ for $f, g : n \to p$ iff $i_n \cdot f \leq i_n \cdot g$ for all $i \in [n]$. Moreover, the composition operation preserves the suprema of nonempty directed sets. When $f : n \to n+p$, then $f^\dagger$ is the least solution of the equation (1). See [7,35] for details. Continuous theories have been generalized to *Park theories* in [17]. Every Park theory is a Conway theory.

Example 8. Let Σ be a ranked alphabet and let T be the theory $\Sigma\mathrm{TR}$, or the theory $\Sigma\mathrm{tr}$. Let the ideal $D(T)$ consist of those morphisms $f : n \to p$ in T whose components $i_n \cdot f$, $i \in [n]$, are proper trees. It is known that for each $f : n \to n+p$ in $D(T)$, the equation (1) has a unique solution in the variable $\xi : n \to p$. Denoting this unique solution by $f^\dagger$, T becomes a partial Conway theory. Moreover, if Σ_0 is not empty, so that there is at least one morphism in $T(1,0)$, then for any choice of a morphism $\perp : 1 \to 0$, the partial dagger operation can be *uniquely* extended to a totally defined dagger operation such that T becomes a Conway theory (in fact an iteration theory). See [5] and [16], or [7].

Example 9. Suppose that Σ contains a single letter $\perp$ that has rank 0. Then any scalar morphism in $\Sigma\mathrm{TR}$ is either a distinguished morphism, or a morphism $\perp_{1,p} = \perp \cdot 0_p : 1 \to p$. Given $f : n \to n+p$, it holds that $f^\dagger = f^n \cdot \langle \perp_{n,p}, \mathbf{1}_p \rangle$, where $\perp_{n,p} = \langle \perp_{1,p}, \ldots, \perp_{1,p} \rangle : n \to p$, $f^0 = \mathbf{1}_n \oplus 0_p$ and $f^{k+1} = f \cdot \langle f^k, 0_n \oplus \mathbf{1}_p \rangle$. Let $\perp\mathrm{TR}$ denote this Conway theory. It is known that $\perp\mathrm{TR}$ is an initial Conway theory (and an initial iteration theory).

Example 10. Let Θ' be the theory whose morphisms $n \to p$ are the partial functions $[n] \to [p]$. Composition is function composition and the distinguished morphisms are defined as in the theory Θ. For each $n, p \geq 0$, let $\perp_{n,p}$ denote the totally undefined partial function $[n] \to [p]$. For each $\rho : n \to n + p$, define $\rho^\dagger = \rho^n \cdot \langle \perp_{n,p}, 1_p \rangle$. Then Θ' is an iteration theory isomorphic to $\perp\mathrm{TR}$. Thus, Θ' is also an initial Conway theory.

Example 11. Each nontrivial Conway theory T contains a subtheory isomorphic to Θ'. Given a partial function $\rho : [n] \to [p]$, let us define the corresponding *partial base morphism* in T as the morphism $\langle f_1, \ldots, f_n \rangle$, where for each $i \in [n]$, $f_i = j_p$ if $i\rho = j$ is defined, and $f_i = \perp \cdot 0_p = \perp_{1,p}$, if $i\rho$ is not defined. Here, $\perp = 1_1{}^\dagger$.

When a matrix theory $\mathbf{Mat}_S$ is a Conway theory, the dagger operation determines a *star operation* mapping a matrix $A : n \to n$ to a matrix $A^* : n \to n$ by

$$A^* = (A, 1_n)^\dagger.$$

In particular, S is equipped with a star operation $* : S \to S$. The equational properties of the dagger operation are then reflected by corresponding properties of the star operation. For example, the fixed point identity corresponds to the identity $A^* = AA^* + 1_n$, $A : n \to n$, and the simplified double dagger identity corresponds to the identity $(A + B)^* = A^*(BA^*)^*$, where $A, B : n \to n$. A similar fact holds for those partial matrix theories $T = \mathbf{Mat}_S$ in which $D(T)$ is a *two-sided ideal*, i.e., $D(T)$ is closed under (finite sums and) composition with any matrix on either side.

Similarly, when a matricial theory $\mathbf{Matr}_{S,V}$ is a Conway theory, then the dagger operation determines a star operation on the underlying matrix theory as well as an *omega operation* mapping a matrix $A : n \to n$ in $\mathbf{Mat}_S$ to a vector in V^n, and thus also a star operation on S and a omega operation $S \to V$. For details, see [7].

Remark 2. An *iteration theory* [7] is any theory satisfying all identities true of continuous theories, or equivalently, the tree theories $\Sigma\mathrm{TR}$ or $\Sigma\mathrm{tr}$ with a total dagger operation as in Example 8. Iteration theories can be axiomatized by the Conway identities and an identity associated with each finite (simple) group, cf. [18]. All of the Conway theories given above are in fact iteration theories. The free iteration theories may be described as the theories $\Sigma\mathrm{tr}$ of regular trees. The free Conway theories have been described in [3]. It is decidable in polynomial time whether an identity holds in all iteration theories, whereas the equational theory of Conway theories is PSPACE-complete, cf. [3]. In [7], partial iteration theories are also defined. Every partial iteration theory is a partial Conway theory.

4   A Kleene Theorem

Let T be a partial dagger theory, T_0 a subtheory of T, and let A be a set of scalar morphisms in $D(T)$. We write $A(T_0)$ for the set of morphisms $\langle f_1, \ldots, f_n \rangle : n \to p$,

$n, p \geq 0$ such that each f_i is the composition of a morphism in A with a morphism in T_0. In particular, $0_p \in A(T_0)$ for all $p \geq 0$. Note that if T_0 is T then $A(T_0)$ is the least ideal in T containing the morphisms in A, and if A is the set of scalar morphisms in $D(T)$, then $A(T_0) = D(T)$ for every subtheory T_0 of T.

We say that (T_0, A) is *dagger compatible*, if for each $\alpha : n \to s + n + p$ in T_0 and $a : s \to s + n + p$ in $A(T_0)$, $s, n, p \geq 0$,

$$\alpha \cdot \langle a^\dagger, \mathbf{1}_{n+p} \rangle \in D(T) \implies \alpha \cdot \langle a, 0_s \oplus \mathbf{1}_{n+p} \rangle \in A(T_0).$$

This condition is clearly fulfilled in a partial dagger theory T if (T_0, A) is *strongly dagger compatible*:

1. For all $\alpha : n \to p \in T_0$ and $a : p \to g \in A(T_0)$, $\alpha \cdot a \in A(T_0)$, i.e., when $A(T_0)$ is closed under left composition with T_0-morphisms.
2. If $\alpha \cdot \langle f, \mathbf{1}_p \rangle \in D(T)$ for some $\alpha : n \to m + p \in T_0$ and $f : m \to p \in D(T)$, then $\alpha = \beta \oplus 0_p$ for some $\beta : n \to m$ in T_0.

Indeed, if these conditions hold and $\alpha \cdot \langle a^\dagger, \mathbf{1}_{n+p} \rangle \in D(T)$ for some $\alpha : n \to s + n + p$ in T_0 and $a : s \to s + n + p$ in $A(T_0)$, then there exists $\beta : n \to s$ in T_0 with $\alpha = \beta \oplus 0_{n+p}$. Thus, $\alpha \cdot \langle a, 0_s \oplus \mathbf{1}_{n+p} \rangle = \beta \cdot a$ is in $A(T_0)$.

Of course, it suffices to require the above conditions when the morphism α is scalar.

Remark 3. When (T_0, A) is dagger compatible, then for every $\alpha : n \to n + p$ in T_0, if $\alpha \cdot \langle 0^\dagger_{n+p}, \mathbf{1}_{n+p} \rangle = \alpha \cdot \langle 0_{n+p}, \mathbf{1}_{n+p} \rangle = \alpha$ is in $D(T)$, then it is in $A(T_0)$.

Below, when we write that (T_0, A) *is a basis*, we will mean that T_0 is a subtheory of T and A is a set of scalar morphisms in $D(T)$.

Definition 4. *A presentation $n \to p$ of dimension s over a basis (T_0, A) is an ordered pair:*

$$D = (\alpha, a) : n \to p,$$

where $\alpha : n \to s + p$ is in T_0 and $a : s \to s + p$ is in $A(T_0)$.
   The behavior of D is the following morphism in T:

$$|D| = \alpha \cdot \langle a^\dagger, \mathbf{1}_p \rangle : n \to p.$$

Definition 5

a) *Let $D = (\alpha, a) : n \to p$ and $E = (\beta, b) : m \to p$ be presentations of dimension s and r, respectively. We define*

$$\langle D, E \rangle = (\gamma, c) : n + m \to p$$

as the presentation of dimension $s + r$, where

$$\gamma = \langle \alpha \cdot (\mathbf{1}_s \oplus 0_r \oplus \mathbf{1}_p), 0_s \oplus \beta \rangle$$
$$c = \langle a \cdot (\mathbf{1}_s \oplus 0_r \oplus \mathbf{1}_p), 0_s \oplus b \rangle.$$

b) Let $D = (\alpha, a) : n \to p$ and $E = (\beta, b) : p \to q$ be presentations of dimension s and r, respectively. Let us define

$$D \cdot E = (\gamma, c) : n \to q$$

as the presentation of dimension $s + r$, where

$$\gamma = \alpha \cdot (\mathbf{1}_s \oplus \beta)$$
$$c = \langle a \cdot (\mathbf{1}_s \oplus \beta), 0_s \oplus b \rangle$$

c) Let $D = (\alpha, a) : n \to n+p$ be a presentation of dimension s with $|D| \in D(T)$. Suppose that (T_0, A) is dagger compatible. Then we define

$$D^\dagger = (\beta, b) : n \to p$$

as the presentation of dimension $s + n$, where

$$\beta = (0_s \oplus \mathbf{1}_n \oplus 0_p)$$
$$b = \langle a, \alpha \cdot \langle a, 0_s \oplus \mathbf{1}_{n+p} \rangle \rangle.$$

And if $T_0 \subseteq D(T)$ is closed under dagger, then we define

$$D^\dagger = (\beta, b) : n \to p$$

as the presentation of dimension s, where

$$\beta = (\alpha \cdot (\langle 0_n \oplus \mathbf{1}_s, \mathbf{1}_n \oplus 0_s \rangle \oplus \mathbf{1}_p))^\dagger$$
$$b = a \cdot \langle \mathbf{1}_s \oplus 0_p, \beta, 0_s \oplus \mathbf{1}_p \rangle.$$

Note that if $T_0 \subseteq D(T)$ then $T = D(T)$, so that T is a Conway theory.

Lemma 1. Let T be a partial Conway theory with basis (T_0, A). Then

a) for each presentations $D : n \to p$, $E : p \to q$ we have $|D| \cdot |E| = |D \cdot E|$,
b) for each presentations $D : n \to p$ and $E : m \to p$ we have $\langle |D|, |E| \rangle = |\langle D, E \rangle|$, and
c) if (T_0, A) is dagger compatible, or when $T_0 \subseteq D(T)$ is closed under dagger, then for each presentation $D : n \to n + p$ such that $|D|$ is in $D(T)$, we have $|D|^\dagger = |D^\dagger|$.

Proof. Although the proofs of a) and b) are the same as the proofs of the corresponding facts on pages 450–452 in [7], we include them for the reader's convenience.

Proof of a).

$$\begin{aligned}
|\langle D, E \rangle| &= \gamma \cdot \langle c^\dagger, \mathbf{1}_p \rangle \\
&= \gamma \cdot \langle a^\dagger, b^\dagger, \mathbf{1}_p \rangle \\
&= \langle \alpha \cdot \langle a^\dagger, \mathbf{1}_p \rangle, \beta \cdot \langle b^\dagger, \mathbf{1}_p \rangle \rangle \\
&= \langle |D|, |E| \rangle.
\end{aligned}$$

In the second line, we used the pairing and parameter identities.

Proof of b).

$$\begin{aligned}
|D \cdot E| &= \gamma \cdot \langle c^\dagger, \mathbf{1}_q \rangle \\
&= \gamma \cdot \langle (a \cdot (\mathbf{1}_s \oplus \beta))^\dagger \cdot \langle b^\dagger, \mathbf{1}_q \rangle, b^\dagger, \mathbf{1}_q \rangle \\
&= \alpha \cdot (\mathbf{1}_s \oplus \beta) \cdot \langle a^\dagger \cdot \beta \cdot \langle b^\dagger, \mathbf{1}_q \rangle, b^\dagger, \mathbf{1}_q \rangle \\
&= \alpha \cdot \langle a^\dagger, \mathbf{1}_p \rangle \cdot \beta \cdot \langle b^\dagger, \mathbf{1}_q \rangle \\
&= |D| \cdot |E|.
\end{aligned}$$

In the second and third lines, we used the pairing and parameter identities.

Proof of c). First suppose that (T_0, A) is dagger compatible. Using the first definition of $D^\dagger$ we have:

$$\begin{aligned}
|D^\dagger| &= \beta \cdot \langle b^\dagger, \mathbf{1}_p \rangle \\
&= (0_s \oplus \mathbf{1}_n \oplus 0_p) \cdot \langle b^\dagger, \mathbf{1}_p \rangle \\
&= (0_s \oplus \mathbf{1}_n) \cdot b^\dagger \\
&= (0_s \oplus \mathbf{1}_n) \cdot \langle a, \alpha \cdot \langle a, 0_s \oplus \mathbf{1}_{n+p} \rangle \rangle^\dagger \\
&= (\alpha \cdot \langle a, 0_s \oplus \mathbf{1}_{n+p} \rangle \cdot \langle a^\dagger, \mathbf{1}_{n+p} \rangle)^\dagger \\
&= (\alpha \cdot \langle a^\dagger, \mathbf{1}_{n+p} \rangle)^\dagger \\
&= |D|^\dagger.
\end{aligned}$$

Here, we used the pairing identity in the fifth equation and the fixed point identity in the sixth equation.

Next, suppose that $T_0 \subseteq D(T)$ is closed under dagger. Then, using the second definition of $D^\dagger$,

$$\begin{aligned}
|D^\dagger| &= \beta \cdot \langle b^\dagger, \mathbf{1}_p \rangle \\
&= \beta \cdot \langle (a \cdot \langle \mathbf{1}_s \oplus 0_p, \beta, 0_s \oplus \mathbf{1}_p \rangle)^\dagger, \mathbf{1}_p \rangle \\
&= \beta \cdot \langle (a^\dagger \cdot \langle \beta, 0_s \oplus \mathbf{1}_p \rangle)^\dagger, \mathbf{1}_p \rangle \\
&= (\beta \cdot \langle a^\dagger, 0_n \oplus \mathbf{1}_p \rangle)^\dagger \\
&= ((\alpha \cdot (\langle 0_n \oplus \mathbf{1}_s, \mathbf{1}_n \oplus 0_s \rangle \oplus \mathbf{1}_p))^\dagger \cdot \langle a^\dagger, 0_n \oplus \mathbf{1}_p \rangle)^\dagger \\
&= (\alpha \cdot (\langle 0_n \oplus \mathbf{1}_s, \mathbf{1}_n \oplus 0_s \rangle \oplus \mathbf{1}_p) \cdot \langle \mathbf{1}_n \oplus 0_p, a^\dagger, 0_n \oplus \mathbf{1}_p \rangle)^\dagger \\
&= (\alpha \cdot \langle a^\dagger, \mathbf{1}_n \oplus 0_p, 0_n \oplus \mathbf{1}_p \rangle)^\dagger \\
&= (\alpha \cdot \langle a^\dagger, \mathbf{1}_{n+p} \rangle)^\dagger \\
&= |D|^\dagger.
\end{aligned}$$

Here, we used the parameter identity in the third line, the composition identity in the fourth line, and in the fifth line the identity

$$(f^\dagger \cdot \langle g, 0_n \oplus \mathbf{1}_p \rangle)^\dagger = (f \cdot \langle \mathbf{1}_n \oplus 0_p, g, 0_n \oplus \mathbf{1}_p \rangle)^\dagger,$$

$f : n \to n + m + p$, $g : m \to n + p$ that holds in all Conway theories.

$\square$

Lemma 2. *Let T be a partial Conway theory with basis (T_0, A). Then every T_0-morphism and every morphism in A is the behavior of some presentation.*

Proof. Indeed, when $\alpha : n \to p$ is in T_0, then $\alpha = |D_\alpha|$ for the presentation $D_\alpha = (\alpha, 0_p)$, and when $a : 1 \to p$ in A, then $a = |D_a|$ where $D_a = (1_{1+p}, 0_1 \oplus a)$. The latter fact requires the left zero identity.   □

Using the previous lemmas we obtain the following Kleene type theorem for partial Conway theories.

Theorem 1. *Let T be a partial Conway theory with basis (T_0, A). Suppose that either (T_0, A) is dagger compatible or $T_0 \subseteq D(T)$ is closed under dagger. Then a morphism f belongs to the least partial Conway subtheory of T containing T_0 and A iff f is the behavior of some presentation over (T_0, A).*

Proof. The necessity of our claim follows from the previous two lemmas. Suppose now that $f : n \to p$ is the behavior of a presentation $(\alpha, a) : n \to p$ over (T_0, A). Let T' denote the least partial Conway subtheory of T containing T_0 and A, so that $D(T') = T' \cap D(T)$. Since $a \in D(T')$, $a^\dagger \in T'$. Since T' is a subtheory containing T_0, it follows that $f = \alpha \cdot \langle a^\dagger, 1_p \rangle \in T'$.   □

Corollary 1. *Let T be a partial Conway theory with basis (T_0, A). Suppose that either (T_0, A) is dagger compatible or $T_0 \subseteq D(T)$ is closed under dagger. Then the following are equivalent:*

a) *$T_0 \cup A$ generates T, so that every morphism can be constructed from the morphisms in $T_0 \cup A$ by the theory operations and dagger.*
b) *For each morphism f in T there is a presentation over (T_0, A) whose behavior is f.*
c) *For each scalar morphism f in T there is a presentation over (T_0, A) whose behavior is f.*

Remark 4. Let T be a partial Conway theory with basis (T_0, A) such that every morphism in T is the behavior of a presentation over (T_0, A). Suppose that $D(T)$ is closed with respect to left composition by T_0-morphisms and if $\alpha \cdot \langle f, 1_p \rangle$ is in $D(T)$, where $\alpha : n \to s + p \in T_0$ and $f : s \to p \in D(T)$, then $\alpha = \beta \oplus 0_p$ for some $\beta : n \to s$. Then $D(T)$ is closed with respect to left composition by every T-morphism.

Indeed, suppose that $f : n \to p$ and $g : p \to q$ in T with $g \in D(T)$. By assumption, there exist presentations (α, a) and (β, b) with $|(\alpha, a)| = f$ and $|(\beta, b)| = g$. Since $g \in D(T)$, there exists some γ in T_0 with $\beta = \gamma \oplus 0_p$. Thus, $g = \gamma \cdot b^\dagger$, and we conclude that

$$f \cdot g = \alpha \cdot \langle a^\dagger, 1_p \rangle \cdot \gamma \cdot b^\dagger = \alpha \cdot \langle a^\dagger \cdot \gamma \cdot b^\dagger, \gamma \cdot b^\dagger \rangle$$

is a morphism in $D(T)$.

The above assumptions hold if T is generated by $T_0 \cup A$ and (T_0, A) is strongly dagger compatible.

Indeed, suppose that (T_0, A) is strongly dagger compatible. Then $D(T)$ is closed with respect to left composition with T_0 morphisms. To prove this, suppose that $f : n \to p$ in $D(T)$. Since (T_0, A) is strongly dagger compatible, there exist $\alpha : n \to s \in T_0$ and $a : s \to s + p \in A(T_0)$ with $f = \alpha \cdot a^\dagger$. Let $\beta : m \to n$ be a T_0-morphism. Then $\beta \cdot f = \beta \cdot (\alpha \cdot a^\dagger) = (\beta \cdot \alpha \cdot a) \cdot \langle a^\dagger, \mathbf{1}_p \rangle$ is in $D(T)$.

4.1   Grove Theories

In this subsection, we consider theories T equipped with additional constants $+ : 1 \to 2$ and $\# : 1 \to 0$. Given the constant $+$, let us define a *sum operation* as follows. For all morphisms $f, g : 1 \to p$ in T, we define

$$f + g = + \cdot \langle f, g \rangle.$$

When $f, g : n \to p$, then let

$$f + g = \langle (1_n \cdot f) + (1_n \cdot g), \ldots, (n_n \cdot f) + (n_n \cdot g) \rangle.$$

Moreover, using the constant $\#$, we define $0_{1,p} = \# \cdot 0_p$ and $0_{n,p} = \langle 0_{1,p}, \ldots, 0_{1,p} \rangle$, for all $n, p \geq 0$. Note that by definition $+ = 1_2 + 2_2$ and $\# = 0_{1,0}$. Below we will follow the convention that composition has higher priority than the sum operation.

Definition 6. *A* grove theory *[7] is a theory equipped with the constants* $+ : 1 \to 2$ *and* $\# : 1 \to 0$ *satisfying the following equations:*

$$1_2 + 2_2 = 2_2 + 1_2$$
$$(1_3 + 2_3) + 3_3 = 1_3 + (2_3 + 3_3)$$
$$1_1 + 0_{1,1} = 1_1.$$

A grove theory morphism $\varphi : T \to T'$ *between the grove theories* T *and* T' *is a theory morphism preserving the constants, i.e.* $+\varphi = +$ *and* $\#\varphi = \#$. *We say that the grove theory* T *is a* subgrove theory *of the grove theory* T' *if* T *is a subtheory of* T' *with the same constants* $+$ *and* $\#$.

It follows that for each $n, p \geq 0$, $(T(n, p), +, 0_{n,p})$ is a commutative monoid. Moreover,

$$(f + g) \cdot h = (f \cdot h) + (g \cdot h)$$
$$0_{m,n} \cdot f = 0_{m,p}$$

for all $f, g : n \to p$ and $h : p \to q$. Note that distributivity on the left need not hold. If T and T' are grove theories and $\varphi : T \to T'$ is a grove theory morphism, then $(f + g)\varphi = f\varphi + g\varphi$ and $0_{n,p}\varphi = 0_{n,p}$ for all $n, p \geq 0$ and $f, g : n \to p$. Similarly, if T' is a subgrove theory of T, then T' contains the zero morphisms $0_{n,p}$ of T and is closed under the sum operation of T.

Examples of grove theories include all matrix theories $\mathbf{Mat}_S$ and all matricial theories $\mathbf{Matr}_{S,V}$, see Examples 2 and 3. In $\mathbf{Mat}_S$, the morphism $+$ is the matrix

$(1,1): 1 \to 2$ and $\#$ is the unique matrix $1 \to 0$. In $\mathbf{Matr}_{S,V}$, $+ = ((1,1);0)$ and $\# = (;0)$.

A grove theory which is a (partial) Conway theory is a *(partial) Conway grove theory*. A morphism of (partial) Conway grove theories is both a grove theory morphism and a (partial) Conway theory morphism. A *(partial) Conway subgrove theory* T' of a (partial) Conway grove theory T is a subgrove theory that is a (partial) Conway subtheory of T.

Proposition 3. *Suppose that T is a partial Conway grove theory with basis (T_0, A). If T_0 is a subgrove theory, then the set of behaviors of presentations over (T_0, A) contains the morphisms $0_{n,p}$ and is closed under the sum operation.*

Proof. Every morphism in T_0 is the behavior of some presentation over (T_0, A). Since $\#$ and $+$ are in T_0, this fact applies to these morphisms. □

An important special case of Theorem 1 concerns partial Conway grove theories T with a basis (T_0, A) such that T_0 is a matrix theory.

Corollary 2. *Suppose that T is a partial Conway grove theory with basis (T_0, A) such that T_0 is a matrix theory. Suppose that one of the following two conditions holds:*

1. *For all $x: 1 \to p$ in T_0 and $f: 1 \to p \in D(T)$, if $x + f \in D(T)$ then $x = 0_{1,p}$. Moreover, for all $x: 1 \to 1 \in T_0$ and $a, b: 1 \to p \in A(T_0)$, $x \cdot a \in A(T_0)$ and $a + b \in A(T_0)$.*
2. *For every $x: 1 \to 1 \in T_0$, x^* is defined and belongs to T_0.*

Then a morphism $n \to p$ belongs to the least partial Conway subgrove theory of T containing T_0 and A iff it is the behavior of some presentation over (T_0, A).

Proof. Suppose that for all $x: 1 \to p$ in T_0 and $f: 1 \to p \in D(T)$, if $x + f \in D(T)$ then $x = 0_{1,p}$. Then if $\alpha \cdot \langle a, 1_p \rangle$ belongs to $D(T)$, for some $\alpha: n \to s + p$ in T_0 and $a: s \to p \in D(T)$, then $\alpha = \beta \oplus 0_p$ for some $\beta: n \to s$. Since T_0 is a matrix theory, $A(T_0)$ is closed under left composition with T_0-morphisms iff for each $p \geq 0$, the set of morphisms $1 \to p$ in T_0 is closed under sum and left composition with morphisms $1 \to 1$ in T_0. The second condition is equivalent to requiring that $T_0 \subseteq D(T)$ and T_0 is closed under dagger, or to the condition that α^* exists and is in T_0 for all $\alpha: n \to n$ in T_0. □

Note that the condition that for all $x: 1 \to p \in T_0$ and $f: 1 \to p$ in $A(T_0)$, if $x + f \in D(T)$ then $x = 0_{1,p}$ holds whenever each (scalar) morphism of T can be written in at most one way as the sum of a (scalar) T_0-morphism and a (scalar) morphism in $D(T)$.

5   Applications

In this section we present several applications of Theorem 1.

5.1   Trees

Suppose that Σ is a ranked alphabet and consider the partial Conway theory $T = \Sigma\text{TR}$. Equipped with the ideal $D(T)$ determined by the proper trees, ΣTR is a partial iterative theory and thus a partial Conway theory. Let T_0 be the subtheory determined by those trees not containing any vertex labeled by a letter of Σ, and let A be the collection of all atomic trees corresponding to the letters of Σ. Every tree in T_0 may be considered as a function in the initial theory Θ. Then $A(T_0)$ is the ideal of all proper trees and $(T_0, A(T_0))$ is a strongly dagger compatible basis. A presentation $(\alpha, a) : n \to p$ of dimension s is nothing but a *flowchart scheme* $n \to p$ over Σ, cf. [7]. We can write $a : s \to s + p$ in a unique way

$$a = \langle \sigma_1 \cdot \rho_1, \ldots, \sigma_s \cdot \rho_s \rangle \tag{2}$$

where each σ_i is in Σ_{n_i} for some $n_i \geq 0$, and each base morphism ρ_i corresponds to some function $[n_i] \to [s+p]$, also denoted ρ_i. Such a scheme is a finite, directed, ordered graph whose vertices are the integers in the set $[s+p]$. A vertex $i \in [s]$ is labeled σ_i and has n_i linearly ordered outgoing edges so that the jth edge leads to the vertex $j\rho_i$. Each vertex $s + i$ with $i \in [p]$ is labeled x_i, the ith variable in the set $\{x_1, \ldots, x_p\}$. The base morphism $\alpha : n \to s+p$ corresponds to a function, also denoted α, that picks the ith begin vertex $i\alpha$ for each $i \in [n]$. The behavior of (α, a) is the tree $t = (t_1, \ldots, t_n) : n \to p$ obtained by unfolding the flowchart scheme. It is known that a tree is the unfolding of a flowchart scheme iff it is regular. Thus the Kleene theorem asserts that a tree can be constructed from the atomic trees corresponding to the letters in Σ by the theory operations and (scalar) dagger iff it is the behavior of a scheme.

When Σ_0 contains the letter $\perp$, $T = \Sigma\text{TR}$ can be turned in a unique way into a Conway theory with a totally defined dagger operation such that $1_1{}^\dagger = \perp$, see Example 9. Thus, $D(T) = T$. Accordingly, we may choose T_0 to be the subtheory of all trees not having any vertex labeled by a symbol in Σ other than $\perp$. Then T_0 is closed under dagger, in fact T_0 is uniquely isomorphic to the theory Θ' of Example 10. The isomorphism $\Theta' \to T_0$ maps a partial function $\rho : [n] \to [p]$ to the tree $n \to p$ whose ith component is the variable $x_{i\rho}$ if $i\rho$ is defined and the tree $\perp$ otherwise, for all $i \in [n]$. Consider a presentation $(\alpha, a) : n \to p$ of dimension s over (T_0, A), where A is the collection of all atomic trees corresponding to the letters of Σ other than $\perp$. Here, α may be viewed as a partial function $[n] \to [s + p]$ and a is given as in (2), where now each ρ_i corresponds to a partial function $[n_i] \to [s + p]$. When $j\rho_i$ is undefined, for some $j \in [n_i]$, then the jth outgoing edge of vertex i leads to the extra vertex labeled $\perp$. Similarly, when $j\alpha$ is undefined for some $j \in [n]$, then this means that the jth begin vertex is the vertex labeled $\perp$. The behavior of the scheme is again the unfolding of the scheme. Since these unfoldings are again the regular trees, the Kleene theorem asserts that a tree can be constructed from the letters in Σ (other than $\perp$) by the theory operations and the total (scalar) dagger operation iff it is regular.

5.2 Synchronization Trees

Suppose that Σ is an alphabet. A *synchronization tree* $t : 1 \rightarrow p$ over Σ is an at most countable directed tree whose edges are labeled by the letters in the set $A \cup \{\text{ex}_1, \ldots, \text{ex}_p\}$, where the ex_i are referred to as the *exit symbols*. It is required that whenever an edge is labeled ex_i, for some i, then its target is a leaf. A morphism between trees $1 \rightarrow p$ preserves the root, the edges and the labeling. We identify isomorphic trees. A synchronization tree $n \rightarrow p$ over Σ is an n-tuple $(t_1, \ldots, t_n)$ of synchronization trees $1 \rightarrow p$ over Σ.

Synchronization trees over Σ form a category ST_Σ with composition defined in the following way. Suppose that $t : 1 \rightarrow p$ and $t' = (t'_1, \ldots, t'_p) : p \rightarrow q$. Then $t \cdot t'$ is the synchronization tree obtained from t by replacing each edge labeled ex_i for some $i \in [p]$ by a copy of t_i. When $t = (t_1, \ldots, t_n) : n \rightarrow p$, $t \cdot t'$ is defined as the tree $(t_1 \cdot t', \ldots, t_n \cdot t')$. With the trees $1 \rightarrow n$, $n \geq 0$, having a single edge labeled by an exit symbol as distinguished morphisms, ST_Σ is a theory. Let $+$ denote the tree $1 \rightarrow 2$ with two edges, an edge labeled ex_1, and an edge labeled ex_2, and let $\# : 1 \rightarrow 0$ be the empty tree having a single vertex and no edges. Equipped with these constants, ST_Σ is a grove theory. For each n, p, each component of $0_{n,p}$ is the empty tree. When $t, t' : 1 \rightarrow p$, $t + t'$ is the tree $1 \rightarrow p$ obtained by taking the disjoint union of t and t' and merging the roots. When $t, t' : n \rightarrow p$, $i_n \cdot (t + t') = i_n \cdot t + i_n \cdot t'$, for all $i \in [n]$. For more details, we refer to [7].

Let $T = \mathrm{ST}_\Sigma$ and define $D(T)$ to be the ideal determined by the *guarded trees* having no exit edge originating in the root. It is known that for each $t : n \rightarrow n+p$, the fixed point equation $\xi = t \cdot \langle \xi, 1_p \rangle$ has a unique solution, denoted $t^\dagger$. When each component of t is *finitely branching*, then the same holds for $t^\dagger$. Thus, T equipped with the ideal $D(T)$ is a partial iterative theory and hence a partial Conway grove theory.

Let T_0 denote the subtheory determined by the finitely branching synchronization trees with no edge labeled in Σ. Then T_0 is isomorphic to the matrix theory $\mathbf{Mat}_\mathbb{N}$ for the semiring $\mathbb{N}$ of natural numbers. Let A denote the collection of all trees $1 \rightarrow 1$ corresponding to letters σ in Σ that have a single path consisting of two edges, labeled σ and ex_1, respectively. Then a presentation $D : 1 \rightarrow p$ over (T_0, A) of dimension s is an ordered pair (α, a) consisting of a row matrix α of dimension $s + p$ over $\mathbb{N}$ and a morphism $a = \langle a_1 \cdot \rho_1, \ldots, a_s \cdot \rho_s \rangle : s \rightarrow s + p$ where each a_i is in Σ and each ρ_i is a row matrix over $\mathbb{N}$ of dimension $s + p$:

$$\alpha = (\alpha_j)_{j \in [s+p]}$$
$$\rho_i = (\rho_{i,j})_{j \in [s+p]} .$$

Such a presentation $(\alpha, a) : 1 \rightarrow p$ of dimension s determines and is determined by a finitely branching *transition system* whose set of states is $[s+p]$ together with an *external exit state* ex and a *begin state* b. For a pair of states $(i, j) \in [s] \times [s+p]$, there are $\rho_{i,j}$ transitions labeled a_j from state i to state j. In addition, there are $\rho_{i,s+j}$ edges labeled ex_j from state i to the external exit state ex, for all $j \in [p]$. Finally, for each $j \in [s]$ there are α_j edges labeled a_j from b to state j, and for each $j \in [p]$, there are α_{s+j} edges labeled ex_j from b to the external exit state ex. The behavior of (α, a) is the unfolding of this transition system from the begin state b.

Now (T_0, A) is a strongly dagger compatible basis and in this setting the Kleene theorem is the assertion that a tree $1 \to p$ can be constructed from the trees corresponding to the letters in Σ and the empty tree by the theory operations, sum, and (scalar) dagger applied to guarded trees iff it is the unfolding of a finitely branching transition system $1 \to p$. These trees are exactly the finitely branching regular trees having a finite number of subtrees.

When T_0 is the subtheory of all synchronization trees not having any edge labeled in Σ, then T_0 is isomorphic to $\mathbf{Mat}_{\mathbb{N}_\infty}$ for the semiring $\mathbb{N}_\infty = (\mathbb{N} \cup \{\infty\}, +, \cdot, 0, 1)$ obtained from $\mathbb{N}$ by adding a point at infinity with the usual operations. The dagger operation defined on the guarded trees can be (uniquely) extended to all trees $t : n \to n + p$ in such a way that ST_Σ becomes a Conway grove theory with $+^\dagger = \infty \cdot \mathbf{1}_1$ being the tree $1 \to 1$ that has a countably infinite number of edges leaving the root, each labeled ex_1, cf. [7]. Now T_0 is closed under dagger, and corresponds to the star operation defined on $\mathbb{N}_\infty$ by $0^* = 1$ and $n^* = \infty$ for all $n \in \mathbb{N}_\infty$, $n \neq 0$. Let A denote the collection of all trees corresponding to the letters in Σ. Then a presentation corresponds to a transition system as before, but now α and the ρ_i are row matrices over $\mathbb{N}_\infty$. The behavior is obtained in the same way. Using the second part of Theorem 1, we conclude that a tree $1 \to p$ can be constructed from the trees corresponding to the letters in Σ by the theory operations, sum and (scalar) dagger iff it is the unfolding of a transition system. These are exactly the regular synchronization trees.

5.3   Bisimulation

Let Σ be an alphabet and consider the Conway grove theory ST_Σ. For $t, t' : 1 \to p$, define $t \sim t'$ iff t and t' are *bisimilar*, i.e., when there is a bisimulation between them [27,30]. For trees $t = (t_1, \dots, t_n) : n \to p$ and $t' = (t_1', \dots, t_n') : n \to p$, let $t \sim t'$ iff $t_i \sim t_i'$ for all i. Then the relation $\sim$ is an equivalence relation on each hom-set $T(n, p)$ preserved by all operations including dagger. Thus, we can form the quotient Conway grove theory of bisimilarity equivalence classes.

Suppose now that T is the quotient partial Conway theory of ST_Σ with respect to the relation $\sim$. We identify each letter in Σ with the bisimilarity equivalence class of the corresponding tree. Let A denote the collection of all these equivalence classes.

Let T_0 be the subtheory determined by the equivalence classes of those trees having no edge labeled in Σ, so that T_0 may be identified with the theory $\mathbf{Mat}_{\mathbb{B}}$ of matrices over the boolean semiring $\mathbb{B}$. The transition system corresponding to a presentation $1 \to p$ over (T_0, A) is defined in the same way as in Subsection 5.2 but without any parallel edges labeled by the same symbol. The behavior of the presentation is the bisimulation equivalence class of its unfolding.

The Kleene theorem asserts that a bisimilarity equivalence class of a tree $1 \to p$ can be constructed from the equivalence classes corresponding to the letters in Σ by the theory operations, sum and (scalar) dagger iff it is the behavior of a transition system. It is known that these behaviors are the bisimilarity equivalence classes of the regular synchronization tees. For more details, see [7].

5.4   Weighted Tree Automata

Suppose that S is a semiring and Σ is a ranked alphabet. A function $s : T_\Sigma(X_p) \to S$ is called a (finite) *tree series* [4,12,21] with coefficients in S, sometimes denoted as a formal sum

$$\sum_{t \in T_\Sigma(X_p)} (s,t)t.$$

Tho *support* of s is the set of all trees mapped to a non-zero element of $S.$ Let $S\langle\!\langle T_\Sigma(X_p)\rangle\!\rangle$ stand for the set of all such series. Note that each $\sigma \in \Sigma_p$ has a corresponding series in $S\langle\!\langle T_\Sigma(X_p)\rangle\!\rangle$ that maps the atomic tree corresponding to σ to 1 and all other trees to 0.

We can form a theory $S\langle\!\langle \Sigma\mathrm{Term}\rangle\!\rangle$ whose morphisms $n \to p$ are all n-tuples of series in $S\langle\!\langle T_\Sigma(X_p)\rangle\!\rangle$. Composition is defined in the following way, cf. [8].

Let $s : 1 \to p$ and $r = (r_1,\ldots,r_p) : p \to q$, and consider a tree $u \in T_\Sigma(X_q)$. Write u in all possible ways as

$$u = \hat{u} \cdot \langle u_1, \ldots, u_k \rangle \tag{3}$$

where $\hat{u} \in T_\Sigma(X_k)$ has exactly one leaf labeled x_i for each $i \in [k]$ and the label sequence (from left to right) of these leaves is $x_1 \ldots x_k$, and where $u_1, \ldots, u_k \in T_\Sigma(X_q)$. Note that there are a finite number of such decomposition. Now for each possible decomposition (3), and for each base morphism $\rho : k \to p$, consider the product

$$(s, \hat{u} \cdot \rho)(r_{1\rho}, u_1) \cdots (r_{k\rho}, u_k),$$

where we have identified ρ with the corresponding function $[k] \to [p]$ as usual. Finally, $(s \cdot r, u)$ is the sum of all these products over all possible decompositions of u and all possible choices of ρ. When $s = (s_1, \ldots, s_n) : n \to p$, define $s \cdot r = (s_1 \cdot r, \ldots, s_n \cdot r)$. For each $i \in [n]$, $n \geq 0$, the distinguished morphism $i_n : 1 \to n$ is the series which maps x_i to 1 and all other trees in $T_\Sigma(X_n)$ to 0. Let $+$ denote the series $1 \to 2$ that maps x_1 and x_2 to 1 and all other trees in $T_\Sigma(X_2)$ to 0, and let $\#$ stand for the series $1 \to 0$ that maps all trees in $T_\Sigma(X_0) = T_\Sigma(\emptyset)$ to 0. Equipped with these constants, $S\langle\!\langle \Sigma\mathrm{Term}\rangle\!\rangle$ is a grove theory. The sum operation determined by the constant $+$ is the pointwise sum, so that

$$(s + s', t) = (s, t) + (s', t)$$

for all $s, s' : 1 \to p$ and $t \in T_\Sigma(X_p)$, and

$$s + s' = (s_1 + s_1', \ldots, s_n + s_n')$$

for all $s = (s_1, \ldots, s_n) : n \to p$ and $s' = (s_1', \ldots, s_n') : n \to p$.

Consider the theory $T = S\langle\langle\Sigma\text{Term}\rangle\rangle$. Call $s : 1 \to p$ *proper* if $(s, x_i) = 0$ for all $x_i \in X_p$. Moreover, call $s : n \to p$ proper if $i_n \cdot s$ is proper for all $i \in [n]$. The proper morphisms form an ideal $D(T)$, and for every proper $s : n \to n + p$, the equation $\xi = s \cdot \langle\xi, 1_p\rangle$ has a unique solution in the set of morphisms $n \to p$. Thus we have a partial iterative grove theory and a partial Conway grove theory.

Now let T_0 denote the subtheory determined by those series that map every proper tree to 0. Clearly, T_0 may be identified with the theory $\mathbf{Mat}_S$. In particular, each element s of S may be identified with the series $1 \to 1$ that maps x_1 to s and all other trees to 0. Let A denote the collection of all series whose support is finite and includes only trees of the form $\sigma \cdot \rho$, where $\sigma \in \Sigma$ and ρ is base. Note that $A(T_0) = A$ and (T_0, A) is a strongly dagger compatible basis. A presentation $D = (\alpha, a) : 1 \to p$ of weight s over (T_0, A) may be viewed as a (variant of a) *weighted tree automaton*, see [4,21]. Indeed, each component a_i of a is a series $1 \to s + p$ in A, and α is a row matrix over S of dimension $s + p$. The corresponding weighted tree automaton has $[s + p]$ as its set of states, with $s + 1, \ldots, s + p$ being the initial states corresponding to the variables $x_1, \ldots, x_p$. For a letter $\sigma \in \Sigma_k$ and states $i_1, \ldots, i_k$ and i, there is a transition from $(i_1, \ldots, i_k)$ to i labeled σ and having weight $(a_i, \sigma(x_{i_1}, \ldots, x_{i_k}))$ if this value is not 0. The row matrix α determines the final weight of each state. The initial weights of the states $s + 1, \ldots, s + p$ are all 1, whereas the initial weight of any state in $[s]$ is 0. The behavior of D is the tree series recognized by the corresponding weighted tree automaton. Thus, a tree series is recognizable iff it can be constructed from the the series corresponding to the letters of Σ and the series corresponding to the elements of S using the theory operations, sum and dagger. (The dagger operation may be replaced by a generalized star operation, see [20].)

5.5 Partial Conway Semirings

Following [10], we define a *partial Conway semiring* to be a semiring S equipped with a distinguished two-sided ideal I and a star operation $* : I \to S$ such that

$$(a + b)^* = a^*(ba^*)^*, \quad a, b \in I$$
$$(ab)^* = 1 + a(ba)^*b, \quad a \in I \text{ or } b \in I.$$

The star operation can be extended to square matrices over I using the following well-known matrix formula (which corresponds to the pairing identity as explained in [7]):

$$\begin{pmatrix} A & B \\ C & D \end{pmatrix}^* = \begin{pmatrix} A^* + A^*B(D + CA^*B)^*CA^* & A^*B(D + CA^*B)^* \\ (D + CA^*B)^*CA^* & (D + CA^*B)^* \end{pmatrix}, \quad (4)$$

where A and D are square matrices. (There are several equivalent formulas, see [7].) The star operation in turn gives rise to a dagger operation:

$$\begin{pmatrix} A & B \end{pmatrix}^\dagger = A^*B$$

where A is an $n \times n$ matrix and B is an $n \times p$ matrix over I. Note that the dagger operation determines the star operation by

$$M^* = \left(M \ M \right)^\dagger + \mathbf{1}_n,$$

for all $n \times n$ matrices M over I.

Let $T = \mathbf{Mat}_S$ and denote by $D(T)$ the ideal of those matrices all of whose entries are in I. Then T, equipped with the above dagger operation defined on morphisms $n \to n + p$ in $D(T)$, $n, p \geq 0$, is a partial Conway theory, called a *partial Conway matrix theory*. When $I = S$, T is a *Conway matrix theory*.

Let $A \subseteq I$ and S_0 be a subsemiring of S. An *automaton* over (S_0, A) is a triple (α, M, β) where $\alpha \in S_0^{1 \times n}$, $M \in (S_0 A)^{n \times n}$ and $\beta \in S_0^{n \times 1}$, where $S_0 A$ is the set of all finite linear combinations of elements of A with coefficients in S_0. The behavior of (α, M, β) is $\alpha M^* \beta$. Corollary 2 gives the following result, cf. [7,10]:

Theorem 2. *Suppose that either $S_0 \subseteq I$ is closed under star, or that whenever $x + a \in I$ for some $x \in S_0$ and $a \in D(T)$, then $x = 0$. Then an element s of S is the behavior of some automaton over (S_0, A) iff s can be generated from $S_0 \cup A$ by the rational operations of $+, \cdot$ and star.*

We note that if $S_0 \subseteq I$ then $1 \in I$ and $I = S$, so that S is a Conway semiring. When S is the power series semiring $S_0 \langle\!\langle \Sigma^* \rangle\!\rangle$, for some alphabet Σ, I is the ideal of proper series and A is the collection of all series associated with the letters in Σ, this is Schützenberger's theorem, see [31,32] or [25]. If in addition S_0 is $\mathbb{B}$ with star operation $0^* = 1^* = 1$, then $S_0 \langle\!\langle \Sigma^* \rangle\!\rangle$ may be identified with the usual Conway semiring of all subsets of Σ^*, and we have Kleene's classical theorem [24].

5.6 Partial Conway Semiring-Semimodule Pairs

In [19], a *partial Conway semiring-semimodule pair* is defined as a semiring-semimodule pair (S, V) equipped with a two-sided ideal $I \subseteq S$ and star and *omega power* operations $^* : I \to S$ and $^\omega : I \to V$ such that S is a partial Conway semiring and

$$(a + b)^\omega = (a^* b)^\omega + (a^* b)^* a^\omega, \quad a, b \in I$$
$$(ab)^\omega = a(ba)^\omega, \quad a \in I \text{ or } b \in I.$$

A *Conway semiring-semimodule pair* [7] is a partial Conway semiring-semimodule pair (S, V) with S as distinguished ideal.

Suppose that (S, V) is a partial Conway semiring-semimodule pair with distinguished two-sided ideal I. Then, as shown in [7,19], the omega power operation can be extended to square matrices over I. Using star and omega power, we can define dagger by

$$\left(A \ B \ ; \ v \right)^\dagger = \left(A^* B \ ; \ A^\omega + A^* v \right)$$

where $A \in I^{n \times n}$, $B \in I^{n \times p}$ and $v \in V^n$. Note that the dagger operation in turn determines both the star and the omega power operations, since $A^\omega = (A; 0)^\dagger$, for all square matrices A.

It is shown in [7] that when (S, V) is a Conway semiring-semimodule pair, then $\mathbf{Matr}_{S,V}$ is a Conway theory, called a *Conway matricial theory*. The same argument proves that when (S, V) is a Conway semiring-semimodule pair, with distinguished two-sided ideal I, then $\mathbf{Matr}_{S,V}$ is a *partial Conway matricial theory* with distinguished ideal the set of those morphisms $(A; v) : n \to p$ such that A is a matrix over I.

Suppose now that S_0 is a subsemiring of S, A is a subset of I. Then a *Büchi automaton* over (S_0, A) is a triple (α, M, k), where $\alpha \in S_0^{1 \times n}$, $M \in (S_0 A)^{n \times n}$ and $k \leq n$. The behavior of (α, M, k) is αM^{ω_k}, where if $M = \begin{pmatrix} A & B \\ C & D \end{pmatrix}$ such that A is $k \times k$ and D is $(n-k) \times (n-k)$, then $M^{\omega_k} = \begin{pmatrix} (A + BD^*C)^\omega \\ D^*C(A + BD^*C)^\omega \end{pmatrix}$. Using Corollary 2, we have the following result, see [19,22]:

Theorem 3. *Suppose that (S, V) is a partial Conway semiring-semimodule pair with distinguished two-sided ideal I. Let S_0 be a subsemiring of S and $A \subseteq I$. Suppose that either $S_0 \subseteq I$ is closed under star, or that $x + a \in I$ with $x \in S_0$ and $a \in I$ implies that $x = 0$. Then $v \in V$ is the behavior of a Büchi-automaton over (S_0, I) iff v can be generated from $S_0 \cup A$ by the rational operations of $+, \cdot$, star and omega power.*

References

1. de Bakker, J.W., Scott, D.: A theory of programs. IBM Seminar, Vienna (1969)
2. Bekić, H.: Definable operations in genaral algebras and the theory of automata and flowcharts. In: Bekic, H. (ed.) Programming Languages and their Definition. LNCS, vol. 177, pp. 30–55. Springer, Heidelberg (1984)
3. Bernátsky, L., Ésik, Z.: Semantics of flowchart programs and the free Conway theories. Theoretical Informatics and Applications, RAIRO 32, 35–78 (1998)
4. Berstel, J., Reutenauer, C.: Recognizable formal power series on trees. Theoretical Computer Science 18, 115–148 (1982)
5. Bloom, S.L., Elgot, C.C., Wright, J.B.: Solutions of the iteration equation and extension of the scalar iteration operation. SIAM J. Computing 9, 26–45 (1980)
6. Bloom, S.L., Elgot, C., Wright, J.B.: Vector iteration in pointed iterative theories. SIAM J. Computing 9, 525–540 (1980)
7. Bloom, S.L., Ésik, Z.: Iteration Theories. Springer, Heidelberg (1993)
8. Bloom, S.L., Ésik, Z.: An extension theorem with an application to formal tree series. J. of Automata, Languages and Combinatorics, 145–185 (2003)
9. Bloom, S.L., Ésik, Z.: Axiomatizing rational power series over natural numbers. Information and Computation 207, 793–811 (2009)
10. Bloom, S.L., Ésik, Z., Kuich, W.: Partial Conway and iteration semirings. Fundamenta Informaticae 86, 19–40 (2008)
11. Bloom, S.L., Ginali, S., Rutledge, J.D.: Scalar and vector iteration. J. Comput. System Sci. 14, 251–256 (1977)

12. Bozapalidis, S., Louscou-Bozapalidou, O.: The rank of a formal tree power series. Theor. Comput. Sci. 27, 211–215 (1983)
13. Elgot, C.C.: Monadic computation and iterative algebraic theories. In: Logic Colloquium 1973, Bristol. Studies in Logic and the Foundations of Mathematics, vol. 80, pp. 175–230. North-Holland, Amsterdam (1975)
14. Elgot, C.C.: Matricial theories. J. Algebra 42, 391–421 (1976)
15. Ésik, Z.: Identities in iterative and rational algebraic theories. Computational Linguistics and Computer Languages XIV, 183–207 (1980)
16. Ésik, Z.: On generalized iterative algebraic theories. Computational Linguistics and Computer Languages XV, 95–110 (1982)
17. Ésik, Z.: Completeness of Park induction. Theoret. Comput. Sci. 177, 217–283 (1997)
18. Ésik, Z.: Group axioms for iteration. Information and Computation 148, 131–180 (1999)
19. Ésik, Z.: Partial Conway and iteration semiring-semimodule pairs. In: Kuich, W., Rahonis, G. (eds.) Bozapalidis Festschrift. LNCS, vol. 7020, pp. 72–93. Springer, Heidelberg (2011)
20. Ésik, Z., Hajgató, T.: Iteration grove theories with applications. In: Bozapalidis, S., Rahonis, G. (eds.) CAI 2009. LNCS, vol. 5725, pp. 227–249. Springer, Heidelberg (2009)
21. Ésik, Z., Kuich, W.: Formal tree series. J. Autom. Lang. Comb. 8, 219–285 (2003)
22. Ésik, Z., Kuich, W.: A semiring-semimodule generalization of ω-regular languages, Parts 1 and 2. J. Automata, Languages, and Combinatorics 10, 203–242, 243–264 (2005)
23. Golan, J.: Semirings and Their Applications. Kluwer Academic Publishers, Dordrecht (1999)
24. Kleene, S.C.: Representation of events in nerve nets and finite automata. In: Automata Studies, pp. 3–42. Princeton University Press, Princeton (1956)
25. Kuich, W., Salomaa, A.: Semirings, Automata, Languages. Springer, Heidelberg (1986)
26. Lawvere, F.W.: Functorial semantics of algebraic theories. Proc. Nat. Acad. Sci. U.S.A. 50, 869–872 (1963)
27. Milner, R.: Communication and Concurrency. Prentice-Hall, Englewood Cliffs (1989)
28. Niwinski, D.: Equational μ-calculus. In: Skowron, A. (ed.) SCT 1984. LNCS, vol. 208, pp. 169–176. Springer, Heidelberg (1985)
29. Niwinski, D.: On fixed-point clones (extended abstract). In: Kott, L. (ed.) ICALP 1986. LNCS, vol. 226, pp. 464–473. Springer, Heidelberg (1986)
30. Park, D.M.R.: Fixed point induction and proofs of program properties. In: Michie, D., Meltzer, B. (eds.) Machine Intelligence, vol. 5, pp. 59–78. Edinburgh Univ. Press, Edinburgh (1970)
31. Schützenberger, M.P.: On the definition of a family of automata. Information and Computation 4, 245–270 (1961)
32. Schützenberger, M.P.: On a theorem of R. Jungen. Proc. American Mathematical Society 13, 885–890 (1962)
33. Simpson, A., Plotkin, G.: Complete axioms for categorical fixed-point operators. In: 15th Annual IEEE Symposium on Logic in Computer Science, Santa Barbara, CA, pp. 30–41. IEEE Comput. Soc. Press, Los Alamitos (2000)
34. Plotkin, G.: Domains. University of Edinburgh, Edinburgh (1983)
35. Wright, J.B., Thatcher, J.W., Wagner, E.G., Goguen, J.A.: Rational algebraic theories and fixed-point solutions. In: 17th Annual Symposium on Foundations of Computer Science, Houston, Tex, pp. 147–158. IEEE Comput. Soc., Long Beach (1976)

Rational Transformations and a Kleene Theorem for Power Series over Rational Monoids

Ina Fichtner and Christian Mathissen

Institut für Informatik, Universität Leipzig, Germany
{fichtner,mathissen}@informatik.uni-leipzig.de

Abstract. In this paper we consider transformations on formal power series and extend well-known results in terms of homomorphisms to rational functions. Using these results we prove a Kleene-Schützenberger Theorem for formal power series over rational monoids. It extends a result of Sakarovitch.

1 Introduction

Kleene's seminal result of the coincidence of rational and recognizable subsets of a finitely generated free monoid has been extended in several directions. Let us mention the two extensions this paper is concerned with.

The first is the characterisation of weighted automata in terms of rational expressions, a quantitative version of Kleene's result, which was shown by Schützenberger already in 1961 [19].

The second example is contained in a paper from Sakarovitch from 1987 [16]. As monoids are clearly one of the most fundamental structures in theoretical computer science, Sakarovitch investigated the natural question for which classes of monoids Kleene's coincidence of recognizable and rational subsets holds. He identified the class of so-called rational monoids for which this property holds true. These monoids possess descriptions in terms of finite state transducers and therefore an operation which is easy to compute.

Note that already Eilenberg [9] defined weighted automata and rational expressions not only over free but over arbitrary monoids. Hence it appears natural to raise the question whether one can combine Schützenberger's and Sakarovitch's result into a Kleene result for formal power series over rational monoids. This is the question this paper is devoted to.

We will generalize the results mentioned above and show that a Kleene Theorem holds true if the underlying semiring is a principal ideal domain or a locally finite semifield. Unfortunately we are not able to answer whether the characterization holds for arbitrary semirings. This difficulty may already indicate that our proof is not a straightforward adaption of Sakarovitch's methods and requires different concepts making use of the characteristics of the underlying semiring. In fact the authors conjecture that there are semirings where the characterization is not true.

Let us outline the structure of the paper and the methods we use for our main result. Our proof will be a reduction to the result of Schützenberger, i.e. to the

W. Kuich and G. Rahonis (Eds.): Bozapalidis Festschrift, LNCS 7020, pp. 94–111, 2011.

Kleene Theorem for formal power series over free monoids. Therefore we have to look at transformations of formal power series. Recall, that it is well known that rational and recognizable languages over the free monoid are closed under homomorphic images and homomorphic preimages. In fact this property does not only hold for homomorphisms but also for the more general class of rational functions. We will prove similar closure properties for formal power series in Section 4. These results enable us directly to show the easier part of our main result, namely that all recognizable formal power series over rational monoids are indeed rational.

The converse direction is more difficult and requires us to develop algebraic concepts in Section 3. One of these concepts is inspired by the syntactic ideals of Reutenauer [15] which were further developed by Symeon Bozapalidis et al. [6,5,4]. The other concept is similar to the syntactic congruence for tree series considered by Borchardt et al. [3,10]. These concepts constitute the part of the proof where we have to use the restrictions that our semiring is either a principal ideal domain or a locally finite semifield.

Finally, in the last section the results are plugged together to prove our main result.

2 Recognizable and Rational Series

In this section we recall some basics for monoids, semirings and from the theory of weighted automata. In particular we recall some basic closure properties which we will use in the sequel. For a more detailed treatment we recommend to consult [9,18,2,17].

Preliminaries. *In the following let M be a monoid.* We call a tuple $(m_1, \ldots, m_k)$ with $m_i \in M \setminus \{1\}$ for all $1 \leq i \leq k$ a factorization of $m_1 \cdots \cdot m_k$. Later on we will need the following basic property. Suppose that Σ is a finite generating set of M and let $\eta : \Sigma^* \to M$ be the natural homomorphism. Then, if we assume that $1 \notin \Sigma$, it is equivalent to say that each $m \in M$ admits only finitely many factorizations or to say that $\eta^{-1}(m)$ is finite for all $m \in M$. Indeed, assume that $\eta^{-1}(m)$ is finite. For a factorization $(m_1, \ldots, m_k)$ of $m \in M$ we have that the Cartesian product

$$\prod_{1 \leq i \leq k} \eta^{-1}(m_i)$$

is non-empty and contains only factorizations of elements in $\eta^{-1}(m)$. On the other hand, for two different factorizations $(m_1, \ldots, m_k), (m_1', \ldots, m_\ell')$ of m we have

$$\prod_{1 \leq i \leq k} \eta^{-1}(m_i) \cap \prod_{1 \leq i \leq \ell} \eta^{-1}(m_i') = \emptyset.$$

Hence, since $\eta^{-1}(m)$ is finite and since each $w \in \Sigma^*$ admits only finitely many factorizations, m admits only finitely many factorizations too.

Conversely, assume that $\eta^{-1}(m)$ is infinite. Any two different elements $w = a_1 \ldots a_k, w' = a'_1 \ldots a'_\ell \in \eta^{-1}(m) \in \Sigma^*$ yield two different factorizations $(a_1, \ldots, a_k)$ and $(a'_1, \ldots, a'_\ell)$ of m. Hence there are infinitely many factorizations of m.

Let us now turn to semirings. A *semiring* K is an algebraic structure $(K, +, \cdot, 0, 1)$ such that $(K, +, 0)$ is a commutative monoid, $(K, \cdot, 1)$ is a monoid, multiplication distributes over addition and 0 is absorbing, i.e. $k \cdot 0 = 0 = 0 \cdot k$ for all elements $k \in K$. Moreover, we assume that $0 \neq 1$. In case the multiplication is commutative, we call K *commutative*. Examples for semirings comprise the natural numbers $(\mathbb{N}, +, \cdot, 0, 1)$ as well as the tropical semiring $(\mathbb{N} \cup \{\infty\}, \min, +, \infty, 0)$ and the arctic semiring $(\mathbb{N} \cup \{-\infty\}, \max, +, -\infty, 0)$ which are used to model problems in operations research. Important examples are also the probabilistic semiring $([0, 1], \max, \cdot, 0, 1)$ and the semiring of formal languages $(\mathscr{P}(\Delta^*), \cup, \cap, \emptyset, \Delta^*)$. We call a semiring *locally finite* if any finitely generated subsemiring is finite. For example any Boolean algebra, the min-max semiring $(\mathbb{R}_+ \cup \{\infty\}, \max, \min, 0, \infty)$ and the fuzzy semiring $([0, 1], \max, \min, 0, 1)$ are each locally finite. We denote by $\mathbb{B}$ the 2-valued Boolean algebra $(\{0, 1\}, \vee, \wedge, 0, 1)$ and refer to it as the Boolean semiring.

Further on, we will need the structure of a semifield. We call a semiring K a *semifield* if each element $k \in K \setminus \{0\}$ has a multiplicative inverse, denoted k^{-1}.

For a semiring K and a finite set Q, we denote by $K^{Q \times Q}$ the set of all $Q \times Q$-matrices over K. The set $K^{Q \times Q}$ together with the usual matrix multiplication forms a monoid. By $K^{1 \times Q}$ and $K^{Q \times 1}$ we denote the set of all row resp. column matrices. *Now, let in the following K be a semiring.*

Assume that K is commutative. A K-*semimodule* $\mathscr{M}$ is a commutative monoid $(\mathscr{M}, +, 0)$ together with a scalar multiplication $\cdot : K \times \mathscr{M} \to \mathscr{M}$ such that for all $k, l \in K$ and $m, n \in \mathscr{M}$ we have

$$k \cdot (m + n) = k \cdot m + k \cdot n, \qquad (k + l) \cdot m = k \cdot m + l \cdot m,$$
$$(k \cdot l) \cdot m = k \cdot (l \cdot m),$$
$$1 \cdot m = m, \qquad\qquad 0 \cdot m = 0.$$

Observe that from these axioms we get: $k \cdot 0 = k \cdot (0 \cdot 0) = (k \cdot 0) \cdot 0 = 0 \cdot 0 = 0$ for all $k \in K$. If K is a ring, then $\mathscr{M}$ is called a K-*module*. A module having only finitely generated submodules is called *Noetherian*. Clearly, every commutative ring is a module over itself. In the latter case, the submodules are the ideals as considered in classical algebra and the finitely generated submodules are precisely the finitely generated ideals. A commutative ring is *Noetherian* if it is a Noetherian module over itself. A commutative ring without zero-divisors having the property that any ideal can in fact be generated by a single element is a called a *principal ideal domain*.

A *formal power series* is a mapping $S : M \to K$. The image of S at $m \in M$ will be denoted by (S, m). The series is then usually denoted as a formal sum $S = \sum_{m \in M} (S, m).m$. The *support* $\text{supp}(S)$ of S is the set $\{m \in M \mid (S, m) \neq 0\}$. If $\text{supp}(S)$ is finite, then S is called a *polynomial*. The class of formal power series on M with coefficients in K is denoted by $K\langle\langle M \rangle\rangle$. Polynomials are collected in

$K\langle M \rangle$. For some $L \subseteq M$ we denote by $\mathbb{1}_L$ the characteristic series of L, i.e. the formal power series that assumes 1 for $m \in L$ and 0 otherwise. By 0 we denote the series that assumes 0 everywhere.

Let us define some operations on formal power series. To this aim let $S, S' \in K\langle\!\langle \Sigma^* \rangle\!\rangle$ and let $k \in K$. First, we define the scalar product $k \cdot S$ which is given by $(k \cdot S, m) = k \cdot (S, m)$ for all $m \in M$. Furthermore have the pointwise sum $S + S'$ of S and S', i.e. $(S + S', m) = (S, m) + (S, m')$ for all $m \in M$. Similarly we define the pointwise product $S \odot S'$ of S and S', denoted by $\odot$, i.e. $(S \odot S', m) = (S, m) \cdot (S, m')$ for all $m \in M$. If the monoid has the property that each $m \in M$ admits only finitely many factorizations, then we can define additional operations. In this case we define the Cauchy product $S \cdot S'$ of S and S' by letting for all $m \in M$

$$(S \cdot S', m) = \sum_{\substack{n, n' \in M \\ nn' = m}} (S, n) \cdot (S', n').$$

Last we define the star S^* of S. Again we have to assume that each $m \in M$ admits only finitely many factorizations and, moreover, that $(S, 1) = 0$. In this case we define:

$$(S^*, m) = \sum_{\substack{m_1, \ldots, m_k \in M \\ m_1 \ldots m_k = m}} (S, m_1) \cdot \ldots \cdot (S, m_k)$$

Rational series are those series that can be constructed from polynomials using the pointwise sum, the Cauchy product and the star (the star only applied to series with $(S, 1) = 0$). The set of rational formal power series on M over K is denoted by $K^{\mathrm{rat}}\langle\!\langle M \rangle\!\rangle$. Note that over $\mathbb{B}$, the mapping supp gives a bijection between rational series and rational languages. Moreover, observe that $K\langle\!\langle M \rangle\!\rangle$, $K\langle M \rangle$ and $K^{\mathrm{rat}}\langle\!\langle M \rangle\!\rangle$ form semirings.

Definition 2.1 (cf. [18]). *A weighted finite automaton over the monoid M and the semiring K is a quadruple $\mathfrak{A} = (Q, \lambda, \mu, \varrho)$, where*

- *Q is a non-empty finite set of states,*
- *$\mu : M \to K^{Q \times Q}$ is a monoid homomorphism, and*
- *$\lambda \in K^{1 \times Q}$, $\varrho \in K^{Q \times 1}$.*

The automaton *computes* a formal power series $\|\mathfrak{A}\| \in K\langle\!\langle M \rangle\!\rangle$, given by

$$(\|\mathfrak{A}\|, m) = \lambda \cdot \mu(m) \cdot \varrho \quad \text{for all } m \in M.$$

Series computed by weighted automata are called *recognizable* and form the set $K^{\mathrm{rec}}\langle\!\langle M \rangle\!\rangle$. Again, over $\mathbb{B}$, the mapping supp gives a bijection between recognizable series and recognizable languages. Moreover, by standard automaton constructions it follows that if $L \subseteq M$ is recognizable and if $S \in K^{\mathrm{rec}}\langle\!\langle M \rangle\!\rangle$, then $\mathbb{1}_L \in K^{\mathrm{rec}}\langle\!\langle M \rangle\!\rangle$ and $S \odot \mathbb{1}_L \in K^{\mathrm{rec}}\langle\!\langle M \rangle\!\rangle$. Clearly, $K^{\mathrm{rec}}\langle\!\langle M \rangle\!\rangle$ is also closed under scalar product and pointwise sum.

In the case that the underlying monoid is the free monoid Σ^*, the monoid homomorphism μ is determined by the image $\mu(\Sigma)$ and hence by a set $E \subseteq Q \times \Sigma \times K \times Q$ of transitions which can naturally be interpreted as a directed graph with vertex set Q and edges labeled with elements from $\Sigma \times K$. Therefore we sometimes call the quadruple (Q, λ, E, ϱ) a weighted automaton. The set of paths from a state p to state q with label $w = a_1 \ldots a_n$ will be indicated by $q \overset{w}{\leadsto}_\mathfrak{A} q$. If $\lambda_p \neq 0$ and $\varrho_q \neq 0$, then we call a path $p \in q \overset{w}{\leadsto}_\mathfrak{A} q$ successful. We define the weight $\mathsf{weight}(p)$ of p by multiplying the weights of the composed transitions. The coefficient of the determined formal power series $\|\mathfrak{A}\|$ for a string w is then the sum of weights over all successful paths for w multiplied with the corresponding values of λ and ϱ.

Transformations of Formal Power Series. It is well known that the image of a rational subset of a monoid under a homomorphism is again rational. This holds also for recognizable subsets only if we require that the homomorphism is surjective. For a recognizable subset, however, it is known that its preimage under a homomorphism is again recognizable.

We now state similar closure properties for rational and recognizable formal power series. To this aim let M and N be monoids and $\beta : M \to N$ be a function. First, we define the transformation $\beta^{-1} : K\langle\!\langle N \rangle\!\rangle \to K\langle\!\langle M \rangle\!\rangle$. If $T \in K\langle\!\langle N \rangle\!\rangle$, then $\beta^{-1}(T) \in K\langle\!\langle M \rangle\!\rangle$ is given by letting

$$(\beta^{-1}(T), m) = (T, \beta(m)) \qquad \text{for all } m \in M.$$

If β has the property that $\beta^{-1}(n)$ is finite for all $n \in N$, then we also define the transformation $\bar{\beta} : K\langle\!\langle M \rangle\!\rangle \to K\langle\!\langle N \rangle\!\rangle$ as follows. If $S \in K\langle\!\langle M \rangle\!\rangle$, then the formal power series $\bar{\beta}(S) \in K\langle\!\langle N \rangle\!\rangle$ is given by letting

$$(\bar{\beta}(S), n) = \sum_{\beta(m)=n} (S, m) \quad \text{for all } n \in N.$$

If β is a homomorphism, then these transformations on formal power series preserve recognizability resp. rationality under certain circumstances:

Proposition 2.2 (cf. [18]). *Let $\beta : M \to N$ be a homomorphism. Moreover, let $S : K\langle\!\langle M \rangle\!\rangle$ and let $T \in K\langle\!\langle N \rangle\!\rangle$.*

(a) If T is recognizable, then $\beta^{-1}(T)$ is recognizable.
(b) If $M = \Sigma^*$ for some finite set Σ, β has the property that $\beta^{-1}(n)$ is finite for all $n \in N$ and if S is rational, then $\bar{\beta}(S)$ is rational.

Let us mention another transformation. Let Σ be a finite monoid and let $\Delta \subseteq \Sigma$. Moreover, let $\beta : \Sigma^* \to \Delta^*$ be a homomorphism such that $\beta^{-1}(w)$ is finite for all $w \in \Delta^*$. In [7] the transformation $\tilde{\beta} : K\langle\!\langle \Sigma^* \rangle\!\rangle \to K\langle\!\langle \Delta^* \rangle\!\rangle$ was introduced. It is given for $S \in K\langle\!\langle \Sigma^* \rangle\!\rangle$ by letting

$$(\tilde{\beta}(S), v) = \sum_{w \in \beta^{-1}(v)} (S, vw), \quad \text{for all } v \in \Delta^*.$$

Now, in [7] it was shown that $\tilde{\beta}$ preserves recognizability if K is commutative or locally finite.

In Section 4 we will extend the closure properties stated above to a larger class of functions than homomorphisms, namely to functions computed by rational transducers. But first, in the next section, we will investigate under which conditions the converse of Proposition 2.2(a) holds. This will be crucial in the proof of our main result. For this we will need some algebraic concepts.

3 Algebraic Characterizations of Weighted Automata

Let $S : M \to K$. We may extend S linearly to $S : K\langle M \rangle \to K$. Now let $\sim_S = \{(P_1, P_2) \in K\langle M \rangle \times K\langle M \rangle \mid (S, uP_1v) = (S, uP_2v) \text{ for all } u, v \in M\}$. It is not hard to see that this is a congruence with respect to $k\cdot$, $+$ and $\cdot$ for all $k \in K$. Let $\sim$ be any congruence contained in the kernel $\ker(S) = \{(P_1, P_2) \in K\langle M \rangle \times K\langle M \rangle \mid (S, P_1) = (S, P_2)\}$ and let $(P_1, P_2) \in \sim$. Then $(uP_1v, uP_2v) \in \sim$ for any $u, v \in M$ as $\sim$ is a congruence. Therefore, we have $(S, uP_1v) = (S, uP_2v)$ for all $u, v \in M$. This shows that $\sim \subseteq \sim_S$ and, hence, that $\sim_S$ is the coarsest congruence fully contained in $\ker(S)$. We define $\mathscr{A}_S = K\langle M \rangle / \sim_S$, the *syntactic algebra* of S. We say $\mathscr{A}_S$ is of *finite rank* if it is finitely generated as a semimodule.

Let us note that the concept of the syntactic algebra for formal power series goes back to Reutenauer. He considered a similar notion of a syntactic ideal for formal power series over free monoids and rings [15]. Symeon Bozapalidis et al. developed this concept further to the syntactic ideal of a tree series over fields [6,5,4].

We now define a second relation $\equiv_S$ similar to the syntactic congruences defined in [3,10]. For this let $m_1 \equiv_S m_2$ iff there exists $k \in K \setminus \{0\}$ such that for all $u, v \in M$ we have $(S, um_1v) = k \cdot (S, um_2v)$. Again it is left to the reader to verify that in case K is a semifield the relation $\equiv_S$ is a monoid congruence.

We will use the congruence relations $\sim_S$ and $\equiv_S$ to characterize recognizable series.

Proposition 3.1. *Let K be a be a principal ideal domain, let M be a finitely generated monoid and let $S \in K\langle\langle M \rangle\rangle$. Then S is recognizable iff $\mathscr{A}_S$ is of finite rank.*

Proof. (*If*). Let $\mathscr{A}_S$ be of finite rank finite. It is easy to see that it is a torsion free module and hence a finitely generated free module over K (cf. e.g. [13, Thm. 7.3]). Let $\{m_1, \ldots, m_n\}$ be a set which freely generates $\mathscr{A}_S$. Let φ be the natural epimorphism $\varphi : K\langle M \rangle \to \mathscr{A}_S$. For all $1 \le i, j \le n$ and $m \in M$ there are $\mu(m)_{i,j} \in K$ such that $m_i \cdot \varphi(m) = \sum_j \mu(m)_{ij} m_j$. This defines a mapping $\mu : M \to K^{n \times n}$. We show that μ is a homomorphism. Indeed, for all i we have

$$\sum_j \mu(mm')_{i,j} m_j = m_i \cdot \varphi(mm') = m_i \cdot \varphi(m)\varphi(m') = \left(\sum_k \mu(m)_{i,k} m_k\right)\varphi(m') =$$

$$= \sum_k (\mu(m)_{i,k} \sum_j \mu_{k,j}(m') m_j) = \sum_{k,j} \mu(m)_{i,k} \mu(m')_{k,j} m_j.$$

Since $m_1, \ldots, m_n$ freely generate $\mathscr{A}_S$, linear combinations are unique and we conclude that $\mu(mm') = \mu(m)\mu(m')$, as required. Moreover, there are $\lambda_i \in K$ ($1 \leq i \leq n$) such that $\varphi(1) = \sum_i \lambda_i m_i$. This defines $\lambda \in K^{1 \times n}$.

Now, let X be a finite set generating M and let $\alpha : X^* \to M$ be the natural homomorphism. We show by induction on the length of $w \in X^*$ that $\varphi(\alpha(w)) = \sum_{i,j} \lambda_i \mu(\alpha(w))_{ij} m_j$. For the empty word this is clear by definition. Now for the induction step suppose the claim holds for u and let $w = ua$. Then $\varphi(\alpha(ua)) = \varphi(\alpha(u)) \cdot \varphi(\alpha(a)) = (\sum_{i,k} \lambda_i \mu(u)_{ik} m_k) \cdot \varphi(\alpha(a)) = \sum_{i,k} \lambda_i \mu(u)_{ik}(m_k \cdot \varphi(\alpha(a))) = \sum_{i,k,j} \lambda_i \mu(\alpha(u))_{ik} \mu(\alpha(a))_{kj} m_j = \sum_{i,j} \lambda_i \mu(\alpha(ua))_{ij} m_j$. Select $P_i \in K\langle M \rangle$ ($1 \leq i \leq n$) such that $\varphi(P_i) = m_i$. Then we have $\varphi(m) = \varphi(\sum_i \lambda_i \mu(m)_{ij} P_j)$ for all $m \in M$. Let $\varrho \in K^{n \times 1}$ be given by $\varrho_i = (S, P_i)$. Observe that this is well-defined. Since $\ker(\varphi) \subseteq \ker(S)$ we have $(S, m) = (S, \sum_i \lambda_i \mu(m)_{ij} P_j) = \sum_{i,j} \lambda_i \mu(m)_{ij} \varrho_j$.

(*Only if*). Let $\mu : M \to K^{n \times n}$ be the transition function of a weighted automaton computing S. We extend μ linearly to $\mu : K\langle M \rangle \to K^{n \times n}$. Since K is a principal ideal domain it is Noetherian and hence $K^{n \times n}$ is a Noetherian K-module (see [13, Proposition X.1.4]). Thus, the K-submodule $\mu(K\langle M \rangle) \subseteq K^{n \times n}$ is of finite rank. Since $\ker(\mu) \subseteq \ker(S)$, we have $\ker(\mu) \subseteq \sim_S$. This shows that $\mathscr{A}_S = K\langle M \rangle / \sim_S$ is of finite rank, too.   □

Proposition 3.2. *Let K be a locally finite semifield, let M be a finitely generated monoid and let $S \in K\langle\langle M \rangle\rangle$. Then S is recognizable iff $\equiv_S$ has finite index.*

Proof. (*If*). The proof is similar to the proof of [10, Lemma 3.33] with small changes in the details. We include it here for the sake of completeness. Let us assume $\equiv_S$ has finite index. We will construct an automaton $\mathfrak{A} = (Q, \lambda, \mu, \varrho)$ computing S. Let $Q = M / \equiv_S$. Now for each equivalence class $[m]$ ($m \in M$) fix some normal form $\mathsf{nf}([m]) \in [m]$. Furthermore, for each $m \in M$ let $k(m) = 1$ if $(S, umv) = 0$ for all $u, v \in M$, otherwise let $k(m)$ be the unique element in K such that $(S, u\mathsf{nf}([m])v) = k(m) \cdot (S, umv)$ for all $u, v \in M$. Moreover, let $L_S = \{m \in M \mid (S, umv) = 0 \text{ for all } u, v \in M\}$. Now define for all $m, u, v \in M$

$$\mu(m)_{[u],[v]} = \begin{cases} (S', u)^{-1}(S', um) & \text{if } [v] = [um] \text{ and } um \notin L_S \\ 0 & \text{otherwise.} \end{cases}$$

Moreover for all $m \in M$, let $\lambda([m]) = k(1)$ if $[m] = [1]$ and let $\lambda([m]) = 0$ otherwise, and, let $\varrho([m]) = (S, \mathsf{nf}(m))$. It remains to show that μ is well defined, that μ is in fact a homomorphism and that $\|\mathfrak{A}\| = S$. Let us start by showing that μ is well defined.

Indeed, let $u, v, u', v' \in M$ such that $u \equiv_S u'$ and $v \equiv_S v'$. Since $\equiv_S$ is a congruence we get $um \equiv_S v$ iff $u'm \equiv_S v'$ and since L_S is either empty or a congruence class we also have $um \in L_S$ iff $u'm \in L_S$. Thus, let us assume that $v \equiv_S um$ and $um \notin L_S$. Then there are $n, n' \in M$ such that $(S, numn') \neq 0$. We conclude that

$$(S, numn') = k(u) \cdot (S, \mathsf{nf}(u)mn')$$

$$(S, numn') = k(um) \cdot (S, \mathsf{nnf}(um)n')$$

and similarly that

$$(S, nu'mn') = k(u') \cdot (S, \mathsf{nnf}(u)mn')$$
$$(S, nu'mn') = k(u'm) \cdot (S, \mathsf{nnf}(um)n').$$

Now $(S, numn') \neq 0$ implies that $(S, nu'mn') \neq 0$ and hence we can infer that $k(u)^{-1}k(um) = k(u')^{-1}k(u'm)$ as required.

Next, we show that μ is a homomorphism. Let $m, m', u, v \in M$. By definition of μ we have

$$\sum_{[n] \in Q} \mu(m)_{[u],[n]} \mu(m')_{[n],[v]} = \mu(m)_{[u],[um]} \mu(m')_{[um],[v]}.$$

Hence we need to show that $\mu(mm')_{[u],[v]} = \mu(m)_{[u],[um]} \mu(m')_{[um],[v]}$. If $v \not\equiv_S umm'$ or if $umm' \in L_S$ then both sides of the equation equal 0. Otherwise we have

$$\mu(mm')_{[u],[v]} = k(u)^{-1}k(umm')$$
$$= k(u)^{-1}k(um)k(um)^{-1}k(umm') = \mu(m)_{[u],[um]} \mu(m')_{[um],[v]}.$$

Last we show that $\|\mathfrak{A}\| = S$. In fact, for all $m \in M$ we have

$$(\|\mathfrak{A}\|, m) = \sum_{[u],[v] \in Q} \lambda_{[u]} \mu(m)_{[u],[v]} \varrho_{[v]} = \sum_{[v] \in Q} k(1)\mu(m)_{[1],[v]} \varrho_{[v]}$$
$$= k(1)\mu(m)_{[1],[m]} \varrho_{[m]} = k(1)k(1)^{-1}k(m)(S, \mathsf{nf}(m)) = (S, m)$$

(*Only if*). Let $\mathfrak{A} = (Q, \lambda, \mu, \varrho)$ be a weighted automaton computing S. Since M is finitely generated, since K is locally finite and since Q is finite we get $\mu(M) \subseteq K^{Q \times Q}$ is finite. Clearly $\ker(\mu) \subseteq \equiv_S$ and hence $\equiv_S$ has finite index. □

From our algebraic characterization we get the following corollary which will be crucial in the proof of our main result. In fact this corollary is the reason why we considered the congruences $\sim_S$ and $\equiv_S$.

Corollary 3.3. *Let K be a principal ideal domain or let K be a locally finite semifield. Let M be a finitely generated monoid and let $\beta : M \to N$ be a surjective homomorphism. Moreover, let $S \in K\langle\!\langle N \rangle\!\rangle$. If $\beta^{-1}(S) : M \to K$ is recognizable, then so is S.*

Proof. Let $\beta^{-1}(S)$ be recognizable. Let us first consider the case where K is a principal ideal domain. By Proposition 3.1 $\mathscr{A}_{\beta^{-1}(S)}$ is of finite rank. We may extend β linearly to $\beta : K\langle N \rangle \to K\langle M \rangle$. Now we get $P_1 \sim_{\beta^{-1}(S)} P_2$ iff $\beta(P_1) \sim_S \beta(P_2)$. Indeed,

$$P_1 \sim_{\beta^{-1}(S)} P_2 \Leftrightarrow (S, \beta(uP_1v)) = (S, \beta(uP_2v)) \text{ for all } u, v \in N$$

$$\Leftrightarrow (S, \beta(u)\beta(P_1)\beta(v)) = (S, \beta(u)\beta(P_2)\beta(v)) \text{ for all } u, v \in N$$
$$\Leftrightarrow \beta(P_1) \sim_S \beta(P_2).$$

Where in the last equivalence we used that β is surjective. Thus we get that $\mathscr{A}_{\beta^{-1}(S)}$ is isomorphic to $\mathscr{A}_S$. We conclude from Proposition 3.1 that S is recognizable.

Let us now consider the case where K is a locally finite commutative semifield. This case is similar but uses the congruence $\equiv_{\beta^{-1}(S)}$. In fact, by Proposition 3.2 $\equiv_{\beta^{-1}(S)}$ has finite index. Similarly as in the first case we conclude that $u \equiv_{\beta^{-1}(S)} v$ iff $\beta(u) \equiv_S \beta(v)$ for all $u, v \in N$. Hence $\equiv_S$ has finite index. Applying Proposition 3.2 again shows that S must be recognizable.   □

4   String and Series Transducer

In this section we consider a particular kind of weighted automata over free monoids, namely weighted automata over the semiring $K^{\mathrm{rat}}\langle\!\langle \Delta^* \rangle\!\rangle$ for some finite alphabet Δ. These automata are called series transducers. If K is the two-valued Boolean algebra $\mathbb{B}$, then they are also known as string transducers.

As we have seen in the last section, if β is a homomorphism, then the transformations β^{-1} and $\bar{\beta}$ preserve recognizability and rationality. In this section we will extend this result to functions β computed by transducers. Some of these results will be important in order to prove our main result; all of them, however, are interesting in their own right.

Let us start with a quite intuitive definition for string transducers and after that turn to the formal definition of series transducers.

Definition 4.1. *A (string) transducer* $\mathfrak{T} = (\Sigma, \Delta, Q, q_-, Q_+, E)$ *consists of a finite input alphabet* Σ, *a finite output alphabet* Δ, *a finite set* Q *of states, an initial state* $q_- \in Q$, *a set of final states* $Q_+ \subseteq Q$ *and a finite set of transitions* E *such that* $E \subseteq Q \times \Sigma^* \times \Delta^* \times Q$.

Let $w \in \Sigma^*$ and let $u \in \Delta^*$. The transducer $\mathfrak{T}$ computes a so called *rational transduction* $\|\mathfrak{T}\| : \Sigma^* \to \mathscr{P}(\Delta^*)$, where $u \in \|\mathfrak{T}\|(w)$ iff there exists a successful path (i.e. a path from q_- to Q_+) with input w and output u. Let us remark that a function $\tau : \Sigma^* \to \mathscr{P}(\Delta^*)$ is a rational transduction if and only if $\{(w, u) \in \Sigma^* \times \Delta^* \mid u \in \tau(w)\}$ is a rational subset of the monoid $\Sigma^* \times \Delta^*$ [1]. If a rational transduction τ has the property that for all $w \in \Sigma^*$ the set $\tau(w)$ has cardinality at most one, then we consider τ as a partial function from Σ^* to Δ^* and call it a *rational function*. Note that for example any homomorphism is a rational function.

A string transducer or a weighted automaton over the free monoid in general is called *unambiguous* if any word is the input of at most one successful path. Clearly, unambiguous transducers compute rational functions. On the other hand, any rational function can be realized by an unambiguous transducer as the next proposition states. For a proof of it see [1].

Proposition 4.2 ([9]). *Let* $\tau : \Sigma^* \to \Delta^*$ *be a rational function such that* $\tau(\varepsilon) = \varepsilon$ *for the empty word* ε. *Then there exists an unambiguous transducer* $\mathfrak{T} = (\Sigma, \Delta, Q, q_-, Q_+, E)$ *with* $E \subseteq Q \times \Sigma \times \Delta^* \times Q$ *and* $Q_+ = \{q_+\}, q_+ \neq q_-$ *computing* τ.

Next we turn to a generalization of string transducers, so called series transducers. The basic definitions and notations in the context of series transducers are in [11].

Definition 4.3. *A series transducer* $\mathfrak{T} = (Q, \mu, q_0, P)$ *is given by a finite set* Q *of states, a homomorphism* $\mu : \Sigma^* \to \left(K^{rat}\langle\!\langle \Delta^* \rangle\!\rangle\right)^{Q \times Q}$, *an initial state* $q_0 \in Q$ *and a finite state vector* $P \in (K^{rat}\langle\!\langle \Delta^* \rangle\!\rangle)^{Q \times 1}$.

The homomorphism μ is called *regulated* if it has the property that there exists a $k \geq 1$ such that, for all $w \in \Sigma^*$ with $|w| \geq k$ and for all $p, q \in Q$ we have $(\mu(w), \varepsilon)_{p,q} = 0$. In this case one can easily conclude that $(\mu(w)_{p,q}, v) = 0$ for all $v \in \Delta^*$ with $|w| \geq k(|v| + 1)$. If μ is regulated we call $\mathfrak{T}$ a *regulated* series transducer. We call $\mathfrak{T}$ *polynomial* if $\mu : \Sigma^* \to (K\langle \Delta^* \rangle)^{Q \times Q}$ and $P \in (K\langle \Delta^* \rangle)^{Q \times 1}$.

Now we extend the homomorphism μ to a mapping $\mu : K\langle\!\langle \Sigma^* \rangle\!\rangle \to \left(K^{rat}\langle\!\langle \Delta^* \rangle\!\rangle\right)^{Q \times Q}$ by letting

$$\mu(S) = \sum_{w \in \Sigma^*} (S, w)\mu(w) \quad \text{for all } S \in K\langle\!\langle \Sigma^* \rangle\!\rangle, \text{ provided the sum exists.}$$

In fact, this is well defined in case μ is regulated, since

$$\left(\mu(S)_{p,q}, v\right) = \sum_{w \in \Sigma^*} (S, w)\left(\mu(w)_{p,q}, v\right) = \sum_{|w| < k(|v|+1)} (S, w)\left(\mu(w)_{p,q}, v\right).$$

The transducer now realizes the mapping $\|\mathfrak{T}\| : K\langle\!\langle \Sigma^* \rangle\!\rangle \to K\langle\!\langle \Delta^* \rangle\!\rangle$ given by

$$\|\mathfrak{T}\|(S) = \sum_{w \in \Sigma^*} (S, w)\left(\mu(w)P\right)_{q_0} = \left(\mu(S)P\right)_{q_0}.$$

A mapping $\tau : K\langle\!\langle \Sigma^* \rangle\!\rangle \to K\langle\!\langle \Delta^* \rangle\!\rangle$ is called a *(regulated) rational series transduction* if there exists a (regulated) series transducer $\mathfrak{T} = (Q, \mu, q_0, P)$ such that $\tau(S) = \|\mathfrak{T}\|(S)$ for all $S \in K\langle\!\langle \Sigma^* \rangle\!\rangle$.

Let us first mention that regulated series transducers have the important property that they transform rational series into rational series:

Theorem 4.4 ([11]). *Let* K *be a commutative semiring. Let* $\mathfrak{T} = (Q, \mu, q_0, P)$ *be a regulated series transducer. If* $S \in K^{rat}\langle\!\langle \Sigma^* \rangle\!\rangle$ *then* $\|\mathfrak{T}\|(S) \in K^{rat}\langle\!\langle \Delta^* \rangle\!\rangle$.

Transformations of Formal Power Series. We now consider the transformation β^{-1}, $\bar{\beta}$ and $\tilde{\beta}$ in the case that β is a rational function. We will discuss which of these transformations arise as regulated rational series transductions. Since by Theorem 4.4 such series preserve rationality this will extend Proposition 2.2. In Section 5, in order to prove for rational monoids that rational series are recognizable we will then apply the results.

Proposition 4.5. *Let $\beta : \Sigma^* \to \Delta^*$ be a rational function with $\beta(\varepsilon) = \varepsilon$ and $\beta^{-1}(w)$ is finite, for all $w \in \Delta^*$. Then*

$$\bar{\beta} : K\langle\!\langle \Sigma^* \rangle\!\rangle \to K\langle\!\langle \Delta^* \rangle\!\rangle$$

is a regulated rational series transduction.

Proof. Let $\mathfrak{T} = (\Sigma, \Delta, Q, q_0, F, E)$ be a string transducer for β. By Proposition 4.2 we may assume that $\mathfrak{T}$ is unambiguous and $E \subseteq Q \times \Sigma \times \Delta^* \times Q$. We may also assume that each state $q \in Q$ is reachable and co-reachable, which means that there is a path from q_0 to q and that there is a path from q to some $p \in F$. We define $\mu : \Sigma^* \to (K^{\mathrm{rat}}\langle\!\langle \Delta^* \rangle\!\rangle)^{Q \times Q}$ by letting for all $a \in \Sigma$

$$\mu(a)_{p,q} = \begin{cases} \mathbb{1}_{\{w\}} & \text{if } (p,a,w,q) \in E \\ 0 & \text{otherwise.} \end{cases}$$

Moreover, let $(P)_q = \mathbb{1}_{\{\varepsilon\}}$ if $q \in F$ and 0 otherwise. Then μ is a morphism which is regulated. Indeed, since each state is reachable and co-reachable and since $\beta^{-1}(w)$ is finite for all $w \in \Delta^*$ we conclude that there can not be a loop with output ε. Thus there are only finitely many paths with output ε. Let now k be the maximum length of a path with output ε. Then $(\mu(w), \varepsilon)_{p,q} = 0$ for all $w \in \Sigma^*, |w| > k$. Furthermore, since T is unambiguous, we have

$$(\mu(w)P)_{q_0} = \sum_{q \in Q} \mu(w)_{q_0,q} = \sum_{q \in F} \mu(w)_{q_0,q} = \mathbb{1}_{\{\beta(w)\}} \quad \text{for all } w \in \Delta^*. \quad (4.1)$$

Now, consider the regulated series transducer $\mathfrak{T} = (Q, \mu, q_0, P)$, let $S \in K\langle\!\langle \Sigma^* \rangle\!\rangle$ and let $v \in \Delta^*$. Then

$$
\begin{aligned}
(\|\mathfrak{T}\|(S), v) &= \big((\mu(S)P)_{q_0}, v\big) = \sum_{w \in \Delta^*} (S,w)\big((\mu(w)P)_{q_0}, v\big) \\
&\overset{(4.1)}{=} \sum_{w \in \Delta^*} (S,w) \sum_{q \in F} \big(\mu(w)_{q_0,q}, v\big) = \sum_{w \in \Delta^*} (S,w)(\mathbb{1}_{\{\beta(w)\}}, v) \\
&= \sum_{w \in \beta^{-1}(v)} (S,w) = \big(\bar{\beta}(S), v\big).
\end{aligned}
$$

This shows that $\|\mathfrak{T}\| = \bar{\beta}$ and hence that $\bar{\beta}$ is a regulated rational series transduction. $\square$

Next we give an example of transformations that preserve rationality but do not arise as regulated rational series transductions. In [7] the transformation $\tilde{\beta}$ was considered in the case where β is a homomorphism. It was shown that it preserves recognizability. Using the last proposition one can prove similar to the proof of Theorem 3.1 in [7] the following result.

Proposition 4.6. *Let K be a commutative or locally finite semiring, let $\Delta \subseteq \Sigma$ and let $S : \Sigma^* \to K$ be recognizable. Moreover, let $\beta : \Sigma^* \to \Delta^*$ be a rational function such that $\beta^{-1}(\varepsilon) = \{\varepsilon\}$ and $\beta^{-1}(w)$ is finite for all $w \in \Delta^*$. Then $\tilde{\beta}(S) : \Delta^* \to K$ is recognizable.*

However, $\tilde{\beta}$ in general is not a rational series transduction.

Proposition 4.7. *Let $\mathbb{B}$ be the Boolean semiring. Let $|\Sigma| \geq 2$ and let β the identity function on Σ^*. Then $\tilde{\beta} : \mathbb{B}\langle\!\langle \Sigma^* \rangle\!\rangle \rightarrow \mathbb{B}\langle\!\langle \Sigma^* \rangle\!\rangle$ is not a rational series transduction.*

Proof. Suppose for contradiction that $\mathfrak{T} = (Q, \mu, q_0, P)$ is a rational series transducer for $\tilde{\beta}$. For $v, w \in \Sigma^*$ we have

$$
\big(\|\mathfrak{T}\|(w), v\big) = \big((\mu(w)P)_{q_o}, v\big)
$$

$$
= \sum_{w' \in \Sigma^*} (\mathbb{1}_{\{w\}}, w')\big((\mu(w)P)_{q_o}, v\big) = \big(\|\mathfrak{T}\|(\mathbb{1}_{\{w\}}), v\big)
$$

$$
= \big(\tilde{\beta}(\mathbb{1}_{\{w\}}), v\big) = \big(\mathbb{1}_{\{w\}}, vv\big) = \begin{cases} 1 & w = vv \\ 0 & \text{otherwise.} \end{cases}
$$

This is equivalent to say that there is a string transducer $\mathfrak{T}'$ computing a rational transduction

$$
\|\mathfrak{T}'\|(w) = \begin{cases} \{v\} & w = vv \\ \emptyset & \text{otherwise.} \end{cases}
$$

Hence $\big\{(vv, v) \mid v \in \Sigma^*\big\} \subseteq \Sigma^* \times \Sigma^*$ must be a rational set [1], which is a contradiction. $\qquad\square$

Consider now a polynomial series transducer $\mathfrak{T} = (Q, \mu, q_0, P)$. We define the transformation $\|\mathfrak{T}\|^{-1} : K\langle\!\langle \Delta^* \rangle\!\rangle \rightarrow K\langle\!\langle \Sigma^* \rangle\!\rangle$ by letting for all $S \in K\langle\!\langle \Delta^* \rangle\!\rangle$ and all $w \in \Sigma^*$

$$
\big(\|\mathfrak{T}\|^{-1}(S), w\big) = \sum_{v \in \Delta^*} (S, v)(\|\mathfrak{T}\|(w), v)
$$

Note that the right-hand side is a finite sum since $\mathfrak{T}$ is polynomial.

Theorem 4.8 ([11]). *Let K be a commutative semiring. If $\mathfrak{T}$ is a polynomial series transducer, then $\|\mathfrak{T}\|^{-1}$ is a regulated rational series transduction.*

Proposition 4.9. *Let K be commutative, $\beta : \Sigma^* \rightarrow \Delta^*$ be a rational function with $\beta(\varepsilon) = \varepsilon$ and $\beta^{-1}(w)$ is finite, for all $w \in \Delta^*$. Then*

$$
\beta^{-1} : K\langle\!\langle \Delta^* \rangle\!\rangle \rightarrow K\langle\!\langle \Sigma^* \rangle\!\rangle
$$

is a regulated rational series transduction.

Proof. Clearly, by the proof of Proposition 4.5 there exists a polynomial series transducer $\mathfrak{T} = (Q, \mu, q_0, P)$ for $\tilde{\beta} : K\langle\!\langle \Sigma^* \rangle\!\rangle \rightarrow K\langle\!\langle \Delta^* \rangle\!\rangle$. With Theorem 4.8, $\|\mathfrak{T}\|^{-1}$ is a regulated rational series transduction. We compute for $S \in K\langle\!\langle \Delta^* \rangle\!\rangle$ and $w \in \Sigma^*$:

$$
\big(\|\mathfrak{T}\|^{-1}(S), w\big) = \sum_{v \in \Delta^*} (S, v)\big(\|\mathfrak{T}\|(w), v\big) = \sum_{v \in \Delta^*} (S, v)\big(\mathbb{1}_{\{\beta(w)\}}, v\big)
$$

$$
= \big(\beta^{-1}S, w\big)
$$

Thus $\beta^{-1} = \|\mathfrak{T}\|^{-1}$ which is a regulated rational series transduction by Theorem 4.8. $\qquad\square$

The last proposition together with Theorems 4.4 and 4.8 give that $\bar{\beta}$ and β^{-1} preserve rationality if β is as required and if K is commutative. Next we will give a direct automaton-theoretic construction that shows that this holds even if K is not commutative.

Proposition 4.10. *Let $\beta : \Sigma^* \to \Delta^*$ be a rational function with the property that $\beta(\varepsilon) = \varepsilon$ and that $\beta^{-1}(w)$ is finite for all $w \in \Delta^*$.*

(a) If $S : \Sigma^* \to K$ is recognizable, then $\bar{\beta}(S) : \Delta^* \to K$ is recognizable.
(b) If $S : \Delta^* \to K$ is recognizable, then $\beta^{-1}S : \Sigma^* \to K$ is recognizable.

Proof. Throughout the proof let $\mathcal{B} = (\Sigma, \Delta, R, r_-, R_+, T_{\mathcal{B}})$ be a transducer for β. With Proposition 4.2, we may choose $\mathcal{B}$ unambiguous and R_+ as a singleton set $\{r_+\}$. We have $T_{\mathcal{B}} \subseteq R \times \Sigma \times \Delta^* \times R$.

(a). Let $S : \Sigma^* \to K$ be recognizable. We will assume that $(S, \varepsilon) = 0$. Let $\mathfrak{A} = (Q, T_{\mathfrak{A}}, \lambda, \gamma)$ be a weighted automaton for S satisfying $\lambda(q_-) = 1$ (0 otherwise) and $\gamma(q_+) = 1$ (0 otherwise), where $q_-, q_+ \in Q$. We have

$$T_{\mathfrak{A}} \subseteq Q \times \Sigma \times K \times Q.$$

We denote $Z = R \times Q, z_- = (r_-, q_-), z_+ = (r_+, q_+)$ and define:

$$T = \Big\{ ((r,q), a, w, k, (r', q')) \mid (r, a, w, r') \in T_{\mathcal{B}} \text{ and } (q, a, k, q') \in T_{\mathcal{A}} \Big\}. \quad (4.2)$$

Now, let $J = \{(z_1, u, w, k, z_2)(z_1', u', w', k', z_2') \in T^2 \mid z_2 \neq z_1'\} \subseteq T^*$ and consider the following rational subset of T^*.

$$P = \big(\{z_-\} \times \Sigma \times \Delta^* \times K \times Z \big) T^* \cap T^* \big(Z \times \Sigma \times \Delta^* \times K \times \{z_+\} \big) \setminus T^* J T^*.$$

For our construction, we now need the two projections $\pi_3 : T^* \to (\Delta^*, \circ, \varepsilon)$ and $\pi_4 : T^* \to (K, \cdot, 1)$ which are given as the unique homomorphic extensions of

$$\pi_3(z, \sigma, \delta, k, z') = \delta, \quad \pi_4(z, \sigma, \delta, k, z') = k \quad \text{for all } (z, \sigma, \delta, k, z') \in T.$$

Clearly, π_4 is a recognizable series over T^*, since it is homomorphic. Hence, $\pi_4 \odot \mathbb{1}_P$ is recognizable. Now consider (4.2) and note that, for a word $w \in \Delta^*$, elements in $\pi_3^{-1}(w) \cap P$ arise by pairing for each $u \in \beta^{-1}(w)$, successful paths for u in $\mathcal{B}$ (exactly one path, since $\mathcal{B}$ is unambiguous) with the successful paths for u in $\mathfrak{A}$ (finitely many).

Now, Proposition 2.2 implies $\overline{\pi_3}(\pi_4 \odot \mathbb{1}_P) \in K^{\mathrm{rec}}\langle\!\langle \Delta^* \rangle\!\rangle$. Let $w \in \Delta^*$. Then

$$(\overline{\pi_3}(\pi_4) \odot \mathbb{1}_P), w) = \sum_{\substack{p \in P \\ \pi_3(p) = w}} \pi_4(p) = \sum_{\substack{p_{\mathcal{B}} \in r_- \xrightarrow{u}_{\mathcal{B}} r_+ \\ \beta(u) = w}} \left(\sum_{p_{\mathfrak{A}} \in q_- \xrightarrow{u}_{\mathfrak{A}} q_+} \mathrm{weight}(p_{\mathfrak{A}}) \right)$$

Since $\mathcal{B}$ is unambiguous we continue

$$= \sum_{\substack{u \in \Sigma^* \\ \beta(u) = w}} \left(\sum_{p_{\mathfrak{A}} \in q_- \xrightarrow{u}_{\mathfrak{A}} q_+} \mathrm{weight}(p_{\mathfrak{A}}) \right) = \sum_{u \in \Sigma^*, \beta(u) = w} (S, u) = (\bar{\beta}(S), w).$$

Hence $\overline{\pi}_3(\pi_4 \odot \mathbb{1}_P) = \bar{\beta}(S)$ and thus $\bar{\beta}(S)$ is recognizable.

(b). The proof for this part is very similar to part (a). Let $S : \Delta^* \to K$ be recognizable. Again we assume that $(S, \varepsilon) = 0$. Let $\mathfrak{A} = (Q, T_{\mathfrak{A}}, \lambda, \gamma)$ be a weighted automaton for S satisfying $\lambda(q_-) = 1$ (0 otherwise) and $\gamma(q_+) = 1$ (0 otherwise), where $q_-, q_+ \in Q$. Let $\mu_{\mathfrak{A}} : \Delta^* \to K^{Q \times Q}$ be the corresponding monoid morphism. Again denote $Z = R \times Q, z_- = (r_-, q_-), z_+ = (r_+, q_+)$ and define this time:

$$T = \Big\{ ((r, q), a, w, k, (r', q')) \mid (r, a, w, r') \in T_{\mathcal{B}} \text{ and } k = \mu_{\mathfrak{A}}(w)_{q, q'} \Big\}.$$

Now, as before let $J = \{(z_1, u, w, k, z_2)(z_1', u', w', k', z_2') \in T^2 \mid z_2 \neq z_1'\} \subseteq T^*$ and consider the rational subset

$$P = \big(\{z_-\} \times \Sigma \times \Delta^* \times K \times Z\big)T^* \cap T^*\big(Z \times \Sigma \times \Delta^* \times K \times \{z_+\}\big) \setminus T^*JT^*.$$

For our construction, we now need the two projections $\pi_2 : T^* \to (\Sigma^*, \circ, \varepsilon)$ and $\pi_4 : T^* \to (K, \cdot, 1)$ which are given as the homomorphic extensions of

$$\pi_2(z, \sigma, \delta, k, z') = \sigma, \quad \pi_4(z, \sigma, \delta, k, z') = k \quad \text{for all } (z, \sigma, \delta, k, z') \in T.$$

Now, Proposition 2.2 implies $\overline{\pi}_2(\pi_4 \odot \mathbb{1}_P) \in K^{\text{rec}} \langle\!\langle \Delta^* \rangle\!\rangle$. Since $\mathcal{B}$ is unambiguous, we get for $w \in \Sigma^*$:

$$(\overline{\pi}_2(\pi_4 \odot \mathbb{1}_P), w) = \sum_{\substack{p \in P \\ \pi_2(p) = w}} \pi_4(p) = (\beta^{-1}S, w).$$

Hence $\beta^{-1}S$ is recognizable. □

Having considered $\bar{\beta}$ and $\tilde{\beta}$ for rational functions β, one can also consider rational transductions in general. We extend the definition of $\bar{\beta}$ in the following sense: Let $\beta : \Sigma^* \to \mathscr{P}(\Delta^*)$ be a rational transduction such that $\{w \mid v \in \tau(w)\}$ is finite for all $v \in \Delta^*$ and define $\bar{\beta} : K\langle\!\langle \Sigma^* \rangle\!\rangle \to K\langle\!\langle \Delta^* \rangle\!\rangle$ by setting for $S \in K\langle\!\langle \Sigma^* \rangle\!\rangle$

$$(\bar{\beta}(S), v) = \sum_{w : v \in \beta(w)} (S, w), \quad \text{for all } v \in \Delta^*.$$

This definition coincides with the considered function $\bar{\beta}$ above, in case β is a rational function. Similarly one can define $\tilde{\beta}$ for rational transductions with finite preimages. It remains open whether the transformations $\bar{\beta}$ and $\tilde{\beta}$ in this general case preserve rationality.

5 Formal Power Series over Rational Monoids

In this section we turn to a particular class of monoids – the so-called rational monoids. For this class Sakarovitch showed that the rational and recognizable subsets coincide. In our main theorem we will extend this result to series.

Definitions. We start by giving the necessary definitions for our results. For more details see [16,14] and also [8,12].

Let M be a monoid. A *generating system* of M is a pair (X, α) where X is a set and $\alpha : X^* \to M$ is a surjective homomorphism. The *kernel* of α, i.e. the relation $\ker(\alpha) = \{(v, w) \in X^* \times X^* \mid \alpha(v) = \alpha(w)\}$, is a congruence relation on X^*. An idempotent function $\beta : X^* \to X^*$ is a *description* of M for (X, α) if $\ker(\beta) = \ker(\alpha)$. A monoid is called *rational* if it has a description which is a rational function.

We can think of $\beta(v)$ as a normal form of the word v. We define a new operation $w_1 \circ w_2 = \beta(w_1 w_2)$ on $\beta(X^*)$. We then have $X^*/\ker(\beta) \cong M \cong (\beta(X^*), \circ, \beta(\varepsilon))$. Note, that $\beta(X^*) \subseteq X^*$ is rational, since a rational function transforms every rational language again into a rational language.

Theorem 5.1 ([16]). *Let M be a rational monoid and let $L \subseteq M$. Then Kleene's Theorem holds, i.e. L is rational iff it is recognizable.*

Let us fix for the rest of this section a rational monoid M, a generating system (X, α) and a rational description β. Furthermore, we make the following assumption:

$$\alpha^{-1}(m) \text{ is finite for all } m \in M \tag{5.1}$$

This is equivalent to assume that every congruence class of an element in X^* induced by β is finite. Note that this implies that 1_M has no proper factorisation, i.e. $m_1, m_2 \neq 1_M \Rightarrow m_1 m_2 \neq 1_M$. Hence we may also assume that $\alpha^{-1}(1_M) = \{\varepsilon\}$. Note that finite monoids are rational but do not fulfil our assumption. Schützenberger's Theorem holds in this case anyway.

Example 5.2. Consider the monoid $\{a, b, c\}^*$ and the congruence C induced by the equation $ab = bc$. Then $\{a, b, c\}^*/C$ is a rational monoid which meets our assumption. The monoid $\{a, b, c\}^*/C$ is a divisibility monoid. In [12], it was shown that a divisibility monoid is rational if and only if it satisfies Kleene's Theorem if and only if it is width-bounded. All width-bounded divisibility monoids meet our assumption (5.1).

Example 5.3. Consider the monoid $\{a, b\}^*$ and the congruence C induced by the equation $aab = bba$. Again this is a rational monoid which meets our assumption (5.1). In [16] this monoid was named Fibonacci monoid.

A Kleene-Schützenberger Theorem. As promised we will now prove that recognizable and rational series over rational monoids coincide under certain conditions. In a first step we show that over rational monoids, recognizable series are closed under the rational operations. Let $S \in K\langle M \rangle$ be a polynomial. By Theorem 5.1 we have that $\{m\}$ is recognizable for all $m \in M$. Thus $\mathbb{1}_{\{m\}} \in K^{\mathrm{rec}}\langle\langle M \rangle\rangle$. Since $K^{\mathrm{rec}}\langle\langle M \rangle\rangle$ is closed under pointwise sum and scalar product, we conclude that $S = \sum_{m \in \mathrm{supp}S}(S, m) \cdot \mathbb{1}_{\{m\}}$, is recognizable. Hence we have:

Lemma 5.4. *Let K be a semiring and let M be a rational monoid. Then $K\langle M \rangle \subseteq K^{\mathrm{rec}}\langle\langle M \rangle\rangle$.*

The next result deals with the Cauchy product.

Proposition 5.5. *Let K be a principal ideal domain or let K be a locally finite semifield. If $S_1, S_2 \in K^{rec}\langle\!\langle M \rangle\!\rangle$ then $S_1 \cdot S_2 \in K^{rec}\langle\!\langle M \rangle\!\rangle$.*

Proof. Since α is a homomorphism, the functions $T_i = (\alpha^{-1}S_i) \odot \mathbb{1}_{\beta(X^*)}$ ($i = 1, 2$) define recognizable formal power series on X^* (Proposition 2.2). Now, let $T = T_1 \cdot T_2$ be the product of these two series. In other words

$$T : X^* \to K; \quad w \mapsto \sum_{\substack{w_1 w_2 = w \\ w_i \in \beta(X^*)}} (T_1, w_1) \cdot (T_2, w_2).$$

Then, applying the Kleene-Schützenberger Theorem for free monoids, we get $T \in K^{rec}\langle\!\langle X^* \rangle\!\rangle$. Proposition 4.10 gives $\beta^{-1}(\bar\beta(T)) \in K^{rec}\langle\!\langle X^* \rangle\!\rangle$. Now, we compute:

$$\left(\beta^{-1}(\bar\beta(T)), w\right) = \left(\bar\beta(T), \beta(w)\right) = \sum_{\beta(u)=w} (T, u) = \sum_{\substack{\beta(u_1 u_2)=w \\ u_i \in \beta(X^*)}} (T_1, u_1) \cdot (T_2, u_2)$$

$$= \sum_{u_1 \circ u_2 = w} (T_1, u_1) \cdot (T_2, u_2) = \left(S_1 \cdot S_2, \alpha(w)\right)$$

This proves that $\beta^{-1}(\bar\beta(T)) = \alpha^{-1}(S_1 \cdot S_2)$. Using Corollary 3.3 we conclude that $S_1 \cdot S_2$ is recognizable. $\square$

Proposition 5.6. *Let K be a principal ideal domain or let K be a locally finite semifield. Let $S \in K^{rec}\langle\!\langle M \rangle\!\rangle$. Then $S^* \in K^{rec}\langle\!\langle M \rangle\!\rangle$.*

Proof. The proof can be done analogously to the proof of Proposition 5.5. Consider $T = \alpha^{-1}(S) \odot \mathbb{1}_{\beta(X^*)}$ which is recognizable. As in the proof of Proposition 5.5 we conclude that $\beta^{-1}(\bar\beta(T^*))$ is recognizable. Since $\beta^{-1}(\bar\beta(T^*)) = \alpha^{-1}(S^*)$ we get from Corollary 3.3 that S^* is recognizable. $\square$

Combining Lemma 5.4 and Propositions 5.5, 5.6 we obtain:

Theorem 5.7. *Let K be a principal ideal domain or let K be a locally finite semifield. Let M be a rational monoid satisfying assumption (5.1). Then $K^{rat}\langle\!\langle M \rangle\!\rangle \subseteq K^{rec}\langle\!\langle M \rangle\!\rangle$.*

We now prove the opposite direction, that is that recognizable series over rational monoids are rational. Recall, that by our assumption (5.1) we have $\alpha^{-1}(m)$ is finite for all $m \in M$. Since $\ker(\alpha) = \ker(\beta)$, all elements in $\alpha^{-1}(m)$ are mapped to the same element under β. In the proof of the next result we will denote this element by $\mathsf{nf}(m)$.

Theorem 5.8. *Let K be a semiring and let M be a rational monoid satisfying assumption (5.1). Then $K^{rec}\langle\!\langle M \rangle\!\rangle \subseteq K^{rat}\langle\!\langle M \rangle\!\rangle$.*

Proof. Let $S \in K^{\text{rec}} \langle\!\langle M \rangle\!\rangle$. Again, we set $T = (\alpha^{-1} S) \odot \mathbb{1}_{\beta(X^*)} \in K^{\text{rec}} \langle\!\langle X^* \rangle\!\rangle$. Hence $T \in K^{\text{rat}} \langle\!\langle X^* \rangle\!\rangle$ by Schützenberger's Theorem. Using Proposition 2.2 we conclude $\bar{\alpha}(T) \in K^{\text{rat}} \langle\!\langle M \rangle\!\rangle$ and compute for $m \in M$:

$$(\bar{\alpha}(T), m) = \sum_{\substack{w \in X^* \\ \alpha(w) = m}} (T, w) = \sum_{w \in \beta(X^*), \alpha(w) = m} ((\alpha^{-1} S), w)$$

$$= \sum_{w \in \beta(X^*), \alpha(w) = m} ((S, \alpha(w)) = \sum_{\mathsf{nf}(m)} \big(S, \alpha(\mathsf{nf}(m))\big) = \big(S, m\big)$$

Thus S is rational as required.   □

Putting Theorem 5.7 and Theorem 5.8 together we obtain our main result.

Theorem 5.9. *Let K be a principal ideal domain or let K be a locally finite semifield. Let M be a rational monoid satisfying assumption* (5.1). *Then we have*

$$K^{rec} \langle\!\langle M \rangle\!\rangle = K^{rat} \langle\!\langle M \rangle\!\rangle.$$

Conclusion. We showed a Kleene-Schützenberger result for formal power series over rational monoids. The part showing that all recognizable series are rational is valid for all semirings. The other directions is based on Corollary 3.3 for which we needed the restrictions that the underlying semiring is either a principal ideal domain or a locally finite semifield.

It remains open to relax this restriction further and show whether this result also holds for a bigger class of semirings. The authors conjecture that it does not hold for all semirings in general. However, unfortunately we do not have a counterexample which remains as another open question.

References

1. Berstel, J.: Transductions and Context-Free Languages. B. G. Teubner, Stuttgart (1979)
2. Berstel, J., Reutenauer, C.: Rational Series and Their Languages. EATCS Monographs on Theoretical Computer Science, vol. 12. Springer, Heidelberg (1988)
3. Borchardt, B.: The Myhill-Nerode theorem for recognizable tree series. In: Ésik, Z., Fülöp, Z. (eds.) DLT 2003. LNCS, vol. 2710, pp. 146–158. Springer, Heidelberg (2003)
4. Bozapalidis, S.: Effective construction of the syntactic algebra of a recognizable series on trees. Acta Informatica 28(4), 351–363 (1991)
5. Bozapalidis, S., Alexandrakis, A.: Représentations matricielles des séries d'arbre reconnaissables. Theoretical Informatics and Applications 23(4), 449–459 (1989)
6. Bozapalidis, S., Louscou-Bozapalidou, O.: The rank of a formal tree power series. Theoretical Computer Science 27, 211–215 (1983)
7. Droste, M., Zhang, G.: On transformations of formal power series. Information and Computation 184(2), 369–383 (2003)
8. Droste, M., Kuske, D.: Recognizable languages in divisibility monoids. Mathematical Structures in Computer Science 11(6), 743–770 (2001)

9. Eilenberg, S.: Automata, Languages, and Machines, vol. A. Academic Press, London (1974)
10. Fülöp, Z., Vogler, H.: Weighted tree automata and tree series transducers. In: Droste, M., Kuich, W., Vogler, H. (eds.) Handbook of Weighted Automata. EATCS Monographs on Theoretical Computer Science, ch. 9. Springer, Heidelberg (2009)
11. Kuich, W., Salomaa, A.: Semirings, Automata, Languages. EATCS Monographs on Theoretical Computer Science, vol. 5. Springer, Heidelberg (1986)
12. Kuske, D.: Contributions to a Trace Theory beyond Mazurkiewicz Traces. PhD Thesis, Technische Universität Dresden (2000)
13. Lang, S.: Algebra. Springer, Heidelberg (2002)
14. Pelletier, M., Sakarovitch, J.: Easy multiplications II. Extensions of rational semigroups. Information and Computation 88(1), 18–59 (1990)
15. Reutenauer, C.: Séries formelles et algèbres syntactiques. Journal of Algebra 66, 448–483 (1980)
16. Sakarovitch, J.: Easy multiplications I. The realm of Kleene's theorem. Information and Computation 74(3), 173–197 (1987)
17. Sakarovitch, J.: Elements of Automata Theory. Cambridge University Press, Cambridge (2009)
18. Salomaa, A., Soittola, M.: Automata-Theoretic Aspects of Formal Power Series. Texts and Monographs in Computer Science. Springer, Heidelberg (1978)
19. Schützenberger, M.: On the definition of a family of automata. Information and Control 4, 245–270 (1961)

Equational Weighted Tree Transformations with Discounting[*]

Zoltán Fülöp[1] and George Rahonis[2]

[1] Department of Computer Science, University of Szeged,
Árpád tér 2., H-6720 Szeged, Hungary
`fulop@inf.u-szeged.hu`
[2] Department of Mathematics, Aristotle University of Thessaloniki,
54124 Thessaloniki, Greece
`grahonis@math.auth.gr`

Dedicated to Symeon Bozapalidis on the occasion of his retirement.

Abstract. We consider systems of equations of polynomial weighted tree transformations over the max-plus (or: arctic) semiring $\mathbb{R}_{max} = (\mathbb{R}_+ \cup \{-\infty\}, \max, +, -\infty, 0)$. We apply discounting with a parameter $0 \le d < 1$ in order to guarantee the existence of the least solution, called least d-solution, of such systems. We compute least d-solutions under u-substitution mode, where $u = [IO]$ or $u = OI$. We define a weighted relation over $\mathbb{R}_{max}$ to be u-d-equational, if it is a component of the least u-d-solution of such a system of equations in a pair of algebras. We mainly focus on u-d-equational weighted tree transformations which are equational relations obtained by considering the least u-d-solutions in pairs of term algebras. We also introduce u-d-equational weighted tree languages over $\mathbb{R}_{max}$. We characterize u-d-equational weighted tree transformations in terms of weighted tree transformations defined by weighted d-bimorphisms, which are bimorphisms from d-recognizable weighted tree languages. Finally, we prove that a weighted relation is u-d-equational if and only if it is, roughly speaking, the morphic image of a weighted u-d-equational tree transformation.

Keywords: equational semantics, recognizable weighted tree languages with discounting, [IO]- and OI-equational weighted tree transformations with discounting, weighted bimorphisms with discounting.

1 Introduction

Weighted tree transformations are defined as the semantics of machines called weighted tree transducers [33,21,25]. The essence of weighing is that weights are associated to pairs of trees of a tree transformation. The weights are taken from a semiring [29,17] or other appropriate algebraic structure. This kind of weighting

---

[*] This research was financially supported by the TÁMOP-4.2.1/B-09/1/KONV-2010-0005 program of the Hungarian National Development Agency.

W. Kuich and G. Rahonis (Eds.): Bozapalidis Festschrift, LNCS 7020, pp. 112–145, 2011.

makes it possible to investigate tree transformations not only from a qualitative but also from a quantitative point of view. For instance, we can compute the probability if that an output tree is a translation of a given input tree, cf. [31]. There are two kinds of semantics, the initial algebra one [33,21,35,36], which is suitable to make precise mathematical reasoning, and the rewriting one [26,24], which has importance in practical applications like natural language processing [31,30,38]. The classical theory of tree transducers [19,27,28], which we call the unweighted case sometimes, is reobtained as the particular "weighted theory" in which the weighting structure is the Boolean semiring.

Recently an equational definition of (unweighted) tree transformations was given in [6] and of weighted tree transformations over continuous and commutative semirings in [7]. The authors introduced the concept of a system of equations of weighted tree transformations with variables. Since the weighting semiring is continuous, the space of the potential solutions of the system becomes a complete poset and the system can be realized as a continuous mapping over that space. Hence, the classical fixpoint theorem assures that the least solution of the system exists [42]. A weighted tree transformation is then defined to be equational if it is a component of the least fixpoint of a system of equations of weighted tree transformations. We note that the above approach is the generalization of the equational definition of recognizable tree languages given in [27,28] and that the idea comes from [40], in which the equational definition of a recognizable subset of an arbitrary algebra was given. Based on the same idea, recognizable weighted tree languages were defined as least solutions of systems of equations of weighted tree languages with variables over continuous and commutative semirings, cf. [5,22,32]. We also note that several other papers in the literature deal with different interpretations of the equational approach of [40], see [20,12,13] for instance.

There are semirings, for instance the max-plus semiring $\mathbb{R}_{\max} = (\mathbb{R}_+ \cup \{-\infty\}, \max, +, -\infty, 0)$, in which the supremum of an infinite set may not exist. Hence such a semiring is not continuous and we cannot apply the fixpoint theorem [42] to provide the least solution of a system of equations of weighted tree transformations over $\mathbb{R}_{\max}$. However, we can guarantee the existence of the least solution by applying discounting, i.e., an appropriate multiplication with a discounting parameter $0 \leq d < 1$. In fact, discounting is a common strategy to face problems arising on systems with non-terminating behavior. Among others, it is used in economic mathematics, in Markov decision processes, and in game theory (cf. [14,23,41]). The method was adapted for weighted automata over infinite words by Droste and Kuske in [18]. In [9,10] further properties of weighted automata with discounting over infinite words were investigated. A weighted MSO-logic with discounting has been introduced in [15] and a Büchi-Elgot-type characterization of infinitary recognizable series with discounting has been established. Kuich [34] proved Kleene theorems for weighted automata with discounting acting on finite and infinite words over Conway semirings. In [16] the authors investigated weighted automata with discounting over semirings and finitely generated graded monoids. Recently, in [39] weighted tree automata with discounting over commutative semirings were considered and a Kleene- and a Büchi-Elgot-type characterization was obtained for this kind of automata.

In this paper we consider systems of equations of weighted tree transformations over the semiring $\mathbb{R}_{\max}$. We also consider a discounting parameter $0 \leq d < 1$ and generalize *[IO]*- and *OI*-substitution of weighted tree languages (cf. [8] and [5,32], resp.) to *[IO]-d-* and *OI-d-*substitution of weighted relations over the direct product of two algebras in weighted tree transformations with variables. Then, for $u = [IO], OI$ we introduce u-d-equational weighted relations and, in particular, u-d-equational weighted tree transformations in the following way. We consider systems of equations of weighted tree transformations with variables. Such a system (E) consists of $n \geq 1$ equations of the form $x_i = \rho_i$, where $\rho_i \in \mathbb{R}_{\max}\langle T_\Sigma(X_n) \times T_\Delta(X_n)\rangle$ is a weighted tree transformation of finite support (i.e., a polynomial) over the ranked alphabets Σ, Δ; the variable set $X_n = \{x_1, \ldots, x_n\}$; and the commutative semiring $\mathbb{R}_{\max}$ for every $1 \leq i \leq n$. For any algebras $\mathcal{A} = (A, \Sigma)$ and $\mathcal{B} = (B, \Delta)$ and discounting parameter $0 \leq d < 1$, a u-d-solution of (E) in $(\mathcal{A}, \mathcal{B}, \mathbb{R}_{\max})$ is a tuple $(\theta_1, \ldots, \theta_n)$ of weighted relations over A, B, and $\mathbb{R}_{\max}$, i.e., an element of the poset $\mathbb{R}_{\max}\langle\langle A \times B\rangle\rangle^n$, such that

$$\theta_i = \rho_i \left[\theta_1, \ldots, \theta_n\right]_u^d,$$

for all $1 \leq i \leq n$, where the expression on the right-hand side means the u-d-substitution of $(\theta_1, \ldots, \theta_n)$ with d-discounting in ρ_i. At the same time, the system (E) induces a sequence $(\theta_{1,k}, \ldots, \theta_{n,k})_{k\geq0}$, called the u-d-approximation sequence of (E), in the space $\mathbb{R}_{\max}\langle A \times B\rangle^n$ of polynomials over $\mathbb{R}_{\max}$ which is defined as follows. For every $1 \leq i \leq n$, we define $(\theta_{i,k})_{k\geq0}$ such that

$$\theta_{i,0} = \widetilde{-\infty}, \text{ and } \theta_{i,k+1} = \rho_i \left[\theta_{1,k}, \ldots, \theta_{n,k}\right]_u^d, \text{ for } k \geq 0.$$

We show that the u-d-approximation sequence is bounded and its limit is the least u-d-solution of (E) (Theorem 1). A weighted relation in $\mathbb{R}_{\max}\langle\langle A \times B\rangle\rangle$ is u-d-equational if it appears as a component of the least u-d-solution of a system (E) of equations of weighted tree transformations in $(\mathcal{A}, \mathcal{B}, \mathbb{R}_{\max})$.

A weighted tree transformation $\tau \in \mathbb{R}_{\max}\langle\langle T_\Sigma \times T_\Delta\rangle\rangle$ is u-d-equational, if it is a component of the least u-d-solution of a system (E) of equations of weighted tree transformations in $(T_\Sigma, T_\Delta, \mathbb{R}_{\max})$, where T_Σ and T_Δ are the corresponding term algebras over Σ and Δ, respectively. In our paper, we focus on u-d-equational weighted tree transformations. We give a sufficient condition for the existence and uniqueness of the u-d-solution of a system of equations of weighted tree transformations in the corresponding term algebras (Theorem 2). We characterize u-d-equational weighted tree transformations in terms of weighted d-bimorphisms of [25,37]. In fact, we show that the class of *[IO]*-equational (resp. *OI*-equational) weighted tree transformations coincides with the class of weighted tree transformations defined by ultimately complete d-bimorphisms (resp. ultimately complete linear d-bimorphisms) (Theorem 4). Finally, we establish a Mezei-Wright like relationship [40] between u-d-equational weighted tree transformations and u-d-equational weighted relations. Namely we show that a weighted relation is u-d-equational if and only if it is, roughly speaking, the morphic image of a u-d-equational weighted tree transformation (Theorem 6). We note that the corresponding results, for a continuous semiring and without discounting, were obtained in [7].

The paper is organized as follows. In Section 2, we introduce the necessary notions and notation. In Section 3, we define [IO]- and OI-substitution of weighted relations with d-discounting in weighted tree transformations with variables, and in Section 4, we prove some technical results concerning series and substitutions in weighted tree transformations with d-discounting. In Section 5, we define the concept of a system of equations of weighted tree transformations and of [IO]-d- and OI-d-equational weighted relations and tree transformations. In Section 6, we give the characterization of [IO]-d-equational and of OI-d-equational weighted tree transformations in terms of weighted d-bimorphisms. In Section 7, we prove the Mezei Wright like characterization of equational weighted relations.

2 Preliminaries

2.1 General Notation

We denote by $\mathbb{N}$ the set of nonnegative integers and by $\mathbb{R}_+$ the set of nonnegative real numbers. The usual multiplication in $\mathbb{R}_+$ will be denoted by concatenation.

Let V be a set. We set $id(V) = \{(a,a) \mid a \in V\}$. For every $n \geq 1$ and $1 \leq i \leq n$, we denote the ith component of a vector $\mathbf{a} \in V^n$ by a_i, hence $\mathbf{a} = (a_1, \ldots, a_n)$. For $n = 0$, we define $V^n = \{(\ )\}$ (even if $V = \emptyset$), where $(\ )$ is the empty vector. Let $1 \leq i_1 < \ldots < i_k \leq n$, and $a_1, \ldots, a_k \in V$. We introduce a notation for the set of those elements of V^n which have a_j as their i_jth component, $j = 1, \ldots, k$. Namely, we set

$$V^n|_{(i_1,a_1)\ldots(i_k,a_k)} = \{(b_1, \ldots, b_n) \in V^n \mid b_{i_1} = a_1, \ldots, b_{i_k} = a_k\}.$$

A *partially ordered set* (for short: poset) is a pair $(V, \leq)$, where V is a set and $\leq$ is a *partial order*, i.e., a reflexive, antisymmetric, and transitive relation on V. We will write just V for $(V, \leq)$. Let $(a_i)_{i \in I}$ be a family of elements of V. If its least upper bound exists in V, then we denote it by $\sup_{i \in I} a_i$. An ω-chain in V is a family $(a_n)_{n \geq 0}$ such that $a_0 \leq a_1 \leq \ldots$.

2.2 Semirings and Σ-Algebras

A *semiring* $(S, +, \cdot, 0, 1)$ is an algebraic structure such that $(S, +, 0)$ is a commutative monoid, $(S, \cdot, 1)$ is a monoid, $0 \neq 1$, the multiplication $\cdot$ distributes over addition $+$ from both sides, and $0 \cdot k = k \cdot 0 = 0$ for every $k \in S$. If no confusion arises, then we denote the semiring simply by S. Then S is called *commutative* if the monoid $(S, \cdot, 1)$ is commutative.

In this paper we work with the semiring $\mathbb{R}_{\max} = (\mathbb{R}_+ \cup \{-\infty\}, \max, +, -\infty, 0)$, which is called *max-plus semiring* or *arctic semiring*. Note that $\mathbb{R}_{\max}$ is commutative. In the following we often identify $\mathbb{R}_{\max}$ with its carrier set $\mathbb{R}_+ \cup \{-\infty\}$. We extend max for finite subsets of $\mathbb{R}_{\max}$ and we consider the supremum sup of arbitrary subsets of $\mathbb{R}_{\max}$ provided it exists, with the understanding that $\max \emptyset = \sup \emptyset = -\infty$.

A *ranked alphabet* is a pair (Σ, rk) (simply denoted by Σ) where Σ is a finite set and $rk : \Sigma \to \mathbb{N}$ is the rank function. As usual, we set $\Sigma_k = \{\sigma \in \Sigma \mid rk(\sigma) = k\}$ for every $k \geq 0$.

A *Σ-algebra* is a pair $\mathcal{A} = (A, \Sigma^{\mathcal{A}})$ where A is a nonempty set, called the domain set of $\mathcal{A}$, and $\Sigma^{\mathcal{A}}$ is a family $(\sigma^{\mathcal{A}} \mid \sigma \in \Sigma)$ of operations on A such that for every $k \geq 0$ and $\sigma \in \Sigma_k$, we have $\sigma^{\mathcal{A}} : A^k \to A$. If no confusion arises, then sometimes we drop $\mathcal{A}$ from $\Sigma^{\mathcal{A}}$ and $\sigma^{\mathcal{A}}$ in what follows. Given a further Σ-algebra $\mathcal{B} = (B, \Sigma)$, a *$\Sigma$-algebra morphism from $\mathcal{A}$ to $\mathcal{B}$* is a mapping $H : A \to B$ such that $H\left(\sigma^{\mathcal{A}}(a_1, \ldots, a_k)\right) = \sigma^{\mathcal{B}}\left(H(a_1), \ldots, H(a_k)\right)$ for $\sigma \in \Sigma_k$, $k \geq 0$, and $a_1, \ldots, a_k \in A$. In particular, $H\left(\sigma^{\mathcal{A}}\right) = \sigma^{\mathcal{B}}$ for every $\sigma \in \Sigma_0$.

2.3   Series and Weighted Relations

Let A be a set. A *series over A and $\mathbb{R}_{\max}$* (or $(A, \mathbb{R}_{\max})$) is a mapping $\eta : A \to \mathbb{R}_{\max}$. For every $a \in A$, we write (η, a) for the value $\eta(a)$ and refer to it as the *coefficient* (or *weight*) of a in η. The *support* of η is the set $\mathrm{supp}(\eta) = \{a \in A \mid (\eta, a) \neq -\infty\}$. If $\mathrm{supp}(\eta)$ is finite, then η is called a *polynomial over* $(A, \mathbb{R}_{\max})$. We write a polynomial η as the formal maximum $\max\{k_1.a_1, \ldots, k_n.a_n\}$, where $\mathrm{supp}(\eta) = \{a_1, \ldots, a_n\}$ and $k_i = (\eta, a_i)$ for every $1 \leq i \leq n$. In case $n = 1$ we write just $\eta = k_1.a_1$. We denote by $\mathbb{R}_{\max}\langle\!\langle A \rangle\!\rangle$ and $\mathbb{R}_{\max}\langle A \rangle$ the class of all series and of all polynomials over $(A, \mathbb{R}_{\max})$, respectively. A series $\eta \in \mathbb{R}_{\max}\langle\!\langle A \rangle\!\rangle$ is *bounded* if there is a $K \in \mathbb{R}_{\max}$ such that $(\eta, a) \leq K$ for every $a \in A$. In this case we also write $\eta \leq K$. The class of all bounded series is denoted by $\mathbb{R}_{\max}^b\langle\!\langle A \rangle\!\rangle$. Obviously, we have $\mathbb{R}_{\max}\langle A \rangle \subseteq \mathbb{R}_{\max}^b\langle\!\langle A \rangle\!\rangle$. For every $k \in \mathbb{R}_{\max}$, we denote by $\widetilde{k}$ the *constant series* defined by $(\widetilde{k}, a) = k$ for every $a \in A$.

Let $\eta, \eta_1, \eta_2 \in \mathbb{R}_{\max}\langle\!\langle A \rangle\!\rangle$ and $k \in \mathbb{R}_{\max}$. The *maximum* $\max(\eta_1, \eta_2)$ *of η_1 and η_2* and the *scalar sum $k + \eta$ of η with k* are the series in $\mathbb{R}_{\max}\langle\!\langle A \rangle\!\rangle$ which are defined such that, for every $a \in A$, we have $(\max(\eta_1, \eta_2), a) = \max\{(\eta_1, a), (\eta_2, a)\}$ and $(k + \eta, a) = k + (\eta, a)$.

Next we make $\mathbb{R}_{\max}\langle\!\langle A \rangle\!\rangle$ a poset by equipping it with the partial order defined in the obvious way: for $\eta_1, \eta_2 \in \mathbb{R}_{\max}\langle\!\langle A \rangle\!\rangle$ we let $\eta_1 \leq \eta_2$ iff $(\eta_1, a) \leq (\eta_2, a)$ for every $a \in A$. A family $(\eta_j)_{j \in I}$ of elements of $\mathbb{R}_{\max}\langle\!\langle A \rangle\!\rangle$ is *bounded* if there is a $K \in \mathbb{R}_{\max}$ such that $\eta_j \leq K$ for every $j \in I$. We remark that a family $(\eta_j)_{j \in I}$ of bounded series in $\mathbb{R}_{\max}^b\langle\!\langle A \rangle\!\rangle$ is not necessarily bounded (because the family of upper bounds $(K_j)_{j \in I}$ may not be bounded in $\mathbb{R}_{\max}$). However, if $(\eta_j)_{j \in I}$ is bounded, then $\sup_{j \in I} \eta_j$ exists in the poset $\mathbb{R}_{\max}\langle\!\langle A \rangle\!\rangle$ and $(\sup_{j \in I} \eta_j, a) = \sup_{j \in I}(\eta_j, a)$ for all $a \in A$. In the particular case that $(\eta_j)_{j \geq 0}$ is a bounded ω-chain, we write $\lim_{j \to \infty} \eta_j$ for $\sup_{j \geq 0} \eta_j$. We can extend the partial order on $\mathbb{R}_{\max}\langle\!\langle A \rangle\!\rangle$ to $\mathbb{R}_{\max}\langle\!\langle A \rangle\!\rangle^n$ componentwise for every $n \geq 1$. We call a family $(\eta_{1,j}, \ldots, \eta_{n,j})_{j \in I}$ in $\mathbb{R}_{\max}\langle\!\langle A \rangle\!\rangle^n$ bounded if $(\eta_{i,j})_{j \in I}$ is bounded for every $1 \leq i \leq n$. If $(\eta_{1,j}, \ldots, \eta_{n,j})_{j \in I}$ is bounded, then $\sup_{j \in I}(\eta_{1,j}, \ldots, \eta_{n,j}) = (\sup_{j \in I} \eta_{1,j}, \ldots, \sup_{j \in I} \eta_{n,j})$.

Finally, let B be a further set. A *weighted relation over* A, B, *and* $\mathbb{R}_{\max}$ (or $(A, B, \mathbb{R}_{\max})$) is a series over $(A \times B, \mathbb{R}_{\max})$.

2.4   Weighted Tree Languages and Tree Transformations

In the rest of the paper Σ, Δ, and Γ will denote ranked alphabets which contain at least one nullary symbol.

Let V be a finite set with $V \cap \Sigma = \emptyset$. The set $T_\Sigma(V)$ of finite trees over Σ and V is defined by induction to be the smallest set T such that (i) $\Sigma_0 \cup V \subseteq T$ and (ii) $\sigma(s_1, \ldots, s_k) \in T$ for every $k \geq 1$, $\sigma \in \Sigma_k$, and $s_1, \ldots, s_k \in T$. We write T_Σ for $T_\Sigma(\emptyset)$. Note that $T_\Sigma \neq \emptyset$ since $\Sigma_0 \neq \emptyset$. For every $V' \subseteq V$, we define $\Sigma(V') = \{\sigma(v_1, \ldots, v_k) \mid k \geq 0, \sigma \in \Sigma_k, \text{ and } v_1, \ldots, v_k \in V'\}$.

A particular Σ-algebra is the *term algebra* $\mathcal{T}_\Sigma(V) = (T_\Sigma(V), \Sigma)$ of all trees over Σ and V, where $\sigma^{\mathcal{T}_\Sigma(V)}(s_1, \ldots, s_k) = \sigma(s_1, \ldots, s_k)$ for every $k \geq 0$, $\sigma \in \Sigma_k$, and $s_1, \ldots, s_k \in T_\Sigma(V)$. In fact, it is the *free Σ-algebra generated by V* in the class of all Σ-algebras, i.e., for every Σ-algebra $\mathcal{A}$, any mapping $h : V \to A$ extends uniquely to a Σ-algebra morphism $H : T_\Sigma(V) \to A$. If $V = \emptyset$, then we denote the unique morphism from $\mathcal{T}_\Sigma$ to $\mathcal{A}$ by $H_\mathcal{A}$ and we abbreviate $H_\mathcal{A}(s)$ by $s_\mathcal{A}$ for $s \in T_\Sigma$.

Any subset of $T_\Sigma(V)$ is called a *tree language* and any relation of the form $S \subseteq T_\Sigma(V) \times T_\Delta(V)$ is called a *tree transformation*.

Let $X = \{x_1, x_2, \ldots\}$ be a countably infinite set of *variables*, which is disjoint from any ranked alphabet considered in the paper. We set $X_n = \{x_1, \ldots, x_n\}$ for $n \geq 0$, hence $X_0 = \emptyset$. Let $s \in T_\Sigma(X_n)$ be a tree. The *height* $ht(s) \in \mathbb{N}$ *of s* and the *set* $var(s) \subseteq X_n$ of variables in s is defined by induction as follows. If $s \in \Sigma_0 \cup X_n$, then $ht(s) = 0$ and $var(s) = \{s\}$ for $s \in X_n$ and $var(s) = \emptyset$ for $s \in \Sigma_0$. If $s = \sigma(s_1, \ldots, s_k)$ for some $k \geq 1$, $\sigma \in \Sigma_k$, and $s_1, \ldots, s_k \in T_\Sigma(X_n)$, then $ht(s) = 1 + \max\{ht(s_i) \mid 1 \leq i \leq k\}$, and $var(s) = \bigcup_{i=1}^{k} var(s_i)$. We denote by $|s|_{x_i}$ the number of occurrences of x_i in s for all $1 \leq i \leq n$. Then s is called $(X_n\text{-})linear$ (resp. $(X_n\text{-})nondeleting$) if $|s|_{x_i} \leq 1$ (resp. $|s|_{x_i} \geq 1$) for every $1 \leq i \leq n$. A subset $L \subseteq T_\Sigma(X_n)$ is *linear* (resp. *nondeleting*), if each $s \in L$ is linear (resp. nondeleting). A pair $(s, t) \in T_\Sigma(X_n) \times T_\Delta(X_n)$ is *linear* (resp. *nondeleting*) if both s and t are linear (resp. nondeleting). Furthermore, it is called *variable symmetric* (resp. *variable identical*) if $|s|_{x_i} = |t|_{x_i}$ for all $1 \leq i \leq n$ (resp. $var(s) = var(t)$). Of course, if (s, t) is variable symmetric, then it is variable identical. We lift these concepts to an arbitrary tree transformation $R \subseteq T_\Sigma(X_n) \times T_\Delta(X_n)$ in the obvious way.

Let now $\Xi = \{\xi_1, \xi_2, \ldots\}$ be another set of variables, which is disjoint from any ranked alphabet considered in the paper, and let $\Xi_n = \{\xi_1, \ldots, \xi_n\}$ for every $n \geq 0$. We define tree substitution. For this, let $V \subseteq X$ or $V \subseteq \Xi$, let $s, s_1, \ldots, s_n \in T_\Sigma(V)$ and $v_1, \ldots, v_n$ be pairwise different elements of V. We denote by $s(s_1/v_1, \ldots, s_n/v_n)$ the tree which we obtain by substituting simultaneously s_i for every occurrence of v_i in s for every $1 \leq i \leq n$. In particular, we abbreviate $s(s_1/x_1, \ldots, s_n/x_n)$ by $s(s_1, \ldots, s_n)$.

A *tree homomorphism from* Σ *to* Δ is a family $(h_k)_{k\geq 0}$ of mappings h_k : $\Sigma_k \to T_\Delta(\Xi_k)$. Such a tree homomorphism is called *linear* (for short *l*) (resp. *nondeleting* or *complete*, for short *c*) if for every $k \geq 1$ and $\sigma \in \Sigma_k$ the tree $h_k(\sigma)$ is Ξ_k-linear (resp. Ξ_k-nondeleting).

For every finite set V, the tree homomorphism $(h_k)_{k\geq 0}$ induces a mapping h : $T_\Sigma(V) \to T_\Delta(V)$ defined inductively in the following way. For every $s \in T_\Sigma(V)$ we let

- $h(s) = s$ if $s \in V$, and
- $h(s) = h_k(\sigma)(h(s_1)/\xi_1, \ldots, h(s_k)/\xi_k)$ if $s = \sigma(s_1, \ldots, s_k)$ with $k \geq 0$, $\sigma \in \Sigma_k$, and $s_1, \ldots, s_k \in T_\Sigma(V)$.

As usual, we also call the induced mapping h tree homomorphism. We will use the fact without reference that the class of all tree homomorphisms is closed under composition [19]. We denote by $\mathcal{H}$ the class of all tree homomorphisms and, for any combination w of l and c we denote by w-$\mathcal{H}$ the class of w-tree homomorphisms. Finally, a pair (h, h') of tree homomorphisms $h : T_\Gamma(X_n) \to T_\Sigma(X_n)$ and $h' : T_\Gamma(X_n) \to T_\Delta(X_n)$ is called *ultimately nondeleting* (or *ultimately complete*) if $var(h_k(\gamma)) \cup var(h'_k(\gamma)) = \{\xi_1, \ldots, \xi_k\}$ for every $\gamma \in \Gamma_k$. We denote by $uc(\mathcal{H}, \mathcal{H})$ the class of all ultimately complete pairs of tree homomorphisms.

Next we introduce weighted tree languages and tree transformations. A series $\varphi \in \mathbb{R}_{\max}\langle\langle T_\Sigma(X_n)\rangle\rangle$ is called a *weighted tree language over* Σ, X_n, *and* $\mathbb{R}_{\max}$ (or $(\Sigma, X_n, \mathbb{R}_{\max})$). In case $n = 0$ we call φ a weighted tree language over $(\Sigma, \mathbb{R}_{\max})$. We say that φ is *linear* if $\mathrm{supp}(\varphi)$ is linear.

A *weighted relation* $\tau \in \mathbb{R}_{\max}\langle\langle T_\Sigma(X_n) \times T_\Delta(X_n)\rangle\rangle$ is called a *weighted tree transformation over* Σ, Δ, X_n, *and* $\mathbb{R}_{\max}$ (or $(\Sigma, \Delta, X_n, \mathbb{R}_{\max})$). In case $n = 0$ we call τ a weighted tree transformation over $(\Sigma, \Delta, \mathbb{R}_{\max})$. We say that τ is linear (resp. variable symmetric, variable identical) if $\mathrm{supp}(\tau)$ is linear (resp. variable symmetric, variable identical).

Finally, we recall weighted tree automata [1,25] and recognizable weighted tree languages with d-discounting, where $0 \leq d < 1$ is a *discounting parameter* [18,39]. A *weighted tree automaton* (wta for short) *over* Γ *and* $\mathbb{R}_{\max}$ is a triple $\mathcal{M} = (Q, \mu, \nu)$ where Q is a finite set of *states*, $\mu = (\mu_k)_{k\geq 0}$ is a family of *transition mappings* $\mu_k : Q^k \times \Gamma_k \times Q \to \mathbb{R}_{\max}$, and $\nu : Q \to \mathbb{R}_{\max}$ is the *root weight mapping*.

We define the mapping $h^d_\mu : T_\Gamma \to \mathbb{R}^Q_{\max}$ by induction as follows. For every $q \in Q$, we let

(i) $h^d_\mu(\sigma)_q = \mu_0(\varepsilon, \sigma, q)$ for every $\sigma \in \Gamma_0$, and
(ii) $h^d_\mu(\sigma(s_1, \ldots, s_k))_q = \max_{q_1, \ldots, q_k \in Q} \big(d/k(h^d_\mu(s_1)_{q_1} + \ldots + h^d_\mu(s_k)_{q_k}) +$
$$\mu_k((q_1, \ldots, q_k), \sigma, q))$$

for every $k \geq 1$, $\sigma \in \Gamma_k$, and $s_1, \ldots, s_k \in T_\Gamma$.

The *weighted tree language* $\|\mathcal{M}\|^d \in \mathbb{R}_{\max}\langle\langle T_\Gamma\rangle\rangle$ recognized by $\mathcal{M}$ with d-discounting is defined for every $s \in T_\Gamma$ by

$$\left(\|\mathcal{M}\|^d, s\right) = \max_{q \in Q}\left(h^d_\mu(s)_q + \nu(q)\right).$$

It is easy to see that $\|\mathcal{M}\|^d$ is bounded. In fact, we can show by induction on s that $h_\mu^d(s)_q \leq M\left(\sum_{i=0}^{ht(s)} d^i\right)$ for every $s \in T_\Gamma$ and $q \in Q$, where $M = \max\{\mu_k((q_1, \ldots, q_k), \sigma, q) \mid k \geq 0, \sigma \in \Gamma_k, \text{ and } q, q_1, \ldots, q_k \in Q\}$. Hence $h_\mu^d(s)_q \leq M/(1-d)$ and thus $\|\mathcal{M}\|^d \leq M/(1-d) + N$ where $N = max\{\nu(q) \mid q \in Q\}$.

A weighted tree language $\varphi \in \mathbb{R}_{\max}^b \langle\!\langle T_\Gamma \rangle\!\rangle$ is called d-$recognizable$ if there is a wta $\mathcal{M}$ over Γ and $\mathbb{R}_{\max}$ such that $\varphi = \|\mathcal{M}\|^d$.

We call $\mathcal{M}$ $finite$-$state$ $normalized$, if there is a $q \in Q$ such that $\nu(q) = 0$ and for every $p \neq q$ we have $\nu(p) = -\infty$. In this case we write $\mathcal{M} = (Q, \mu, q)$. We recall the following result (cf. [39], Lemma 1).

Lemma 1. For every wta $\mathcal{M}$, there is a finite-state normalized wta $\mathcal{M}'$ such that $\|\mathcal{M}\|^d = \|\mathcal{M}'\|^d$ for every $0 \leq d < 1$.

3   Substitutions with Discounting in Polynomial Weighted Tree Languages

In the rest of the paper $\mathcal{A} = (A, \Sigma)$ and $\mathcal{B} = (B, \Delta)$ will denote an arbitrary Σ- and a Δ-algebra, respectively. Moreover, d a discounting parameter with $0 \leq d < 1$.

In this section we generalize IO- and OI-substitution of tree languages [20] and of weighted tree languages [8,5,32] by introducing the $[IO]$- and the OI-substitution with d-discounting of weighted relations of $\mathbb{R}_{\max}\langle\!\langle A \times B \rangle\!\rangle$ in polynomial weighted tree languages. The same concept without discounting and for a continuous semiring was introduced in [7]. We begin with some elementary concepts.

Let $h : X_n \to A$ be any mapping with $h(x_i) = a_i, 1 \leq i \leq n$. For every $s \in T_\Sigma(X_n)$, we denote $H(s)$ by $s(a_1, \ldots, a_n)_\mathcal{A}$ and call it the $evaluation$ of s at $(a_1, \ldots, a_n)$ in $\mathcal{A}$. Hence $\sigma()_\mathcal{A} = \sigma^\mathcal{A}$ for every $\sigma \in \Sigma_0$.

We will need another kind of evaluation. Let $s \in T_\Sigma(X_n)$ with $\lambda_i = |s|_{x_i}$ and $\mathbf{a}^{(i)} = \left(a_1^{(i)}, \ldots, a_{\lambda_i}^{(i)}\right) \in A^{\lambda_i}$ for every $1 \leq i \leq n$. The OI-$evaluation$ of s at $(\mathbf{a}^{(1)}, \ldots, \mathbf{a}^{(n)})$ in $\mathcal{A}$ is denoted by $s\left(\mathbf{a}^{(1)}, \ldots, \mathbf{a}^{(n)}\right)_\mathcal{A}$ and is defined as follows.

(i) If $s = x_i$, then $s\left(\mathbf{a}^{(1)}, \ldots, \mathbf{a}^{(n)}\right)_\mathcal{A} = a_1^{(i)}$. (Note that in this case $\lambda_i = 1$, hence $\mathbf{a}^{(i)} = (a_1^{(i)})$, and $\mathbf{a}^{(j)} = ()$ for $j \neq i$).

(ii) If $s = \sigma(s_1, \ldots, s_k)$ for some $k \geq 0$ and $s_1, \ldots, s_k \in T_\Sigma(X_n)$, then let $\lambda_{i,1} = |s_1|_{x_i}, \ldots, \lambda_{i,k} = |s_k|_{x_i}$ and let $\mathbf{a}^{(i,1)}, \ldots, \mathbf{a}^{(i,k)}$ be the unique decomposition of the vector $\mathbf{a}^{(i)}$ into components of dimension $\lambda_{i,1}, \ldots, \lambda_{i,k}$, respectively, for every $1 \leq i \leq n$. (Note that $\lambda_i = \lambda_{i,1} + \ldots + \lambda_{i,k}$.) Then let

$$s\left(\mathbf{a}^{(1)}, \ldots, \mathbf{a}^{(n)}\right)_\mathcal{A} = \sigma^\mathcal{A}\left(s_1\left(\mathbf{a}^{(1,1)}, \ldots, \mathbf{a}^{(n,1)}\right)_\mathcal{A}, \ldots, s_k\left(\mathbf{a}^{(1,k)}, \ldots, \mathbf{a}^{(n,k)}\right)_\mathcal{A}\right).$$

If no confusion arises, then sometimes we drop $\mathcal{A}$ from $s(a_1, \ldots, a_n)_\mathcal{A}$ and $s\left(\mathbf{a}^{(1)}, \ldots, \mathbf{a}^{(n)}\right)_\mathcal{A}$.

Now let $s \in T_\Sigma(X_n)$ and $\eta_1, \ldots, \eta_n \in \mathbb{R}^b_{\max}\langle\!\langle A \rangle\!\rangle$ be bounded series with upper bounds $K_1, \ldots, K_n \in \mathbb{R}_{\max}$, respectively. The *[IO]-substitution with d-discounting* (simply *[IO]-d-substitution*) of $\eta_1, \ldots, \eta_n$ in s is the series denoted by $s[\eta_1, \ldots, \eta_n]^d_{[IO]}$ and defined by case distinction as follows. If $s \in T_\Sigma$, then $s[\eta_1, \ldots, \eta_n]^d_{[IO]} = 0.s_{\mathcal{A}}$ and if $s = x_i$, then $s[\eta_1, \ldots, \eta_n]^d_{[IO]} = \eta_i$. Otherwise, we let

$$\left(s[\eta_1, \ldots, \eta_n]^d_{[IO]}, a \right) = d/k \left(\sup_{\substack{a_1, \ldots, a_k \in A \\ s(b_1, \ldots, b_n) = a}} \left((\eta_{i_1}, a_1) + \ldots + (\eta_{i_k}, a_k) \right) \right),$$

for every $a \in A$, where $var(s) = \{x_{i_1}, \ldots, x_{i_k}\}$, and for every $a_1, \ldots, a_k \in A$, the sequence $b_1, \ldots, b_n$ is an arbitrary element of $A^n|_{(i_1, a_1) \ldots (i_k, a_k)}$. Note that $\left(s[\eta_1, \ldots, \eta_n]^d_{[IO]}, a \right) \leq \max\{K_1, \ldots, K_n, (d/k)(K_1 + \ldots + K_n)\}$. Hence the *[IO]-d-substitution* of series is well-defined because the supremum in the right-hand side of the defining equation exists and it is independent of the choice of the sequences $b_1, \ldots, b_n$.

The *OI-substitution with d-discounting* (simply *OI-d-substitution*) of $\eta_1, \ldots, \eta_n$ in s is the series $s[\eta_1, \ldots, \eta_n]^d_{OI}$ defined as follows. If $s \in T_\Sigma$, then $s[\eta_1, \ldots, \eta_n]^d_{OI} = 0.s_{\mathcal{A}}$, and if $s = x_i$, then $s[\eta_1, \ldots, \eta_n]^d_{OI} = \eta_i$. Otherwise, we define

$$\left(s[\eta_1, \ldots, \eta_n]^d_{OI}, a \right) = d/\lambda \left(\sup_{\substack{\mathbf{a}^{(i)} \in A^{\lambda_i}, 1 \leq i \leq n \\ s(\mathbf{a}^{(1)}, \ldots, \mathbf{a}^{(n)}) = a}} \left(\left(\eta_1, \mathbf{a}^{(1)} \right) + \ldots + \left(\eta_n, \mathbf{a}^{(n)} \right) \right) \right)$$

for every $a \in A$, where $\lambda_i = |s|_{x_i}$, $\mathbf{a}^{(i)} = \left(a^{(i)}_1, \ldots, a^{(i)}_{\lambda_i} \right) \in A^{\lambda_i}$, and $(\eta_i, \mathbf{a}^{(i)}) = \left(\eta_i, a^{(i)}_1 \right) + \ldots + \left(\eta_i, a^{(i)}_{\lambda_i} \right)$ if $\lambda_i \geq 1$ and $(\eta_i, \mathbf{a}^{(i)}) = 0$ if $\mathbf{a}^{(i)} = ()$ for every $1 \leq i \leq n$. Finally, $\lambda = \sum_{1 \leq i \leq n} \lambda_i$. Clearly, we have $(s[\eta_1, \ldots, \eta_n]^d_{OI}, a) \leq \max\{K_1, \ldots, K_n, (d/\lambda)(\lambda_1 K_1 + \ldots + \lambda_n K_n)\}$, hence the *OI-d-substitution* is well-defined.

We note that both for *[IO]* and *OI*, in case $s = x_i$ we do not apply any discounting. This property of the substitution will be used strongly in the proof of Lemma 9.

Finally, for every polynomial weighted tree language $\varphi \in \mathbb{R}_{\max}\langle T_\Sigma(X_n) \rangle$ and $u=[IO]$, *OI* the *u-d-substitution* of $\eta_1, \ldots, \eta_n$ in φ is the series

$$\varphi[\eta_1, \ldots, \eta_n]^d_u = \max_{s \in \text{supp}(\varphi)} \left((\varphi, s) + s[\eta_1, \ldots, \eta_n]^d_u \right).$$

Lemma 2. If $\varphi \in \mathbb{R}_{\max}\langle T_\Sigma(X_n) \rangle$ is a linear weighted tree language, then $\varphi[\eta_1, \ldots, \eta_n]^d_{[IO]} = \varphi[\eta_1, \ldots, \eta_n]^d_{OI}$ for every $\eta_1, \ldots, \eta_n \in \mathbb{R}^b_{\max}\langle\!\langle A \rangle\!\rangle$.

Proof. It follows from the definition of the u-d-substitution in polynomial weighted tree languages and the fact that $s\,[\eta_1,\ldots,\eta_n]^d_{[IO]} = s\,[\eta_1,\ldots,\eta_n]^d_{OI}$ for every linear tree s.

Now, let $(s,t) \in T_\Sigma(X_n) \times T_\Delta(X_n)$ for some $n \geq 0$ with $\lambda_i = |s|_{x_i}$, $\mu_i = |t|_{x_i}$, and $m_i = \max\{\lambda_i,\mu_i\}$ for every $1 \leq i \leq n$. Furthermore, let $\mathbf{v}^{(i)} \in (A \times B)^{m_i}$ for every $1 \leq i \leq n$.

The *OI-evaluation* of (s,t) at $(\mathbf{v}^{(1)},\ldots,\mathbf{v}^{(n)})$ in $(\mathcal{A},\mathcal{B})$ is denoted by $(s,t)\left(\mathbf{v}^{(1)},\ldots,\mathbf{v}^{(n)}\right)_{(\mathcal{A},\mathcal{B})}$ and is defined by

$$(s,t)\left(\mathbf{v}^{(1)},\ldots,\mathbf{v}^{(n)}\right)_{(\mathcal{A},\mathcal{B})} = \left(s\left(\mathbf{a}^{(1)},\ldots,\mathbf{a}^{(n)}\right)_{\mathcal{A}},\, t\left(\mathbf{b}^{(1)},\ldots,\mathbf{b}^{(n)}\right)_{\mathcal{B}}\right),$$

where $\mathbf{v}^{(i)} = \left(\left(a^{(i)}_1, b^{(i)}_1\right),\ldots,\left(a^{(i)}_{m_i}, b^{(i)}_{m_i}\right)\right)$, $\mathbf{a}^{(i)} = \left(a^{(i)}_1,\ldots,a^{(i)}_{\lambda_i}\right)$, and $\mathbf{b}^{(i)} = \left(b^{(i)}_1,\ldots,b^{(i)}_{\mu_i}\right)$ for every $1 \leq i \leq n$.

Next, we define *[IO]*- and *OI*-substitutions with d-discounting of bounded weighted relations in pairs of terms using evaluations of pairs of terms. Let $(s,t) \in T_\Sigma(X_n) \times T_\Delta(X_n)$ and $\theta_1,\ldots,\theta_n \in \mathbb{R}^b_{\max}\langle\!\langle A \times B\rangle\!\rangle$ be bounded weighted relations. The *[IO]-substitution with d-discounting* (simply *[IO]-d-substitution*) of $\theta_1,\ldots,\theta_n$ in (s,t) is the weighted relation $(s,t)\,[\theta_1,\ldots,\theta_n]^d_{[IO]}$ defined as follows. If $(s,t) \in T_\Sigma \times T_\Delta$, then $(s,t)\,[\theta_1,\ldots,\theta_n]^d_{[IO]} = 0.(s_\mathcal{A}, t_\mathcal{B})$, and if $(s,t) = (x_i,x_i)$, then $(s,t)\,[\theta_1,\ldots,\theta_n]^d_{[IO]} = \theta_i$. Otherwise, we define

$$\left((s,t)\,[\theta_1,\ldots,\theta_n]^d_{[IO]},(a,b)\right) =$$

$$d/k\left(\sup_{\substack{(a_1,b_1),\ldots,(a_k,b_k)\in A\times B \\ (s(c_1,\ldots,c_n),t(d_1,\ldots,d_n))=(a,b)}} ((\theta_{i_1},(a_1,b_1)) + \ldots + (\theta_{i_k},(a_k,b_k)))\right),$$

for every $(a,b) \in A \times B$, where $var(s) \cup var(t) = \{x_{i_1},\ldots,x_{i_k}\}$, and for every $(a_1,b_1),\ldots,(a_k,b_k) \in A \times B$, the sequence $(c_1,d_1),\ldots,(c_n,d_n) \in A \times B$ is an arbitrary element of $(A \times B)^n\,|_{(i_1,(a_1,b_1))\ldots(i_k,(a_k,b_k))}$.

The *OI-substitution with d-discounting* (simply *OI-d-substitution*) of $\theta_1,\ldots,\theta_n$ in (s,t) is the weighted relation $(s,t)\,[\theta_1,\ldots,\theta_n]^d_{OI}$ defined in the following way. If $(s,t) \in T_\Sigma \times T_\Delta$, then $(s,t)\,[\theta_1,\ldots,\theta_n]^d_{OI} = 0.(s_\mathcal{A}, t_\mathcal{B})$, and if $(s,t) = (x_i,x_i)$, then $(s,t)\,[\theta_1,\ldots,\theta_n]^d_{OI} = \theta_i$. Otherwise

$$\left((s,t)\,[\theta_1,\ldots,\theta_n]^d_{OI},(a,b)\right) =$$

$$d/m\left(\sup_{\substack{\mathbf{v}^{(i)}\in(A\times B)^{m_i},1\leq i\leq n \\ (s,t)\left(\mathbf{v}^{(1)},\ldots,\mathbf{v}^{(n)}\right)=(a,b)}} \left(\left(\theta_1,\mathbf{v}^{(1)}\right) + \ldots + \left(\theta_n,\mathbf{v}^{(n)}\right)\right)\right)$$

for every $(a, b) \in A \times B$, where $\lambda_i = |s|_{x_i}$, $\mu_i = |t|_{x_i}$, $m_i = \max\{\lambda_i, \mu_i\}$, $\mathbf{v}^{(i)} = \left(\left(a_1^{(i)}, b_1^{(i)}\right), \ldots, \left(a_{m_i}^{(i)}, b_{m_i}^{(i)}\right)\right)$, and $\left(\theta_i, \mathbf{v}^{(i)}\right) = \left(\theta_i, \left(a_1^{(i)}, b_1^{(i)}\right)\right) + \ldots + \left(\theta_i, \left(a_{m_i}^{(i)}, b_{m_i}^{(i)}\right)\right)$ if $m_i \geq 1$, and $\left(\theta_i, \mathbf{v}^{(i)}\right) = 0$ if $\mathbf{v}^{(i)} = ()$ for every $1 \leq i \leq n$. Finally, $m = \sum_{1 \leq i \leq n} m_i$.

It is easy to see that both the *[IO]-d-* and the *OI-d-*substitution of weighted relations is well-defined.

Remark 1. If $x_i \notin var(s) \cup var(t)$, then

$$(s, t) [\theta_1, \ldots, \theta_i, \ldots, \theta_n]_u^d = (s, t) [\theta_1, \ldots, \theta, \ldots, \theta_n]_u^d$$

for every $\theta \in \mathbb{R}_{\max}^b \langle\!\langle A \times B \rangle\!\rangle$ and u=[IO],OI.

Proof. The case *[IO]* is clear because, by definition, neither θ_i nor θ contributes to the values on the corresponding side of the equation. In case *OI*, we have $m_i = 0$. Hence $\mathbf{v}^{(i)} = ()$ and $\left(\theta_i, \mathbf{v}^{(i)}\right) = \left(\theta, \mathbf{v}^{(i)}\right) = 0$. Thus we get the same value on both sides of the equation.

In this paper we will mainly be interested in *[IO]-d-* and *OI-d-*substitutions of bounded weighted tree transformations over $(\Sigma, \Delta, \mathbb{R}_{\max})$ in pairs (s, t) of terms in $T_\Sigma(X_n) \times T_\Delta(X_n)$. In this particular case we evaluate (s, t) in $(\mathcal{T}_\Sigma, \mathcal{T}_\Delta)$ and thus the *[IO]-* and *OI*-evaluation becomes an *[IO]-* and *OI*-substitution of trees, respectively. Next we give an example of such substitutions.

Example 1. (cf. [6], Example 4) Let $\sigma \in \Sigma_3$, $\delta \in \Delta_2$, and $(s, t) = (\sigma(x_1, x_1, x_3), \delta(x_3, x_1))$. Moreover, let $\theta_1 = \max\{1.(s_1, t_1), 2.(s'_1, t'_1)\}$, $\theta_2 = \widetilde{-\infty}$, and $\theta_3 = 1.(s_3, t_3)$ be weighted tree transformations over $(\Sigma, \Delta, \mathbb{R}_{\max})$. Then

$$(s, t) [\theta_1, \theta_2, \theta_3]_{[IO]}^d = (d/2) \max\{2.(\sigma(s_1, s_1, s_3), \delta(t_3, t_1)), 3.(\sigma(s'_1, s'_1, s_3), \delta(t_3, t'_1))\}$$

and

$$(s, t) [\theta_1, \theta_2, \theta_3]_{OI}^d = (d/3) \max\{3.(\sigma(s_1, s_1, s_3), \delta(t_3, t_1)), 4.(\sigma(s_1, s'_1, s_3), \delta(t_3, t_1)),$$
$$5.(\sigma(s'_1, s'_1, s_3), \delta(t_3, t'_1)), 4.(\sigma(s'_1, s_1, s_3), \delta(t_3, t'_1))\}.$$

However, for $(s', t) = (\sigma(x_1, x_1, x_2), \delta(x_3, x_1))$, we have

$$(s', t) [\theta_1, \theta_2, \theta_3]_{[IO]}^d = (s', t) [\theta_1, \theta_2, \theta_3]_{OI}^d = \widetilde{-\infty}.$$

Finally, we define the *[IO]-d-* and the *OI-d-*substitution of bounded weighted relations in polynomial weighted tree transformations. For every $\tau \in \mathbb{R}_{\max}\langle T_\Sigma(X_n) \times T_\Delta(X_n)\rangle$ and u=[IO],OI, we define the *u-d-substitution of* $\theta_1, \ldots, \theta_n$ *in* τ by

$$\tau [\theta_1, \ldots, \theta_n]_u^d = \max_{(s,t) \in \mathrm{supp}(\tau)} \left((\tau, (s, t)) + (s, t) [\theta_1, \ldots, \theta_n]_u^d\right).$$

Lemma 3. If $\tau \in \mathbb{R}_{\max}\langle\, T_\Sigma\,(X_n) \times T_\Delta\,(X_n)\,\rangle$ is linear, then

$$\tau\,[\theta_1,\ldots,\theta_n]^d_{[IO]} = \tau\,[\theta_1,\ldots,\theta_n]^d_{OI}$$

for every $\theta_1,\ldots,\theta_n \in \mathbb{R}^b_{\max}\langle\!\langle A \times B \rangle\!\rangle$.

Proof. It follows from the definition of u-d-substitution in polynomial weighted tree transformations and the fact that $(s,t)\,[\theta_1,\ldots,\theta_n]^d_{[IO]} = (s,t)\,[\theta_1,\ldots,\theta_n]^d_{OI}$ if (s,t) is linear.

Remark 2. The *[IO]*-d- and the *OI*-d-substitution of weighted relations in weighted tree transformations are monotone, i.e., if $\tau \le \tau'$ and $\theta_i \le \theta'_i$ for every $1 \le i \le n$, then $\tau\,[\theta_1,\ldots,\theta_n]^d_u \le \tau'\,[\theta'_1,\ldots,\theta'_n]^d_u$ for $u = [IO],\ OI$.

Proof. It trivially follows from the definition of u-d-substitution of weighted relations in weighted tree transformations.

In the rest of the paper we shall omit, for the sake of simplicity, the superscript d in the notation of substitutions.

4 Preliminary Results

In this section, we prove some technical results concerning series and substitutions in weighted tree transformations which we will use in proving our main results. The proof of the following statement is obvious.

Remark 3. For every bounded family $(\theta_i)_{i \in I}$ in $\mathbb{R}_{\max}\langle\!\langle A \rangle\!\rangle$ and $k \in \mathbb{R}_{\max}$, the family $(k + \theta_i)_{i \in I}$ is also bounded and we have $k + (\sup_{i \in I} \theta_i) = \sup_{i \in I}(k + \theta_i)$. In particular, if $(\theta_i)_{i \ge 0}$ is a bounded ω-chain, then $(k + \theta_i)_{i \ge 0}$ is a bounded ω-chain and $k + (\lim_{i \to \infty} \theta_i) = \lim_{i \to \infty}(k + \theta_i)$.

Lemma 4. Let $(\theta_i)_{i \ge 0}$ be a bounded ω-chain in $\mathbb{R}_{\max}\langle\!\langle A \rangle\!\rangle$. Then $\lim_{i \to \infty}(\sup_{a \in A}(\theta_i, a)) = \sup_{a \in A}(\lim_{i \to \infty}(\theta_i, a))$.

Proof. By our assumptions the family $(\sup_{a \in A}(\theta_i, a))_{i \ge 0}$ is a bounded ω-chain in $\mathbb{R}_{\max}$, hence $\lim_{i \to \infty}(\sup_{a \in A}(\theta_i, a))$ exists. Now, for every $a \in A$ and $i \ge 0$ we have $(\theta_i, a) \le \sup_{a \in A}(\theta_i, a)$, hence $\lim_{i \to \infty}(\theta_i, a) \le \lim_{i \to \infty}(\sup_{a \in A}(\theta_i, a))$. Then the last approximation implies $\sup_{a \in A}(\lim_{i \to \infty}(\theta_i, a)) \le \lim_{i \to \infty}(\sup_{a \in A}(\theta_i, a))$. The reverse approximation can be proven similarly, starting with the one $(\theta_i, a) \le \lim_{i \to \infty}(\theta_i, a)$.

Lemma 5. Let $(s,t) \in T_\Sigma\,(X_n) \times T_\Delta\,(X_n)$ and $1 \le i \le n$ be such that $i \in var(s) \cup var(t)$. Moreover, let $\theta_1,\ldots,\theta_{i-1},\theta_{i+1},\ldots,\theta_n \in \mathbb{R}^b_{\max}\langle\!\langle A \times B \rangle\!\rangle$ be bounded series, and $(\eta_j \mid j \ge 0)$ a bounded ω-chain of weighted relations in $\mathbb{R}_{\max}\langle\!\langle A \times B \rangle\!\rangle$. Then

$$(s,t)\left[\theta_1,\ldots,\lim_{j \to \infty}\eta_j,\ldots,\theta_n\right]_{[IO]} = \lim_{j \to \infty}(s,t)\,[\theta_1,\ldots,\eta_j,\ldots,\theta_n]_{[IO]},$$

where the limit occurs in the ith argument.

Proof. If $(s,t) \in id(X_n)$ or $(s,t) \in T_\Sigma \times T_\Delta$, then our claim follows by definition. Next, let $(s,t) \in T_\Sigma(X_n) \times T_\Delta(X_n) \setminus (id(X_n) \cup T_\Sigma \times T_\Delta)$ with $var(s) \cup var(t) = \{x_{i_1}, \ldots, x_{i_k}\}$ and assume that $i = i_m$. In the following computation, for every $(a_1, b_1) \ldots, (a_k, b_k) \in A \times B$, the sequence $(c_1, d_1), \ldots, (c_n, d_n) \in A \times B$ is an arbitrary element of $(A \times B)^n|_{(i_1,(a_1,b_1))\ldots(i_k,(a_k,b_k))}$. Then, for every $(a,b) \in A \times B$, we have

$$\left((s,t)[\theta_1, \ldots, \lim_{j\to\infty} \eta_j, \ldots, \theta_n]_{[IO]}, (a,b) \right) =$$

$$d/k \left(\sup_{\substack{(a_1,b_1),\ldots,(a_k,b_k)\in A\times B \\ (s(c_1,\ldots,c_n),t(d_1,\ldots,d_n))=(a,b)}} \left((\theta_{i_1},(a_1,b_1)) + \ldots + (\lim_{j\to\infty} \eta_j,(a_m,b_m)) + \ldots + (\theta_{i_k},(a_k,b_k)) \right) \right) =$$

$$d/k \left(\sup_{\substack{(a_1,b_1),\ldots,(a_k,b_k)\in A\times B \\ (s(c_1,\ldots,c_n),t(d_1,\ldots,d_n))=(a,b)}} \left((\theta_{i_1},(a_1,b_1)) + \ldots + \lim_{j\to\infty}(\eta_j,(a_m,b_m)) + \ldots + (\theta_{i_k},(a_k,b_k)) \right) \right) =$$

$$d/k \left(\sup_{\substack{(a_1,b_1),\ldots,(a_k,b_k)\in A\times B \\ (s(c_1,\ldots,c_n),t(d_1,\ldots,d_n))=(a,b)}} \lim_{j\to\infty} \left((\theta_{i_1},(a_1,b_1)) + \ldots + (\eta_j,(a_m,b_m)) + \ldots + (\theta_{i_k},(a_k,b_k)) \right) \right) =$$

$$d/k \left(\lim_{j\to\infty} \left(\sup_{\substack{(a_1,b_1),\ldots,(a_k,b_k)\in A\times B \\ (s(c_1,\ldots,c_n),t(d_1,\ldots,d_n))=(a,b)}} \left((\theta_{i_1},(a_1,b_1)) + \ldots + (\eta_j,(a_m,b_m)) + \ldots + (\theta_{i_k},(a_k,b_k)) \right) \right) \right) =$$

$$\lim_{j\to\infty}(s,t)[\theta_1, \ldots, \eta_j, \ldots, \theta_n]_{[IO]}$$

where in the third equality we use Remark 3 and in the fourth one Lemma 4.

Next we investigate weighted tree transformations defined by means of bounded weighted tree languages and pairs of tree homomorphisms. For this, let $h : T_\Gamma(X_n) \to T_\Sigma(X_n)$ and $h' : T_\Gamma(X_n) \to T_\Delta(X_n)$ be a pair of tree homomorphisms. We define the mapping

$$\langle h, h' \rangle : \mathbb{R}^b_{\max}\langle\!\langle T_\Gamma(X_n) \rangle\!\rangle \to \mathbb{R}^b_{\max}\langle\!\langle T_\Sigma(X_n) \times T_\Delta(X_n) \rangle\!\rangle$$

by letting

$$((\langle h, h' \rangle (\varphi), (s, t)) = \sup_{\substack{u \in T_\Gamma(X_n) \\ \langle h, h' \rangle (u) = (s, t)}} (\varphi, u)$$

for every weighted tree language $\varphi \in \mathbb{R}^b_{\max}\langle\!\langle T_\Gamma(X_n) \rangle\!\rangle$ and $(s, t) \in T_\Sigma(X_n) \times T_\Delta(X_n)$, where $\langle h, h' \rangle (u) = (h(u), h'(u))$ for every $u \in T_\Gamma(X_n)$.

By the above definition we immediately see that any upper bound of φ is an upper bound of $\langle h, h' \rangle (\varphi)$, hence $\langle h, h' \rangle (\varphi) \in \mathbb{R}^b_{\max}\langle\!\langle T_\Sigma(X_n) \times T_\Delta(X_n) \rangle\!\rangle$. Moreover, that the subsequent result holds.

Remark 4. The mapping $\langle h, h' \rangle : \mathbb{R}^b_{\max}\langle\!\langle T_\Gamma(X_n) \rangle\!\rangle \rightarrow \mathbb{R}^b_{\max}\langle\!\langle T_\Sigma(X_n) \times T_\Delta(X_n) \rangle\!\rangle$ is monotonic.

Lemma 6. Let $(\varphi_i)_{i \in I}$ be a bounded family of weighted tree languages in $\mathbb{R}_{\max}\langle\!\langle T_\Gamma(X_n) \rangle\!\rangle$. Furthermore, let $k \in \mathbb{R}_{\max}$, $\varphi \in \mathbb{R}^b_{\max}\langle\!\langle T_\Gamma(X_n) \rangle\!\rangle$, and pairs of tree homomorphisms $h : T_\Gamma(X_n) \rightarrow T_\Sigma(X_n)$ and $h' : T_\Gamma(X_n) \rightarrow T_\Delta(X_n)$. Then, we have

$$\langle h, h' \rangle \left(\sup_{i \in I} \varphi_i \right) = \sup_{i \in I} \langle h, h' \rangle (\varphi_i) \text{ and } \langle h, h' \rangle (k + \varphi) = k + \langle h, h' \rangle (\varphi).$$

Proof. We prove only the first equality as follows. For every $(s, t) \in T_\Sigma(X_n) \times T_\Delta(X_n)$, we have

$$\left(\langle h, h' \rangle \left(\sup_{i \in I} \varphi_i \right), (s, t) \right) = \sup_{\substack{u \in T_\Gamma(X_n) \\ \langle h, h' \rangle (u) = (s, t)}} \left(\sup_{i \in I} \varphi_i, u \right) = \sup_{\substack{u \in T_\Gamma(X_n) \\ \langle h, h' \rangle (u) = (s, t)}} \sup_{i \in I} (\varphi_i, u)$$

$$= \sup_{i \in I} \left(\sup_{\substack{u \in T_\Gamma(X_n) \\ \langle h, h' \rangle (u) = (s, t)}} (\varphi_i, u) \right) = \sup_{i \in I} (\langle h, h' \rangle (\varphi_i), (s, t)).$$

Lemma 7. Let $\psi \in \mathbb{R}_{\max}\langle \Gamma(X_n) \cup X_n \rangle$ and $\varphi_1, \ldots, \varphi_n \in \mathbb{R}^b_{\max}\langle\!\langle T_\Gamma(X_n) \rangle\!\rangle$ be bounded weighted tree languages and $h : T_\Gamma(X_n) \rightarrow T_\Sigma(X_n)$ and $h' : T_\Gamma(X_n) \rightarrow T_\Delta(X_n)$ be an ultimately complete pair of tree homomorphisms. (a) Then

$$\langle h, h' \rangle \left(\psi[\varphi_1, \ldots, \varphi_n]_{[IO]} \right) = \langle h, h' \rangle (\psi) \left[\langle h, h' \rangle (\varphi_1), \ldots, \langle h, h' \rangle (\varphi_n) \right]_{[IO]}.$$

(b) If ψ, h, and h' are linear, then

$$\langle h, h' \rangle (\psi[\varphi_1, \ldots, \varphi_n]_{OI}) = \langle h, h' \rangle (\psi) \left[\langle h, h' \rangle (\varphi_1), \ldots, \langle h, h' \rangle (\varphi_n) \right]_{OI}.$$

Proof. (a) Let us abbreviate $\Gamma(X_n) \cup X_n$ by $\Gamma(X_n)^\cup$. First we show that for every $u \in \Gamma(X_n)^\cup$

$$\langle h, h' \rangle \left(u[\varphi_1, \ldots, \varphi_n]_{[IO]} \right) = \langle h, h' \rangle (u) \left[\langle h, h' \rangle (\varphi_1), \ldots, \langle h, h' \rangle (\varphi_n) \right]_{[IO]}.$$

It is clear for $u \in X_n$. Hence, let $u \in \Gamma(X_n)$ with $var(u) = \{x_{i_1}, \ldots, x_{i_k}\}$. In the next computation, for every $u_1, \ldots, u_k \in T_\Gamma(X_n)$, the sequence $v_1, \ldots, v_n \in T_\Gamma(X_n)$ is an arbitrary element of $(T_\Gamma(X_n))^n \mid_{(i_1, u_1)\ldots(i_k, u_k)}$. Moreover, for every $(s_1, t_1), \ldots, (s_k, t_k) \in T_\Sigma(X_n) \times T_\Delta(X_n)$, the sequence $(\overline{s_1}, \overline{t_1}), \ldots, (\overline{s_n}, \overline{t_n})$ is an arbitrary element of $(T_\Sigma(X_n) \times T_\Delta(X_n))^n \mid_{(i_1, (s_1, t_1))\ldots(i_k, (s_k, t_k))}$. Then for every $(s, t) \in T_\Sigma(X_n) \times T_\Delta(X_n)$ we have

$$\left(\langle h, h' \rangle \left(u[\varphi_1, \ldots, \varphi_n]_{[IO]}\right), (s, t)\right) =$$

$$d/k \left(\sup_{\substack{u_1, \ldots, u_k \in T_\Gamma(X_n) \\ \langle h, h' \rangle (u(v_1, \ldots, v_n)) = (s, t)}} \left((\varphi_{i_1}, u_1) + \ldots + (\varphi_{i_k}, u_k)\right) \right) =$$

$$d/k \left(\sup_{\substack{u_1, \ldots, u_k \in T_\Gamma(X_n) \\ \langle h, h' \rangle (u)(\langle h, h' \rangle (v_1), \ldots, \langle h, h' \rangle (v_n)) = (s, t)}} \left((\varphi_{i_1}, u_1) + \ldots + (\varphi_{i_k}, u_k)\right) \right) =$$

$$d/k \left(\sup_{\substack{u_1, \ldots, u_k \in T_\Gamma(X_n) \\ \langle h, h' \rangle (u_1) = (s_1, t_1), \ldots, \langle h, h' \rangle (u_k) = (s_k, t_k) \\ \langle h, h' \rangle (u)\left((\overline{s_1}, \overline{t_1}), \ldots, (\overline{s_n}, \overline{t_n})\right) = (s, t)}} \left((\varphi_{i_1}, u_1) + \ldots + (\varphi_{i_k}, u_k)\right) \right) =$$

$$d/k \left(\sup_{\substack{(s_1, t_1), \ldots, (s_k, t_k) \in T_\Sigma(X_n) \times T_\Delta(X_n) \\ \langle h, h' \rangle (u)\left((\overline{s_1}, \overline{t_1}), \ldots, (\overline{s_n}, \overline{t_n})\right) = (s, t)}} \left(\left(\sup_{\substack{u_1 \in T_\Gamma(X_n) \\ \langle h, h' \rangle (u_1) = (s_1, t_1)}} (\varphi_{i_1}, u_1) \right) + \ldots + \left(\sup_{\substack{u_k \in T_\Gamma(X_n) \\ \langle h, h' \rangle (u_k) = (s_k, t_k)}} (\varphi_{i_k}, u_k) \right) \right) \right) =$$

$$d/k \left(\sup_{\substack{(s_1, t_1), \ldots, (s_k, t_k) \in T_\Sigma(X_n) \times T_\Delta(X_n) \\ \langle h, h' \rangle (u)\left((\overline{s_1}, \overline{t_1}), \ldots, (\overline{s_n}, \overline{t_n})\right) = (s, t)}} \left((\langle h, h' \rangle (\varphi_{i_1}), (s_1, t_1)) + \ldots + (\langle h, h' \rangle (\varphi_{i_k}), (s_k, t_k))\right) \right) =$$

$$\left(\langle h, h' \rangle (u) [\langle h, h' \rangle (\varphi_1), \ldots, \langle h, h' \rangle (\varphi_n)]_{[IO]}, (s, t)\right) =$$

where the last equality holds because (h, h') is an ultimately complete pair of tree homomorphisms, hence $var(h(u)) \cup var(h'(u)) = \{x_{i_1}, \ldots, x_{i_k}\}$. Moreover, we have

$$\langle h, h' \rangle \left(\psi[\varphi_1, \ldots, \varphi_n]_{[IO]} \right)$$

$$= \langle h, h' \rangle \left(\max_{u \in \Gamma(X_n)^{\cup}} \left((\psi, u) + u[\varphi_1, \ldots, \varphi_n]_{[IO]} \right) \right)$$

$$= \max_{u \in \Gamma(X_n)^{\cup}} \left((\psi, u) + \langle h, h' \rangle \left(u[\varphi_1, \ldots, \varphi_n]_{[IO]} \right) \right)$$

$$= \max_{u \in \Gamma(X_n)^{\cup}} \left((\psi, u) + \langle h, h' \rangle (u)[\langle h, h' \rangle (\varphi_1), \ldots, \langle h, h' \rangle (\varphi_n)]_{[IO]} \right)$$

$$= \sup_{(s,t) \in T_{\Sigma}(X_n) \times T_{\Delta}(X_n)} \left(\left(\max_{\substack{u \in \Gamma(X_n)^{\cup} \\ \langle h, h' \rangle (u) = (s,t)}} (\psi, u) \right) + \right.$$
$$\left. (s, t)[\langle h, h' \rangle (\varphi_1), \ldots, \langle h, h' \rangle (\varphi_n)]_{[IO]} \right)$$

$$= \max_{(s,t) \in \mathrm{supp}(\langle h, h' \rangle (\psi))} \left((\langle h, h' \rangle (\psi), (s, t)) + \right.$$
$$\left. (s, t)[\langle h, h' \rangle (\varphi_1), \ldots, \langle h, h' \rangle (\varphi_n)]_{[IO]} \right)$$

$$= \langle h, h' \rangle (\psi)[\langle h, h' \rangle (\varphi_1), \ldots, \langle h, h' \rangle (\varphi_n)]_{[IO]},$$

where at the second equality we use Lemma 6 and the third equality is justified above.

(b) By Lemma 2 we have

$$\langle h, h' \rangle \left(\psi[\varphi_1, \ldots, \varphi_n]_{[IO]} \right) = \langle h, h' \rangle \left(\psi[\varphi_1, \ldots, \varphi_n]_{OI} \right)$$

since ψ is linear, and by Lemma 3 we get

$$\langle h, h' \rangle (\psi) [\langle h, h' \rangle (\varphi_1), \ldots, \langle h, h' \rangle (\varphi_n)]_{[IO]} = \langle h, h' \rangle (\psi) [\langle h, h' \rangle (\varphi_1), \ldots, \langle h, h' \rangle (\varphi_n)]_{OI}$$

since ψ, h, and h' are linear. Hence, we conclude the proof of (b) by (a) of this lemma.

5   Equational Weighted Tree Transformations with Discounting

In this section we introduce systems of equations of weighted tree transformations with finite support. We show that the least u-d-solution of such systems exists both for $u = [IO]$ and $u = OI$ in any pair of algebras. We define a weighted relation to be u-d-equational if it is a component of the least u-d-solution of a system of equations of weighted tree transformations. Then we give some basic relationships between classes of u-d-equational weighted relations. We focus on u-d-equational weighted tree transformations, which are equational relations obtained by considering the least u-d-solutions in pairs of term algebras. We give sufficient conditions for the existence and uniqueness of the $[IO]$-d- and the OI-d-solution of a system of equations of weighted tree transformations. Finally, we recall systems of equations of weighted tree languages, and associate a system of equations of weighted tree transformations with a system of equations of weighted tree languages and a pair of tree homomorphisms.

A *system of equations of weighted tree transformations* over Σ, Δ, X_n, and $\mathbb{R}_{\max}$ (or over $(\Sigma, \Delta, X_n, \mathbb{R}_{\max})$) is a system

$$\text{(E)}\quad x_1 = \rho_1, \ldots, x_n = \rho_n,$$

where $\rho_i \in \mathbb{R}_{\max}\langle T_\Sigma(X_n) \times T_\Delta(X_n)\rangle$, i.e., ρ_i is a polynomial over $(\Sigma, \Delta, X_n, \mathbb{R}_{\max})$ for every $1 \leq i \leq n$. Moreover, we require that for every $1 \leq i, j \leq n$ if $(x_j, x_j) \in \mathrm{supp}(\rho_i)$, then $(\rho_i, (x_j, x_j)) = 0$. The system (E) is called *linear* (resp. *variable symmetric, variable identical*) if ρ_i is linear (resp. variable symmetric, variable identical) for every $1 \leq i \leq n$.

Let $u = [IO]$ or $u = OI$. An *u-d-solution* of (E) in $(\mathcal{A}, \mathcal{B}, \mathbb{R}_{\max})$ is an n-tuple $(\theta_1, \ldots, \theta_n)$ of weighted relations in $\mathbb{R}_{\max}^b \langle\!\langle A \times B \rangle\!\rangle^n$ such that

$$\theta_i = \rho_i[\theta_1, \ldots, \theta_n]_u \text{ for every } 1 \leq i \leq n.$$

(Note that $\rho_i[\theta_1, \ldots, \theta_n]_u$ also depends on d in the known manner, cf. Section 3.) Furthermore, it is called the *least u-d-solution* of (E) in $(\mathcal{A}, \mathcal{B}, \mathbb{R}_{\max})$ if $(\theta_1, \ldots, \theta_n) \leq (\theta'_1, \ldots, \theta'_n)$ for every u-d-solution $(\theta'_1, \ldots, \theta'_n)$ of (E) in $(\mathcal{A}, \mathcal{B}, \mathbb{R}_{\max})$.

We will show that the least u-d-solution of (E) exists. For this, we define the *u-d-approximation sequence* of (E) to be the family $(\theta_{1,k}, \ldots, \theta_{n,k})_{k \geq 0}$ in $\mathbb{R}_{\max}\langle A \times B\rangle^n$ such that

$$\theta_{i,0} = \widetilde{-\infty}, \text{ and } \theta_{i,k+1} = \rho_i[\theta_{1,k}, \ldots, \theta_{n,k}]_u, \text{ for } 1 \leq i \leq n \text{ and } k \geq 0.$$

We can see easily that the u-d-approximation sequence of (E) is an ω-chain. In fact, we prove by induction that $\theta_{i,k} \leq \theta_{i,k+1}$ for every $1 \leq i \leq n$ and $k \geq 0$. For $k = 0$ it is obvious by definition. Then we have

$$\theta_{i,k+1} = \rho_i[\theta_{1,k}, \ldots, \theta_{n,k}]_{[IO]} \leq \rho_i[\theta_{1,k+1}, \ldots, \theta_{n,k+1}]_{[IO]} = \theta_{i,k+2}$$

where the inequality holds by the induction hypothesis and Remark 2.

We can also show that the u-d-approximation sequence of (E) is bounded.

Lemma 8. For every system

$$\text{(E)}\quad x_1 = \rho_1, \ldots, x_n = \rho_n,$$

of equations of weighted tree transformations, $u = [IO]$, OI, and $1 \leq i \leq n$, the family $(\theta_{i,k})_{k \geq 0}$ is bounded.

Proof. Let $M = \max\{(\rho_i, (s,t)) \mid (s,t) \in \mathrm{supp}(\rho_i), 1 \leq i \leq n\}$.

Case *[IO]*: We show by induction on k that $\theta_{i,k} \leq M/(1-d)$ for every $1 \leq i \leq n$ and $k \geq 0$. For $k = 0$ it is clear by definition. Next, for every $(a,b) \in A \times B$ and $k \geq 1$ we have

$$\left(\theta_{i,k}, (a,b) \right) = \left(\rho_i \left[\theta_{1,k-1}, \ldots, \theta_{n,k-1} \right]_{[IO]}, (a,b) \right) = \max \left\{ SUP, MX_1, MX_2 \right\}$$

where

$$
SUP = \sup_{\substack{(s,t)\in\mathrm{supp}(\rho_i)\backslash(id(X_n)\cup T_\Sigma \times T_\Delta) \\ var(s)\cup var(t)=\{x_{j_1},\ldots,x_{j_p}\} \\ (a_1,b_1),\ldots,(a_p,b_p)\in A\times B \\ (s(c_1,\ldots,c_n),t(d_1,\ldots,d_n))=(a,b)}} \left((\rho_i,(s,t)) + d/p \left(\begin{matrix} (\theta_{j_1,k-1},(a_1,b_1)) + \ldots \\ + (\theta_{j_p,k-1},(a_p,b_p)) \end{matrix} \right) \right)
$$

$$
\leq \sup_{\substack{(s,t)\in\mathrm{supp}(\rho_i)\backslash(id(X_n)\cup T_\Sigma \times T_\Delta) \\ var(s)\cup var(t)=\{x_{j_1},\ldots,x_{j_p}\} \\ (a_1,b_1),\ldots,(a_p,b_p)\in A\times B \\ (s(c_1,\ldots,c_n),t(d_1,\ldots,d_n))=(a,b)}} \left(M + (d/p)\left(pM/(1-d) \right) \right)
$$

$$
= M + d\left(M/(1-d) \right) = M/(1-d),
$$

and for every $(a_1, b_1), \ldots, (a_p, b_p) \in A \times B$, the sequence $(c_1, d_1), \ldots, (c_n, d_n)$ is an arbitrary element of $(A \times B)^n |_{(i_1,(a_1,b_1))\ldots(i_p,(a_p,b_p))}$. Moreover,

$$
MX_1 = \max_{(x_j,x_j)\in\mathrm{supp}(\rho_i)} \left((\rho_i,(x_j,x_j)) + \left((x_j,x_j)\left[\theta_{1,k-1},\ldots,\theta_{n,k-1}\right]_{[IO]}, (a,b) \right) \right)
$$

$$
= \max_{(x_j,x_j)\in\mathrm{supp}(\rho_i)} \left(\theta_{j,k-1},(a,b) \right) \leq M/(1-d)
$$

and

$$
MX_2 = \max_{(s,t)\in\mathrm{supp}(\rho_i)\cap T_\Sigma \times T_\Delta} \left((\rho_i,(s,t)) + \left((s,t)\left[\theta_{1,k-1},\ldots,\theta_{n,k-1}\right]_{[IO]}, (a,b) \right) \right)
$$

$$
= \max_{(s,t)\in\mathrm{supp}(\rho_i)\cap T_\Sigma \times T_\Delta} \left((\rho_i,(s,t)) + (0.(s_A,t_B),(a,b)) \right) \leq M.
$$

Altogether, this proves that $(\theta_{i,k}, (a,b)) \leq M/(1-d)$.

Case *OI*: We prove again that $\theta_{i,k} \leq M/(1-d)$ for every $1 \leq i \leq n$ and $k \geq 0$. For $k = 0$ it is clear by definition. Next for every $(a,b) \in A \times B$ and $k \geq 1$ we have

$$\left(\theta_{i,k}, (a,b) \right) = \left(\rho_i \left[\theta_{1,k-1}, \ldots, \theta_{n,k-1} \right]_{OI}, (a,b) \right) = \max\{ SP, NX_1, NX_2 \},$$

where

$$SP = \sup_{\substack{(s,t)\in\mathrm{supp}(\rho_i)\setminus(id(X_n)\cup T_\Sigma\times T_\Delta) \\ m_j=\max\{|s|_{x_j},|t|_{x_j}\},1\leq j\leq n \\ m=\sum_{j=1}^n m_j \\ \mathbf{v}^{(j)}\in(A\times B)^{m_j},1\leq j\leq n \\ (s,t)(\mathbf{v}^{(1)},\ldots,\mathbf{v}^{(n)})=(a,b)}} \left(\left(\rho_i,(s,t)\right)+d/m\left(\begin{array}{c}(\theta_{1,k-1},\mathbf{v}^{(1)})+\cdots\\+(\theta_{n,k-1}\mathbf{v}^{(n)})\end{array}\right)\right)$$

$$\leq \sup_{\substack{(s,t)\in\mathrm{supp}(\rho_i)\setminus(id(X_n)\cup T_\Sigma\times T_\Delta) \\ m_j=\max\{|s|_{x_j},|t|_{x_j}\},1\leq j\leq n \\ m=\sum_{j=1}^n m_j \\ \mathbf{v}^{(j)}\in(A\times B)^{m_j},1\leq j\leq n \\ (s,t)(\mathbf{v}^{(1)},\ldots,\mathbf{v}^{(n)})=(a,b)}} \left(M+(d/m)\left(m\left(M/(1-d)\right)\right)\right)$$

$$= M+d\left(M/(1-d)\right) = M/(1-d),$$

$$NX_1 = \max_{(x_j,x_j)\in\mathrm{supp}(\rho_i)}\left(\left(\rho_i,(x_j,x_j)\right)+(\theta_{j,k-1},(a,b))\right)\leq M/(1-d),$$

and

$$NX_2 = \max_{(s,t)\in\mathrm{supp}(\rho_i)\cap T_\Sigma\times T_\Delta}\left(\left(\rho_i,(s,t)\right)+(0.(s_\mathcal{A},t_\mathcal{B}),(a,b))\right)\leq M.$$

Again, we obtain that $(\theta_{i,k},(a,b))\leq M/(1-d)$.

Now we are able to show that the last u-d-solution of (E) exists.

Theorem 1. *Let*

$$\text{(E)}\qquad x_1=\rho_1,\ldots,x_n=\rho_n,$$

be a system of equations of weighted tree transformations, $u = [\mathrm{IO}]$ *or* $u = \mathrm{OI}$, *and let* $(\theta_{1,k},\ldots,\theta_{n,k})_{k\geq0}$ *the* u-d-*approximation sequence of* (E). *Then* $\lim_{k\to\infty}(\theta_{1,k},\ldots,\theta_{n,k})$ *exists and it is the least* u-d-*solution of* (E) *in* $(\mathcal{A},\mathcal{B},\mathbb{R}_{\max})$.

Proof. We prove the case $u=[\mathrm{IO}]$ because the proof of the other case is similar.

Since $(\theta_{1,k},\ldots,\theta_{n,k})_{k\geq0}$ is an ω-chain and by Lemma 8 each component of it is a bounded family, $\lim_{k\to\infty}(\theta_{1,k},\ldots,\theta_{n,k})$ exists and each component of it is bounded. We let $(\theta_1,\ldots,\theta_n)=\lim_{k\to\infty}(\theta_{1,k},\ldots,\theta_{n,k})$ and show that $(\theta_1,\ldots,\theta_n)$ is the least $[\mathrm{IO}]$-d-solution of (E). For every $1\leq i\leq n$ and $(a,b)\in A\times B$, we have

$$(\theta_i,(a,b)) = \left(\lim_{k\to\infty}\theta_{i,k},(a,b)\right) = \lim_{k\to\infty}(\theta_{i,k},(a,b))$$

$$= \lim_{k\to\infty}\left(\rho_i\left[\theta_{1,k-1},\ldots,\theta_{n,k-1}\right]_{[\mathrm{IO}]},(a,b)\right)$$

$$= \lim_{k\to\infty}\left(\max\{SUP,MX_1,MX_2\}\right)$$

$$= \max\left\{\lim_{k\to\infty}SUP,\lim_{k\to\infty}MX_1,\lim_{k\to\infty}MX_2\right\},$$

where SUP, MX_1, and MX_2 are defined in the proof of Lemma 8. Then we have

$$\lim_{k\to\infty} SUP =$$

$$\sup_{\substack{(s,t)\in\text{supp}(\rho_i)\setminus(id(X_n)\cup T_\Sigma\times T_\Delta) \\ var(s)\cup var(t)=\{x_{j_1},\ldots,x_{j_p}\} \\ (a_1,b_1),\ldots,(a_p,b_p)\in A\times B \\ (s(c_1,\ldots,c_n),t(d_1,\ldots,d_n))=(a,b)}} \left((\rho_i,(s,t)) + d/p \left(\begin{array}{c} \lim_{k\to\infty} \left(\theta_{j_1,k-1},(a_1,b_1)\right) + \ldots \\ + \lim_{k\to\infty} \left(\theta_{j_p,k-1},(a_p,b_p)\right) \end{array} \right) \right) =$$

$$\sup_{\substack{(s,t)\in\text{supp}(\rho_i)\setminus(id(X_n)\cup T_\Sigma\times T_\Delta) \\ var(s)\cup var(t)=\{x_{j_1},\ldots,x_{j_p}\} \\ (a_1,b_1),\ldots,(a_p,b_p)\in A\times B \\ (s(c_1,\ldots,c_n),t(d_1,\ldots,d_n))=(a,b)}} \left((\rho_i,(s,t)) + d/p \left(\begin{array}{c} \left(\lim_{k\to\infty}\theta_{j_1,k-1},(a_1,b_1)\right) + \ldots \\ + \left(\lim_{k\to\infty}\theta_{j_p,k-1},(a_p,b_p)\right) \end{array} \right) \right) =$$

$$\sup_{\substack{(s,t)\in\text{supp}(\rho_i)\setminus(id(X_n)\cup T_\Sigma\times T_\Delta) \\ var(s)\cup var(t)=\{x_{j_1},\ldots,x_{j_p}\} \\ (a_1,b_1),\ldots,(a_p,b_p)\in A\times B \\ (s(c_1,\ldots,c_n),t(d_1,\ldots,d_n))=(a,b)}} \left((\rho_i,(s,t)) + d/p \left((\theta_{j_1},(a_1,b_1)) + \ldots + (\theta_{j_p},(a_p,b_p)) \right) \right)$$

where the first equality holds by Lemma 4. Moreover

$$\lim_{k\to\infty} MX_1 = \max_{(x_j,x_j)\in\text{supp}(\rho_i)} \left(\lim_{k\to\infty} (\theta_{j,k-1},(a,b)) \right)$$

$$= \max_{(x_j,x_j)\in\text{supp}(\rho_i)} ((\theta_j,(a,b))),$$

and

$$\lim_{k\to\infty} MX_2 = \max_{(s,t)\in\text{supp}(\rho_i)\cap T_\Sigma\times T_\Delta} ((\rho_i,(s,t)) + (0.(s_A,t_B),(a,b))).$$

Hence, we get

$$(\theta_i,(a,b)) = \left(\rho_i\,[\theta_1,\ldots,\theta_n]_{[IO]},(a,b) \right)$$

which proves that $(\theta_1,\ldots,\theta_n)$ is an $[IO]$-d-solution of (E).

Now assume that $\left(\theta'_1,\ldots,\theta'_n\right)$ is another $[IO]$-d-solution of (E). We show by induction on k that $(\theta_{i,k},(a,b)) \leq (\theta'_i,(a,b))$ for every $1\leq i\leq n$, $(a,b)\in A\times B$, and $k\geq 0$. For $k=0$ it is clear by definition. Furthermore, we have that

$$(\theta_{i,k+1},(a,b)) = \left(\rho_i\,[\theta_{1,k},\ldots,\theta_{n,k}]_{[IO]},(a,b) \right)$$

$$\leq \left(\rho_i\,[\theta'_1,\ldots,\theta'_n]_{[IO]},(a,b) \right) = (\theta'_i,(a,b))$$

where the first inequality holds by the induction hypothesis and Remark 2. This finishes the proof of the theorem.

A weighted relation $\theta \in \mathbb{R}^b_{\max}\langle\!\langle A \times B \rangle\!\rangle$ is called u-d-*equational* (resp. l-u-d-*equational*, vs-u-d-*equational*, and vi-u-d-*equational*) if it is a component of the least u-d-solution in $(\mathcal{A}, \mathcal{B}, \mathbb{R}_{\max})$ of a system (resp. linear, variable symmetric, and variable identical system) of equations of weighted tree transformations over $(\Sigma, \Delta, X_n, \mathbb{R}_{\max})$. We will denote by $EQUA^d_u$ (resp. l-$EQUA^d_u$, vs-$EQUA^d_u$, and vi-$EQUA^d_u$) the class of all u-d-equational (resp. l-u-d-equational, vs-u-d-equational, and vi-u-d-equational) weighted relations.

Next we show some equalities between certain classes of equational weighted relations. The first one immediately follows from Lemma 3.

Corollary 1. l-$EQUA^d_{[IO]} = l$-$EQUA^d_{OI}$.

In the following, we show that OI-d-equational weighted relations are the same as l-OI-d-equational ones. For this, we introduce the notion of the rank of a system (E) of equations of weighted tree transformations. Firstly, we recall (cf. [6]) that the rank of a pair $(s, t) \in T_\Sigma(X_n) \times T_\Delta(X_n)$ is given by $rk((s,t)) = card(\{j \mid 1 \le j \le n, |s|_{x_j} > 1 \text{ or } |t|_{x_j} > 1\})$. Then, the *rank of the system* (E) is given by $rk(E) = \sum_{i=1}^n rk(\rho_i)$, where $rk(\rho_i) = \sum_{(s,t)\in\text{supp}(\rho_i)} rk((s,t))$.

Lemma 9. $EQUA^d_{OI} = l$-$EQUA^d_{OI}$.

Proof. We apply the same technique as in the proof of Lemma 12 in [6], and show that $EQUA^d_{OI} \subseteq l$-$EQUA^d_{OI}$. For this, let

$$\text{(E)} \quad x_1 = \rho_1, \ldots, x_n = \rho_n,$$

be a system of equations of weighted tree transformations over $(\Sigma, \Delta, X_n, \mathbb{R}_{\max})$ such that the least OI-d-solution $(\varphi_1, \ldots, \varphi_n)$ of (E) exists. We effectively construct a linear system (F) of weighted tree transformations over $(\Sigma, \Delta, X_{n+m}, \mathbb{R}_{\max})$ for some $m \ge 0$ such that the least OI-d-solution of (F) exists and the least OI-d-solution of (E) is the first n components of the least OI-d-solution of (F).

If (E) is linear, then obviously (F)=(E). Otherwise, there is an $1 \le i_0 \le n$ such that $\text{supp}(\rho_{i_0})$ contains a nonlinear pair $(s, t) \in T_\Sigma(X_n) \times T_\Delta(X_n)$. This means that for some $1 \le j \le n$, $\lambda = |s|_{x_j}$, and $\mu = |t|_{x_j}$ we have $mx = \max\{\lambda, \mu\} > 1$.

We first transform (E) to a system

$$\text{(E')} \quad x_1 = \rho'_1, \ldots, x_{n+mx} = \rho'_{n+mx},$$

such that $rk(E') = rk(E) - 1$, the least OI-d-solution of (E') exists, and $(\varphi_1, \ldots, \varphi_n)$ is the first n components of the least OI-d-solution of (E'). Let the first n equations of (E') be those of (E) except that

$$(\rho'_{i_0}, (u, v)) = \begin{cases} (\rho_{i_0}, (s, t)) & \text{if } (u, v) = (s', t') \\ -\infty & \text{if } (u, v) = (s, t) \\ (\rho_{i_0}, (u, v)) & \text{otherwise,} \end{cases}$$

where we obtain (s', t') by replacing the occurrences of x_j in s and t from left to right by $x_{n+1}, \ldots, x_{n+\lambda}$ and by $x_{n+1}, \ldots, x_{n+\mu}$, respectively. Moreover, let $\rho'_{n+k} = 0.(x_j, x_j)$ for every $1 \le k \le mx$.

To prove the statement concerning the least OI-d-solution of (E) and of (E'), we observe that the first n components of any element of the OI-d approximation sequence of (E') form an element of the OI-d approximation sequence of (E). Moreover, any element of the OI-d approximation sequence of (E) is the first n components of some element of the OI-d approximation sequence of (E'). Hence, by Theorem 1 and the fact that $(\varphi_1, \ldots, \varphi_n)$ is the least OI-d-solution of (E), the least OI-d-solution of (E') exists, and $(\varphi_1, \ldots, \varphi_n)$ is its first n components.

Since $rk(E') = rk(E) - 1$, we can obtain the desired linear system (F) by applying the above procedure a finite number of times.

Corollary 2. $EQUA_{OI}^d = l\text{-}EQUA_{[IO]}^d$.

If the system (F) in the proof of Lemma 9 is variable symmetric, then the linear system (F) will also be variable symmetric. Hence we obtain the following result.

Corollary 3. $vs\text{-}EQUA_{OI} = l\text{-}vs\text{-}EQUA_{OI}$.

In the rest of the paper we will focus on u-d-equational weighted tree transformations over $(\Sigma, \Delta, \mathbb{R}_{\max})$, i.e., u-d-equational weighted relations in $\mathbb{R}_{\max}\langle\!\langle T_\Sigma \times T_\Delta \rangle\!\rangle$ for $u = [IO]$, OI. They are obtained by considering the least u-d-solution of systems (E) of equations of weighted tree transformations over $(\Sigma, \Delta, X_n, \mathbb{R}_{\max})$ in $(T_\Sigma, T_\Delta, \mathbb{R}_{\max})$. For the sake of simplicity, we call a u-d-solution (resp. the least u-d-solution) of such an (E) in $(T_\Sigma, T_\Delta, \mathbb{R}_{\max})$ just a u-d-solution (resp. the least u-d-solution) of (E). We define u-d-equational (resp. l-u-d-equational, vs-u-d-equational, and vi-u-d-equational) weighted tree transformations as we defined the corresponding concepts for weighted relations. We will denote by $EQUT_u^d$ (resp. $l\text{-}EQUT_u^d$, $vs\text{-}EQUT_u^q$, and $vi\text{-}EQUT_u^d$) the class of all u-d-equational (resp. l-u-d-equational, vs-u-d-equational, and vi-u-d-equational) weighted tree transformations.

In the next theorem we give sufficient conditions for the existence and uniqueness of the $[IO]$-d- and the OI-d-solution of a system of equations of weighted tree transformations. Hereby, we generalize the corresponding result for systems of equations of weighted tree languages obtained in [4], Proposition 6.1, cf. also [25], Lemma 3.37. For this, we define a system $x_i = \rho_i$, $1 \leq i \leq n$ of equations of weighted tree transformations to be *proper* if $supp(\rho_i) \subseteq (T_\Sigma(X_n) \setminus X_n) \times (T_\Delta(X_n) \setminus X_n)$ for every $1 \leq i \leq n$.

Theorem 2. *Any proper and variable identical (resp. variable symmetric) system of equations of weighted tree transformations has a unique $[IO]$-d-solution (resp. OI-d-solution).*

Proof. Let

$$\text{(E)} \quad x_1 = \rho_1, \ldots, x_n = \rho_n,$$

be a system of equations of weighted tree transformations over $(\Sigma, \Delta, X_n, \mathbb{R}_{\max})$. Let $(\tau_1, \ldots, \tau_n)$ be an $[IO]$-d-solution of (E). Then, for every $(u, v) \in T_\Sigma \times T_\Delta$, we have

$$(\tau_i, (u, v)) = (\rho_i [\tau_1, \ldots, \tau_n]_{[IO]}, (u, v)) =$$

$$\max_{\substack{(s,t)\in \mathrm{supp}(\rho_i) \\ var(s)=var(t)=\{x_{i_1},\ldots,x_{i_k}\} \\ (s_1,t_1),\ldots,(s_k,t_k)\in T_\Sigma \times T_\Delta \\ (s(\overline{s_1},\ldots,\overline{s_n}),t(\overline{t_1},\ldots,\overline{t_n}))=(u,v)}} ((\rho_i, (s,t)) + d/k \, ((\tau_{i_1}, (s_1,t_1)) + \ldots + (\tau_{i_k}, (s_k,t_k)))) =$$

$$\max_{\substack{(s,t)\in \mathrm{supp}(\rho_i) \\ var(s)=var(t)=\{x_{i_1},\ldots,x_{i_k}\} \\ (s_1,t_1),\ldots,(s_k,t_k)\in (sub(u)\setminus\{u\})\times(sub(v)\setminus\{v\}) \\ (s(\overline{s_1},\ldots,\overline{s_n}),t(\overline{t_1},\ldots,\overline{t_n}))=(u,v)}} \left((\rho_i, (s,t)) + d/k \left(\begin{array}{c} (\tau_{i_1}, (s_1,t_1)) + \ldots \\ + (\tau_{i_k}, (s_k,t_k)) \end{array} \right) \right),$$

where for every $(s_1, t_1), \ldots, (s_k, t_k) \in T_\Sigma \times T_\Delta$, the sequence $(\overline{s_1}, \overline{t_1}), \ldots, (\overline{s_n}, \overline{t_n})$ is an arbitrary element of $(T_\Sigma \times T_\Delta)^n |_{(i_1,(s_1,t_1))\ldots(i_k,(s_k,t_k))}$. In the first summation, we can write $var(s) = var(t) = \{x_{i_1}, \ldots, x_{i_k}\}$ because (E) is variable identical, and the third equality is justified by the fact that (E) is proper. Hence $(\tau_i, (u, v))$ is uniquely determined by ρ_i and by the values of the τ_j's on pairs, of which the components are proper subtrees of u and v, respectively.

Next, let $(\tau_1, \ldots, \tau_n)$ be an OI-d-solution of (E) and $(u, v) \in T_\Sigma \times T_\Delta$. Then

$$(\tau_i, (u, v)) = (\rho_i [\tau_1, \ldots, \tau_n]_{OI}, (u, v)) =$$

$$\max_{\substack{(s,t)\in \mathrm{supp}(\rho_i) \\ |s|_{x_i}=|t|_{x_i}=\lambda_i, 1\leq i\leq n \\ m=\sum_{1\leq i\leq n}\lambda_i \\ \mathbf{v}^{(i)}\in(T_\Sigma\times T_\Delta)^{\lambda_i}, 1\leq i\leq n \\ (s(\mathbf{s}^{(1)},\ldots,\mathbf{s}^{(n)}),t(\mathbf{t}^{(1)},\ldots,\mathbf{t}^{(n)}))=(u,v)}} \left((\rho_i, (s,t)) + d/m \left(\begin{array}{c} (\tau_1, \mathbf{v}^{(1)}) + \ldots \\ + (\tau_n, \mathbf{v}^{(n)}) \end{array} \right) \right) =$$

$$\max_{\substack{(s,t)\in \mathrm{supp}(\rho_i) \\ |s|_{x_i}=|t|_{x_i}=\lambda_i, 1\leq i\leq n \\ m=\sum_{1\leq i\leq n}\lambda_i \\ \mathbf{v}^{(i)}\in((sub(u)\setminus\{u\})\times(sub(v)\setminus\{v\}))^{\lambda_i}, 1\leq i\leq n \\ (s(\mathbf{s}^{(1)},\ldots,\mathbf{s}^{(n)}),t(\mathbf{t}^{(1)},\ldots,\mathbf{t}^{(n)}))=(u,v)}} \left((\rho_i, (s,t)) + d/m \left(\begin{array}{c} (\tau_1, \mathbf{v}^{(1)}) + \ldots \\ + (\tau_n, \mathbf{v}^{(n)}) \end{array} \right) \right),$$

where the vectors $\mathbf{s}^{(i)}$ and $\mathbf{t}^{(i)}$ are made of the first and the second components of $\mathbf{v}^{(i)}$, respectively (cf. page 121, OI-evaluation). The second equality follows from that (E) is variable symmetric, and the third from that (E) is proper. Hence again, $(\tau_i, (u, v))$ is uniquely determined by ρ_i and by the values of the τ_j's on pairs, of which the components are proper subtrees of u and v, respectively.

On the other hand, the above calculations can be used as defining equalities for the $[IO]$-d- and the OI-d-solution, respectively, provided the conditions of the theorem hold.

Finally, we introduce equational weighted tree languages with discounting. A *system of equations of weighted tree languages over* Γ, X_n, *and* $\mathbb{R}_{\max}$ (or ($\Gamma, X_n, \mathbb{R}_{\max}$)) is a system

$$(G) \quad x_1 = \psi_1, \ldots, x_n = \psi_n,$$

where $\psi_i \in \mathbb{R}_{\max}\langle T_\Gamma(X_n) \rangle$ for every $1 \le i \le n$. The system (G) is called *linear* (resp. *simple*) if ψ_i is linear (resp. $\psi_i \in \mathbb{R}_{\max}\langle \Gamma(X_n) \rangle$) for every $1 \le i \le n$. The concept of the least *OI-d-solution* of (G) can be defined as for systems of equations of weighted tree transformations using *OI-d-substitution* of weighted tree languages rather than that of weighted tree transformations. Obviously, the least *OI-d-solution* of (G) is an n-tuple $(\varphi_1, \ldots, \varphi_n) \in \mathbb{R}_{\max}^b \langle\!\langle T_{\Gamma'} \rangle\!\rangle^n$ of weighted tree languages. Similarly to weighted tree transformations, a weighted tree language is called *OI-d-equational* (resp. *l-OI-d-equational, s-OI-d-equational*) if it is a component of the least *OI-d-solution* of a system (resp. linear, simple) of equations of weighted tree languages.

In the following we relate recognizable weighted tree languages with *d-discounting* and *s-OI-d-equational* weighted tree languages. For this, we recall the following concept, cf. [25]. A simple system (G) of equations of weighted tree languages and a finite-state normalized wta $\mathcal{M} = (Q, \mu, x_1)$ are *related* if $Q = X_n$, and for every $1 \le i \le n$, $k \ge 0, \sigma \in \Sigma_k$, and $x_{i_1}, \ldots, x_{i_k} \in X_n$ we have

$$(\rho_i, \sigma(x_{i_1}, \ldots, x_{i_k})) = \mu_k((x_{i_1}, \ldots, x_{i_k}), \sigma, x_i).$$

It is obvious that, given a simple system (G) of equations of weighted tree languages, we can construct a finite-state normalized wta $\mathcal{M}$ such that (G) and $\mathcal{M}$ related, and vice versa.

Lemma 10. If (G) and $\mathcal{M}$ are related, then $(\varphi_1, \ldots, \varphi_n)$ is the least *OI-d-solution* of (G), where $(\varphi_i, s) = h_\mu^d(s)_{x_i}$ for every $1 \le i \le n$.

Proof. Let us consider a tree $\sigma(s_1, \ldots, s_k)$ for some $k \ge 0$, $\sigma \in \Gamma_k$, and $s_1, \ldots, s_k \in T_\Gamma$. Then we have

$$(\rho_i[\varphi_1, \ldots, \varphi_n]_{OI}, \sigma(s_1, \ldots, s_k)) =$$

$$\left(\sup_{u \in T_\Sigma(X_n)} ((\rho_i, u) + u[\varphi_1, \ldots, \varphi_n]_{OI}), \sigma(s_1, \ldots, s_k) \right) =$$

$$\sup_{u \in T_\Sigma(X_n)} ((\rho_i, u) + (u[\varphi_1, \ldots, \varphi_n]_{OI}, \sigma(s_1, \ldots, s_k))) =$$

$$\max_{x_{i_1}, \ldots, x_{i_k} \in X_n} \{(\rho_i, \sigma(x_{i_1}, \ldots, x_{i_k})) + (\sigma(x_{i_1}, \ldots, x_{i_k})[\varphi_1, \ldots, \varphi_n]_{OI}, \sigma(s_1, \ldots, s_k))\} =$$

$$\max_{x_{i_1}, \ldots, x_{i_k} \in X_n} \left\{ (\rho_i, \sigma(x_{i_1}, \ldots, x_{i_k})) + d/k \left(\sum_{j=1}^{k} (\varphi_{i_j}, s_j) \right) \right\} =$$

$$\max_{x_{i_1},\ldots,x_{i_k}\in X_n}\left\{(\rho_i,\sigma(x_{i_1},\ldots,x_{i_k}))+d/k\left(\sum_{j=1}^{k}h_\mu^d(s_j)_{x_{i_j}}\right)\right\}=$$

$$\max_{x_{i_1},\ldots,x_{i_k}\in X_n}\left\{d/k\left(\sum_{j=1}^{k}h_\mu^d(s_j)_{x_{i_j}}\right)+\mu_k((x_{i_1},\ldots,x_{i_k}),\sigma,x_i)\right\}=$$

$$h_\mu^d(\sigma(s_1,\ldots,s_k))_{x_i}=(\varphi_i,\sigma(s_1,\ldots,s_k)).$$

The third equality is justified by the fact that for every decomposition of the form $u(\mathbf{v}^{(1)},\ldots,\mathbf{v}^{(n)})$ of $\sigma(s_1,\ldots,s_k)$ such that $(\rho_i,u)\neq-\infty$, we have that $u=\sigma(x_{i_1},\ldots,x_{i_k})$ for some $x_{i_1},\ldots,x_{i_k}\in X_n$. The fourth one holds by the definition of the *OI-d*-substitution, the fifth one by the definition of φ_{i_j}, and the sixth one because (G) and $\mathcal{M}$ are related. This proves that $(\varphi_1,\ldots,\varphi_n)$ is an *OI-d*-solution of (G). Moreover, it is the least solution, because, by an obvious adaptation of Theorem 2, (G) has a unique *d*-solution.

Theorem 3. *A weighted tree language is d-recognizable if and only if it is s-OI-d-equational.*

Proof. Let φ be a *d*-recognizable weighted tree langu9age. We may assume that $\varphi=\|\mathcal{M}\|^d$ for some finite-state normalized wta $\mathcal{M}=(Q,\mu,x_1)$ with $Q=X_n$. In particular, $(\varphi,s)=h_\mu^d(s)_{x_1}$ for every $s\in T_\Sigma$. Now consider the simple equation system (G) of weighted tree languages such that (G) and $\mathcal{M}$ are related. By Lemma 10, φ is the first component of the least *d*-OI-solution of (G), hence it is *s-OI-d*-equational. The other direction can be proved similarly.

Lemma 11. *If a weighted tree language* $\varphi\in\mathbb{R}_{\max}\langle\!\langle T_\Gamma\rangle\!\rangle$ *is d-recognizable, then it is a component of the least OI-d-solution of a linear system*

$$x_1=\psi_1,\ldots,x_n=\psi_n,$$

of equations of weighted tree languages over $(\Gamma,X_n,\mathbb{R}_{\max})$ *with* $\psi_i\in\mathbb{R}_{\max}\langle\Gamma(X_n)\cup X_n\rangle$, $1\leq i\leq n$.

Proof. By Theorem 3 there is a simple system (G) of equations of weighted tree languages over $(\Gamma,X_n,\mathbb{R}_{\max})$ such that φ is a component of the least *OI-d*-solution of (G). We apply (the adapted version of) the linearization algorithm appearing in the proof of Lemma 9 to (G). Then we obtain a system with the desired properties.

Now, let

$$(\text{G})\quad x_1=\psi_1,\ldots,x_n=\psi_n,$$

be a system of weighted tree languages over $(\Gamma,X_n,\mathbb{R}_{\max})$, and $h:T_\Gamma\to T_\Sigma$ and $h':T_\Gamma\to T_\Delta$ be tree homomorphisms. The *system of equations of tree transformations over* $(\Sigma,\Delta,X_n,\mathbb{R}_{\max})$ *associated with* (G), h and h' is the system

$$\langle h,h'\rangle\,(\text{G})\quad x_1=\langle h,h'\rangle\,(\psi_1),\ldots,x_n=\langle h,h'\rangle\,(\psi_n).$$

Lemma 12. Let

$$\text{(G)} \quad x_1 = \psi_1, \ldots, x_n = \psi_n,$$

be a linear system of equations of weighted tree languages over $(\Gamma, X_n, \mathbb{R}_{\max})$ such that $\psi_i \in \mathbb{R}_{\max}\langle \Gamma(X_n) \cup X_n \rangle$ for every $1 \leq i \leq n$ and $(\varphi_1, \ldots, \varphi_n)$ be its least *OI-d*-solution. Moreover, let $h : T_\Gamma \to T_\Sigma$ and $h' : T_\Gamma \to T_\Delta$ be an ultimately complete pair of tree homomorphisms.

(a) Then the least *[IO]-d*-solution of $\langle h, h' \rangle$ (G) is $(\langle h, h' \rangle (\varphi_1), \ldots, \langle h, h' \rangle (\varphi_n))$.
(b) If in addition the tree homomorphisms h and h' are linear, then the least *OI-d*-solution of $\langle h, h' \rangle$ (G) is $(\langle h, h' \rangle (\varphi_1), \ldots, \langle h, h' \rangle (\varphi_n))$.

Proof. First we prove (a) as follows. We have

$$\langle h, h' \rangle (\varphi_i) = \langle h, h' \rangle (\psi_i[\varphi_1, \ldots, \varphi_n]_{OI})$$
$$= \langle h, h' \rangle (\psi_i[\varphi_1, \ldots, \varphi_n]_{[IO]})$$
$$= \langle h, h' \rangle (\psi_i) [\langle h, h' \rangle (\varphi_1), \ldots, \langle h, h' \rangle (\varphi_n)]_{[IO]},$$

for every $1 \leq i \leq n$, where the second equality follows from Lemma 2 and the fact that (G) is linear, and the third one from Lemma 7(a). Hence $(\langle h, h' \rangle (\varphi_1), \ldots, \langle h, h' \rangle (\varphi_n))$ is an *[IO]-d*-solution of $\langle h, h' \rangle$ (G).

Now assume that $(\zeta_1, \ldots, \zeta_n)$ is another *[IO]-d*-solution of $\langle h, h' \rangle$ (G). By Theorem 1 we may assume that $(\varphi_1, \ldots, \varphi_n) = \lim_{k \to \infty} ((\varphi_{1,k}, \ldots, \varphi_{n,k}))$, where $(\varphi_{1,k}, \ldots, \varphi_{n,k})_{k \geq 0}$ is the *[IO]-d*-approximation sequence of (E).

We show by induction that, for every $1 \leq i \leq n$ and $k \geq 0$, we have $\langle h, h' \rangle (\varphi_{i,k}) \leq \zeta_i$. For $k = 0$ this is true by definition. Then, for every $k \geq 0$, we have

$$\langle h, h' \rangle (\varphi_{i,k+1}) = \langle h, h' \rangle (\psi_i[\varphi_{1,k}, \ldots, \varphi_{n,k}]_{OI})$$
$$= \langle h, h' \rangle (\psi_i[\varphi_{1,k}, \ldots, \varphi_{n,k}]_{[IO]})$$
$$= \langle h, h' \rangle (\psi_i) [\langle h, h' \rangle (\varphi_{1,k}), \ldots, \langle h, h' \rangle (\varphi_{n,k})]_{[IO]}$$
$$\leq \langle h, h' \rangle (\psi_i) [\zeta_1, \ldots, \zeta_n]_{[IO]} = \zeta_i,$$

where the second equality follows from Lemma 2 and the fact that (G) is linear, and the third one from Lemma 7(a). Finally, the inequality holds by the induction hypothesis and Remark 2. Hence, by Remark 4, we get $\langle h, h' \rangle (\varphi_i) \leq \zeta_i$ for every $1 \leq i \leq n$, therefore $(\langle h, h' \rangle (\varphi_1), \ldots, \langle h, h' \rangle (\varphi_n))$ is the least *[IO]-d*-solution of $\langle h, h' \rangle$ (G).

Next we prove (b). We have

$$\langle h, h' \rangle (\varphi_i) = \langle h, h' \rangle (\psi_i[\varphi_1, \ldots, \varphi_n]_{OI})$$
$$= \langle h, h' \rangle (\psi_i) [\langle h, h' \rangle (\varphi_1), \ldots, \langle h, h' \rangle (\varphi_n)]_{OI}$$

where the second equality follows from Lemma 7(b) because ψ_i is linear and (h, h') is an ultimately complete pair of linear tree homomorphisms. Thus, $(\langle h, h' \rangle (\varphi_1), \ldots, \langle h, h' \rangle (\varphi_n))$ is an *OI-d*-solution of $\langle h, h' \rangle$ (G). We complete the proof in the same way as in (a), where again we use Lemma 7(b).

6   Characterizations in Terms of Weighted Bimorphisms

In this section we give a characterization of the classes $EQUT^d_{[IO]}$, $EQUT^d_{OI}$, $vi\text{-}EQUT^d_{[IO]}$, and $vs\text{-}EQUT^d_{OI}$ of d-equational weighted tree transformation classes in terms of weighted d-bimorphisms. The concept of an (unweighted) bimorphisms for tree languages was introduced in [2,3], and of weighted bimorphisms for weighted tree languages in [25,37]. Moreover, we show that in fact the class $vi\text{-}EQUT^d_{[IO]}$ is the closure of $vs\text{-}EQUT^d_{OI}$ under complete tree homomorphisms.

A *weighted d-bimorphism* (over Γ, Σ, Δ, and $\mathbb{R}_{\max}$) is a triple (h, φ, h'), where $\varphi \in \mathbb{R}_{\max}\langle\langle T_\Gamma \rangle\rangle$ is a d-recognizable weighted tree language, and $h : T_\Gamma \to T_\Sigma$ and $h' : T_\Gamma \to T_\Delta$ are the input and the output tree homomorphism, respectively. The weighted tree transformation computed by (h, φ, h') is $\langle h, h'\rangle(\varphi)$. For any combinations w_1 and w_2 of l, c, we denote by $B^d(w_1\text{-}\mathcal{H}, w_2\text{-}\mathcal{H})$ the class of all weighted tree transformations computed by weighted d-bimorphisms with input tree homomorphism of type w_1 and output tree homomorphism of type w_2. Furthermore, we denote by $B^d(uc(\mathcal{H}, \mathcal{H}))$ (resp. $B^d(uc(l\text{-}\mathcal{H}, l\text{-}\mathcal{H}))$) the class of all weighted tree transformations computed by weighted d-bimorphisms whose input and output tree homomorphism constitute an ultimately complete pair of tree homomorphisms (resp. linear tree homomorphisms). We characterize the mentioned four classes as follows.

Theorem 4

(a) $EQUT^d_{[IO]} = B^d(uc(\mathcal{H}, \mathcal{H}))$ (a1) $vi\text{-}EQUT^d_{[IO]} = B(c\text{-}\mathcal{H}, c\text{-}\mathcal{H})$

(b) $EQUT^d_{OI} = B^d(uc(l\text{-}\mathcal{H}, l\text{-}\mathcal{H}))$ (b1) $vs\text{-}EQUT^d_{OI} = B(lc\text{-}\mathcal{H}, lc\text{-}\mathcal{H})$

Proof. (a) and (a1): First we prove the inclusions from right to left. In case (a) let $\tau = \langle h, h'\rangle(\varphi)$ for some d-recognizable weighted tree language $\varphi \in \mathbb{R}_{\max}\langle\langle T_\Gamma \rangle\rangle$, and the pair of tree homomorphisms $h : T_\Gamma \to T_\Sigma$ and $h' : T_\Gamma \to T_\Delta$ is ultimately complete. If $\varphi = \widetilde{-\infty}$, then obviously $\tau = \widetilde{-\infty} \in EQUT^d_{[IO]}$.

Otherwise, by Lemma 11, we can assume that φ is a component of the least OI-d-solution $(\varphi_1, \ldots, \varphi_n)$ of a linear system

$$\text{(G)} \quad x_1 = \psi_1, \ldots, x_n = \psi_n,$$

of weighted tree languages over $(\Gamma, X_n, \mathbb{R}_{\max})$ which satisfies $\psi_i \in \mathbb{R}_{\max}\langle\Gamma(X_n)\cup X_n\rangle$ for every $1 \leq i \leq n$. Furthermore (h, h') is an ultimately complete pair of tree homomorphisms, hence by Lemma 12(a) we get that $(\langle h, h'\rangle(\varphi_1), \ldots, \langle h, h'\rangle(\varphi_n))$ is the least $[IO]$-d-solution of $\langle h, h'\rangle$ (G). This implies that τ is a component of the least $[IO]$-d-solution of $\langle h, h'\rangle$ (G), hence $\tau \in EQUT^d_{[IO]}$.

In case (a1) the tree homomorphisms h and h' are nondeleting (and thus the pair (h, h') is ultimately complete). This implies that the system $\langle h, h'\rangle$ (G) is variable identical. Hence we get that $\tau \in vi\text{-}EQUT_{[IO]}$.

Now we prove the inclusion from left to right. In case (a) let $\tau \in \mathbb{R}_{\max}\langle\!\langle T_\Sigma \times T_\Delta\rangle\!\rangle$ be a component of the least *[IO]-d*-solution of a system

$$(\text{E}) \quad x_1 = \rho_1, \ldots, x_n = \rho_n,$$

of equations of weighted tree transformations over $(\Sigma, \Delta, X_n, \mathbb{R}_{\max})$.

For every $1 \leq i \leq n$ and every $(s,t) \in \text{supp}(\rho_i)$, we specify a new symbol $\sigma_{s,t}$ with rank $m = |var(s) \cup var(t)|$. Let Γ be the ranked alphabet consisting of all such symbols. Consider the system of equations

$$(\text{G}) \quad x_1 = \psi_1, \ldots, x_n = \psi_n,$$

of weighted tree languages over $(\Gamma, X_n, \mathbb{R}_{\max})$, where

$$\text{supp}(\psi_i) = \{\sigma_{s,t}(x_{i_1}, \ldots, x_{i_m}) \mid (s,t) \in \text{supp}(\rho_i), var(s) \cup var(t) = \{x_{i_1}, \ldots, x_{i_m}\},$$
$$1 \leq i_1 < \ldots < i_m \leq n\}$$

for every $1 \leq i \leq n$, and

$$(\psi_i, \sigma_{s,t}(x_{i_1}, \ldots, x_{i_m})) = (\rho_i, (s,t)).$$

Clearly, the system (G) is linear and simple. Now consider the tree homomorphisms $h : T_\Gamma \to T_\Sigma$ and $h' : T_\Gamma \to T_\Delta$ determined by $h_m(\sigma_{s,t}) = s(\xi_1/x_{i_1}, \ldots, \xi_m/x_{i_m})$ and $h'_m(\sigma_{s,t}) = t(\xi_1/x_{i_1}, \ldots, \xi_m/x_{i_m})$ for every $m \geq 0$ and $\sigma_{s,t} \in \Gamma_m$. Obviously, $\langle h, h'\rangle(\sigma_{s,t}(x_{i_1}, \ldots, x_{i_m})) = (s,t)$, and hence $\langle h, h'\rangle(\psi_i) = \rho_i$ for every $1 \leq i \leq n$, i.e., $(\text{E}) = \langle h, h'\rangle(\text{G})$. Moreover, by definition, (h, h') is an ultimately complete pair of tree homomorphisms, hence by Lemma 12(a) we obtain that $\tau = \langle h, h'\rangle(\varphi)$, where φ is a component of the least *OI-d*-solution of (G). Since φ is a *d*-recognizable series, our proof is completed.

The proof of (a1) is analogous, we just add the following. Since the system (E) is variable identical, the tree homomorphisms h and h' are nondeleting. Hence again Lemma 12(a) holds and we conclude $\tau \in B(c\text{-}\mathcal{H}, c\text{-}\mathcal{H})$, as required.

(b) and (b1): The proof of the inclusion from right to left in both cases is the same as in (a), except that we write *OI* for *[IO]* and Lemma 12(b) for Lemma 12(a). Moreover, in case (b1) we observe that the system $\langle h, h'\rangle(\text{G})$ is variable symmetric because the tree homomorphisms h and h' are linear and nondeleting.

For the proof of the other inclusion in case (b), we adapt the corresponding proof in (a) in the following way. Let us write *OI* for *[IO]*. By Lemma 9, we can assume that (E) is linear. This yields that (h, h') is an ultimately complete pair of linear tree homomorphisms. Then we can finish the proof by writing Lemma 12(b) for Lemma 12(a).

In case (b1), by Corollary 3, we can assume that (E) is linear. This yields that the tree homomorphisms h and h' are linear (and nondeleting).

Finally, we show that the class of *vi-[IO]-d*-equational weighted tree transformations is the closure of the class of *vs-OI-d*-equational weighted tree transformations under nondeleting tree homomorphisms. We will need the subsequent notation.

Let $(s,t) \in T_\Sigma \times T_\Delta$ and $g : T_\Sigma \to T_{\Sigma'}$ and $g' : T_\Delta \to T_{\Delta'}$ two tree homomorphisms. We set $\langle g, g' \rangle ((s,t)) = (g(s), g'(t))$. Furthermore, for every weighted tree transformation $\tau \in \mathbb{R}_{\max}^d \langle\langle T_\Sigma \times T_\Delta \rangle\rangle$, we define the weighted tree transformation $\langle g, g' \rangle (\tau)$ over $(\Sigma', \Delta', \mathbb{R}_{\max})$, such that

$$(\langle g, g' \rangle (\tau), (u,v)) = \sup_{\substack{(s,t) \in T_\Sigma \times T_\Delta \\ \langle g, g' \rangle ((s,t)) = (u,v)}} (\tau, (s,t))$$

for every $(u,v) \in T_{\Sigma'} \times T_{\Delta'}$. For a class $\mathcal{C}$ of bounded weighted tree transformations over $\mathbb{R}_{\max}$, we let

$$\langle c\text{-}\mathcal{H}, c\text{-}\mathcal{H} \rangle (\mathcal{C}) = \{ \langle g, g' \rangle (\tau) \mid \tau \in \mathcal{C}, \tau \in \mathbb{R}_{\max}^b \langle\langle T_\Sigma \times T_\Delta \rangle\rangle, \text{ and } g : T_\Sigma \to T_{\Sigma'}$$
$$\text{and } g' : T_\Delta \to T_{\Delta'} \text{ are nondeleting tree homomorphisms} \}.$$

Theorem 5. $\langle c\text{-}\mathcal{H}, c\text{-}\mathcal{H} \rangle \left(vs\text{-}EQUT_{OI}^d \right) = vi\text{-}EQUT_{[IO]}^d.$

Proof. We observe that $\langle c\text{-}\mathcal{H}, c\text{-}\mathcal{H} \rangle (B(lc\text{-}\mathcal{H}, lc\text{-}\mathcal{H})) = B(c\text{-}\mathcal{H}, c\text{-}\mathcal{H})$. The proof follows from the following obvious facts. The composition of a complete tree homomorphism and a linear and complete tree homomorphism is a complete tree homomorphism. Moreover, any complete tree homomorphism appears as the composition of a complete tree homomorphism and a linear and complete tree homomorphism (which may be, e.g, the complete tree homomorphism itself and the identity mapping, respectively). Then the statement follows from Theorem 4(a1) and (b1). □

7   A Mezei-Wright Like Relationship

In this section, we give a Mezei-Wright type result which relates u-d-equational weighted tree transformations and u-d-equational weighted relations both for $u=[IO]$ and $u=OI$. Namely, we show that a weighted relation is u-d-equational if and only if it is, roughly speaking, the morphic image of a u-d-equational weighted tree transformation. First we recall a preparatory result from [6].

Lemma 13. (cf. [6], Lemma 29) Let $s \in T_\Sigma(X_n)$ for some $n \geq 0$ with $|s|_{x_i} = \lambda_i$, $1 \leq i \leq n$. Moreover, let $s_1, \ldots, s_n \in T_\Sigma$ and $\mathbf{s}^{(i)} = \left(s_1^{(i)}, \ldots, s_{\lambda_i}^{(i)} \right) \in T_\Sigma^{\lambda_i}$ for every $1 \leq i \leq n$. Then

(a) $H_{\mathcal{A}} (s (s_1, \ldots, s_n)) = s (H_{\mathcal{A}}(s_1), \ldots, H_{\mathcal{A}}(s_n))$ and
(b) $H_{\mathcal{A}} (s (\mathbf{s}^{(1)}, \ldots, \mathbf{s}^{(n)})) = s (H_{\mathcal{A}}(\mathbf{s}^{(1)}), \ldots, H_{\mathcal{A}}(\mathbf{s}^{(n)})),$

where $H_{\mathcal{A}} (\mathbf{s}^{(i)}) = \left(H_{\mathcal{A}} \left(s_1^{(i)} \right), \ldots, H_{\mathcal{A}} \left(s_{\lambda_i}^{(i)} \right) \right)$ for every $1 \leq i \leq n$.

Next we define the mapping $H_{(\mathcal{A}, \mathcal{B})} : \mathbb{R}_{\max}^b \langle\langle T_\Sigma \times T_\Delta \rangle\rangle \to \mathbb{R}_{\max}^b \langle\langle A \times B \rangle\rangle$ such that for any $\tau \in \mathbb{R}_{\max} \langle\langle T_\Sigma \times T_\Delta \rangle\rangle$ and $(a,b) \in A \times B$, we have

$$(H_{(\mathcal{A}, \mathcal{B})}(\tau), (a,b)) = \sup_{\substack{(s,t) \in T_\Sigma \times T_\Delta \\ H_{(\mathcal{A}, \mathcal{B})}((s,t)) = (a,b)}} (\tau, (s,t)),$$

where $H_{(\mathcal{A},\mathcal{B})}((s,t)) = (H_{\mathcal{A}}(s), H_{\mathcal{B}}(t))$ for all $(s,t) \in T_{\Sigma} \times T_{\Delta}$. We note that $H_{(\mathcal{A},\mathcal{B})}$ is monotonic. We will need the subsequent lemmas.

Lemma 14. Let $(\tau_n)_{n \geq 0}$ be an ω-chain in $\mathbb{R}_{\max}\langle\!\langle T_{\Sigma} \times T_{\Delta} \rangle\!\rangle$ such that the set $\{(\tau_n, (s,t)) \mid n \geq 0, (s,t) \in T_{\Sigma} \times T_{\Delta}\}$ is bounded in $\mathbb{R}_{\max}$. Then $\lim_{n \to \infty} (H_{(\mathcal{A},\mathcal{B})}(\tau_n))$ exists in $\mathbb{R}_{\max}\langle\!\langle A \times B \rangle\!\rangle$ and $\lim_{n \to \infty} (H_{(\mathcal{A},\mathcal{B})}(\tau_n)) = H_{(\mathcal{A},\mathcal{B})}(\lim_{n \to \infty} \tau_n)$.

Proof. By our assumption and by the definition of $H_{(\mathcal{A},\mathcal{B})}$, the set $\{(H_{(\mathcal{A},\mathcal{B})}(\tau_n), (a,b)) \mid n \geq 0, (a,b) \in A \times B\}$ is bounded in $\mathbb{R}_{\max}$, hence $\lim_{n \to \infty} (H_{(\mathcal{A},\mathcal{B})}(\tau_n))$ exists. Moreover, for every $(a,b) \in A \times B$ we have

$$\left(\lim_{n \to \infty} (H_{(\mathcal{A},\mathcal{B})}(\tau_n)), (a,b) \right) = \lim_{n \to \infty} (H_{(\mathcal{A},\mathcal{B})}(\tau_n), (a,b)) =$$

$$\lim_{n \to \infty} \left(\sup_{\substack{(s,t) \in T_{\Sigma} \times T_{\Delta} \\ H_{(\mathcal{A},\mathcal{B})}((s,t))=(a,b)}} (\tau_n, (s,t)) \right) =$$

$$\sup_{\substack{(s,t) \in T_{\Sigma} \times T_{\Delta} \\ H_{(\mathcal{A},\mathcal{B})}((s,t))=(a,b)}} \left(\lim_{n \to \infty} (\tau_n, (s,t)) \right) = \sup_{\substack{(s,t) \in T_{\Sigma} \times T_{\Delta} \\ H_{(\mathcal{A},\mathcal{B})}((s,t))=(a,b)}} \left(\lim_{n \to \infty} \tau_n, (s,t) \right)$$

$$= \left(H_{(\mathcal{A},\mathcal{B})} \left(\lim_{n \to \infty} \tau_n \right), (a,b) \right),$$

where the third equality holds by Lemma 4.

Lemma 15. For every $n \geq 0$, $\tau \in \mathbb{R}_{\max}\langle T_{\Sigma}(X_n) \times T_{\Delta}(X_n) \rangle$, $\tau_1, \ldots, \tau_n \in \mathbb{R}_{\max}^b\langle\!\langle T_{\Sigma} \times T_{\Delta} \rangle\!\rangle$, and $u = [IO]$, OI, we have

$$H_{(\mathcal{A},\mathcal{B})}(\tau[\tau_1, \ldots, \tau_n]_u) = \tau [H_{(\mathcal{A},\mathcal{B})}(\tau_1), \ldots, H_{(\mathcal{A},\mathcal{B})}(\tau_n)]_u.$$

Proof. We show the equality for the *[IO]*-d-substitution only because the proof for the *OI*- one is similar. Firstly, we prove that

$$H_{(\mathcal{A},\mathcal{B})}\left((s,t)[\tau_1, \ldots, \tau_n]_{[IO]} \right) = (s,t)[H_{(\mathcal{A},\mathcal{B})}(\tau_1), \ldots, H_{(\mathcal{A},\mathcal{B})}(\tau_n)]_{[IO]}$$

for every $(s,t) \in T_{\Sigma}(X_n) \times T_{\Delta}(X_n)$. Let $var(s) \cup var(t) = \{x_{i_1}, \ldots, x_{i_k}\}$. Then for every $(a,b) \in A \times B$ we have

$$\left(H_{(\mathcal{A},\mathcal{B})} \left((s,t) \left[\tau_1, \ldots, \tau_n \right]_{[IO]} \right), (a,b) \right)$$

$$= d/k \left(\sup_{\substack{(s_i,t_i)\in T_\Sigma\times T_\Delta,\, 1\le i\le k \\ H_{(\mathcal{A},\mathcal{B})}(s(u_1,\ldots,u_n),t(v_1,\ldots,v_n))=(a,b)}} \left((\tau_{i_1},(s_1,t_1)) + \ldots + (\tau_{i_k},(s_k,t_k)) \right) \right)$$

$$= d/k \left(\sup_{\substack{(s_i,t_i)\in T_\Sigma\times T_\Delta,\, 1\le i\le k \\ (H_{\mathcal{A}}(s(u_1,\ldots,u_n)),H_{\mathcal{B}}(t(v_1,\ldots,v_n)))=(a,b)}} \left((\tau_{i_1},(s_1,t_1)) + \ldots + (\tau_{i_k},(s_k,t_k)) \right) \right)$$

$$= d/k \left(\sup_{\substack{(s_i,t_i)\in T_\Sigma\times T_\Delta,\, 1\le i\le k \\ (s(H_{\mathcal{A}}(u_1),\ldots,H_{\mathcal{A}}(u_n)),t(H_{\mathcal{B}}(v_1),\ldots,H_{\mathcal{B}}(v_n)))=(a,b)}} \right.$$
$$\left. \left((\tau_{i_1},(s_1,t_1)) + \ldots + (\tau_{i_k},(s_k,t_k)) \right) \right)$$

$$= d/k \left(\sup_{\substack{(s_i,t_i)\in T_\Sigma\times T_\Delta,\, 1\le i\le k \\ (s,t)\left(H_{(\mathcal{A},\mathcal{B})}(u_1,v_1),\ldots,H_{(\mathcal{A},\mathcal{B})}(u_1,v_n)\right)=(a,b)}} \right.$$
$$\left. \left((\tau_{i_1},(s_1,t_1)) + \ldots + (\tau_{i_k},(s_k,t_k)) \right) \right)$$

$$= (s,t) \left[H_{(\mathcal{A},\mathcal{B})}(\tau_1), \ldots, H_{(\mathcal{A},\mathcal{B})}(\tau_n) \right]_{[IO]}$$

where for every $((s_1,t_1),\ldots,(s_k,t_k))$, the sequence $((u_1,v_1),\ldots,(u_n,v_n))$ is an arbitrary element of $(T_\Sigma \times T_\Delta)^n|_{(i_1,(s_1,t_1))\ldots(i_k,(s_k,t_k))}$. Moreover, at the third equality we use Lemma 13(a). Finally, we have

$$H_{(\mathcal{A},\mathcal{B})} \left(\tau \left[\tau_1, \ldots, \tau_n \right]_{[IO]} \right)$$

$$= H_{(\mathcal{A},\mathcal{B})} \left(\max_{(s,t)\in\mathrm{supp}(\tau)} \left((\tau,(s,t)) + (s,t) \left[\tau_1, \ldots, \tau_n \right]_{[IO]} \right) \right)$$

$$= \max_{(s,t)\in\mathrm{supp}(\tau)} \left((\tau,(s,t)) + H_{(\mathcal{A},\mathcal{B})} \left((s,t) \left[\tau_1, \ldots, \tau_n \right]_{[IO]} \right) \right)$$

$$= \max_{(s,t)\in\mathrm{supp}(\tau)} \left((\tau,(s,t)) + (s,t) \left[H_{(\mathcal{A},\mathcal{B})}(\tau_1), \ldots, H_{(\mathcal{A},\mathcal{B})}(\tau_n) \right]_{[IO]} \right)$$

$$= \tau \left[H_{(\mathcal{A},\mathcal{B})}(\tau_1), \ldots, H_{(\mathcal{A},\mathcal{B})}(\tau_n) \right]_{[IO]}.$$

Now we are ready to state and prove the mentioned Mezei-Wright like correspondence (cf. [40], Theorem 5.5) between u-d-equational weighted relations and u-d-equational weighted tree transformations for $u = [IO], OI$.

Theorem 6. *Let $u = [\text{IO}]$ or $u=\text{OI}$. A weighted relation $\theta \in \mathbb{R}_{\max}\langle\!\langle A \times B \rangle\!\rangle$ is u-d-equational iff there exists a u-d-equational weighted tree transformation $\tau \in \mathbb{R}_{\max}\langle\!\langle T_\Sigma \times T_\Delta \rangle\!\rangle$ such that $H_{(\mathcal{A},\mathcal{B})}(\tau) = \theta$.*

Proof. Assume first that θ is u-d-equational. Then there is a system

$$(\text{E}) \quad x_1 = \rho_1, \ldots, x_n = \rho_n,$$

of equations of weighted tree transformations over $(\Sigma, \Delta, X_n, \mathbb{R}_{\max})$ such that θ is a component of its least u-d-solution $(\theta_1, \ldots, \theta_n)$ in $(\mathcal{A}, \mathcal{B}, \mathbb{R}_{\max})$. Let $(\tau_1, \ldots, \tau_n)$ be the least u-d-solution of (E). We show that $H_{(\mathcal{A},\mathcal{B})}(\tau_i) = \theta_i$ for every $1 \leq i \leq n$. By Lemma 15 we have

$$H_{(\mathcal{A},\mathcal{B})}(\tau_i) = H_{(\mathcal{A},\mathcal{B})}\left(\rho_i\left[\tau_1, \ldots, \tau_n\right]_u\right) = \rho_i\left[H_{(\mathcal{A},\mathcal{B})}\left(\tau_1\right), \ldots, H_{(\mathcal{A},\mathcal{B})}\left(\tau_n\right)\right]_u,$$

i.e., $\left(H_{(\mathcal{A},\mathcal{B})}\left(\tau_1\right), \ldots, H_{(\mathcal{A},\mathcal{B})}\left(\tau_n\right)\right)$ is a u-d-solution of (E) in $(\mathcal{A}, \mathcal{B}, \mathbb{R}\max)$. We show that in fact it is the least u-d-solution of (E). For this let

$$(\tau_1, \ldots, \tau_n) = \lim_{k \to \infty}(\tau_{1,k}, \ldots, \tau_{n,k}),$$

where $(\tau_{1,k}, \ldots, \tau_{n,k})_{k \geq 0}$ is the u-d-approximation sequence of (E). We show by induction that, for every $1 \leq i \leq n$ and $k \geq 0$, we have $H_{(\mathcal{A},\mathcal{B})}(\tau_{i,k}) \leq \theta_i$. For $k = 0$ this is true by definition. Then, for every $k \geq 0$, we have

$$H_{(\mathcal{A},\mathcal{B})}(\tau_{i,k+1}) = H_{(\mathcal{A},\mathcal{B})}\left(\rho_i\left[\tau_{1,k}, \ldots, \tau_{n,k}\right]_u\right)$$
$$= \rho_i\left[H_{(\mathcal{A},\mathcal{B})}\left(\tau_{1,k}\right), \ldots, H_{(\mathcal{A},\mathcal{B})}\left(\tau_{n,k}\right)\right]_u$$
$$\leq \rho_i\left[\theta_1, \ldots, \theta_n\right]_u = \theta_i,$$

where the second equality holds by Lemma 15 and the inequality holds by the induction hypothesis and Remark 2. By Lemma 14 we conclude $H_{(\mathcal{A},\mathcal{B})}(\tau_i) = \theta_i$ for every $1 \leq i \leq n$ and this proves one half of our theorem. The other direction can be proved similarly.

Acknowledgment. The authors are grateful to Andreas Maletti for his valuable remarks on an earlier version of this paper.

References

1. Alexandrakis, A., Bozapalidis, S.: Weighted grammars and Kleene's theorem. Inform. Process. Lett. 24, 1–4 (1987)
2. Arnold, A., Dauchet, M.: Bi-transductions de forets. In: Michaelson, S., Milner, R. (eds.) Proc. 3rd Int. Coll. Automata, Languages, and Programming, pp. 74–86. Edinburgh University Press, Edinburgh (1976)
3. Arnold, A., Dauchet, M.: Morphismes et bimorphismes d'arbes. Theoret. Comput. Sci. 20, 33–93 (1982)
4. Berstel, J., Reutenauer, C.: Recognizable formal power series on trees. Theoret. Comput. Sci. 18, 115–148 (1982)

5. Bozapalidis, S.: Equational elements in additive algebras. Theory of Comput. Syst. 32, 1–33 (1999)
6. Bozapalidis, S., Fülöp, Z., Rahonis, G.: Equational tree transformations. Theoret. Comput. Sci. 412, 3676–3692 (2011)
7. Bozapalidis, S., Fülöp, F., Rahonis, G.: Equational weighted tree transformations (submitted, 2011)
8. Bozapalidis, S., Rahonis, G.: On the closure of recognizable tree series under tree homomorphisms. J. Autom. Lang. Comb. 10, 185–202 (2005)
9. Chatterjee, K., Doyen, L., Henzinger, T.A.: Quantitative languages. In: Kaminski, M., Martini, S. (eds.) CSL 2008. LNCS, vol. 5213, pp. 385–400. Springer, Heidelberg (2008)
10. Chatterjee, K., Doyen, L., Henzinger, T.A.: Composition and alternation for weighted automata. EPLF Technical Report MTC-REPORT-2008-004
11. Comon, H., Dauchet, M., Gilleron, R., Jacquema, F., Lugiez, D., Tison, S., Tommasi, M.: Tree Automata Techniques and Applications, http://tata.gforge.inria.fr/
12. Courcelle, B.: Equivalences and transformations of regular systems – Applications to recursive program schemes and grammars. Theoret. Comput. Sci. 42, 1–122 (1986)
13. Courcelle, B.: Basic notions of universal algebra for language theory and graph grammars. Theoret. Comput. Sci. 163, 1–54 (1996)
14. de Alfaro, L., Henzinger, T.A., Majumda, R.: Discounting the future in systems theory. In: Baeten, J.C.M., Lenstra, J.K., Parrow, J., Woeginger, G.J. (eds.) ICALP 2003. LNCS, vol. 2719, pp. 1022–1037. Springer, Heidelberg (2003)
15. Droste, M., Rahonis, G.: Weighted automata and weighted logics with discounting. Theoret. Comput. Sci. 410, 3481–3494 (2009)
16. Droste, M., Sakarovitch, J., Vogler, H.: Weighted automata with discounting. Inform. Process. Lett. 108, 23–28 (2008)
17. Droste, M., Kuich, W.: Semirings and formal power series. In: Droste, M., Kuich, W., Vogler, H. (eds.) Handbook of Weighted Automata. EATCS Monographs in Theoretical Computer Science, pp. 3–28. Springer, Heidelberg (2009)
18. Droste, M., Kuske, D.: Skew and infinitary formal power series. Theoret. Comput. Sci. 366, 189–227 (2006)
19. Engelfriet, J.: Bottom-up and top-down tree transformations - a comparison. Math. Systems Theory 9, 198–231 (1975)
20. Engelfriet, J., Schmidt, E.M.: IO and OI. I.;II. J. Comput. System Sci. 15, 328–353 (1977); 16, 67–99 (1978)
21. Engelfriet, J., Fülöp, Z., Vogler, H.: Bottom-up and top-down tree series transformations. J. Autom. Lang. Comb. 7, 11–70 (2002)
22. Ésik, Z., Kuich, W.: Formal tree series. J. Autom. Lang. Comb. 8, 219–285 (2003)
23. Filar, J., Vrieze, K.: Competitive Marcov Decision Processes. Springer, Heidelberg (1997)
24. Fülöp, Z., Maletti, A., Vogler, H.: Backward and forward application of extended tree series transformations. Fund. Inform. 112, 1–39 (2011)
25. Fülöp, Z., Vogler, H.: Weighted tree automata and tree transducers. In: Droste, M., Kuich, W., Vogler, H. (eds.) Handbook of Weighted Automata. EATCS Monographs in Theoretical Computer Science, pp. 313–404. Springer, Heidelberg (2009)
26. Fülöp, Z., Vogler, H.: Weighted tree transducers. J. Autom. Lang. Comb. 9, 31–54 (2004)
27. Gécseg, F., Steinby, M.: Tree Automata. Akadémiai Kiadó, Budapest (1984)

28. Gécseg, F., Steinby, M.: Tree languages. In: Rozenberg, G., Salomaa, A. (eds.) Handbook of Formal Languages, vol. III, pp. 1–68. Springer, Heidelberg (1997)
29. Golan, J.S.: Semirings and their Applications. Kluwer Academic Publishers, Dordrecht (1999)
30. Graehl, J., Knight, K., May, J.: Training tree transducers. Computational Linguistics 34, 391–427 (2008)
31. Knight, K., Graehl, J.: An overview of probabilistic tree transducers for natural language processing. In: Gelbukh, A. (ed.) CICLing 2005. LNCS, vol. 3406, pp. 1–24. Springer, Heidelberg (2005)
32. Kuich, W.: Formal power series over trees. In: Bozapalidis, S. (ed.) Proceedings of DLT 1997, pp. 61–101. Aristotle University of Thessaloniki, Thessaloniki (1998)
33. Kuich, W.: Tree transducers and formal tree series. Acta Cybernet. 14, 135–149 (1999)
34. Kuich, W.: On skew formal power series. In: Bozapalidis, S., Kalampakas, A., Rahonis, G. (eds.) Proceedings of CAI 2005, pp. 7–30 (2005)
35. Maletti, A.: Relating tree series transducers and weighted tree automata. Internat. J. Found. Comput. Sci. 16, 723–741 (2005)
36. Maletti, A.: Compositions of tree series transformations. Theoret. Comput. Sci. 366, 248–271 (2006)
37. Maletti, A.: Compositions of extended top-down tree transducers. Inform. and Comput. 206, 1187–1196 (2008)
38. Maletti, A., Graehl, J., Hopkins, M., Knight, K.: The power of extended top-down tree transducers. SIAM J. Comput. 39(2), 410–430 (2008)
39. Mandrali, E., Rahonis, G.: Recognizable tree series with discounting. Acta Cybernet. 19, 411–439 (2009)
40. Mezei, J., Wright, J.B.: Algebraic automata and context-free sets. Inform. Control 11, 3–29 (1967)
41. Shapley, L.S.: Stochastic games. Roc. National Acad. of Sciences 39, 1095–1100 (1953)
42. Wechler, W.: Universal Algebra for Computer Scientists. EATCS Monographs on Theoretical Computer Science, vol. 25. Springer, Heidelberg (1992)

Quantum Automata Theory – A Review*

Mika Hirvensalo

Department of Mathematics, University of Turku, FIN-20014, Turku, Finland and
TUCS – Turku Centre for Computer Science
`mikhirve@utu.fi`

Abstract. In the first part of this survey paper, the notions of finite
automata and regular languages are reviewed from various points of view.
The middle part contains an introduction to the Hilbert space formalism
of finite-level quantum systems, and the final part is a presentation of the
most notable quantum finite automata models introduced up to date.

1 Finite Automata

The theory of finite automata is one of the cornerstones of theoretical computer
science. Finite automata were introduced in 1940's and 1950's via a series of
papers: notable ones include those of McCulloch and Pitts [31], Kleene [28],
Mealy [32], Moore [33], and Rabin and Scott [37]. The notion of finite automaton,
as well as the other notions of this section, are formally defined in the next
section, but to understand the meaning and importance of finite automata, it
may be necessary to consider various points of view.

Finite automata are theoretical models for real-time computing with a finite
memory. By real-time computing we mean here that the automaton reads its
input once, and gives the answer immediately when the whole input is read.
Such a model is evidently an important object of research by itself, but seeing
finite automata merely as computing machines gives only a part of the picture.
More points of view arise from the language theory: As a language accepted by
a finite automaton we understand the set of inputs the automaton "permits"
in a sense defined later. It turns out that the languages accepted by finite automata are exactly regular languages, which can be built from finite languages
by using concatenation, union, and Kleene star. The third point of view arises
from formal power series: The supports of power series of rational functions are
exactly regular languages. The fourth point of view is connected to the monoids:
Rational languages are exactly the languages having a finite syntactic monoid.

Listed as such, the aforementioned viewpoints are merely mathematically
provable equivalences, but the most important point lies among them: finite
automata and languages they accept have many faces. It is true that finite
automata have various applications from compiling and parsing [2] to image
compression [15], [16], but the applications only are hardly the propelling force
which has made finite automata an interesting research object for researchers
over the decades. Instead, the significance of finite automata most likely arises

---

* Dedicated to Symeon Bozapalidis on his very special occasion.

W. Kuich and G. Rahonis (Eds.): Bozapalidis Festschrift, LNCS 7020, pp. 146–167, 2011.

from the fact that they, and the languages they determine are mathematically extremely fascinating objects: Regular languages are closed under union, intersection, complementation, concatenation and Kleene star, and all the closure properties can be proven true constructively from the automata point of view. Multiple characterizations for a single object from different viewpoints almost always enriches mathematics, and here the theory of finite automata serves as an exemplar of elegance.

1.1 Formal Definitions

The literature on automata theory is very rich, and it is certainly possible to exhaust all pages (and far more) of a short article like this by only listing all notable work on the topic, hence there is no point in trying to do so. Instead, we just mention a classic work by Eilenberg [19], and another recommendable presentation by Sheng Yu [43] from language theory viewpoint. The same minimalistic line will be followed when introducing automata theory notions: only those of major importance to this article will be presented.

This presentation does not follow literally any particular source, but the definitions presented in this section are generally recognized anyway.

Definition 1. *An alphabet Σ is a finite set. Taking concatenation as the operation, Σ^* denotes the free monoid generated by Σ. Elements of Σ^* are called* words. *The neutral element of Σ^* is denoted by 1 and called* the empty word. *The* length *of a word w is defined as $|w| = 0$, if $w = 1$ is the empty word, and n, if $w = a_1 \ldots a_n$, where $a_i \in \Sigma$. For any word $w = a_1 a_2 \ldots a_n$, its* mirror image *is defined as $w^R = a_n \ldots a_2 a_1$. Any subset of Σ^* is called a* language over Σ.

Definition 2 (Regular languages). *A language $L \subseteq \Sigma^*$ is* regular, *if 1) L is finite or 2) L is obtained from regular languages L_1 and L_2 by either union $L_1 \cup L_2$, concatenation $L_1 L_2 = \{w_1 w_2 \mid w_1 \in L_1, w_2 \in L_2\}$, or Kleene star: $L_1^* = \{w_1 w_2 \ldots w_k \mid k \geq 0, w_i \in L_1\}$.*

Any language constructed by using the above rules can be proven regular simply by showing the derivation tree. Such an expression is called *regular expression*.

Example 1. The language over alphabet $\Sigma = \{a, b\}$ consisting of words which contain an even number of letters a or at least one b is regular, as it can be presented as a regular expression $(\Sigma^* a \Sigma^* a \Sigma^*)^* \cup \Sigma^* b \Sigma^*$.

Definition 3 (DFA). *A deterministic finite automaton (DFA) $\mathcal{F}$ is a quintuple*

$$\mathcal{F} = (Q, \Sigma, \delta, q_I, F),$$

where Q is a finite set of states, *Σ an* alphabet, *$\delta : Q \times \Sigma \to Q$ the* transition function[1], *$q_I \in Q$ the* initial state, *and $F \subseteq Q$ the set of* final states.

---

[1] In this article, we always assume the automata *complete*, meaning that the transition function is total. This can always be achieved by adding an extra state.

The transition function δ can be extended into a function $\delta : Q \times \Sigma^* \to Q$ by $\delta(q, 1) = q$ and if $w = aw_1$, where $a \in \Sigma$ and $w_1 \in \Sigma^*$, then $\delta(q, w) = \delta(\delta(a, q), w_1)$.

Definition 4. *The* Boolean semiring $\mathbb{B} = \{0, 1\}$ *is equipped with obvious multiplication and addition, but* $1 + 1 = 1$.

Definition 5. *A DFA $\mathcal{F}$ computes a function $f_{\mathcal{F}} : \Sigma^* \to \mathbb{B}$ defined as*

$$f_{\mathcal{F}}(w) = \begin{cases} 1 \text{ if } \delta(q_I, w) \in F \\ 0 \text{ otherwise.} \end{cases}$$

Definition 6 (Recognizable Languages). *The language recognized (accepted) by finite automaton $\mathcal{F}$ is*

$$L(\mathcal{F}) = \{w \in \Sigma^* \mid f_{\mathcal{F}}(w) = 1\}.$$

The following characterization is well known (see [19], for instance).

Theorem 1. *A language L is regular if and only if it is recognized by a DFA.*

It turns out that the matrix formalism is very useful for introducing generalizations and variants of DFA. We fix an order $Q = \{q_1, \ldots, q_n\}$ on the state set and for each letter $a \in \Sigma$, define a matrix

$$(M_a)_{ij} = \begin{cases} 1 \text{ if } \delta(q_j, a) = q_i \\ 0 \text{ otherwise.} \end{cases}$$

over the Boolean semiring. From the definition it is clear that M_a has exactly one 1 in each column. The initial vector $\boldsymbol{x} \in \mathbb{B}^n$ is defined so that $\boldsymbol{x}_i = 1$, if q_i is the initial state, and $\boldsymbol{x}_i = 0$ otherwise. The final state vector $\boldsymbol{y} \in \mathbb{B}^n$ is then defined so that $\boldsymbol{y}_i = 1$ if and only if $q_i \in F$. Both $\boldsymbol{x}$ and $\boldsymbol{y}$ are regarded as row vectors.

Now if $\boldsymbol{z}$ is any vector with only one 1 at position k and $\delta(q_k, a) = q_l$, then $(M_a)_{lk} = 1$ is the only nonzero element in the kth column, and hence

$$(M\boldsymbol{z}^T)_m = \sum_{j=1}^{n} (M_a)_{mj} \boldsymbol{z}_j = (M_a)_{mk} = \begin{cases} 1 \text{ if } m = l \\ 0 \text{ otherwise.} \end{cases}$$

This simply means that M_a moves the nonzero element of $\boldsymbol{z}$ from position k to position l, as the automaton moves from state q_k to state q_l when reading letter a. If we now define recursively

$$M_w = \begin{cases} I \text{ if } w = 1, \\ M_a M_{w_1} \text{ if } w = aw_1, \text{ where } a \in \Sigma, \end{cases}$$

it is then clear that $(M\boldsymbol{x}^T)_m = 1$ if and only if $\delta(q_I, w) = q_m$, and we see that

$$f_{\mathcal{F}}(w) = \boldsymbol{y} M_{w^R} \boldsymbol{x}^T. \tag{1}$$

Equation (1) offers a good starting point for generalizations.

1.2 Classical Variants

Definition 7 (NFA). *A nondeterministic finite automaton $\mathcal{N}$ is defined exactly as DFA, but instead of a transition function, the dynamics is determined by a transition relation $\delta \subseteq Q \times \Sigma \times Q$, and there may be many initial states.*

The matrix representation of an NFA is as simple as that of DFA: $(M_a)_{ij} = 1$ if and only if $(q_j, a, q_i) \in \delta$, and the initial vector $\boldsymbol{x}$ has 1 exactly at the positions corresponding to the initial states. The function computed by an NFA $\mathcal{N}$ is

$$f_{\mathcal{N}}(w) = \boldsymbol{y} M_{w^R} \boldsymbol{x}^T,$$

and the languages recognized by NFA are defined as those recognized by DFA.

In principle, nondeterminism does not bring any advantage for the language recognition; the following fact is well-known [19], [43].

Theorem 2. *For each NFA $\mathcal{N}$ with n states there is a DFA $\mathcal{F}$ with at most 2^n states so that $f_{\mathcal{N}} = f_{\mathcal{F}}$.*

Even though NFAs does not bring any advantage for language recognition, the complexity (number of states needed) may change essentially: It is known that for any $n \in \mathbb{N}$, there are n-state NFAs recognizing languages which cannot be recognized by any DFA with less than 2^n states [19].

For a general treatment of probabilistic automata, the reader is advised to consult [36], here the introduction is short:

Definition 8 (PFA). *A probabilistic finite automaton $\mathcal{P}$ is defined as DFA, but instead of transition function, there is a transition probability function $\delta : \Sigma \times Q \times \Sigma \rightarrow [0,1]$, and the initial state is replaced by an initial distribution $\boldsymbol{x} \in \mathbb{R}^n$ so that $\boldsymbol{x}_j \geq 0$ and $\sum_{j=1}^{n} \boldsymbol{x}_j = 1$.*

The matrix form consists now of matrices over $\mathbb{R}$ defined as

$$(M_a)_{ij} = \delta(q_j, a, q_i),$$

which stands for the transition probability: $(M_a)_{ij}$ is the probability that being in state q_j and reading a symbol a, the automaton will enter state q_j. The evident requirement is then that

$$\sum_{i=1}^{n} \delta(q_j, a, q_i) = \sum_{i=1}^{n} (M_a)_{ij} = 1,$$

meaning that each column is a probability distribution. Matrices satisfying these requirements are called *Markov matrices*.

A PFA computes a function $f_{\mathcal{P}} : \Sigma^* \rightarrow [0,1]$ defined as

$$f_{\mathcal{P}}(w) = \boldsymbol{y} M_{w^R} \boldsymbol{x}^T,$$

but as this function is not anymore $\{0,1\}$-valued, it is not so straightforward how to define the language recognized (or accepted) by a PFA. The most evident approach begins with a *cut-point* $\lambda \in [0,1]$ and continues with a definition

$$L_{>\lambda}(\mathcal{P}) = \{w \in \Sigma^* \mid f_{\mathcal{P}}(w) > \lambda\}$$

or

$$L_{\geq\lambda}(\mathcal{P}) = \{w \in \Sigma^* \mid f_{\mathcal{P}}(w) \geq \lambda\}.$$

In other words, $L_{>\lambda}(\mathcal{P})$ (resp. $L_{\geq\lambda}(\mathcal{P})$) consists of words with acceptance probability greater (resp. equal or greater) than λ.

Such languages are no longer necessarily regular [38], [35], [42], but the regularity can be guaranteed by assuming that the cut-point is *isolated*.

Definition 9. *Let $\mathcal{P}$ be a probabilistic automaton and $\epsilon > 0$. A cut-point $\lambda \in (0,1)$ is ϵ-isolated, if $f_{\mathcal{P}}(w) \notin (\lambda - \epsilon, \lambda + \epsilon)$ whenever $w \in \Sigma^*$.*

The notion of isolated cut-point is very desirable for practical reasons: if values $f_{\mathcal{P}}(w)$ can get arbitrarily close to the cut-point, it is difficult in practice to decide whether the automaton accepts w. Indeed, the cases when a final state is reached with a probability of $\frac{1}{2} + \frac{1}{2^{|w|}}$ and $\frac{1}{2} - \frac{1}{2^{|w|}}$ cannot be separated reliably with less than $\Omega(2^{|w|})$ attempts. Unfortunately the isolation of the cut-point should usually emerge intentionally from the construction of the automaton, since, given an automaton $\mathcal{P}$ and cut-point λ, there is no way to determine whether the cut-point is isolated, but it is an undecidable problem [10].

On the other hand, if the cut-point is ϵ-isolated, then only a constant number (depends on ϵ) of runs is enough to determine the acceptance question with probability as close to one as desired.

The following theorem, due to Rabin [38], shows that PFA with isolated cut-point cannot recognize more languages than DFA.

Theorem 3. *Let $\mathcal{P}$ be a probabilistic automaton with n states and one final state.[2] Let also λ be an ϵ-isolated cut-point. Then there exists a DFA with at most $(1 + \frac{1}{\epsilon})^{n-1}$ states recognizing language $L_{\geq\lambda}(\mathcal{P}) = L_{>\lambda}(\mathcal{P})$.*

In [38] Rabin also demonstrated that the probabilistic automata can be more succinct than the deterministic ones. He indeed constructed a sequence L_n of languages and the corresponding cut-points λ_n so that each L_n is accepted by a 2-state PFA with isolated cut-point λ_n, but a DFA recognizing L_n requires at least n states.

Theorem 3 shows that if the cut-point is isolated, then PFAs can be at most exponentially more succinct than DFAs. In [21] R. Freivalds presented a subexponential gap: A sequence of languages L_n accepted by a PFA with n states and fixed cut-point isolation, whereas any DFA accepting L_n requires $\Omega(2^{\sqrt{n}})$ states. Eventually in [22] R. Freivalds proved the existence of a language sequence where the separation between PFA and DFA sizes is exponential.

---

[2] A probabilistic automaton $\mathcal{P}$ can be always translated into $\mathcal{P}_1$ with one more state and only one final state.

2   Syntactic Monoids

The notion of a syntactic monoid was presented by Rabin and Scott in [37] and it brings another aspect to the languages accepted by finite automata. In fact, the main idea connecting automata to monoids has been already presented: Any DFA can be represented as a set of matrices M_a over $\mathbb{B}$. As $\mathbb{B}$ is finite, it is clear that the matrices M_a generate a finite monoid. This can be represented in a bit more abstract form as follows.

Definition 10. *Let L be a language over Σ. We say that words v and u are (syntactically) congruent (with respect to L), denoted $u \sim_L v$, if for all $x, y \in \Sigma^*$ we have*

$$xuy \in L \iff xvy \subset L.$$

It is straightforward to verify that the syntactic congruence is an equivalence relation, and that it is compatible with the concatenation, meaning that if $u_1 \sim_L u_2$ and $v_1 \sim_L v_2$, then also $u_1 v_1 \sim_L u_2 v_2$.[3] This implies that the multiplication on equivalence classes

$$[u] = \{v \in \Sigma^* \mid v \sim_L u\}$$

defined as $[u][v] = [uv]$ is a well-defined operation.

Definition 11. *The syntactic monoid of language L is the quotient*

$$M(L) = \Sigma^* / \sim = \{[u] \mid u \in \Sigma^*\}$$

equipped with operation $[u][v] = [uv]$.

Notice that the notion of syntactic monoid is defined for each language L independently of regularity or other assumptions.

Definition 12. *Let M_1 and M_2 be monoids. Mapping $\varphi : M_1 \to M_2$ is a morphism, if $\varphi(m_1 m_2) = \varphi(m_1)\varphi(m_2)$ holds for all $m_1, m_2 \in M_1$ and $\varphi(1) = 1$.*

Definition 13. *Language L is recognized by a monoid M, if there exists a morphism $\varphi : \Sigma^* \to M$ and a subset $B \subseteq M$ so that $L = \varphi^{-1}(B)$.*

For each language $L \subseteq \Sigma^*$ there are some obvious choices for recognizing monoids. For instance taking $M = \Sigma^*$, φ the identity mapping and $B = L$ gives obviously a recognizing monoid. Another choice is $M = M(L)$ (the syntactic monoid), φ the projection $\varphi(w) = [w]$, and $B = \varphi(L)$.

In a very true sense, the syntactic monoid is the smallest one recognizing L: If N is another monoid recognizing L, then $M(L)$ is a quotient of a submonoid of N. To see this, let $\varphi : \Sigma^* \to N$ be a morphism and $L = \varphi^{-1}(B)$. Now for any words u and v for which $\varphi(u) = \varphi(v)$ we have also $\varphi(xuy) = \varphi(x)\varphi(u)\varphi(y) = \varphi(x)\varphi(v)\varphi(y) = \varphi(xvy)$, and hence $xuy \in L \iff xvy \in L$. This shows that in

---

[3] In general, a *congruence* on an algebraic structure is an equivalence relation $\sim$ compatible with the algebraic operations. For monoids, this means that $\sim$ must satisfy $u \sim v \Rightarrow xuy \sim xvy$.

the first place, relation $\varphi(u) = \varphi(v)$ is a congruence, and on the second hand, that $\varphi(u) = \varphi(v)$ implies $u \sim_L v$, meaning that relation $\sim_L$ is coarser than relation $\varphi(u) = \varphi(v)$. The remaining details are left to reader.

For any monoid M recognizing language L there is a straightforward way to construct an automaton (not necessarily with a finite state set) of recognizing L. In fact, we can define

$$\mathcal{F} = (M, \Sigma, \delta, 1, B),$$

where $\delta(m, a) = m\varphi(a)$ for each $m \in M$ and $a \in \Sigma$. It is then straightforward to see that $\delta(1, w) = \varphi(w)$, hence $\delta(1, w) \in B \iff \varphi(w) \in B$, so the language accepted by this automaton is indeed $L = \varphi^{-1}(B)$.

The aforementioned circumstances justify the following theorem, which was introduced in [37].

Theorem 4. *A language L is recognized by a finite automaton if and only if its syntactic monoid is finite.*

Definition 14. *A language L is called recognizable, if it is recognized by a finite monoid.*

Schützenberger was the first to characterize a highly nontrivial property of regular languages in terms of their syntactic monoids. He demonstrated that a language is *star-free* if and only if its syntactic monoid is aperiodic. For the notions and proofs, see [41]. Decades later, the properties of a certain subclass of quantum automata have been characterized by so-called *forbidden constructions*, which can be naturally interpreted as properties of the syntactic monoids [7].

3   Formal Power Series

We will shortly present the basics of formal power series here. For a detailed exposition, we refer to [30].

Definition 15. *Let Σ be an alphabet. A formal power series over a semiring R is a function $S : \Sigma^* \to R$. It is usual to write S as*

$$S = \sum_{w \in \Sigma^*} S(w)w,$$

and the elements of Σ are understood as (noncommutative) variables.

Example 2.

$$\frac{1}{1 - ab} = 1 + ab + abab + ababab + abababab + \dots$$

is a formal power series over $\mathbb{R}$ so that $S(w) = 1$, if w is of form $(ab)^i$, and $S(w) = 0$ otherwise.

For a PFA $\mathcal{P}$ (DFA can be viewed as a subcase) we define

$$S = \sum_{w \in \Sigma^*} f_{\mathcal{P}}(w)w, \tag{2}$$

and recall that $f_{\mathcal{P}}$ can be represented as $f_{\mathcal{P}}(w) = \boldsymbol{y} M_{w^R} \boldsymbol{x}^T = (\boldsymbol{x}(M_{w^R})^T \boldsymbol{y}^T)^T$. This implies that

$$\sum_{|w|=n} f_{\mathcal{P}}(w)w = (\boldsymbol{x} \sum_{|w|=n} w(M_{w^R})^T \boldsymbol{y}^T)^T = (\boldsymbol{x}(\sum_{a \in \Sigma} aM_a^T)^n \boldsymbol{y}^T)^T$$

$$= (\boldsymbol{x}(M^T)^n \boldsymbol{y}^T)^T = \boldsymbol{y} M^n \boldsymbol{x}^T,$$

where we have denoted $M^T = \sum_{a \in \Sigma} aM_a^T$. Hence

$$S = \sum_{w \in \sigma^*} f_{\mathcal{P}}(w)w = \sum_{n=0}^{\infty} \sum_{|w|=n} f_{\mathcal{P}}(w)w = \boldsymbol{y} \sum_{n=0}^{\infty} M^n \boldsymbol{x}^T = \boldsymbol{y}(1-M)^{-1}\boldsymbol{x}^T,$$

which shows that S is a rational function in variables $a \in \Sigma$. Especially we see that if L is a recognizable language, then

$$S = \sum_{w \in L} w$$

is a rational function. This connection was initially presented by Kleene [28] and Schützenberger [40] (in fact, Schützenberger's theorem is a generalization of the below theorem):

Theorem 5. *A formal power series*

$$S = \sum_{w \in L} w$$

is rational if and only if L is a recognizable language.

4   Formalism of Finite Quantum Systems

Before introducing quantum automata, it is necessary to present the formalism shortly. For more details, see [23].

4.1   Hilbert Space Preliminaries

Mathematical description of finite-level quantum systems is built on *Hilbert spaces* of finite dimension. As an n-dimensional Hilbert space H_n we understand the complex vector space $\mathbb{C}^n$ equipped with *Hermitian inner product* $\langle \boldsymbol{x} \mid \boldsymbol{y} \rangle = x_1^* y_1 + \ldots + x_n^* y_n$. The inner product induces *norm* by $||\boldsymbol{x}|| = \sqrt{\langle \boldsymbol{x} \mid \boldsymbol{x} \rangle}$.

For each $\boldsymbol{x} = (x_1, \ldots, x_n)$ in H_n we define $|\boldsymbol{x}\rangle$ to be a column vector ($n \times 1$-matrix)

$$|\boldsymbol{x}\rangle = \begin{pmatrix} x_1 \\ \vdots \\ x_n \end{pmatrix}$$

and $\langle \boldsymbol{x} | = (x_1^*, \ldots, x_n^*)$ a row vector ($1 \times n$-matrix). $|\boldsymbol{x}\rangle$ is called a *ket*-vector and $\langle \boldsymbol{x} |$ a *bra*-vector. If necessary, we can identify H_n either with column vector space or row vector space. The set of all linear mappings $H_n \to H_n$ is denoted by $L(H_n)$. For $\boldsymbol{x}, \boldsymbol{y} \in H_n$ we define a mapping $|\boldsymbol{x}\rangle\langle \boldsymbol{y} | : H_n \to H_n$ by setting $|\boldsymbol{x}\rangle\langle \boldsymbol{y} \,|\, | \boldsymbol{z}\rangle = \langle \boldsymbol{y} \mid \boldsymbol{z}\rangle \mid \boldsymbol{x}\rangle$. Clearly $|\boldsymbol{x}\rangle\langle \boldsymbol{y} |$ is a linear mapping, and in a special case $\boldsymbol{y} = \boldsymbol{x}$, $||\boldsymbol{x}|| = 1$ we see that $|\boldsymbol{x}\rangle\langle \boldsymbol{x} \,|\, | \boldsymbol{z}\rangle = \langle \boldsymbol{x} \mid \boldsymbol{z}\rangle \mid \boldsymbol{x}\rangle$, meaning that $|\boldsymbol{x}\rangle\langle \boldsymbol{x} |$ is an orthogonal projection onto a one-dimensional subspace spanned by $|\boldsymbol{x}\rangle$. It is straightforward to interpret $|\boldsymbol{x}\rangle\langle \boldsymbol{y} |$ as a Kronecker product: If A is an $r \times s$-matrix and B an $t \times u$-matrix, then $A \otimes B$ is an $rt \times su$-matrix

$$A \otimes B = \begin{pmatrix} a_{11}B & a_{12}B & \ldots & a_{1s}B \\ a_{21}B & a_{22}B & \ldots & a_{2s}B \\ \vdots & \vdots & \ddots & \vdots \\ a_{r1}B & a_{r2}B & \ldots & a_{rs}B \end{pmatrix}.$$

Now

$$|\boldsymbol{x}\rangle \otimes \langle \boldsymbol{y} | = \begin{pmatrix} x_1 \\ x_2 \\ \vdots \\ x_n \end{pmatrix} \otimes (y_1^*, y_2^*, \ldots, y_n^*) = \begin{pmatrix} x_1 y_1^* & x_1 y_2^* & \ldots & x_1 y_n^* \\ x_2 y_1^* & x_2 y_2^* & \ldots & x_2 y_n^* \\ \vdots & \vdots & \ddots & \vdots \\ x_n y_1^* & x_n y_2^* & \ldots & x_n y_n^* \end{pmatrix}$$

is the matrix of mapping $|\boldsymbol{x}\rangle\langle \boldsymbol{y} |$ in the natural basis.

It is worth noticing that if $\{\boldsymbol{x}_1, \ldots, \boldsymbol{x}_n\}$ is an orthonormal basis, then the matrix representation of $| \boldsymbol{x}_i\rangle\langle \boldsymbol{x}_j |$ in this orthonormal basis consists only of zeros but a single one in the intersection of the ith row and the jth column. The *trace* of a mapping is the sum of diagonal elements of the matrix: $\mathrm{Tr}(A) = \sum_{j=1}^n \langle \boldsymbol{x}_j \mid A\boldsymbol{x}_j \rangle$. It can be shown that the trace is independent of the choice of the orthonormal basis $\{\boldsymbol{x}_1, \ldots, \boldsymbol{x}_n\}$.

The mapping A is *positive*, if $\langle \boldsymbol{x} \mid A\boldsymbol{x}\rangle \geq 0$ for each $\boldsymbol{x} \in H_n$. The *adjoint* mapping A^* is defined by condition $\langle \boldsymbol{x} \mid A\boldsymbol{y}\rangle = \langle A^*\boldsymbol{x} \mid \boldsymbol{y}\rangle$, and it is easy to see that the matrix presentation for A^* is obtained from that of A by transposing and taking complex conjugates. Mapping A is *normal*, if $AA^* = A^*A$, *self-adjoint*, if $A^* = A$, and *unitary*, if $A^* = A^{-1}$. All normal mappings have a remarkable representation introduced in the following theorem, whose proof can be found in [27], for instance.

Theorem 6. *For each normal mapping A there is an orthonormal basis $\{\boldsymbol{x}_1, \ldots, \boldsymbol{x}_n\}$ of H_n consisting of the eigenvectors of A so that*

$$A = \lambda_1 |\boldsymbol{x}_1\rangle\langle \boldsymbol{x}_1 | + \ldots + \lambda_n |\boldsymbol{x}_n\rangle\langle \boldsymbol{x}_n |. \tag{3}$$

The numbers λ_i are the eigenvalues of A, and representation (3) is called a spectral representation *of A.*

In terms of matrices, a spectral representation corresponds to a diagonal form. Thus the above theorem states that all normal matrices can be diagonalized unitarily, meaning that there is an orthonormal basis on which the matrix becomes diagonal. Self-adjointness then means that all the eigenvalues in (3) are real, positivity means that they are nonnegative, and unitarity that they lie in the unit circle.

4.2 States and Observables

In this article, we will not define the physical notions *state* or *observable* in a rigorous manner, but they are understood only intuitively.

Definition 16. *As an n-level quantum system we understand a physical system whose mechanics is depicted according to quantum physics and that has exactly n (but no more) perfectly distinguishable values for some observable.*

Definition 17 (States). *The states of an n-level quantum system are described as self-adjoint positive mappings of H_n with unit trace. A matrix representation of a state is called a* density matrix. *H_n is called the* state space *of the system.*

According to Theorem 6, any state S has a presentation

$$S = \lambda_1 \,|\boldsymbol{x}_1\rangle\langle\boldsymbol{x}_1| + \ldots + \lambda_n \,|\boldsymbol{x}_n\rangle\langle\boldsymbol{x}_n|,$$

where $\{\boldsymbol{x}_1, \ldots, \boldsymbol{x}_n\}$ is an orthonormal basis of H_n, $1 = \mathrm{Tr}(S) = \lambda_1 + \ldots + \lambda_n$ (unit trace), and $\lambda_j \geq 0$ (positivity). It is evident that if S_1 and S_2 are states, so is also $\lambda S_1 + (1 - \lambda)S_2$ for any $\lambda \in (0, 1)$, meaning that the state set is *convex*.

Definition 18. *The state S is* pure, *if representation $S = \lambda S_1 + (1 - \lambda)S_2$ with $S_1 \neq S_2$ implies $\lambda \in \{0, 1\}$.*

According to the previous definition, pure states are the *extremals* of the state set, i.e., states that cannot be represented as a convex combination in a nontrivial way. The following theorem is well-known.

Theorem 7. *The state S is pure if and only if $S = |\boldsymbol{x}\rangle\langle\boldsymbol{x}|$ is a projection onto a one-dimensional subspace (recall that then $\|\boldsymbol{x}\| = 1$ must hold).*

Pure states are also called *vector states*, since to describe $S = |\boldsymbol{x}\rangle\langle\boldsymbol{x}|$ it is enough to give $\boldsymbol{x} \in H_n$. It is required that $\|\boldsymbol{x}\| = 1$, but unfortunately this does not fix $\boldsymbol{x}$ uniquely, as any $e^{i\theta}\boldsymbol{x}$ with $\theta \in \mathbb{R}$ also satisfies $\|e^{i\theta}\boldsymbol{x}\| = 1$ and belongs to the subspace generated by $\boldsymbol{x}$. Nevertheless, it is easy to see that all such vector states generate the same state, meaning that $|e^{i\theta}\boldsymbol{x}\rangle\langle e^{i\theta}\boldsymbol{x}| = |\boldsymbol{x}\rangle\langle\boldsymbol{x}|$.

Definition 19 (Observable). *A (sharp) observable of a quantum system is a self-adjoint mapping $H_n \to H_n$.*

As a self-adjoint mapping, any observable A has a spectral representation

$$A = \mu_1 \, |\boldsymbol{y}_1\rangle\langle\boldsymbol{y}_1| + \ldots + \mu_n \, |\boldsymbol{y}_n\rangle\langle\boldsymbol{y}_n|, \tag{4}$$

where $\mu_i \in \mathbb{R}$. The eigenvalues of A are the potential values of observable A. For any set X of real numbers we define

$$E_A(X) = \sum_{\{j|\mu_j \in X\}} |\boldsymbol{y}_j\rangle\langle\boldsymbol{y}_j|,$$

hence $E_A(X)$ is a projection onto the subspace spanned by those vectors $\boldsymbol{y}_j$ whose eigenvalues belong to X.

States and observables are the primary objects of quantum theory, but they have to be connected for a meaningful interpretation. This connection is presented as an axiom referred as to the *minimal interpretation* of quantum mechanics. Quantum mechanics is ultimately a probabilistic theory, meaning that the outcome, when measuring the value of an observable, is not necessarily determined even if the system is in a pure state.

Definition 20 (Minimal Interpretation). *Let notations be as before. If A is an observable and S a state of quantum system, then the probability that the observed value of A is in set X is given by*

$$\mathbb{P}_S(X) = \mathrm{Tr}(SE_A(X)).$$

Example 3. Let H_n be a state space of an n-level quantum system and $\{\boldsymbol{x}_1, \ldots, \boldsymbol{x}_n\}$ an orthonormal basis. Then any unit-length vector $\boldsymbol{y} \in H_n$ can be represented as

$$\boldsymbol{y} = \alpha_1 \boldsymbol{x}_1 + \ldots + \alpha_n \boldsymbol{x}_n,$$

where $|\alpha_1|^2 + \ldots + |\alpha_n|^2 = 1$. Numbers α_i are called *amplitudes* and we say that $\boldsymbol{y}$ is a *superposition* of $\boldsymbol{x}_1, \ldots, \boldsymbol{x}_n$. Let $S = |\boldsymbol{y}\rangle\langle\boldsymbol{y}|$ be a pure state and

$$A = 1\cdot \, |\boldsymbol{x}_1\rangle\langle\boldsymbol{x}_1| + 2\cdot \, |\boldsymbol{x}_2\rangle\langle\boldsymbol{x}_2| + \ldots + n\cdot \, |\boldsymbol{x}_n\rangle\langle\boldsymbol{x}_n| \tag{5}$$

be an observable. Then the probability of observing value k is given by

$$\mathbb{P}_S(j) = \mathrm{Tr}(S \, |\boldsymbol{x}_j\rangle\langle\boldsymbol{x}_j|) = |\alpha_j|^2.$$

This can be interpreted so that if a pure state is expanded using the eigenvectors of observable A, then the coefficient α_j (which is called the amplitude of $\boldsymbol{x}_j$) induces the probability of measuring value j by $\mathbb{P}(j) = |\alpha_j|^2$ (this is known as Born probability rule). In fact, *observing a vector state*

$$\boldsymbol{y} = \alpha_1 \boldsymbol{x}_1 + \ldots + \alpha_n \boldsymbol{x}_n$$

is a common term in quantum computing, and it refers to measuring observable (5). The probability of seeing j as the value of A is then given as $|\alpha_j|^2$, and the usual terminology speaks about "observing $\boldsymbol{x}_j$", which is a synonym for measuring value j.

It may be noted that the numerical values $1, 2, \ldots, n$ as potential values of observable (5) are not important, but can be replaced by an arbitrary set of distinct values.

4.3   Compound Systems

To form a description of two quantum systems with state spaces H_n and H_m, we use the tensor product construction. Hence the state space of the joint system is mn-dimensional space $H_n \otimes H_m$, and the principal objects (states, observables) can be constructed as tensor products.[4]

Especially, if S_1 and S_2 are states of subsystems, then $S_1 \otimes S_2$ is a state of the joint system. Analogously it is possible to construct an observable of the whole system from observables A_1 and A_2 of the subsystems. It is however worth noticing that space $H_n \otimes H_m$ contains much more states than those ones of form $S_1 \otimes S_2$. Indeed, a state S is called *decomposable*, if there is a representation

$$S = \sum p_i S_i^{(1)} \otimes S_i^{(2)},$$

otherwise S is *entangled*.

The subsystem states of a compound system state are defined via statistical basis: We say that S_1 and S_2 are obtained from S via *partial trace*, and are formally defined as

$$S_1 = \mathrm{Tr}_1(S) \iff \mathrm{Tr}(S_1 A) = \mathrm{Tr}(S(A \otimes I))$$

whenever $A \in L(H_n)$ is an observable. The state $S_2 = \mathrm{Tr}_2(S)$ of the second subsystem is defined analogously. For an explicit formula for S_1 and S_2, see [23].

4.4   State Transformations

States and observables are sufficient to give a description of a quantum system at a fixed time. For the purposes of quantum computing, we need to describe how quantum systems change in time. In so-called *Schrödinger picture*, the state depends on the time and the observables remain, whereas the *Heisenberg picture* is built on time-dependent observables. Both representations are mathematically equivalent, and here we choose the usual Schrödinger picture.

The task is then to describe how quantum systems change in time. Recall that a state of a quantum system is a positive, unit-trace mapping in $L(H_n)$. We should then find out the following: if S_1 and S_2 are states of a quantum system, which properties a mapping $V : L(H_n) \rightarrow L(H_n)$ taking S_1 to S_2 should satisfy? As both S_1 and S_2 are states (unit-trace positive mappings), V should preserve the unit trace. In the same spirit, V should preserve positivity. This serves as a good basis when characterizing mappings V: state transformations should be trace-preserving, and positivity-preserving mappings on the state set.

Mapping $V : L(H_n) \rightarrow L(H_n)$ is called *positive*, if $V(S)$ is always a positive mapping when S is. Unfortunately it turns out that trace-preserving property and the positivity are not enough to characterize all acceptable state transformations. Instead, we need to take care of the *environment*, and say that

---

[4] In the matrix representations, tensor products are represented as Kronecker products.

$V : L(H_n) \to L(H_n)$ is *completely positive*, if $V \otimes I$ is a positive mapping in $H_n \otimes H_m \to H_n \otimes H_m$, where I is an identity mapping on any potential environment H_m of H_n.

The proof of the following theorem can be found in [23].

Theorem 8. *The following are equivalent:*

1. *The mapping $V : L(H_n) \to L(H_n)$ is completely positive, trace-preserving mapping.*
2. $V(S) = \sum_{j=1}^{n^2} V_i S V_i^*$, *where $V_i \in L(H_n)$ satisfy $\sum_{j=1}^{n^2} V_i^* V_i = I$.*
3. $V(S) = \mathrm{Tr}_1(U(S \otimes E)U^*)$, *where E is a pure state of the "environment" system and $U \in L(H_n \otimes H_m)$ a unitary mapping.*

Definition 21. *A quantum system is called* closed, *if its state transformations are of form $V(S) = USU^*$, where U is a unitary mapping, i.e., $U^*U = 1$.*

Notice that the "closedness" is a subcase of both conditions 2 and 3 of the previous theorem: When the system is closed, there is only one single mapping $V_1 = U$ in condition 2, and in condition 3, either the "environment system" is nonexistent or $U = U_1 \otimes I$ does not change the environment state space H_m at all.

For all mappings $A, B \in L(H_n)$ it is easy to see that $A \, |\boldsymbol{x}\rangle\langle\boldsymbol{y}| \, B = |A\boldsymbol{x}\rangle\langle B^*\boldsymbol{y}|$ holds. Hence the state transformation on a pure state $|\boldsymbol{x}\rangle\langle\boldsymbol{x}|$ of a closed system is described as follows: $V(|\boldsymbol{x}\rangle\langle\boldsymbol{x}| = U \, |\boldsymbol{x}\rangle\langle\boldsymbol{x}| \, U^* = |U\boldsymbol{x}\rangle\langle U\boldsymbol{x}|$. This means that a vector state $\boldsymbol{x}$ is transformed into $U\boldsymbol{x}$, if the system is closed. A frequently occurring phrase "quantum time evolution is unitary" simply refers to closed systems beginning at a pure state. It is also important to notice that state transformations in a closed quantum system are always reversible: from $U\boldsymbol{x}$ one can always recover $\boldsymbol{x}$ by $U\boldsymbol{x} \mapsto U^{-1}U\boldsymbol{x} = \boldsymbol{x}$.

4.5   Projection Postulate

The case of the measurement theory of quantum mechanics is far from being closed. In fact, the *measurement problem* of quantum mechanics is a profound and fundamental problem for which a satisfactory resolution is not in sight [13].

The *projection postulate* describes the state transformation in a measurement process in an simple way, but there is no easy way to embed the projection postulate into the theory of quantum mechanics consistently. To introduce the projection postulate, consider a vector state

$$\boldsymbol{x} = \alpha_1 \boldsymbol{x}_1 + \ldots + \alpha_n \boldsymbol{x}_n$$

and an observable $A = 1 \cdot |\boldsymbol{x}_1\rangle\langle\boldsymbol{x}_1| + \ldots + n \cdot |\boldsymbol{x}_n\rangle\langle\boldsymbol{x}_n|$. Now the probability of seeing value j is $|\alpha_j|^2$, and, according to the projection postulate, if value j is observed, *then the post-observation state is $\boldsymbol{x}_j$*. Projection postulate, as presented here can be naturally extended to non-pure states, too, but here we will not need such an extension. Notice that in the definition of observable A, numbers 1, 2,

..., n are not important, but can be replaced with any set of distinct numbers (which would then become the potential values of the observable).

In the theory of quantum computing, it is usually possible to avoid referring to the projection postulate, but sometimes using it makes the notations simpler.

5 Finite Quantum Automata

The theory of quantum computing was implicitly launched by Richard Feynman in 1982, when he suggested that it may be impossible to simulate quantum mechanical systems with a classical computer without an exponential slowdown [20]. Quantum computing attracted only little attention in the beginning, but nevertheless, important theoretical works were conducted by David Deutsch [17], [18]. In 1994 Peter Shor succeeded in raising the theory of quantum computing from the margin by introducing his famous quantum algorithms for factoring integers and extracting discrete logarithms in polynomial time [39]. However, the early research on quantum computing was mainly focused on quantum algorithms with unlimited computing space, and the first studies of finite memory quantum computing were presented as late as 1997.

5.1 Early Models

Quantum finite automata (QFA) were introduced in 1997 by A. Kondacs and J. Watrous [29], and independently by C. Moore and J. P. Crutchfield (although the journal version [34] we cite here appeared later). The model of Kondacs and Watrous is frequently referred as to "Measure-Many" model (MM), and that one by Moore and Crutchfield as "Measure-Once" (MO). The models are crucially different, and formally, MO-QFA seems to be the model more faithful to DFA.

More QFA variants have been introduced later, and in the sequel, we will present some of the most notable ones. When presenting the notion of QFA, the matrix formalism is evidently the most useful way to choose. The intuition behind MO-QFA is that the state set of the automaton is a closed quantum mechanical system, and all state transformation are determined by the input letters.

Let $Q = \{q_1, \ldots, q_n\}$ be the state set of the automaton. There is a "canonical" way to introduce a "quantum model" for any classical one, and here it works as follows: We introduce an n-dimensional Hilbert space with an orthonormal basis $\{|q_1\rangle, |q_2\rangle, \ldots, |q_n\rangle\}$. A general state of the automaton is hence a superposition of basis states $|q_i\rangle$:

$$|q\rangle = \alpha_1 |q_1\rangle + \alpha_2 |q_2\rangle + \ldots + \alpha_n |q_n\rangle, \tag{6}$$

where $|\alpha_1|^2 + \ldots + |\alpha_n|^2 = 1$. It is possible to select a model with an initial state $|q_I\rangle$, but also a superposition over all states $|q_i\rangle$ is equally acceptable for an initial state of the automaton.

The dynamics of an MO-automaton is that one of a closed quantum system, meaning that for any input letter a, there is a unitary mapping $U_a : H_n \to H_n$

describing how (6) changes under a read input letter. Finally, if states in F are specified to be final, then the probability of observing

$$|q\rangle = \alpha_1 |q_1\rangle + \ldots + \alpha_n |q_n\rangle$$

in a final state is $\sum_{q \in F} |\alpha_q|^2$. By writing $P = \sum_{q \in F} |q\rangle\langle q|$ we notice that P is a projection onto the final states, and that

$$P|q\rangle = \sum_{q \in F} \alpha_q |q\rangle.$$

Hence the observation probability can be written in form

$$\sum_{q \in F} |\alpha_q|^2 = ||P|q\rangle||^2 .$$

The definition of MO-QFA follows these outlines, but the generalization allows the initial state and the final projection to be chosen more generally.

Definition 22 (MO-QFA). *A measurement-once quantum finite automaton Q with n states over the alphabet Σ is a triplet $(\boldsymbol{x}, \{U_a \mid a \in \Sigma\}, P)$ where $\boldsymbol{x} \in H_n$ is the* initial vector, *$\{U_a \mid a \in \Sigma\}$ is the set of* unitary transition matrices, *and P is the* final projection.

Remark 1. It also is possible to define an MO-QFA as a fivetuple $(Q, \Sigma, \delta, q_I, F)$, where other components are as in the definition of DFA, but $\delta : \Sigma \to L(H_n)$ is a transition function so that each $\delta(a)$ is a unitary mapping in $L(H_n)$. The definition we used here is then obtained from this different one by specifying $\boldsymbol{x} = |q_I\rangle$, $U_a = \delta(a)$, and the final projection as $P = \sum_{f \in F} |f\rangle\langle f|$.

As in the case of probabilistic automata, the primary function of an MO-QFA is to compute a probability for every word $w \in \Sigma^*$, and in the MO-case it is done as follows:

$$f_Q(w) = ||PU_{w^R}\boldsymbol{x}^T||^2 .$$

This is a quantum analogue of the probability computed by PFA. In the article presenting the Moore-Crutchfield model, the authors demonstrated that many properties of classical automata also hold for MO-QFA [34]. For instance, they prove that for any QFA Q, the series

$$\sum_{w \in \Sigma^*} f_Q(w)w$$

is rational, and that the following closure properties hold: Let f and $g : \Sigma^* \to [0,1]$ be functions computed by MO-QFA and $|\alpha|^2 + |\beta|^2 = 1$. Then also $\alpha f + \beta g$ is a function $\Sigma^* \to [0,1]$ computed by an MO-QFA (convexity). Moreover, fg (intersection) and $1 - f$ (complement) are computed by an MO-QFA. The closure under inverse morphism was also demonstrated: if $h : \Sigma_1^* \to \Sigma^*$ is a morphism, then $fh : \Sigma_1 \to [0,1]$ is again a function computed by an MO-QFA. All the aforementioned properties are well-known for regular languages. In [34], Moore and Crutchfield use the compactness of the unit sphere of H_n to prove also a version of the *pumping lemma*. On the other hand that version is very different from the classical pumping lemma: in the MO-QFA case, it is shown that for any $w \in \Sigma^*$ and each $\epsilon > 0$, there exists $k \in \mathbb{N}$ so that for all $u, v \in \Sigma^*$, $\left| f_Q(uw^k v) - f_Q(uv) \right| \leq \epsilon$.

The studies of Moore and Crutchfield can be understood within a generalized notion of a language. Whereas the basic definition simply means a subset of Σ^*, the generalized notion refers to a *fuzzy* subset of Σ^*, i.e., a function $f : \Sigma^* \rightarrow [0,1]$. The traditional notion of a (crisp) language then means that f is actually onto $\{0,1\}$. From any fuzzy language $f : \Sigma^* \rightarrow [0,1]$ it is then possible to obtain a traditional language by discretizing f, and that can be done exactly in the same way as for the probabilistic automata to obtain cut-point languages $L_{>\lambda}(Q)$ and $L_{\geq\lambda}(Q)$.

Now if the cut-point is not isolated, languages recognized by MO-QFA need not to be regular. For a concrete example, the reader is invited to design a two-state MO-QFA Q over the binary alphabet $\Sigma = \{a, b\}$ with the following property: $f_Q(w) = 0$, if $|w|_a = |w|_b$ (the number of as and bs in w coincide), and $f_Q > 0$ otherwise.

On the other hand, it was noted in [1] that the technique of Rabin can be used also for QFA (in fact, Rabin's technique applies to H_n more elegantly than to the classical probability polyhedra) to show that quantum cut-point languages with isolated cut-points are all regular (If not explicitly mentioned otherwise, the language recognition by automata will here and hereafter mean recognition of a cut-point language with an isolated cut-point). Most regularity proofs for isolated cut-point model, including the first one by Rabin [38], are based on the compactness of the state set. It is therefore interesting to notice that Symeon Bozapalidis has shown that the regularity of cut-point languages can be derived also in a different way, which applies for a large class of languages containing those recognized by MO-QFA [11].

For language recognition, the unitarity of the evolution matrices turns out to be an essential restriction. All unitary matrices are invertible, and that may lead to a sophisticated guess that the syntactic monoid of a cut-point language of QFAs should also contain the inverses of its elements. This guess turns out to be true: In [34] Moore and Crutchfield already point out that if the characteristic function of a regular language L equals to f_Q for some quantum automaton Q, then L is a *group language*, meaning that its syntactic monoid is a group. This result was extended to all isolated cut-point languages recognized by MO-QFA in [12].

The aforementioned results show that MO-QFA and their related languages are mathematically very elegant objects: They satisfy a number of closure properties, and the model itself seems to be a satisfactory "quantum counterpart" of a probabilistic model. Unfortunately their language recognition power is very weak: isolated cut-point languages of MO-QFA are all group languages, which is a small subset of all regular languages. This is essentially different from PFA, which are in fact genuine generalizations of DFA and can accept all regular languages (in the cut-point acceptance model). Therefore MO-QFA cannot be seen as generalizations of DFA, but rather as "variants".

The other model, MM-QFA has almost inverse properties: Language recognition power is greater, but the closure properties and mathematical elegancy are weak. In [29] Kondacs and Watrous introduced 1-way and 2-way MM-QFA, and studied mainly their language recognition power in the isolated cut-point model.

Definition 23 (MM-QFA). *A measure-many (1-way) quantum finite automaton Q consists of a state set Q, set of unitary transition mappings $\{U_a \mid a \in \Sigma\}$, and an initial vector $\boldsymbol{x} \in H_n$. The state set Q is divided into disjoint sets of accepting, rejecting, and neutral states: $Q = Q_a \cup Q_r \cup Q_n$.*

To describe the computation of, let first P_a, P_r and P_n be projections onto the subspaces H_a, H_r, and H_n spanned by the accepting, rejecting, and neutral states, respectively.

The computation of a MM-QFA goes as follows: The automaton begins at the state $\boldsymbol{x}$, and the input word w is scanned one letter at time. When letter a is read, transition U_a is applied to the state of the automaton, and then the state is observed to see whether the automaton is in an accepting, rejecting or neutral. This means measuring the value of three-valued observable $A = 1 \cdot P_a + 2 \cdot P_r + 3 \cdot P_n$ (numbers 1, 2, and 3 are not important and can be replaced with any set of distinct numbers). If the state was observed to be accepting (resp. rejecting), then the input word is accepted (resp. rejected), and if the state seen was neutral, then the computation goes on, and the next input letter is read. The post-observation state after each observation is determined according to the projection postulate, meaning that the state

$$\sum_{q \in Q_a} \alpha_q \,|q\rangle + \sum_{q \in Q_r} \alpha_q \,|q\rangle + \sum_{q \in Q_n} \alpha_q \,|q\rangle \qquad (7)$$

collapses into

$$\frac{1}{\sqrt{\mathbb{P}(a)}} \sum_{q \in Q_a} \alpha_q \,|q\rangle, \qquad \frac{1}{\sqrt{\mathbb{P}(r)}} \sum_{q \in Q_r} \alpha_q \,|q\rangle, \quad \text{or} \quad \frac{1}{\sqrt{\mathbb{P}(n)}} \sum_{q \in Q_n} \alpha_q \,|q\rangle,$$

according to which type of state was seen. Here $\mathbb{P}(a) = \sum_{q \in Q_a} |\alpha_q|^2$, $\mathbb{P}(r) = \sum_{q \in Q_r} |\alpha_q|^2$, and $\mathbb{P}(n) = \sum_{q \in Q_n} |\alpha_q|^2$ are the probabilities of seeing the automaton in state (7) in an accepting, rejecting, or neutral state, respectively. It is assumed that the input word is surrounded by special *endmarkers* which do not belong to the actual input alphabet, and that when reading the right endmarker, the automaton cannot any more stay in a neutral state, but must decide whether the word is accepted or rejected. Unlike the other models previously presented in this article, the MM-QFA can accept or reject a word without reading all letters of it.

The computational process is more complicated than for MO-QFA, but nevertheless, an MM-QFA Q also computes a function $f_Q : \Sigma^* \to [0, 1]$ (the acceptance probability for each word $w \in \Sigma^*$). It was noted already in [29], that in the isolated cut-point model, all languages recognized by MM-QFA are also regular. This follows by applying the technique of Rabin [37] or Bozapalidis [11]. However, even MM-QFAs with isolated cut-point are not powerful enough to recognize all regular languages. In fact, it was noted in [29] that even the language $L = \{a, b\}^*a$ cannot be recognized with a MM-QFA with isolated cut-point.

In this paper, we are not going to treat 2-way finite automata capable of scanning the input various times, but it may be worth emphasizing that in [29], the authors demonstrated that 2-way MM-QFA with isolated cut-point can indeed recognize all regular languages, and even more: The non-regular language $\{a^n b^n \mid n \in \mathbb{N}\}$ can be accepted with a 2-way MM-QFA with an isolated cut-point. This may appear somewhat surprising, as 2-way DFA are known equally as powerful as ordinary DFAs.

Subsequent studies have revealed rather strange behaviours of MM-QFA. It was shown in [3], that if MM-QFA are required to work it a high probability (at least $\frac{7}{9}$), then there is a classical *reversible automaton* doing the same job (see [3] for the definitions). But for weaker correctness probabilities, MM-QFA are more powerful than the classical automata. In [3] it was also demonstrated that in some cases, MM-QFA can be exponentially smaller than PFA recognizing the same language. The aforementioned probability $\frac{7}{9}$ was subsequently improved to $0.7726\ldots$ in [6].

A direction of [3] was followed in [4], where the authors constructed a hierarchy of languages recognizable by MM-QFA with isolated cut-point, but whose potential isolation tends to zero.

More troublesome news for MM-QFA were brought forth in [5], where it was shown that the class of languages recognized by MM-QFA is not closed under union, intersection, or under any genuinely binary Boolean operation. Very interestingly, the authors of [5] also launched the study of so-called *forbidden constructions*. Forbidden constructions are properties of the graph of the minimal DFA for the language in question which prohibit the language to be accepted by an MM-QFA (with isolated cut-point, of course).

The forbidden constructions can be translated into the properties of syntactic monoids, and this course of research was followed in [7]. However, any good description to syntactic monoids of languages recognized by MM-QFA is not known.

From the aforementioned description it is obvious that MM-QFA is not very elegant model mathematically, but more satisfactory models have been introduced subsequently.

5.2 Latvian QFA

In [7] the authors introduced another variant of quantum automata called *Latvian QFA*. A Latvian QFA $\mathcal{Q}$ is a sixtuple $\mathcal{Q} = (Q, \Sigma, \{U_a \mid a \in \Sigma\}, \{M_a \mid a \in \Sigma\}, q_I, F)$, Q and Σ are as for DFA, U_a is an unitary transition associated to letter a, and M_a is a measurement determined by an orthogonal decomposition $H_n = V_1^a \oplus \ldots \oplus V_k^a$. The automaton starts in state $|q_I\rangle$, then reads the input one letter at time. When reading the input letter a, transition U_a is applied, and then the measurement M_a is performed. The procedure continues until the right endmarker is read. It is required that the measurement associated to the right endmarker projects either to the subspace generated by the final states or to its orthogonal complement, the subspace generated by non-final states. The probability function computed by the automaton is then the probability that a final state is seen.

In [7] it is demonstrated that Latvian QFA have far more elegant closure properties than MM-QFA. Indeed, languages recognized by Latvian QFA are closed under union, intersection, complement, and inverse morphishms. It was also shown that the languages recognized by Latvian automata are exactly those whose syntactic monoid is of wreath product form $J * G$, where J is a $\mathcal{J}$-trivial monoid and G group (for the definitions, see [7]). As form $J * G$ does not cover all finite monoids, this characterization shows also that even the Latvian QFA are not sufficient to recognize all regular languages.

To enrich the model, Bertoni & al. introduced a QFA model with a control language, where also the state is measured after each transition, and the acceptance depends on whether the sequence of the measurement results belong to the control language [9], thus obtaining an QFA model capable of recognizing a set of regular languages closed under Boolean operations. In [14] Ciamarra introduced a reversibility construction using extra space to provide a model capable of accepting all regular languages.

5.3   Open Quantum Automata

We have seen that sometimes QFAs can provide some advantage over DFA or PFA, when the efficiency is measured in the number of states needed to recognize a language. On the other hand, many models of QFA are unfortunately restricted very heavily, implying that they cannot recognize even all regular languages. The reason for the restrictions has been correctly located by the aforementioned authors: Unitary time evolution is always reversible, and the real-time computation of 1-way finite automaton does not allow any reversibility construction: It was shown by Charles Bennett that all computation can be made reversible [8], but the price to be paid for this is to introduce an extra memory to write the history of the computation. Such a construction is not applicable to finite automata.

On the other hand, the state transformations in quantum systems need not to be reversible, but reversibility is just a property of closed quantum systems. In the classical theory of computation irreversibility (and subsequent information loss) is perfectly acceptable, and hence there is no reason to assume quantum systems closed when studied in the context of computability. Therefore, the most general QFA model should be naturally based on open system state transitions (completely positive, trace-preserving mapping, see definitions in section 4.4).

Such a model was introduced in [24] and [25], and its properties were studied in [26].

Definition 24. *A QFA with open time evolution (or shortly open QFA) is a quintuple $\mathcal{Q} = (Q, \Sigma, \delta, q_I, F)$, where Q and Σ are as before, q_I is the initial state, and $F \subseteq Q$ is the set of final states, and $\delta : \Sigma \to L(L(H_n))$ is a transition function associating to each letter $a \in \Sigma$ a trace-preserving completely positive mapping $L(H_n) \to L(H_n)$, $\delta(a) = V_a$.*

The computation of an open QFA begins in (pure) state $| q_I \rangle \langle q_I |$, and each read input letter changes the state by $V_a : L(H_n) \to L(H_n)$. The acceptance probability of the word is then given by

$$f_{\mathcal{Q}}(w) = \text{Tr}(P V_{w^R} | q_I \rangle \langle q_I |),$$

where $P = \sum_{f \in F} | f \rangle \langle f |$ is the projection onto the subspace generated by the final states.

In [24] it was shown that PFA (and subsequently DFA), is a subcase of open QFA, and in [26] it was shown that MM-QFA (and subsequently MO-QFA), and Latvian QFA can considered as subcases of QFA with open time evolution. Hence it is justified to say that QFAs with open time evolution are the genuine quantum extensions of DFAs.

In [26], it was also demonstrated the functions $f_{\mathcal{Q}} : \Sigma^* \to [0, 1]$ computed by QFA with open time evolution satisfy the same closure properties as those computed by MO-QFA, and that the formal power series

$$\sum_{w \in \Sigma^*} f_{\mathcal{Q}}(w) w$$

is rational. As open QFA is an extension of all other automata models mentioned in this paper, it follows that there are cut-point languages accepted by open QFA which are not regular. The question of regularity with isolated cut-point was not studied in [26]. But as the set of all unit-trace, positive linear mappings $H_n \to H_n$ is evidently compact, the technique of Rabin obviously applies and the isolated cut-point languages accepted by open QFA are evidently regular.

Another possibility to settle the regularity question is to notice that Bozapalidis' theorem [11] evidently covers also the dynamics of open QFAs.

References

1. Ablayev, F., Gainutdinova, A.: On the Lower Bounds for One-Way Quantum Automata. In: Nielsen, M., Rovan, B. (eds.) MFCS 2000. LNCS, vol. 1893, pp. 132–140. Springer, Heidelberg (2000)
2. Aho, A.V., Sethi, R., Ullman, J.D.: Compilers: Principles, Techniques, and Tools. Addison-Wesley, Reading (1986)
3. Ambainis, A., Freivalds, R.: 1-way quantum finite automata: strengths, weaknesses and generalizations. In: Proceedings of the 39th FOCS, pp. 376–383 (1998)
4. Ambainis, A., Bonner, R.F., Freivalds, R., Ķikusts, A.: Probabilities to Accept Languages by Quantum Finite Automata. In: Asano, T., Imai, H., Lee, D.T., Nakano, S.-i., Tokuyama, T. (eds.) COCOON 1999. LNCS, vol. 1627, pp. 174–185. Springer, Heidelberg (1999)

5. Ambainis, A., Ķikusts, A., Valdats, M.: On the class of languages recognizable by 1-way quantum finite automata. In: Ferreira, A., Reichel, H. (eds.) STACS 2001. LNCS, vol. 2010, pp. 75–86. Springer, Heidelberg (2001)

6. Ambainis, A., Ķikusts, A.: Exact results for accepting probabilities of quantum automata. Theoretical Computer Science 295(1), 3–25 (2003)

7. Ambainis, A., Beaudry, M., Golovkins, M., Ķikusts, A., Mercer, M., Thérien, D.: Algebraic Results on Quantum Automata. Theory of Computing Systems 39, 165–188 (2006)

8. Bennett, C.H.: Logical reversibility of computation. IBM Journal of Research and Development 17, 525–532 (1973)

9. Bertoni, A., Mereghetti, C., Palano, B.: Quantum computing: 1-way quantum automata. In: Ésik, Z., Fülöp, Z. (eds.) DLT 2003. LNCS, vol. 2710, pp. 1–20. Springer, Heidelberg (2003)

10. Blondel, V.D., Canterini, V.: Undecidable problems for probabilistic automata of fixed dimension. Theory of Computing systems 36, 231–245 (2003)

11. Bozapalidis, S.: Extending Stochastic and Quantum Functions. Theory of Computing Systems 2, 183–197 (2003)

12. Brodsky, A., Pippenger, N.: Characterizations of 1-Way Quantum Finite Automata. SIAM Journal on Computing 31(5), 1456–1478 (2002)

13. Busch, P., Lahti, P., Mittelstaedt, P.: The quantum theory of measurement. Springer, Heidelberg (1996)

14. Ciamarra, M.: Quantum reversibility and a new model of quantum automaton. Fundamentals of Computation Theory 13, 376–379 (2001)

15. Culik II, K., Kari, J.: Image compression using weighted finite automata. Computers and Graphics 17, 305–313 (1993)

16. Culik II, K., Kari, J.: Image-data compression using edge-optimizing algorithm for WFA inference. Journal of Information Processing and Management 30, 829–838 (1994)

17. Deutsch, D.: Quantum theory, the Church-Turing principle and the universal quantum computer. Proceedings of the Royal Society of London A 400, 97–117 (1985)

18. Deutsch, D.: Quantum computational networks. Proceedings of the Royal Society of London A 425, 73–90 (1989)

19. Eilenberg, S.: Automata, languages, and machines, vol. A. Academic Press, London (1974)

20. Feynman, R.P.: Simulating physics with computers. International Journal of Theoretical Physics 21(6/7), 467–488 (1982)

21. Freivalds, R.: On the growth of the number of states in result of the determinization of probabilistic finite automata. Avtomatika i Vichislitelnaya Tekhnika 3, 39–42 (1982) (Russian)

22. Freivalds, R.: Non-constructive Methods for Finite Probabilistic Automata. International Journal of Foundations of Computer Science 19(3), 565–580 (2008)

23. Hirvensalo, M.: Quantum Computing, 2nd edn. Springer, Heidelberg (2004)

24. Hirvensalo, M.: Some Open Problems Related to Quantum Computing. In: Paun, G., Rozenberg, G., Salomaa, A. (eds.) Current Trends in Theoretical Computer Science – The Challenge of the New Century, vol. 1. World Scientific, Singapore (2004)

25. Hirvensalo, M.: Various Aspects of Finite Quantum Automata. In: Ito, M., Toyama, M. (eds.) DLT 2008. LNCS, vol. 5257, pp. 21–33. Springer, Heidelberg (2008)

26. Hirvensalo, M.: Quantum Automata with Open Time Evolution. International Journal of Natural Computing Research 1, 70–85 (2010)

27. Horn, R.A., Johnson, C.R.: Matrix Analysis. Cambridge University Press, Cambridge (1985)
28. Kleene, S.: Representation of Events in Nerve Nets and Finite Automata. In: Shannon, C., McCarthy, J. (eds.) Automata Studies, pp. 3–41. Princeton University Press, Princeton (1956)
29. Kondacs, A., Watrous, J.: On the power of quantum finite state automata. In: Proceedings of the 38th IEEE Symposium on Foundations of Computer Science, pp. 66–75 (1997)
30. Kuich, W., Salomaa, A.: Semirings, Automata and Languages. EATCS Monographs on Theoretical Computer Science, vol. 5. Springer, Heidelberg (1986)
31. McCulloch, W., Pitts, W.: A logical calculus of the ideas immanent in nervous activity. Bulletin of Mathematical Biophysics 7, 115–133 (1943)
32. Mealy, G.: A Method for Synthesizing Sequential Circuits. Bell Systems Technical Journal 34, 1045–1079 (1955)
33. Moore, E.: Gedanken-experiments on Sequential Machines. In: Shannon, C., Ashby, W. (eds.) Automata Studies, pp. 129–153. Princeton University Press, Princeton (1956)
34. Moore, C., Crutchfield, J.P.: Quantum automata and quantum grammars. Theoretical Computer Science 237(1-2), 275–306 (2000)
35. Paz, A.: Some aspects of probabilistic automata. Information and Control 9, 26–60 (1966)
36. Paz, A.: Introduction to Probabilistic Automata. Academic Press, London (1971)
37. Rabin, M.O., Scott, D.: Finite Automata and Their Decision Problems. IBM Journal of Research and Development 3(2), 114–125 (1959)
38. Rabin, M.O.: Probabilistic Automata. Information and Control 6, 230–245 (1963)
39. Shor, P.W.: Algorithms for quantum computation: discrete log and factoring. In: Proceedings of the 35th Annual Symposium on the Foundations of Computer Science, pp. 20–22 (1994); Physical Review Letters 81(17), 3563–3566 (1998)
40. Schützenberger, M.-P.: On the Definition of a Family of Automata. Information and Control 4, 245–270 (1961)
41. Schützenberger, M.-P.: On finite monoids having only trivial subgroups. Information and Control 8, 190–194 (1965)
42. Turakainen, P.: On Stochastic Languages. Information and Control 12, 304–313 (1968)
43. Yu, S.: Regular Languages. In: Rozenberg, G., Salomaa, A. (eds.) Handbook of Formal Languages. Word, Language, Grammar, vol. 1. Springer, Heidelberg (1997)

Graph Automata: The Algebraic Properties of Abelian Relational Graphoids

Antonios Kalampakas

Technical Institute of Kavala,
Department of Exact Sciences,
65404, Kavala, Greece
akalampakas@gmail.com

Abstract. Automata operating on arbitrary graphs were introduced in
a previous paper by virtue of a particular instance of an abelian relational
graphoid. As it is indicated in the same paper, in order to construct a
graph automaton it is necessary and sufficient that the relations over the
Kleene star of the state set constitute a graphoid. In this respect, various
different versions of graph automata arise corresponding to the specific
relational graphoid that is employed. We prove that the generation of an
abelian graphoid by a set Q implies the partitioning of Q into disjoint
abelian groups and vise versa.

1 Introduction

The last 40 years several types of automata operating on restricted classes of
graphs (rooted, planar, acyclic, etc) have been introduced in the literature (cf.
[1], [9], [4]). Automata on general (hyper)graphs are constructed for the first time
in [6] by utilizing the algebraic properties of graphoids, i.e., magmoids satisfying
the 15 equations of graphs which are specified in [5].

The notion of magmoids, introduced by Arnold and Dauchet (cf. [2],[3]), is
the algebraic structure which is employed to generate graphs from a finite alphabet in a role similar to the one that monoids play for the generation of strings.
Recall that a magmoid is a doubly ranked set $M = (M_{m,n})_{m,n \geq 0}$ equipped with
two operations $\circ$ and $\square$ which are associative, unitary and mutually coherent
in a canonical way. Engelfriet and Vereijken, in [7], proved that the set of (hyper)graphs $GR(\Sigma)$, with hyperedges labeled over a finite doubly ranked alphabet
Σ, can be organized into a magmoid with $\circ$ being the graph product and $\square$ the
graph sum, and that every graph can be built from a specific finite set of elementary graphs D, together with the elements of Σ, by using the two operations
of product and sum (Theorem 7 of [7]). In this construction every hypergraph is
represented by an infinite number of expressions; this ambiguity was settled by
determining a *finite* set of equations $\mathcal{E}$, involving the elements of D and Σ, with
the property that two expressions represent the same hypergraph, if and only if,
one can be transformed into the other through these equations (cf. [5]). Deriving
from this result, a *graphoid* $\mathbf{M}$ is defined to be a magmoid with a designated set

W. Kuich and G. Rahonis (Eds.): Bozapalidis Festschrift, LNCS 7020, pp. 168–182, 2011.
© Springer-Verlag Berlin Heidelberg 2011

of elements that satisfy the equations $\mathcal{E}$. Hence $GR(\Sigma)$ can be structured into a graphoid by virtue of the previously mentioned set D of elementary graphs.

Given a set Q, the *relational magmoid* over Q is constructed by defining the operations of composition and sum on the set of all relations from Q^m to Q^n, $m, n \geq 0$. This set is structured into a *relational graphoid* over Q, by specifying a set D of relations that satisfy the equations $\mathcal{E}$. A relational graphoid is called abelian when a particular relation of D consists of all the transpositions in Q. In [6] graph automata, over a state set Q, were introduced by virtue of a specific abelian relational graphoid over Q and by exploiting the fact that $GR(\Sigma)$ is the free graphoid generated by Σ. As it is indicated by this construction, different kinds of graphoids produce graph automata with diverse operation and recognizability capacity.

We show that all abelian relational graphoids, generated from a given set, can be determined by virtue of the following characterization. A set Q generates an abelian relational graphoid, if and only if, Q is partitioned into disjoint abelian groups. In other words it is proved that the structuring of $Rel(Q)$ into a relational graphoid results in the partitioning of the set Q, and the endowment of every class with the group axioms.

2   Magmoids

Recall that a doubly ranked set - or doubly ranked alphabet - $(A_{m,n})_{m,n \in \mathbb{N}}$ is a set A together with a function $rank : A \to \mathbb{N} \times \mathbb{N}$, where $\mathbb{N}$ is the set of natural numbers. For $m, n \in \mathbb{N}$, $A_{m,n}$ is the set $\{a \in A \mid rank(a) = (m, n)\}$. In what follows we will drop the subscript $m, n \in \mathbb{N}$ and denote a doubly ranked set simply by $(A_{m,n})$. A *semi-magmoid* is a doubly ranked set $M = (M_{m,n})$ equipped with two operations

$$\circ : M_{m,n} \times M_{n,k} \to M_{m,k}, \qquad m, n, k \geqslant 0$$

$$\Box : M_{m,n} \times M_{m',n'} \to M_{m+m',n+n'}, \qquad m, n, m', n' \geqslant 0$$

which are associative in the obvious way and satisfy the distributivity law

$$(f \circ g) \Box (f' \circ g') = (f \Box f') \circ (g \Box g')$$

whenever all the above operations are defined. A *magmoid* is a semi-magmoid $M = (M_{m,n})$, equipped with a sequence of constants $e_n \in M_{n,n}$ $(n \geqslant 0)$, called units, such that

$$e_m \circ f = f = f \circ e_n, \quad e_0 \Box f = f = f \Box e_0$$

for all $f \in M_{m,n}$ and all $m, n \geqslant 0$, and the additional condition

$$e_m \Box e_n = e_{m+n}, \qquad \text{for all } m, n \geqslant 0$$

holds. Notice that, due to the last equation, the element e_n ($n \geq 2$) is uniquely determined by e_1. From now on e_1 will be simply denoted by e.

The sets $Rel_{m,n}(Q)$ of all relations from Q^m to Q^n

$$Rel_{m,n}(Q) = \{R \mid R \subseteq Q^m \times Q^n\}$$

can be structured into a magmoid with $\circ$ being the usual relation composition, while the operation $\square$ is defined as follows: for $R \in Rel_{m,n}(Q)$ and $S \in Rel_{m',n'}(Q)$

$$R \square S = \{(u_1 u_2, v_1 v_2) \mid (u_1, v_1) \in R \text{ and } (u_2, v_2) \in S)\},$$

where $u_1 \in Q^m$, $u_2 \in Q^{m'}$, $v_1 \in Q^n$, $v_2 \in Q^{n'}$. Notice that $Q^0 = \{\varepsilon\}$, where ε is the empty word of Q^*. The units of this magmoid are given by $e_0 = \{(\varepsilon, \varepsilon)\}$ and

$$e = \{(g, g) \mid g \in Q\}. \tag{1}$$

We denote by $Rel(Q) = (Rel_{m,n}(Q))$ the magmoid constructed in this way and call it the *relational magmoid of* Q.

Let Σ be a doubly ranked alphabet. *We denote by* $SM(\Sigma) = (SM_{m,n}(\Sigma))$ *the smallest doubly ranked set satisfying the next items:*

- $\Sigma_{m,n} \subseteq SM_{m,n}(\Sigma)$ for all $m, n \geq 0$,
- if $p \in SM_{m,n}(\Sigma)$ and $q \in SM_{n,k}(\Sigma)$ then their horizontal concatenation $pq \in SM_{m,k}(\Sigma)$,
- if $p \in SM_{m,n}(\Sigma)$ and $p' \in SM_{m',n'}(\Sigma)$ then their vertical concatenation $\dfrac{p}{p'} \in SM_{m+m',n+n'}(\Sigma)$.

Let $\sim = (\sim_{m,n})$ be the doubly ranked equivalence on $SM(\Sigma)$, compatible with horizontal and vertical concatenation, generated by the relations

$$\begin{matrix} p_1\, p_1' \\ \\ p_2\, p_2' \end{matrix} \sim \begin{matrix} p_1 & p_1' \\ p_2 & p_2' \end{matrix}$$

for all p_i, p_i' of suitable ranks. The quotient

$$SM(\Sigma)/\sim = (SM_{m,n}(\Sigma)/\sim_{m,n})$$

is denoted by $smag(\Sigma)$ and obviously is a semi-magmoid. The elements of $smag_{m,n}(\Sigma)$ are called (m,n)-*patterns* or *patterns of rank* (m,n). They are analogous with the unsorted abstract dags of [9], [10] and [4]; for another formalization see also [8]. Subsets of $smag(\Sigma)$ are called *pattern languages*. In order to avoid confusion in the pattern calculus instead of $\dfrac{p}{p'}$ we write $\begin{pmatrix} p \\ p' \end{pmatrix}$. Actually $smag(\Sigma)$ is the *free* semi-magmoid generated by Σ as confirms the next result.

Proposition 1. *For every semi-magmoid* $M = (M_{m,n})$ *and every doubly ranked function* $f : \Sigma \to M$, *there exists a unique morphism of semi-magmoids* $\bar{f} : smag(\Sigma) \to M$ *making the following triangle commutative.*

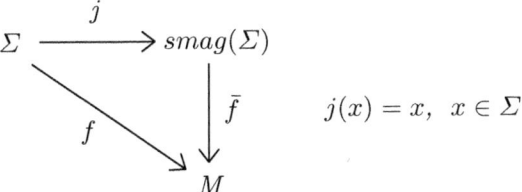

$$j(x) = x, \quad x \in \Sigma$$

Actually, $\bar{f}$ is given by the clauses,

- $\bar{f}(x) = f(x)$, *for all* $x \in \Sigma$,
- $\bar{f}(p\,q) = \bar{f}(p) \circ \bar{f}(q), \quad \bar{f}\left(\dfrac{p}{p'}\right) = \bar{f}(p) \,\square\, \bar{f}(p')$,

for all $p, q, p' \in smag(\Sigma)$ of suitable rank.

The construction of the free magmoid follows naturally. Let $(e_n)_{n \geq 0}$ be a sequence of symbols not in Σ and denote by $mag(\Sigma)$ the free semi-magmoid $smag(\Sigma \cup \{e_n \mid n \geq 0\})$ divided by the congruence generated by the relations

$$e_m\, p \equiv p \equiv p\, e_n, \qquad \begin{pmatrix} e_0 \\ p \end{pmatrix} \equiv p \equiv \begin{pmatrix} p \\ e_0 \end{pmatrix}, \qquad e_m\, e_n \equiv e_{m+n}$$

for all $m, n \geq 0$ and all patterns p of suitable rank, then $mag(\Sigma)$ clearly constitutes a magmoid which has a universal property analogous to the one stated in Proposition 1, i.e., $mag(\Sigma)$ is the *free* magmoid generated by Σ (cf. [5]).

The set of all hypergraphs can be structured into a magmoid in the following way. Given a finite alphabet X, we denote by X^* the set of all words over X and for every word $w \in X^*$, $|w|$ denotes its length. Formally, a *concrete (m, n)-graph*, with hyperedges labeled over a doubly ranked alphabet $\Sigma = (\Sigma_{m,n})$, is a tuple

$$G = (V, E, s, t, l, begin, end)$$

where

- V is the finite set of nodes,
- E is the finite set of hyperedges,
- $s : E \to V^*$ is the source function,
- $t : E \to V^*$ is the target function,
- $l : E \to \Sigma$ is the labelling function such that $rank(l(e)) = (|s(e)|, |t(e)|)$ for every $e \in E$,
- $begin \in V^*$ with $|begin| = m$ is the sequence of begin nodes and
- $end \in V^*$ with $|end| = n$ is the sequence of end nodes.

Notice that according to this definition vertices can be duplicated in the begin and end sequences of the graph and also at the sources and targets of an edge. For an edge e of a hypergraph G we simply write $rank(e)$ to denote $rank(l(e))$. The specific sets V and E chosen to define a concrete graph G are actually irrelevant. We shall not distinguish between two isomorphic graphs. Hence we have the following definition of an abstract graph. Two concrete (m, n)-graphs

$G = (V, E, s, t, l, begin, end)$ and $G' = (V', E', s', t', l', begin', end')$ over Σ are isomorphic iff there exist two bijections $h_V : V \to V'$ and $h_E : E \to E'$ commuting with source, target, labelling, $begin$ and end in the usual way.

An *abstract* (m, n)-*graph* is defined to be the equivalence class of a concrete (m, n)-graph with respect to isomorphism. We denote by $GR_{m,n}(\Sigma)$ the set of all abstract (m, n)-graphs over Σ. Since we shall mainly be interested in abstract graphs we simply call them graphs except when it is necessary to emphasize that they are defined up to an isomorphism. Any graph $G \in GR_{m,n}(\Sigma)$ having no edges is called a *discrete* (m, n)-*graph*.

If G is an (m, n)-graph represented by $(V, E, s, t, l, begin, end)$ and H is an (n, k)-graph represented by $(V', E', s', t', l', begin', end')$ then their *product* $G \circ H$ is the (m, k)-graph represented by the concrete graph obtained by taking the disjoint union of G and H and then identifying the ith end node of G with the ith begin node of H, for every $i \in \{1, ..., n\}$; also, $begin(G \circ H) = begin(G)$ and $end(G \circ H) = end(H)$.

The *sum* $G \square H$ of arbitrary graphs G and H is their disjoint union with their sequences of begin nodes concatenated and similarly for their end nodes.

For instance let $\Sigma = \{a, b, c\}$, with $rank(a) = (2, 1)$, $rank(b) = (1, 1)$ and $rank(c) = (1, 2)$. In the following pictures, edges are represented by boxes, nodes by dots, and the sources and targets of an edge by directed lines that enter and leave the corresponding box, respectively. The order of the sources and targets of an edge is the vertical order of the directed lines as drawn in the pictures. We display two graphs $G \in GR_{3,4}(\Sigma)$ and $H \in GR_{4,2}(\Sigma)$, where the ith begin node is indicated by b_i, and the ith end node by e_i.

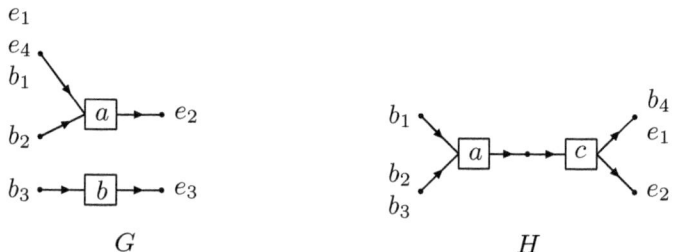

Then their product $G \circ H$ is the $(3, 2)$-graph

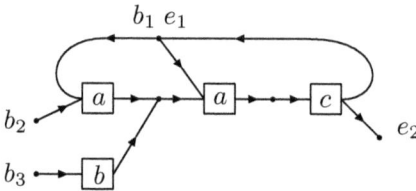

and, their sum $G \square H$ is the $(7, 6)$-graph

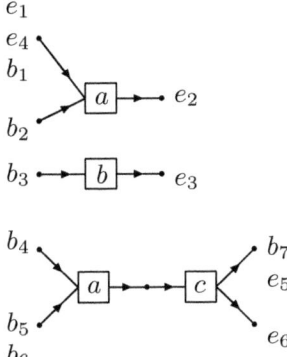

For every $n \in \mathbb{N}$ we denote by E_n the discrete graph of rank (n, n) with nodes $x_1, ..., x_n$ and $begin(E_n) = end(E_n) = x_1 \cdots x_n$; we write E for E_1. Note that E_0 is the empty graph.

It is straightforward to verify that $GR(\Sigma) = (GR_{m,n}(\Sigma))$ with the operations defined above is a magmoid, whose units are the graphs E_n, $n \geq 0$, see Lemma 6 of [7]. Subsets of $GR(\Sigma)$ are referred to as graph languages. The discrete graphs of $GR(\Sigma)$ form obviously a sub-magmoid $DISC$ of $GR(\Sigma)$ and the function sending each graph $G \in GR(\Sigma)$ to its underlying discrete graph is indeed an epimorphism of magmoids

$$disc_\Sigma : GR(\Sigma) \to DISC.$$

Engelfriet and Vereijken proved that, $GR(\Sigma)$ is finitely generated, that is, any graph can be built from a specific finite set of elementary graphs (cf. [7]). More precisely, let us denote by $I_{p,q}$ the discrete (p, q)-graph having a single node x and whose begin and end sequences are $x \cdots x$ (p times) and $x \cdots x$ (q times) respectively. Note that $I_{1,1}$ is equal with E. Let also Π be the discrete $(2, 2)$-graph having two nodes x and y and whose begin and end sequences are xy and yx, respectively. Finally, for every $\sigma \in \Sigma_{m,n}$, we denote again by σ the (m, n)-graph having only one edge and $m + n$ nodes $x_1, \ldots, x_m, y_1, \ldots, y_n$. The edge is labelled by σ, and the begin (resp. end sequence) of the graph is the sequence of sources (resp. targets) of the edge, viz. $x_1 \cdots x_m$ (resp. $y_1 \cdots y_n$).

Now let us introduce the alphabet D, formed by the following five symbols

$$i_{21} : 2 \to 1 \quad i_{01} : 0 \to 1 \quad i_{12} : 1 \to 2 \quad i_{10} : 1 \to 0 \quad \pi : 2 \to 2$$

where $x : m \to n$ indicates that symbol x has first rank m and second rank n, and denote by $mag(\Sigma \cup D)$ the free magmoid generated by the doubly ranked alphabet $\Sigma \cup D$. We denote by

$$val_\Sigma : mag(\Sigma \cup D) \to GR(\Sigma)$$

the unique magmoid morphism extending the function described by the assignments

$$i_{21} \mapsto I_{2,1}, \quad i_{01} \mapsto I_{0,1}, \quad i_{12} \mapsto I_{1,2}, \quad i_{10} \mapsto I_{1,0}, \quad \pi \mapsto \Pi,$$

$$\sigma \mapsto \sigma, \text{ for all } \sigma \in \Sigma, \quad e_n \mapsto E_n, \text{ for all } n \in \mathbb{N}.$$

Theorem 1 (cf. [7]). *The magmoid $GR(\Sigma)$ is generated by the set*

$$\Sigma \cup \{I_{12}, I_{10}, I_{21}, I_{01}, \Pi\}.$$

The previous theorem implies that the morphism val_Σ is a surjection. However, val_Σ is not an injection and in fact, for any given hypergraph, there are infinitely many patterns representing it. This ambiguity was settled by constructing a *finite* set of equations with the property that two patterns represent the same hypergraph, if and only if, one can be transformed into the other through these equations (cf. [5]). More precisely, we denote by $\pi_{n,1}$ the pattern inductively defined by

$$\pi_{1,0} = e, \quad \pi_{n,1} = \binom{\pi_{n-1,1}}{e}\binom{e_{n-1}}{\pi}$$

which will represent the graph associated with the permutation

$$\begin{pmatrix} 1 & 2 & \dots & n+1 \\ 2 \dots & n+1 & 1 \end{pmatrix}$$

interchanging the last n numbers with the first one (see below). Notice that for $n = 1$, $\pi_{1,1} = \pi$. Given a finite doubly ranked alphabet Σ, the set of equations $\mathcal{E}$:

$$\pi\pi = e_2, \quad \binom{e}{\pi}\binom{\pi}{e}\binom{e}{\pi} = \binom{\pi}{e}\binom{e}{\pi}\binom{\pi}{e},$$

$$\binom{e}{i_{21}} i_{21} = \binom{i_{21}}{e} i_{21}, \quad \binom{e}{i_{01}} i_{21} = e,$$

$$\pi i_{21} = i_{21}, \quad \binom{e}{i_{01}} \pi = \binom{i_{01}}{e},$$

$$\binom{\pi}{e}\binom{e}{\pi}\binom{i_{21}}{e} = \binom{e}{i_{21}} \pi,$$

$$i_{12} \binom{e}{i_{12}} = i_{12} \binom{i_{12}}{e}, \quad i_{12} \binom{e}{i_{10}} = e,$$

$$i_{12}\pi = i_{12}, \quad \pi \binom{e}{i_{10}} = \binom{i_{10}}{e},$$

$$\binom{i_{12}}{e}\binom{e}{\pi}\binom{\pi}{e} = \pi \binom{e}{i_{12}},$$

$$i_{12} i_{21} = e, \quad \binom{i_{12}}{e}\binom{e}{i_{21}} = i_{21} i_{12},$$

$$\pi_{m,1} \binom{\sigma}{e} = \binom{e}{\sigma} \pi_{n,1}, \quad \text{where } \sigma \in \Sigma_{m,n}, \ m, n \geq 0,$$

has the following property: for all patterns p and q,

$$val_\Sigma(p) = val_\Sigma(q) \text{ if and only if } p \underset{\mathcal{E}}{=} q.$$

3 Graphoids and Graph Automata

As we have seen in the previous section, the equations $\mathcal{E}$ are satisfied in $GR(\Sigma)$ by replacing π by Π and $i_{\kappa\lambda}$ by $I_{\kappa,\lambda}$. Magmoids with such a property are called *graphoids*. Formally, a graphoid $\mathbf{M} = (M, D)$ consists of a magmoid M with units e_0 and e and a set $D = \{s, d_{01}, d_{21}, d_{10}, d_{12}\}$, where $s \in M_{2,2}$, $d_{01} \in M_{0,1}$, $d_{21} \in M_{2,1}$, $d_{10} \in M_{1,0}$, $d_{12} \in M_{1,2}$ such that the following equations hold:

$$s \circ s = e_2 \quad (2) \qquad (s \,\square\, e) \circ (e \,\square\, s) \circ (s \,\square\, e) = (e \,\square\, s) \circ (s \,\square\, e) \circ (e \,\square\, s) \quad (3)$$

$$(e \,\square\, d_{21}) \circ d_{21} = (d_{21} \,\square\, e) \circ d_{21} \quad (4) \qquad (e \,\square\, d_{01}) \circ d_{21} = e \quad (5)$$

$$s \circ d_{21} = d_{21} \quad (6) \qquad (e \,\square\, d_{01}) \circ s = (d_{01} \,\square\, e)\,, \qquad (7)$$

$$(s \,\square\, e) \circ (e \,\square\, s) \circ (d_{21} \,\square\, e) = (e \,\square\, d_{21}) \circ s, \qquad (8)$$

$$d_{12} \circ (e \,\square\, d_{12}) = d_{12} \circ (d_{12} \,\square\, e) \quad (9) \qquad d_{12} \circ (e \,\square\, d_{10}) = e, \quad (10)$$

$$d_{12} \circ s = d_{12} \quad (11) \qquad s \circ (e \,\square\, d_{10}) = (d_{10} \,\square\, e)\,, \quad (12)$$

$$(d_{12} \,\square\, e) \circ (e \,\square\, s) \circ (s \,\square\, e) = s \circ (e \,\square\, d_{12})\,, \qquad (13)$$

$$d_{12} \circ d_{21} = e \quad (14) \qquad (d_{12} \,\square\, e) \circ (e \,\square\, d_{21}) = d_{21} \circ d_{12} \quad (15)$$

$$s_{m,1} \circ (f \,\square\, e) = (e \,\square\, f) \circ s_{n,1}, \quad \text{for all } f \in mag_{m,n}(\Sigma \cup D), \qquad (16)$$

where $s_{m,1}$ is defined inductively by s analogously with $\pi_{m,1}$ (see Section 2).

We point out that the last equation holds in $GR(\Sigma)$ since it holds for all the letters of the doubly ranked alphabet Σ (cf. [5]). Thus the pair $\mathbf{GR}(\Sigma) = (GR(\Sigma), D)$, where $D = \{\Pi, I_{0,1}, I_{2,1}, I_{1,0}, I_{1,2}\}$ is a graphoid. Given graphoids (M, D) and (M', D'), a magmoid morphism $H : M \to M'$ preserving D-sets, i.e., $H(s) = s'$ and $H(d_{\kappa\lambda}) = d'_{\kappa\lambda}$, is called a morphism of graphoids. Graphoids $\mathbf{Rel}(\mathbf{Q}) = (Rel(Q), D)$ constructed from the magmoid of relations over a given set Q are called *relational graphoids* and a relational graphoid is called abelian when $s = \{(g_1 g_2, g_2 g_1) \mid g_1, g_2 \in Q\}$.

Example 1. One way to construct an abelian relational graphoid is by setting s as above and

$$d_{01} = \{(\varepsilon, g) \mid g \in Q\}, \quad d_{21} = \{(gg, g) \mid g \in Q\},$$

$$d_{10} = \{(g, \varepsilon) \mid g \in Q\}, \quad d_{12} = \{(g, gg) \mid g \in Q\}.$$

This graphoid was introduced in [6] where it was used in the construction of graph automata.

We have already discussed how the set $GR(\Sigma)$ can be structured into a graphoid; in fact it is the free graphoid generated by Σ.

Theorem 2 ([6]). *The doubly ranked function $j : \Sigma \to GR(\Sigma)$, with $j(\sigma) = \sigma$, for all $\sigma \in \Sigma$, has the following universal property: for any graphoid $\mathbf{M} = (M, D)$, $D = \{\eta_0, \eta, s, d_{10}, d_{12}, d_{01}, d_{21}\}$ and any doubly ranked function $f : \Sigma \to M$, there exists a unique morphism of graphoids $\bar{f} : GR(\Sigma) \to \mathbf{M}$ making commutative the following triangle.*

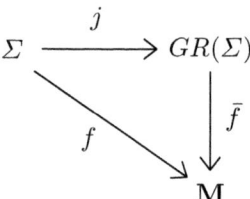

The morphism $\bar{f}$ is defined by the clauses

- $\bar{f}(\sigma) = f(\sigma)$, $\sigma \in \Sigma$,
- $\bar{f}(E_0) = \eta_0$, $\bar{f}(E) = \eta$, $\bar{f}(\Pi) = s$, $\bar{f}(I_{ij}) = d_{ij}$,
- $\bar{f}(G_1 \circ G_2) = \bar{f}(G_1) \circ \bar{f}(G_2)$,
- $\bar{f}(G_1 \,\square\, G_2) = \bar{f}(G_1) \,\square\, \bar{f}(G_2)$,

for all graphs G_1, G_2 of suitable rank.

A *graph homomorphism* $H : GR(\Sigma) \to GR(\Sigma')$ is just a morphism of graphoids. Hence, by virtue of the previous theorem it is completely determined by its values $H(\sigma)$, $\sigma \in \Sigma$. A graph homomorphism $H : GR(\Sigma) \to GR(\Sigma')$ is called a *projection* whenever $H(\Sigma) \subseteq \Sigma'$.

Automata operating on arbitrary (hyper)graphs were introduced in [6]. A *graph automaton* is a structure $\mathcal{A} = (\Sigma, Q, \mathbf{Rel}(Q), \delta_{\mathcal{A}}, I_{\mathcal{A}}, T_{\mathcal{A}})$, where:

- Σ is the doubly ranked alphabet of hyperedge labels;
- Q is the finite set of states;
- $\mathbf{Rel}(Q)$ is a relational graphoid over Q;
- $\delta_{\mathcal{A}} : \Sigma \to \mathbf{Rel}(Q)$ is the doubly ranked transition function;
- $I_{\mathcal{A}}, T_{\mathcal{A}}$ are initial and final rational subsets of Q^*.

According to Theorem 2 the function

$$\delta_{\mathcal{A}} : \Sigma \to \mathbf{Rel}(Q)$$

is uniquely extended into a morphism of graphoids

$$\bar{\delta}_{\mathcal{A}} : GR(\Sigma) \to \mathbf{Rel}(Q).$$

The behavior of $\mathcal{A}$ is given by

$$|\mathcal{A}| = \{F \mid F \in GR_{m,n}(\Sigma),\ \bar{\delta}_{\mathcal{A}}(F) \cap (I_{\mathcal{A}}^{(m)} \times T_{\mathcal{A}}^{(n)}) \neq \emptyset,\ m, n \in \mathbb{N}\}$$

where $I_A^{(m)} = I_A \cap Q^m$ and $T_A^{(n)} = T_A \cap Q^n$. From their construction, graph automata are finite machines due to the fact that the set of equations (2)-(16) is finite. A graph language is called recognizable whenever it is obtained as the behavior of a graph automaton. The class of all such languages over the doubly ranked alphabet Σ is denoted by $Rec(\Sigma)$.

Example 2. Let $\Sigma = \Sigma_{1,1} = \{a, b\}$, and $Q = \{q_1, q_2\}$, the graph automaton

$$\mathcal{A} = (\Sigma, Q, \mathbf{Rel}(Q), \delta_A, I_A, T_A)$$

where $\mathbf{Rel}(Q)$ is as in Example 1, $I_A = \{q_1\}$, $T_A = \{q_1\}$, $\delta_A(a) = \{(q_1, q_2)\}$, $\delta_A(b) = \{(q_2, q_1)\}$, clearly computes the graph language $L \subseteq GR_{1,1}(\Sigma)$, consisting of exactly those graphs that have at least one path labeled $(ab)^*$ from their initial to their final node.

4   The Structure of Abelian Relational Graphoids

From the construction of the graph automata it is evident that the selection of the particular graphoid is crucial in the sense that it determines the behavior of the automaton. In this respect, we determine the properties that a set Q inherits from the structuring of $Rel(Q)$ into a graphoid. In what follows, let Q be a finite set and $\mathbf{Rel}(\mathbf{Q}) = (Rel(Q), D)$, with $D = \{s, d_{01}, d_{21}, d_{10}, d_{12}\}$ an abelian relational graphoid over Q.

By using the fact that for $\mathbf{Rel}(\mathbf{Q})$ it holds

$$e = \{(g, g) \mid g \in Q\} \quad (17) \qquad \text{and} \quad s = \{(g_1 g_2, g_2 g_1) \mid g_1, g_2 \in Q\} \quad (18)$$

we obtain the following two lemmata.

Lemma 1. *i)* If $(h, fg) \in d_{12}$ then $(h, gf) \in d_{12}$.
ii) If $(fg, h) \in d_{21}$ then $(gf, h) \in d_{21}$.

Proof. We combine Equations (11) and (6) with Equation (18).

Lemma 2. *i)* If $(f, ag) \in d_{12}$ or $(f, ga) \in d_{12}$ and $(a, \varepsilon) \in d_{10}$, then $f = g$.
ii) If $(ag, f) \in d_{21}$ or $(ga, f) \in d_{21}$ and $(\varepsilon, a) \in d_{01}$, then $f = g$.

Proof. i) From Lemma 1

$$(f, ag) \in d_{12} \text{ implies } (f, ga) \in d_{12}$$

from this and $(a, \varepsilon) \in d_{10}$ we get

$$(f, g) \in d_{12} \circ (e \,\square\, d_{10})$$

which from Equation (10) gives $(f, g) \in e$ and from (1) we obtain the desired result.

ii) Similarly from using Equation (5).

By this result we obtain the following Lemma.

Lemma 3. *If $(f, fa) \in d_{12}$, with $(a, \varepsilon) \in d_{10}$ and $(f, fb) \subset d_{12}$, with $(b, \varepsilon) \in d_{10}$, then $a = b$.*

Proof. Since $(f, fa) \in d_{12}$ and $(f, fb) \in d_{12}$ we get that

$$(f, fba) \in d_{12} \circ (d_{12} \square e).$$

Now from Equation (9) it must also hold

$$(f, fba) \in d_{12} \circ (e \square d_{12}).$$

Hence there exists an element $c \in Q$ such that $(f, fc) \in d_{12}$ and $(c, ba) \in d_{12}$. This together with $(a, \varepsilon) \in d_{10}$ and Lemma 2 $i)$ gives $c = b$. By the same lemma and $(b, \varepsilon) \in d_{10}$ we get that $c = a$ which concludes the proof.

At this point we are able to define a partition inside Q.

Proposition 2. *The relation $\approx$ defined for $f, g \in Q$ by $f \approx g$ if and only if there exists $a \in Q$ with $(a, \varepsilon) \in d_{10}$ such that*

$$(f, fa) \in d_{12} \text{ and } (g, ga) \in d_{12},$$

is an equivalence.

Proof. By Equations (10) and (1) we deduce that for every $f \in Q$ there exists an element $a \in Q$ such that

$$(f, fa) \in d_{12} \text{ and } (a, \varepsilon) \in d_{10}.$$

Hence $f \approx f$. The relation is obviously reflexive and from Lemma 3 it is easily seen that it is also transitive.

The equivalence $\approx$ defined in the above proposition partitions Q into sets Q_i which, as we shall prove, have the structure of an abelian group with operations induced by d_{21} and d_{12} and units by d_{10} and d_{01}. The following two lemmata are a valuable step towards this direction.

Lemma 4. *If $(a, \varepsilon) \in d_{10}$, then*

i) $(a, aa) \in d_{12}$,
ii) $(aa, a) \in d_{21}$, and
iii) $(\varepsilon, a) \in d_{01}$.

Proof. i) From Equations (10) and (1) there exists an element $b \in Q$ with $(b, \varepsilon) \in d_{10}$ such that $(a, ab) \in d_{12}$ and by Lemma 2 $i)$ we obtain $a = b$.
   ii) From Equation (14) there are elements $f, g \in Q$ such that

$$(a, fg) \in d_{12} \quad \text{and} \quad (fg, a) \in d_{21}.$$

From the first of the above and Equation (11) we have $(a, gf) \in d_{12}$ and this together with $(g, g) \in e$ and $(fg, a) \in d_{21}$ gives

$$(ag, ga) \in (d_{12} \square e) \circ (e \square d_{21}).$$

Hence by virtue of Equation (15) it also holds $(ag, ga) \in d_{21} \circ d_{12}$. This means that there exists an element $k \in Q$ with $(ag, k) \in d_{21}$ and $(k, ag) \in d_{12}$ and this by Lemma 2 gives $k = g$. Thus

$$(ag, g) \in d_{21} \quad \text{and} \quad (g, ag) \in d_{12},$$

or by Lemma 1

$$(ga, g) \in d_{21} \quad \text{and} \quad (g, ga) \in d_{12},$$

hence $(ga, ga) \in d_{21} \circ d_{12}$ and from Equation (15)

$$(ga, ga) \in (d_{12} \,\square\, e) \circ (e \,\square\, d_{21}).$$

From this we deduce that there exists an element $b \in Q$ with

$$(g, gb) \in d_{12} \quad \text{and} \quad (ba, a) \in d_{21}.$$

The second of the above together with $(a, aa) \in d_{12}$ gives

$$(ba, aa) \in d_{21} \circ d_{12}$$

which again by Equation (15) means that

$$(ba, aa) \in (d_{12} \,\square\, e) \circ (e \,\square\, d_{21}).$$

Hence there exists an element k such that

$$(b, ak) \in d_{12} \quad \text{and} \quad (ka, a) \in d_{21}.$$

The first of the above from Lemma 2 gives $k = b$ so the above equation are rewritten as

$$(b, ab) \in d_{12} \quad \text{and} \quad (ba, a) \in d_{21},$$

and from these and Lemma 1 it holds $(b, a) \in d_{12} \circ d_{21}$, which from Equations (14) and (1) gives $b = a$ which concludes the proof.

$iii)$ From Equation (5) there exists an element $b \in Q$ with $(\varepsilon, b) \in d_{01}$ such that $(ab, a) \in d_{21}$ and by using the same argument we used in the end of the previous item we get $a = b$ as wanted.

Lemma 5. *If* $(f, fa) \in d_{12}$, *and* $(a, \varepsilon) \in d_{10}$ *then* $(fa, f) \in d_{21}$.

Proof. From $(f, fa) \in d_{12}$ and $(aa, a) \in d_{21}$ (obtained from Lemma 4) we get

$$(fa, fa) \in (d_{12} \,\square\, e) \circ (e \,\square\, d_{21})$$

and by Equation (15) there exists an element k such that $(fa, k) \in d_{21}$ which by Lemma 4 $iii)$ and Lemma 2 $ii)$ gives $f = k$.

The next lemma guarantees that the group operations are always defined for two elements of the same group.

Lemma 6. *If for $f, g \in Q$, $f \approx g$, then there exists $k \in Q$ such that $(fg, k) \in d_{21}$ and $(k, fg) \in d_{12}$.*

Proof. Let $(f, fa) \in d_{12}$ and $(g, ga) \in d_{12}$ with $(a, \varepsilon) \in d_{10}$. By Lemma 5 $(ag, g) \in d_{21}$ and hence by using $(f, fa) \in d_{12}$ and Equation (15)

$$(fg, fg) \in d_{21} \circ d_{12},$$

which gives the desired result.

The following result states that the operations are closed and defined only for elements of the same group.

Lemma 7. *If for $f, g, k \in Q$, $(fg, k) \in d_{21}$ or $(k, fg) \in d_{12}$, then $f \approx g \approx k$.*

Proof. Let $(fg, k) \in d_{21}$ and $(f, af) \in d_{12}$ with $(a, \varepsilon) \in d_{10}$. Then

$$(fg, ak) \in (d_{12} \,\square\, e) \circ (e \,\square\, d_{21})$$

and by Equation (15) there exists an element $h \in Q$ with $(h, ak) \in d_{12}$. This by Lemma 2 *i)* gives $h = k$ so $(k, ak) \in d_{12}$ and hence $k \approx f$. Similarly we get $k \approx g$. By an argument akin to the above we can also prove the second case.

We are now ready to state the following theorem.

Theorem 3. *If the set Q is structured into an abelian relational graphoid by the set of relations $D = \{s, d_{01}, d_{21}, d_{10}, d_{12}\}$, then Q is partitioned by the equivalence $\approx$ into abelian groups Q_i with operation for $f, g \in Q_i$*

$$f \cdot g = k, \quad \text{where} \quad (fg, k) \in d_{21}$$

and unit the unique element $a \in Q_i$ with the property $(a, \varepsilon) \in d_{10}$.

Proof. First we will show that the operation is well defined for every set Q_i. From Lemma 6 we get that for all $f, g \in Q_i$ there exists an element k with

$$(fg, k) \in d_{21} \quad \text{and} \quad (k, fg) \in d_{12}$$

At this point assume that there exists a second element k' with $(fg, k') \in d_{21}$; from this and the second of the above together with Equation (14) we obtain that $(k, k') \in e$, thus $k = k'$.

From Lemma 7 we see that the operation is closed for every set Q_i.

For every Q_i there is exactly one element $a \in Q_i$ with the property $(a, \varepsilon) \in d_{10}$. Indeed from Proposition 2 and from Lemma 4 *i)* we get that there is at least one such element $a \in Q_i$. Now let $b \in Q_i$ with $(b, \varepsilon) \in d_{10}$. We have $(b, ba) \in d_{12}$ and from $(b, \varepsilon) \in d_{10}$ and Lemma 2 we get $a = b$. This element is the unit of Q_i as asserts Lemma 5.

Associativity is obtained from Equation (4) in a straightforward way.

Now let $f \in Q_i$, with a the unit of Q_i, to prove invertibility we have to show that there exists an element $g \in Q_i$ with $(fg, a) \in d_{21}$. From $(af, f) \in d_{21}$ and $(f, fa) \in d_{12}$ we obtain $(af, fa) \in d_{21} \circ d_{12}$ which from Equation (15) gives

$$(af, fa) \in (d_{12} \,\square\, e) \circ (e \,\square\, d_{21}).$$

This means that there exists an element g with $(a, fg) \in d_{12}$ and $(gf, a) \in d_{21}$ and from Lemma 7 it holds $g \in Q_i$.

Every group Q_i is commutative as it is asserted by Lemma 1 ii).

It is straightforward to prove that the opposite direction of the above theorem also holds, i.e., every partition of a set Q into abelian groups instructs a structuring of Q into an abelian relational graphoid. Hence, we have specified all the possible abelian relational graphoids that can be constructed from a given set. Since the role of the graphoid is instrumental in the construction of a graph automaton, this result actually determines all the possible types of (abelian) graph automata with a given state set Q. Furthermore, it introduces previously unknown graphoids considerably more sophisticated than the existing ones. Indeed, the graphoid of Example 1 that was used for the construction of graph automata in [6] corresponds to the partitioning of the state set into singleton sets each one being the trivial group. Further research directions include the following.

- Determine the function and the behavior of graph automata utilizing the discovered graphoids.
- Compare the new types of graph automata with the present version as well as with other kinds of graph recognizability.
- Examine weather non-abelian relational graphoids exist.

References

1. Arbib, M.A., Give'on, Y.: Algebra automata I: Parallel programming as a prolegomena to the categorical approach. Inform. and Control 12, 331–345 (1968)
2. Arnold, A., Dauchet, M.: Théorie des magmoides. I. RAIRO Inform. Théor. 12, 235–257 (1978)
3. Arnold, A., Dauchet, M.: Théorie des magmoides. II. RAIRO Inform. Théor. 13, 135–154 (1979)
4. Bossut, F., Dauchet, M., Warin, B.: A Kleene theorem for a class of planar acyclic graphs. Inform. and Comput. 117, 251–265 (1995)
5. Bozapalidis, S., Kalampakas, A.: An axiomatization of graphs. Acta Informatica 41, 19–61 (2004)
6. Bozapalidis, S., Kalampakas, A.: Graph Automata. Theoretical Computer Science 393, 147–165 (2008)
7. Engelfriet, J., Vereijken, J.J.: Context-free graph grammars and concatenation of graphs. Acta Informatica 34, 773–803 (1997)

8. Gibbons, J.: An initial-algebra approach to directed acyclic graphs. In: Möller, B. (ed.) MPC 1995. LNCS, vol. 947, pp. 282–303. Springer, Heidelberg (1995)
9. Kamimura, T., Slutzki, G.: Parallel and two-way automata on directed ordered acyclic graphs. Inform. and Control 49, 10–51 (1981)
10. Kamimura, T., Slutzki, G.: Transductions of DAGS and trees. Math. Syst. Theory 15, 225–249 (1982)

A Survey on Picture-Walking Automata[*]

Jarkko Kari and Ville Salo

Department of Mathematics,
University of Turku, FI-20014 Turku, Finland

Abstract. Picture walking automata were introduced by M. Blum and
C. Hewitt in 1967 as a generalization of one-dimensional two-way finite automata to recognize pictures, or two-dimensional words. Several
variants have been investigated since then, including deterministic, non-
deterministic and alternating transition rules; four-, three- and two-way
movements; single- and multi-headed variants; automata that must stay
inside the input picture, or that may move outside. We survey results
that compare the recognition power of different variants, consider their
basic closure properties and study decidability questions.

Keywords: Picture-walking automata, 2-dimensional automata, picture
languages.

1   Introduction

Informally, a picture is a matrix over a finite alphabet, and a picture language
is a set of matrices over the same alphabet. A picture-walking automaton is a
finite state automaton moving on the cells of the given picture according to a
local rule, accepting if it reaches a final state [1].

The theory of picture languages is a branch of formal language theory which
studies natural picture language families and connections between them. Most
of this theory has concentrated on classes having to do with 'finite state', in an
effort to find a natural counterpart for the one-dimensional regular languages.
The classes obtained from picture-walking automata are usually not considered
to be a very natural counterpart due to their rather weak closure properties,
but the recognizable picture languages take this place instead. In fact, many
interesting proofs about picture-walking automata have more of a 'navigational'
than a language-theoretic feel to them.

In the first section, we deal with very basic and natural questions. We only
consider Boolean closure properties and the lattice of inclusions between the
four basic automata classes we define: DFA, NFA, UFA and AFA, that is, deterministic, non-deterministic, universally quantifying and alternating finite state
automata, respectively. We present a slightly more refined view to these questions
by considering unary rectangles and non-unary square pictures separately with
the aim of clarifying the strengths and weaknesses of each approach. In each case

---

[*] Research supported by the Academy of Finland Grant 131558.

W. Kuich and G. Rahonis (Eds.): Bozapalidis Festschrift, LNCS 7020, pp. 183–213, 2011.

we obtain a slightly different set of non-closure properties, which taken together give a rather complete set of results for the class of all pictures.

The second section introduces a very modest-looking change to the definition of an automaton: allowing it to exit the picture. We will see that this change is relevant for AFA, but not for the other classes, although this is not altogether straightforward to prove. In the third and fourth sections, we introduce much greater changes to the definition. In the third section, we restrict the directions in which the automaton is allowed to move. We give a brief introduction to the results known for these classes. In particular, for three-way automata, we give a complete set of Boolean closure properties and complete comparison results between all three- and four-way automata classes, mostly without proofs. In the fourth section we give our automata a finite set of markers they can move around the picture. In Section 5, we review decidability results, and recall connections between 2-dimensional DFA and Minsky machines.

Finally, we mention that a survey on two-dimensional picture-walking automata theory already exists [6], although it's main focus is not on finite state automata but on general Turing machines. Since few papers on picture-walking automata have been published after [6], it is still mostly up-to-date, except for some open problems which have since been solved (many of which are presented here). There is also a good survey of general two-dimensional language theory in the Handbook of Formal Languages [2].

2   Definitions

A *picture* is the two-dimensional analog of a finite word: a not necessarily square matrix over a finite alphabet. We may usually assume a binary alphabet $\Sigma = \{0, 1\}$. We write Σ_*^* for the full language of all matrices over Σ, and define a *picture language* as a subset of a full language. A class of picture languages, usually called a *picture class*, is a collection of picture languages.

For $p \in \Sigma_*^*$ we write $p[(i,j)] = p_{i,j}$ for the contents of the cell in position (i,j) in p, with the usual matrix indexing. We also draw p as a matrix, and thus, for instance, the cell with index $(1,1)$ is considered the top left cell and the first axis is the vertical one, ascending downward. If (i,j) is not a cell of p, we define $p[(i,j)]$ to be a special symbol $\#$ (that is, in practise we index pictures by $\mathbb{Z}^2$). We write $\mathrm{dom}(p)$ for the set of indexes of (non-$\#$) cells in p, called the *domain* of p, and $\mathrm{edom}(p)$ for the cells of $\mathrm{dom}(p)$ and all their neighbors. The width and height of a picture p are denoted by $|p|$ and $\bar{p}$, respectively. The set $\mathrm{edom}(p) - \mathrm{dom}(p)$ is called the *border* of p.

Let us start by defining the main picture-walking automata considered in this survey, and their corresponding picture classes. Our automata have a single head, and they walk on the positions of the picture according to a local rule, accepting if they reach a final state. Existential and universal states can be used to make (perfect) guesses and to check multiple local properties 'simultaneously', respectively.

Definition 1. *An alternating finite state automaton (an AFA) A is a tuple $(Q, \Sigma, E, U, I, F, \delta)$ where Q is the set of* states *partitioned into E and U, the sets of* universal *and* existential *states. The sets $I, F \subset Q$ are called the sets of* initial *and* final *states. The function δ is called the* transition function *or the* local rule *and it has type $\delta : Q \times \Sigma \to 2^{Q \times \{(1,0),(-1,0),(0,1),(0,-1)\}}$.*

If $Q = E$ (and thus $U = \emptyset$), the automaton is said to be non-deterministic *(an NFA), and if $Q = U$ and $|I| = 1$, it is said to be* universally quantifying *(a UFA). If all images of δ are singletons or empty sets and $|I| = 1$, the automaton is called* deterministic *(a DFA). We use the variable XFA to state things for all the automata classes simultaneously.*

Definition 2. *Let $A = (Q, \Sigma, E, U, I, F, \delta)$ be an AFA. An* instantaneous description *(an ID) of A on a picture $p \in \Sigma_*^*$ is a pair $(q, x) \subset Q \times edom(p)$. An ID $l_2 = (q_2, x_2)$ is a successor of another ID $l_1 = (q_1, x_1)$ if $(q_2, x_2 - x_1) \in \delta(q_1, p[x_1])$. We then write $l_1 \to l_2$. For a vertex-labeled tree r, we write $a_1 \to a_2$ if a_2 is a child of a_1 in r, and we write $l(a)$ for the label of node $a \in r$.*

Now, an accepting run *of A on p is a finite tree r labeled with ID's such that*

- *the root of r has its label in $I \times \{(1,1)\}$.*
- *if $a_1 \in r$ is not a leaf and $l(a_1) \in E \times edom(p)$, then*

$$\exists a_2 \in r : a_1 \to a_2 \wedge l(a_1) \to l(a_2).$$

- *if $a_1 \in r$ is not a leaf and $l_1 = l(a_1) \in U \times edom(p)$, then*

$$\forall l_2 : l_1 \to l_2 \implies \exists a_2 \in r : a_1 \to a_2 \wedge l(a_2) = l_2.$$

- *the leaves of r have labels in $F \times edom(p)$.*

We write $\mathcal{L}(A)$ for the set of pictures p over Σ for which there exists an accepting run of A.

The restriction $|I| = 1$ is important for UFA, since otherwise I provides the automaton with existential quantification. For DFA, NFA and UFA, we may leave E and U out of the definition, and for DFA, we take δ to have the type $\delta : Q \times \Sigma \to Q \times \{(1,0),(-1,0),(0,1),(0,-1)\}$, with the obvious meaning.

Note that an accepting run for an NFA is just a sequence of ID's. For UFA, an accepting run can look complicated, but we may – in an obvious way – define non-accepting runs for them, that is, possibly infinite sequences of ID's that prove no accepting run exists. This is possible since *all* choices of transitions must lead to acceptance or a picture is rejected by a UFA. Of course, a non-accepting run exists if and only if the picture is not accepted.

Note that our automata cannot sense if the border of the picture is next to them, but must step on it and read the special symbol # in order to obtain this information. We will also need 1-dimensional automata. These are always two-way automata, that is, they move both left and right. The left and right ends of the input word are again observed by reading a bordering #.

For each automata class XFA, the picture class corresponding to XFA will also be called XFA. That is, asking whether there exists an XFA for a picture language L is equivalent to asking whether $L \in$ XFA. This should not cause confusion.

We use the naming scheme of [11] and [22] for the XFA classes. DFA and NFA were defined already in [1] in 1967 and were simply called 'automata', while the most used names for DFA and NFA seem to have been '2-DA' and '2-NA', respectively. UFA and AFA were discussed at least in [10], as a subset of Turing machines, and were given the names '2-UFA' and '2-AFA' in [3]. Slightly confusingly, we will later write 2XFA to refer to *two-way* automata instead of two-dimensional automata. Two- and three-way restrictions of automata have often been denoted by having 'TR' and 'TW' somewhere in the name of the class, respectively. We do not consider probabilistic automata in this article: see [17] for a incomparability result between probabilistic automata and AFA.

The square pictures and the unary pictures are natural subclasses of the set of all pictures (also called the general pictures). Some negative results (results of the type $L \notin$ CLS for some picture class CLS) are easier to prove in the square world (content results) and some are easier to prove in the unary world (shape results).

Definition 3. *For each automata class XFA, we write XFA_s for the languages of square pictures accepted by some automaton in XFA. We write XFA_u for the unary languages of the XFA automata. Complements of languages in XFA_s and XFA_u are usually taken with respect to the set of all square pictures and all unary pictures, respectively.*

The theory of picture languages has not been primarily concerned with these kinds of automata, but a different model of computation known as recognizability. We will not discuss the recognizable picture languages (REC) in length, but we do show some connections between the two worlds, and obtain some results for picture-walking automata from non-closure properties of REC. For more information on REC, see [2] or [22].

Definition 4. *A Wang tile is an element of C^4 for some set of colors C containing a special element #. A set T of Wang tiles with the same C defines a picture language over T by taking the pictures where neighboring colors match, and #'s occur (only) facing the border of the picture. Such a language is called a* local *picture language. A recognizable picture language is the image of a local picture language through a symbol-to-symbol projection. We denote the class of recognizable picture languages by REC.*

We may also think of an REC grammar (tileset and projection) as 'accepting' languages instead of generating them, by taking a picture p, and assigning 'states' (elements of T) on top of the cells of p, which agree in their colors with the neighboring states. When picture-walking automata are involved, we will, however, avoid the term 'state' in this context, and simply say tiles are assigned on top of the cells.

3   Basic Results on 4-Way Automata

As is often the case in mathematics, the main results in the theory of picture-walking automata tend to be of one of two types: 'positive' or 'negative'. A positive result says a class is a subset of another, or that a language belongs to a class, while a negative result tells us a class is a *proper* subset of another, or that a language *cannot* be accepted by some type of machine.

Many negative results are known for both XFA_u and XFA_s, and taken together, one could say the natural questions for XFA have mostly been answered. However, the intersection of these classes is not understood at all. We repeat the following two conjectures, which were made in [12] in 2004, and are still unanswered.

Conjecture 1. The language of unary squares with prime side length are not in DFA.

Conjecture 2. DFA is a proper subset of NFA when restricted to unary squares.

As we mentioned, by varying either shape *or* content, we can build the beginning of a (proper) polynomial hierarchy and prove basic non-closure properties for all the automata classes involved. We first show positive results for the class of all picture languages and then devote a subsection for negative results in both the case of varying shape with unary alphabet and varying content with square shape. In the unary case, a proper diamond of inclusions is shown to exist between the four classes. As for the square case, proper inclusions are known to hold between DFA_s, NFA_s and AFA_s, but little is known about UFA_s. In each case a slightly different set of Boolean closure properties is obtained.

Figure 3 summarizes the known results. All inclusions drawn are proper, and dashed lines signify incomparable classes. On the right side, UFA_s is known to be between DFA_s and AFA_s, but not between DFA_s and NFA_s.

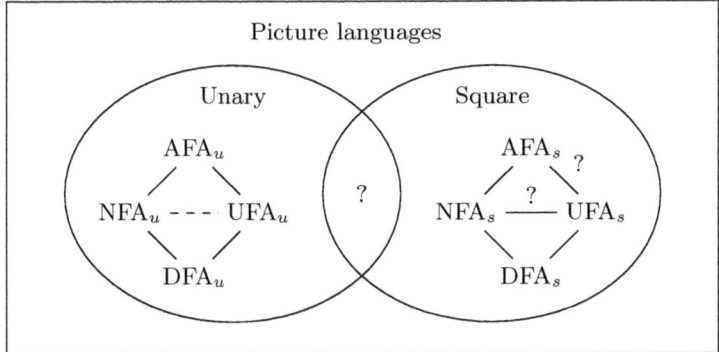

Fig. 1. A diagram of inclusions of the automata classes in the unary and square case

In the following sections, we will see that DFA is closed under complementation, while NFA, UFA and AFA are not. We prove non-closure under complement separately for the unary and square cases for NFA and UFA. Non-closure in the case of all pictures can easily be inferred from either proof. The case of complementing AFA_u is unknown, but we prove non-closure for AFA_s, from which the general case of AFA easily follows. Figure 3 summarizes the closure properties known for the different types of automata, where we write XFA_x for XFA, XFA_u and XFA_s.

	DFA_x	NFA	NFA_u	NFA_s	UFA_x	AFA	AFA_u	AFA_s
¬	Yes	No	No	?	No	No	?	No
∩	Yes	Yes	Yes	Yes	Yes	Yes	Yes	Yes
∪	Yes	Yes	Yes	Yes	?	Yes	Yes	Yes

Fig. 2. Refined table of Boolean closure properties of the XFA classes

Note that for general pictures, all Boolean closure properties except for the closure of UFA under union are known. In [22], a negative answer is conjectured.

Conjecture 3. UFA is not closed under union.

3.1   Positive Results

Obviously, DFA $\subset$ NFA $\subset$ AFA and DFA $\subset$ UFA $\subset$ AFA with not necessarily proper inclusions. It is also clear that all the classes are closed under rotation around the center of the picture, vertical and horizontal flips and matrix transpose. We can also prove natural Boolean closure properties for these classes with simple constructions outlined below.

The following observation has been used in numerous articles, including [3,11,22]. The proof is a direct modification of [23].

Theorem 1. *For every DFA A, there is a DFA A' with $\mathcal{L}(A') = \mathcal{L}(A)$ such that A' halts on every input.* ☐

Corollary 1. *DFA is closed under complementation, that is, DFA = co-DFA.* ☐

Theorem 2. *All the XFA classes are closed under intersection and DFA, NFA and AFA are closed under union.*

Proof. Since an accepting computation is necessarily halting in all branches of the accepting run, intersection can be implemented for any machine by simply testing inclusion in the languages in question one by one. As for union, NFA and AFA can use an existential state, and for a halting DFA (guaranteed by Theorem 1), union can be implemented like intersection. ☐

3.2   Negative Results for Varying Shape and Unary Alphabet

The shape-based approach is more recent than the content-based one, and gives better results for connections between classes. It is based on 1-dimensional automata. The tools we need are a reduction from 2D to 1D (Lemma 1) and a lemma connecting the number of states in a 1-dimensional automata, and the period and threshold of its language (Lemma 2). After this, a single concrete language is enough to separate all the four automata classes. We also obtain non-closure under complementation for NFA_u and UFA_u.

Let A be an XFA running on unary input, with language $L = \mathcal{L}(A)$. For each height h, we may, in a natural way, associate a 1-dimensional XFA A_h accepting the unary language of the corresponding widths. Furthermore, A_h has $O(h)$ states where the invisible constant only depends on A.

Lemma 1. *[11] Let A be an AFA with k_u universal states and k_e existential states recognizing the unary picture language $L \subset \{1\}_*^*$. Then, for each h, the language $L_h = \{1^{|p|} \mid p \in L, \bar{p} = h\}$ is recognized by a one-dimensional two-way AFA A_h with $k_u(h+2)$ universal states and $k_e(h+2)$ existential states.*

Proof. Let $A = (Q, \{1\}, E, U, I, F, \delta)$ and $X = [0, h+1]$. Then

$$A_h = (Q \times X, \{1\}, E \times X, U \times X, I \times \{1\}, F \times X, \delta_h)$$

where δ_h simulates the local rule of A, interpreting the location of A_h as the horizontal location of A, and the X component of the state of A_h as the vertical location of A. It is clear that A_h correctly recognizes L_h.   □

Of course, each A_h accepts a unary regular language, and thus an eventually periodic language. Since the A_h have linearly many states with respect to h, it will be useful to prove a lemma which tells us something about the connection between the number of states in a (one-dimensional) automaton, and the period and threshold of the unary language it accepts.

Lemma 2. *[11,22] Let A be a 1D (two-way) AFA with k states over a unary alphabet with language $L \subset 1^*$. Let $n > k$. Then*

- *if A is an NFA, $1^n \in L \Longrightarrow 1^{n+k!} \in L$.*
- *if A is a UFA, $1^n \in L \Longleftarrow 1^{n+k!} \in L$.*

In particular, if A is a DFA, an equivalence holds, since if the transition relation is a function, we may change between universal and existential states without changing the language.

Proof. First, assume all states are existential, and let $1^n \in L$. Consider a subsequence s of an accepting run r for 1^n, which visits border symbols # at s_1 and $s_{|s|}$ or possibly ends with the automaton halting in the middle of the word (we think of the automaton as being on the left border just before starting its run). We will translate such 'partial runs' to corresponding partial runs for the longer input $1^{n+k!}$, which we then glue together to obtain $1^{n+k!} \in L$.

If a partial run moves from the left border back to the left border, the same partial run can be used for the longer input, and similarly for the right border. If a partial run ends in the middle of the word, it can also directly be translated for the longer word.

If s moves from the left border to the right border, we note that there must be a repetition of states such that the automaton moved some number of steps $0 < l \leq k$ to the right in between. But l divides $k!$, so we may repeat this partial run an additional $\frac{k!}{l}$ times to obtain a corresponding partial run for the longer input. A similar claim holds for partial runs from right to left.

The claim for UFA follows by considering non-accepting runs instead. The fact that these runs can be infinite does not lead to complications.   □

We will now separate NFA and UFA by using the duality between existential and universal quantification found in Lemma 2. For this, we will use the following 'billiard ball language' from [11].

Definition 5. *The billiard ball language $L_{billiard}$ is defined as*

$$L_{billiard} = \{p \in \{1\}_*^* \mid |p| \in \langle \overline{p}, \overline{p} + 1 \rangle\} = \{m\overline{p} + n(\overline{p} + 1) \mid m, n \in (\mathbb{N} \cup 0)\}$$

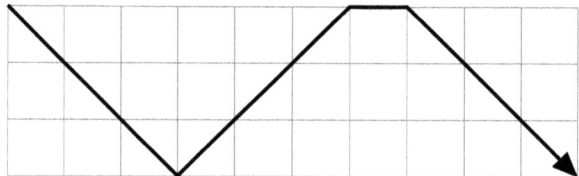

Fig. 3. The movement of an NFA accepting the unary 3x10 picture, which is in $L_{billiard}$

That is, for each height h, the widths must be some linear combination of h and $h + 1$. It is easy to make an NFA accepting this language by having it move to the right diagonally, bouncing off the walls as in Figure 3.2. Just as easily, we can make a UFA accepting its (unary) complement. Using the lemmas we proved, we can also prove the converse claims: a UFA cannot accept $L_{billiard}$ and an NFA cannot accept its complement:

Lemma 3. $L_{billiard} \notin UFA_u$, $L_{billiard}^c \notin NFA_u$

Proof. We will only prove $L_{billiard} \notin UFA_u$, the proof for NFA_u is symmetric. So assume on the contrary that A is a UFA with k states accepting $L_{billiard}$, and for each h let A_h with $\mathcal{L}(A_h) = L_h$ be the 1-dimensional UFA given by Lemma 1.

By looking at the definition of $L_{billiard}$, it is not hard to see that for each h, $h^2 - h - 1 = h(h - 2) + (h - 1)$ is not in L_h, but $h^2 - h$ is. Let h be large enough that $h^2 - h - 1 > k(h + 2)$. Here, $k(h + 2) = k_h$ is the number of states of A_h, and $1^{h^2 - h - 1} \notin L_h$, so by Lemma 2, also $1^{h^2 - h - 1 + mk_h!} \notin L_h$ for all m. But obviously L_h contains every word longer than some threshold t, a contradiction.   □

Corollary 2. $NFA_u \bowtie UFA_u$.　　　　　　　　　　　　　　　　　　□

Corollary 3.

$$DFA_u \subsetneq NFA_u \subsetneq AFA_u$$

and

$$DFA_u \subsetneq UFA_u \subsetneq AFA_u$$

　　　　　　　　　　　　　　　　　　　　　　　　　　　　　　□

Thus, we have built the diamond on the left in Figure 3. As a side product, we obtained that NFA and UFA are not closed under complement, completing the unary part of Figure 3.

Corollary 4. *NFA_u and UFA_u are not closed under complement.*　　　□

3.3　Negative Results for Varying Content and Square Shape

The separation of DFA and NFA on binary squares was done already in [1], and it is based on a pigeonhole argument.

Definition 6. *The language L_{center} consists of square pictures p of odd side length over $\{0, 1\}$ containing a 1 in the middle cell.*

Theorem 3. *L_{center} is in $NFA_s - DFA_s$.*

Proof. It is easy to see that L_{center} is in NFA_s: An NFA for it follows the main diagonal southeast, and can turn northeast at any 1 it sees. It accepts if it reaches the northeast corner. (Checking that the picture has square shape is trivial.)

Now, consider a hypothetical DFA $A = (Q, \Sigma, I, F, \delta)$ for it. We may assume A always leaves the domain of the picture before accepting it, and that it always halts by Theorem 1. To each binary $n \times n$ picture p we can then associate a function f_p characterizing the behavior of A inside the picture p. The function f_p has the type

$$f_p : Q \times E(p) \to Q \times (\mathrm{edom}(p) - \mathrm{dom}(p)),$$

where $E(p)$ is the set of cells of p with a neighbor outside $\mathrm{dom}(p)$, and $f_p((q, x)) = (q', x')$ if from the ID (q, x), A eventually leaves the domain of p in ID (q', x') (this is well-defined by the assumptions we made).

For some constant C depending only on A, we have that for any n there are less than Cn elements in $Q \times E(p)$, and there are less than Cn elements in $Q \times (\mathrm{edom}(p) - \mathrm{dom}(p))$. Therefore, there are less than Cn^{Cn} different functions f_p for a fixed side length n. Of course, there are 2^{n^2} pictures p with side length n.

By taking n large enough that $Cn^{Cn} < 2^{n^2}$, we thus obtain two pictures of side length n with $p[x] \neq q[x]$ for some x, but $f_p = f_q$. Now we can construct two larger pictures p' and q' that agree everywhere except at a subpicture where p' has a translated copy of p and q' has q such that the position x is moved to the middle of the larger pictures. One of these pictures is in L_{center} and the other is not, but A will obviously accept p' if and only if it accepts q', a contradiction. □

By Theorem 1, we then have the following.

Corollary 5. $DFA_s \subsetneq NFA_s$.   □

The proofs involving universal states use non-closure properties of REC. Of course, we will first need to prove some connections between the automata classes and REC. Theorem 5 and its natural corollaries first appeared in [3], and independently in [11] where also Theorem 4 and its natural corollaries were proven. We will only scetch these proofs, more details can be found in [3] and [22]. The proofs are for general pictures.

Theorem 4. $NFA \subset REC$.

Proof. For an NFA, we construct a set of Wang tiles whose valid tilings draw accepting runs of the NFA on top of the picture. This is done by carrying the set of states the NFA can reach at each tile. The top left corner must contain an initial state, and some tile must contain a final state. The problem is we need to make sure a final state can only occur if there is a path to it from the top left corner. To achieve this, every state contained in a tile (except the initial state at the top left corner) will have a single predecessor, and a final state is the predecessor of no state. Then no loops containing a final state can occur, so a chain of successor links ending in a final state must have started at the initial state at the top left cell – and thus represents a valid loopless computation of the NFA.   □

Theorem 5. $co\text{-}AFA \subset REC$.

Proof. For this, we construct a set of Wang tiles whose valid tilings depict failed computations. Again, tiles contain a set of states, and a tile contains a state if there is no accepting computation from that state. The top left tile must contain all the initial states, a universal state must have at least one successor in the neighboring tiles (at least one choice of action leads to a failing computation), and an existential state must have all its successors in neighboring tiles (every possible choice of action leads to a failing computation). No final states may occur. It is then easy to believe that a tiling exists if and only if the automaton doesn't accept the input picture.   □

Next, we will need a concrete language. Let $L_{acyclic}$ be the language of (not necessarily planar) acyclic graphs. Any representation of graphs where the graphs are somehow 'drawn' on the (square) picture will do. We will assume a node can be contained in every cell, an edge can go through multiple cells, edges can cross each other freely, and a large (but fixed) number of them can move through a single cell.

We note that $L_{acyclic}$ is in UFA_s, since an automaton can use universal states to check that every possible run along the forward edges of the graph leads to a dead end. However, a standard pigeonhole argument shows it is not in REC [11]: Split pictures of side length n in two, with a vertical border in the middle, containing n nodes. On both sides, implement the same total order, connecting each node to its immediate successor. For large enough n, there are more partial orders than there are possible tilings of the middle column in tilings accepting the pictures representing the partial orders. Therefore, some two pictures can swap their right sides, necessarily resulting in a picture with a valid tiling, but which contains a cycle.

By the previous paragraph, UFA_s cannot be a subset of NFA_s, or it would also be a subset of REC, which is contradicted by $L_{acyclic} \in UFA_s - REC$. We thus obtain the partial diamond on the right side in Figure 3.

Theorem 6.
$$DFA_s \subsetneq NFA_s \subsetneq AFA_s,$$
$$DFA_s \subsetneq UFA_s$$

and
$$UFA_s \not\subset NFA_s$$

Theorem 7. *UFA_s and AFA_s are not closed under complement.*

Proof. Otherwise,
$$UFA_s = \text{co-}UFA_s \subset REC,$$

which is a contradiction. The case of AFA_s is proved similarly. $\qquad\square$

Using the connection with REC, it is also easy to show certain non-closure properties for all of the classes XFA simultaneously. We outline the idea for an operation that inherently works on general pictures instead of squares: concatenation. Consider the following lemma, which directly follows from Theorem 5.

Lemma 4. *If f is an operation on languages, $L_1, \ldots, L_k \in DFA$ but $f(L_1, \ldots, L_k)^c \notin REC$, then none of the classes XFA are closed under the operation f.* $\qquad\square$

This implies that we can prove certain non-closure properties working completely within REC – and REC is very easy to work with!

Theorem 8. *None of the classes XFA are closed under concatenation.*

Proof. We prove the claim for horizontal concatenation. Consider the language $L_{l=r} = \{p \in \Sigma_*^* \mid p[*, 1] = p[*, |p|]\}$, where $\Sigma = \{a, b, c\}$. Clearly, $L_{l=r}$ is in DFA, and therefore in all of the classes. Consider the language $L = L_{l=r}L_{l=r}L_{l=r}$. If one of the classes XFA were closed under concatenation, then L would be in this class. We show that L is not even in the class AFA. If it were, then L^c would be in co-AFA, and therefore also in REC by Theorem 5. We will show, however, that L^c is outside REC, proving the claim. Suppose, on the contrary, that L^c were in REC.

The language L^c is the the language of pictures p such that

$$\forall i, j \text{ such that } 3 \leq i < j \leq |p| - 2 :$$
$$p[*, i] = p[*, j] \implies p[*, 1] \neq p[*, i-1] \lor p[*, j+1] \neq p[*, |p|] \, ,$$

where $p[*, k]$ means the kth column of p. It looks kind of hard to work with, so we use the many well-known closure properties of REC to simplify it: First we restrict to the subset of pictures with columns alternatingly over $\{c\}$ and $\{a, b\}$ and with additional borders over $\{c\}$. In other words, the rows will be in the regular language given by the expression $c(c(a + b))^*cc$. This restriction is obtained by intersecting with a recognizable language. On this subset of L^c, the constraint simply forbids equal columns over $\{a, b\}$. We then erase the columns over $\{c\}$ using further closure properties to obtain that the simpler language

$$L_{c \neq c} = \{p \mid \not\exists i \neq j : p[*, i] = p[*, j]\}$$

is in REC. But it is proven in [22] that $L_{c \neq c}$ is in fact not in REC, a contradiction. See [22] for the details.   □

4   Moving Outside the Picture

In this section, we think of pictures as being embedded on the plane $\mathbb{Z}^2$ in the position indicated by the indexes. Thus, extending the indexing of matrices, the plane will be drawn with the first axis being the vertical one, and the second the horizontal one, with the first coordinate increasing as we move down, and the second as we move to the right.

So far, we have considered automata that are not allowed to exit the picture they are accepting. It is easy to see that a DFA does not gain any extra strength if it is allowed to exit a picture [21]. The corresponding question of whether NFA are strengthened if they are allowed to exit the picture was solved for pictures of height 1 in [21] in the negative. The question remained open for arbitrary shapes [21,15,11], until a negative answer was given recently in [22]. The solution is based on a theorem from [12] characterizing the languages of non-deterministic automata that are not allowed to enter the picture they are accepting, and we will outline the proof in Section 4.2. Also the case of AFA was solved in [22]. In this case, a simple example shows automata are in fact strengthened if they can exit the picture.

Definition 7. *FNFA is the class of picture languages accepted by NFA that are allowed to exit the pictures they are accepting. FAFA is the corresponding picture class for AFA.*

Formally, this just means redefining the set of ID's as $Q \times \mathbb{Z}^2$ instead of $Q \times$ edom(p). As already mentioned, we will also need the following 'dual' automata.

Definition 8. *ONFA is the class of unary picture languages accepted by NFA that cannot enter the picture they are accepting. ONFA can sense the border of the picture next to them, and they are started at $(0, 1)$, that is, just above the top left cell.*

4.1 AFA Is Not FAFA

In the one-dimensional case, it is well-known that two-way alternating finite state automata accept exactly the regular languages [14]. On the other hand, even two-headed one-way deterministic automata clearly accept non-regular languages. We use the natural embedding of words into pictures by considering the sublanguages of $L_{words} = \{p \in \Sigma_*^* : \bar{p} = 1\}$, and show that even restricted to such languages, alternating finite state automata become stronger if allowed to exit the domain of the picture. We denote by 2HAFA the class of picture languages accepted by 2-headed AFA that are not allowed to exit the picture.

Theorem 9. *AFA $\subsetneq$ FAFA. More precisely,*

$$\emptyset \neq (2HAFA - AFA) \cap L_{words} \subseteq FAFA - AFA.$$

Proof. Clearly AFA $\subseteq$ FAFA. To prove proper inclusion, we will simulate an arbitrary 2-headed AFA on strings, which are represented as $1 \times n$ pictures, using the space above the string to remember the distance between the two heads. This proves the claim, because a two-headed one-dimensional AFA can recognize, for instance, the non-regular language $\{a^n b^n \mid n \in \mathbb{N}\}$, while a one-headed AFA restricted to a string will only recognize (embeddings of) regular languages.

Given a one-dimensional language $L \subseteq \Sigma^*$ recognized by a 2-headed AFA A, we construct a 1-headed FAFA A' recognizing the picture language $L' = \{p \in \Sigma_*^* \mid \bar{p} = 1, p[1, *] \in L\}$.

We may assume A's first head is always to the left of its second head. When the first head of the one-dimensional 2HAFA is at p_1 and its second head is at p_2, the head of A' hovers over the input string at $(1 + p_1 - p_2, p_1)$. The moves of A are directly translatable to moves of A', so the only problem is to read the contents of the cells under the two heads. But an AFA can do this by guessing the contents, and using a universal state to branch two heads that check that the guess was correct. A third head then continues the simulation. □

4.2 NFA Is FNFA

We give a proof of this result based on 'landing sequences' of an FNFA, the information of which states and cells it can reach when it returns to the domain of the picture after exiting it. That is, we simulate the run of an FNFA by an NFA while it stays inside the picture, and when it leaves, we predict its landing without leaving the picture. When considering the behavior of an FNFA outside the domain of the picture, we only need to consider transition functions $\delta : Q \to 2^{Q \times \mathbb{Z}^2}$ where the $\mathbb{Z}^2$ part is the move of the automaton in the transition, which we can assume to be in $\{(0, 1), (0, -1), (1, 0), (-1, 0)\}$. This is because an FNFA reads only unary input outside the picture. We give some further definitions to simplify the following discussion.

Definition 9. *A* run *of an FNFA A on unary input is a sequence* $((q_i, x_i))_i$ *of pairs* (q, x) *where q is a state of A and* $x \in \mathbb{Z}^2$, *such that* $\forall i : \delta(q_i) \ni (q_{i+1}, x_{i+1} - x_i)$ *where* δ *is the transition rule of A.*

We assume an FNFA never accepts outside the picture, since it can always navigate back to the picture in order to accept.

Definition 10. *A* state configuration *is a subset of* $Q \times \mathbb{Z}^2$ *(that is, a set of ID's).*

Definition 11. *Let A be an FNFA with states Q. Then, if X, Y, Z are state configurations, we define*

$$A(X, Y, Z) = \{z \in Z \mid \exists \ run \ r \ of \ A : r \in XY^*z\}$$

referred to as a set of landings *(which is also a state configuration). We also use the syntactic conventions that if a tuple is used in place of one of X, Y, Z, the tuple is enclosed in a singleton set, which is then used as the argument. If one of X, Y, Z is not a subset of* $Q \times \mathbb{Z}^2$, *but a subset of* $\mathbb{Z}^2$, *then its cartesian product with Q is used instead.*

The proof relies crucially on the following Landing Lemma, from [12], which is interesting in its own right. It says that, while moving one step downward, the possible moves to the left and to the right form an eventually periodic set. We call a run that ends right after moving one step downward from the initial position a *south landing run*.

In the proof, we use the following basic definitions and results from automata theory: A *semilinear subset* of a commutative monoid M is a finite union of *linear sets*, which are sets of the form $x + \mathbb{N}_0 x_1 + \cdots + \mathbb{N}_0 x_k$ for $x, x_i \in M$. The Parikh set corresponding to $L \subset \Sigma^*$ is

$$\{x \in \mathbb{N}_0^\Sigma \mid \exists w \in L : \forall a \in \Sigma : |w|_a = x_a\}.$$

Lemma 5 (Landing Lemma). *For all* $s, f \in Q$, *the set*

$$A((s, (-1, 0)), \{(y, x) \in \mathbb{Z}^2 \mid y < 0\}, \{f\} \times (\{0\} \times \mathbb{Z})),$$

considered as a $\mathbb{Z}$-*indexed sequence, is eventually periodic in both directions.*

Proof. Let $s \in \{0, 1\}^{\mathbb{Z}}$ be the corresponding binary sequence containing a 1 in the positions A can reach in state f. To show s_i is eventually periodic in both directions, we construct a PDA accepting the language L_{moves} of words

$$w \in \{(1, 0), (-1, 0), (0, 1), (0, -1)\}^*$$

such that some south landing run r of A has exactly this sequence as its sequence of moves.

The PDA accepts when the stack becomes empty. It originally has one symbol Z on the stack, representing the fact that the NFA starts at height 1. The finite control makes state transitions just as the NFA would, always reading the current move from the input word, rejecting the word if a different move is read. A Z' is pushed on top of the stack whenever the NFA moves up, and a Z' or Z is popped whenever it moves down. However, the automaton may only pop a Z if it is about to enter state f. The stack is not changed when the NFA moves horizontally. It is clear that such a PDA accepts exactly L_{moves}, since when the stack become empty, the NFA being simulated would have entered state f, and would have, altogether, moved one step down from its initial position.

Because this language is context-free, its Parikh set L_p is a semilinear subset of $\mathbb{N}_0^4$ [18] and thus also a semilinear subset of $\mathbb{Z}^4$. But then, by well-known closure properties of semilinear sets,

$$\{r - l \mid (d, u, r, l) \in L_p\}$$

is semilinear in $\mathbb{Z}$, where r and l are the amounts of right and left moves, respectively. But this is exactly the set $\{i \mid s_i = 1\}$, and a semilinear subset of $\mathbb{Z}$ must be eventually periodic in both directions.   □

Of course, symmetric claims hold for landings in other directions than down. The problem of whether NFA = FNFA essentially boils down to proving such a lemma for landings on the border of a rectangle instead of a straight line.

Definition 12. *The* fundamental threshold *and* fundamental period *of an ONFA A are the smallest possible t and q such that its landing sequences in all directions from any state s to any state f all have t as a threshold and q as a period in both directions. More precisely, we let q be the smallest period, and t is then chosen to be the smallest threshold for this particular q.*

The proof we present is based on the following characterization of the class ONFA. This result is also interesting in its own right, and was also proven in [12].

Definition 13. *An eventually periodic subset of $\mathbb{N}^2$ is a set X such that there exist t, q such that*

$$\forall w > t : (h, w) \in X \iff (h, w + q) \in X$$

and

$$\forall h > t : (h, w) \in X \iff (h + q, w) \in X.$$

We state two characterizations of the eventually periodic sets without proof, see [22] or [12].

Lemma 6. *A subset X of $\mathbb{N}^2$ is eventually periodic if and only if there exist t, q such that*

$$\forall w > t : (h, w) \in X \implies (h, w + q) \in X$$

and

$$\forall h > t : (h, w) \in X \implies (h + q, w) \in X.$$

□

Lemma 7. *A subset X of $\mathbb{N}^2$ is eventually periodic if and only if the language $\{0^m 1^n \mid (m, n) \in X\}$ is a regular language.* $\square$

Theorem 10. *The shapes of languages of ONFA are exactly the eventually periodic sets.*

Proof. It is not hard to see that ONFA can accept all the eventually periodic sets using Lemma 7. For the other direction, we will use Lemma 6. We consider a language L' in ONFA accepted by some automaton A, and let $L = \{(\overline{p}, |p|) \mid p \in L'\}$, which uniquely determines the language. We prove both the widths and heights of pictures in the language can be pumped, that is, there is a threshold t and a period q such that

$$\forall (h, w) \in L : h > t \implies (h + q, w) \in L$$

$$\forall (h, w) \in L : w > t \implies (h, w + q) \in L$$

It is enough to find such t and q that work for widths, since a symmetric argument will work for heights, and we can use the larger of the thresholds and a least common period of the periods to find the threshold and the period of the whole language.

Let q' and t' be the fundamental period and fundamental threshold of A. We will prove that $q = (|Q|t')! q'$ and $t = (|Q| + 1) t'$, work as pumping constants for L. We start by naming some areas of the plane. We define the top left ray, the top right ray, the bottom left ray and the bottom right ray as the sets of cells

$$\{(0, -n) \mid n \in \mathbb{N}_0\},$$

$$\{(0, |p| + n + 1) \mid n \in \mathbb{N}_0\},$$

$$\{(\overline{p} + 1, -n) \mid n \in \mathbb{N}_0\},$$

and

$$\{(\overline{p} + 1, |p| + n + 1) \mid n \in \mathbb{N}_0\},$$

respectively. Note that the top and bottom rays do not coincide even if the picture has height 1.

Now consider a run r accepting the picture of shape (h, w), where $w > t$. We split the run into partial runs $r^1, \ldots, r^k$ between the points of intersection with one of the four rays, that is, we partition the run into semiopen intervals by cutting the run at the points where a ray is crossed (we may assume A accepts on a ray).

These partial runs are transformed into a run of the automaton on the larger picture of shape $(h, w + q)$. There is an obvious way to identify the rays of the small picture with the rays of the large picture, and each partial run is transformed into a run that starts and ends at the corresponding cell in the corresponding ray. It is then clear that by gluing together the partial runs obtained for the larger picture, a valid accepting run for the larger picture is obtained, proving $(h, w + q) \in L$.

For partial runs between two cells both in left rays, or two cells both in right rays, it is obvious how to find a corresponding partial run for the larger picture. Now, all we need to do is apply the Landing Lemma 5 to transform the partial runs that move between the two sides of the small picture to such runs on the large picture. For this, consider such a partial run $r^j = u$ on the small picture. By symmetry, we may assume u starts just above a cell of the top left ray, and ends on a cell of the top right ray. We split u further into subpartial runs u^j exactly as we did for r, but this time at the positions where the automaton senses that it is next to the picture. Let J the sequence of indices of u where the automaton senses p.

We now have two cases to consider:

- One of the subpartial runs u^j moves the automaton to the right more than t' cells.
- All subpartial runs u^j move the automaton at most t' to the right.

In the first case, we are done, since the automaton's fundamental threshold has been exceeded, so the subpartial run u^j can be made mq' cells longer to the right, for any $m \in \mathbb{N}$. In particular, it can be made $q = (|Q|t')!q'$ cells longer, proving the claim in this case.

In the latter case, we note that there must be at least than w/t' of these subpartial runs, where we chose $w > t = (|Q| + 1)t'$, and thus $|J| > |Q|$. Then, we may take an ascending subsequence K of J such that

- the sequence of positions of $(u_i)_{i \in K}$ is ascending to the right.
- between each two indices of K the automaton has moved at most t' cells to the right.
- $|K| > |Q|$.

Now note that there must be a repetition of states in $(u_i)_{i \in K}$, say at $a, b \in K, a < b$, and the distance d between u_a and u_b satisfies $1 \le d \le |Q|t'$. Then the partial run of length d between u_a and u_b can be repeated $\frac{q}{d} = \frac{(|Q|t')!q'}{d}$ times to obtain a partial run for the picture $(h, w + q)$.

This means that in both cases, we were able to make the partial run work for a picture q wider, as long as the picture had width at least t, and thus t and q as we chose them are a threshold and period for pumping the widths of pictures in L. □

Let us now give more notation for parts of a picture, and the space around it. Given a picture p, we call the cells (i, j) such that $j < 1$ the west half-plane $H_w(p)$, and similarly we get the east, north and south half-planes $H_e(p)$, $H_n(p)$ and $H_s(p)$. These cover all the positions outside the picture. In what follows, we use a slightly different definition of rays: the top left west ray $R_{tlw}(p)$ is the set $\{(1, x) : x < 1\}$ and similarly we define $R_{tln}(p)$, $R_{trn}(p)$, $R_{tre}(p)$, $R_{bre}(p)$, $R_{brs}(p)$, $R_{bls}(p)$ and $R_{blw}(p)$. The west edge of p is the set $E_w(p)$ of cells on its domain with a left neighbor outside its domain. Similarly we obtain $E_n(p)$, $E_e(p)$ and $E_s(p)$. We define the edge of p as $E(p) = E_w(p) \cup E_n(p) \cup E_e(p) \cup E_s(p)$.

We define the west line of p as $L_w(p) = R_{tln}(p) \cup E_w(p) \cup R_{bls}(p)$, and similarly we get $L_n(p)$, $L_e(p)$ and $L_s(p)$. We define the outside of p as $O(p) = H_w(p) \cup H_e(p) \cup H_n(p) \cup H_s(p)$. Finally, we define $C_m(p)$ as the subset of $E(p)$ of cells at most distance m away from some corner of p.

Given a run of an FNFA starting from just outside the edge of a picture and ending at an edge, staying outside the picture during the run, we call the unique half-plane on which it starts the *initial half-plane*. The part of the run before exiting the initial half-plane for the first time is called the *initial segment* of the run. Similarly, we define the *final half-plane* and the *final segment* of the run. Note that the initial and final segments can partially overlap, or be equivalent if the initial half-plane is never exited.

In what follows, we will use the terms 'landing' and 'computing a state configuration' somewhat informally. Implicitly, one of the following two definitions will then apply.

Definition 14. *Let Q be a finite set of states. For any $Z \subset \mathbb{Z}^2$ and FNFA A with unique initial state s, the set of landings of A from x onto Z is defined as the state configuration c on Z given by*

$$A((s, x), Q \times O(Z), Q \times Z).$$

Let A be an FNFA with state set $Q' \supset Q$ and with a unique initial state $s \in Q' - Q$. We say A computes the state configuration c over Q from x within Z if c is the state configuration

$$A((s, x), (Q' - Q) \times Z, Q \times Z).$$

Now, we can prove the main theorem of this section: NFA $=$ FNFA. First, let us compute the sub-landing sequences 'near the corners' of pictures, assuming the FNFA leaves the picture at one of its corners.

Lemma 8. *Let A be an FNFA with states Q. Then, for every $i \in Q$ and $m > 0$ there exists an NFA A' with states $Q' \supset Q$ such that*

$$\forall p : A((i, (0, 1)), O(p), C(p, m)) = A'((s', (1, 1)), dom(p), Q \times C(p, m))$$

Proof. $C(p, m)$ is a finite set, so A' simply needs to guess one of the landing possibilities $(s, x) \in Q \times C(p, m)$ and check if it's a possible landing of A. If it is, A' finds x and enters s on it. But it's easy to check if (s, x) is a landing of A: There exists an ONFA B that simulates A until it tries to enter the picture, and accepts if it would've entered the correct cell x in the correct state s. The set $\{(h, w) \mid \exists p \in \mathcal{L}(L(B)) : (\overline{p}, |p|) = (h, w)\}$ is eventually periodic, so A' can easily check if p belongs to $\mathcal{L}(B)$. Since $p \in \mathcal{L}(B)$ if and only if (s, x) is a landing of A, we are done. $\square$

Using the previous lemma, we can now characterize *all* landings of runs that start near a corner.

Lemma 9. *Let A be an FNFA with states Q. Then, for every $i \in Q$ there exists an NFA A' with states $Q' \supset Q$ such that*

$$\forall p : A((i, (0, 1)), O(p), E(p)) = A'((s', (1, 1)), dom(p), Q \times E(p))$$

Proof. Let t and q be the fundamental threshold and period of A, respectively, and assume $(s, x) \in Z = A((i, (0, 1)), O(p), E(p))$. First assume $x \in E_w(p)$. Then

$$x \notin C(p, t + q + 1) \implies (s, x + (q, 0)) \in Z$$

by the Landing Lemma 5, since A's run with landing (s, x) must have entered the west half-plane for the last time at some point, and we may thus pump the final segment of the run before landing, by q steps in either direction. By repeating the argument, (s, y) is also in Z for some $y \in C(p, t + q + 1) - C(p, t + 1)$.

Conversely, if (s, y) is a landing of A onto $C(p, t+q+1) - C(p, t+1)$, on the west side, then also all other $(s, y + (mq, 0))$ such that $y + (mq, 0) \in E(p) - C(p, t+1)$ are landings of A, again by the Landing Lemma 5, in particular if y is obtained from pumping x onto $C(p, t+q+1) - C(p, t+1)$, then the landing (s, x) is found by pumping y the same distance in the other direction.

Now, for landings of A on $C(p, t + q + 1)$, we apply Lemma 8. As for other landings, consider again landings on the west side of p. For all landings of A onto (s, x), where $x \in C(p, t + q + 1) - C(p, t + 1)$, we let A' move q steps up or down as many times as it likes before entering s, while staying inside $E(p) - C(p, t+1)$. By the argument of the first paragraph, we obtain all landings of A on E_w this way. The other edges are handled similarly. □

Finally, the result is obtained by handling runs that never exit the initial half-plane separately, and then using Lemma 9 for those that exit it.

Theorem 11. *NFA $=$ FNFA.*

Proof. Given FNFA A, we construct an NFA A' that has states $Q \cup Q'$ where Q are the states of A and Q' are helper states used in the simulation. The NFA A' uses only states of Q when inside p, and has the same transition function as A when restricted to these states. Of course, this means A might try to exit p during its accepting run of p. When it tries to exit p, x being the first cell outside p it would've entered, it instead enters a special search state in Q' that computes the landing sequence of A' from x onto the edges of the picture inside $dom(p)$. We assume A never accepts a picture p while outside $dom(p)$.

If we can accomplish this, then it should be clear that the languages of A and A' are the same. We do not give a formal construction of A', but an informal algorithm for obtaining such an automaton from the behavior of A. So assume A exits p during a run, and let x be the first cell outside p it sees. We may assume x is on the west side of p, since the other cases are symmetric. First, we note that

$$A((s, x), O, E) = A((s, x), H_w, E_w) \cup$$
$$A(A((s, x), H_w, R_{tln}), O, E) \cup$$
$$A(A((s, x), H_w, R_{bls}), O, E)$$

where we write X instead of $X(p)$ for brevity. We explain how A' can compute each of the three parts within $\mathrm{dom}(p)$, in which case the pointwise union is also computable. By symmetry, it's enough to handle $A((s,x),H_w,E_w)$ and $A(A((s,x),H_w,R_{tln}),O,E)$.

Case 1: Computing $A((s,x),H_w,E_w)$ inside $\mathrm{dom}(p)$.

Consider the landing sequence s^1 of A from (s,x) upwards, that is, s_i^1 is the set of landing states on the cell $x+(-i,1)$. By the Landing Lemma 5, this sequence s^1 is eventually periodic. Therefore A' can compute this sequence inside E_w, since the sequence does not depend on the picture. This concludes the first case.

Case 2: Computing $A(A((s,x),H_w,R_{tln}),O,E)$ inside $\mathrm{dom}(p)$.

Let $x=(i,j)$. Then,

$$A((s,x),H_w,R_{tln}) = s^1_{[i+1,\ldots)} = s^2,$$

where s^1 is as in Case 1. Let B_{s^2} be an FNFA with initial state s'' that computes this sequence onto R_{tln}, and then simulates A. That is, B_{s^2} has states $Q \cup Q''$, with Q'' and Q disjoint, and

$$B_{s^2}((s'',(0,1)),Q'' \times R_{tln}, Q \times R_{tln}) = A((s,x),H_w,R_{tln})$$

for the unique initial state s'' of B_{s^2}, and B_{s^2} has the same transition function as A, when restricted to states in Q, which implies

$$B_{s^2}((s'',(0,1)),O,E) = A(A((s,x),H_w,R_{tln}),O,E).$$

By Lemma 9, there is an inner NFA with the same landings as B_{s^2}. Therefore, A' can move up the side of p, determine the sequence s^2, and depending on this sequence, compute the landing sequence of B_{s^2} onto E inside $\mathrm{dom}(p)$. This is possible, because the sequence s^2 is a final segment of the eventually periodic sequence s^1, and thus one of a finite amount of possibilities. This concludes the second case.   □

The techniques presented here do not generalize for 3- or more-dimensional pictures (with the obvious definitions). We end this section with the following conjecture from [22].

Conjecture 4. NFA = FNFA on three-dimensional pictures.

5   Restricting the Directions of Movement

We will now turn to automata for which some directions of movement are forbidden. We define the 2XFA as XFA that cannot use the directions up and left (recall that our automata are started from the top left corner). The classes 3XFA are defined by forbidding only upward moves. Although results will still be used from both the unary and the square worlds, we will only give results for general pictures in this section. We will first collect results from the literature to give the Boolean closure properties for each 3XFA. We then compare the three-way classes with each other and the four-way classes and give a strong connection between 2AFA and a deterministic version of REC.

Analysis of three-way automata is somewhat simpler than that of four-way automata. Interestingly, now 3AFA can be shown to be closed under complement [3], and so can 3DFA [24], while 3NFA and 3UFA still lack this property [11] (proof below). Now our proofs for the other Boolean closure properties break down, and in fact the class 3DFA is not closed under either intersection or union, and 3NFA is not closed under intersection [4]. Also the question for 3UFA corresponding to Conjecture 3 has a positive answer – 3UFA is not closed under union [10]. Of course, 3NFA is closed under union, 3UFA is closed under intersection and 3AFA is still closed under both operations. Thus, we obtain the rather natural situation of Figure 5.

	3DFA	3NFA	3UFA	3AFA
¬	Yes	No	No	Yes
∩	No	No	Yes	Yes
∪	No	Yes	No	Yes

Fig. 4. Boolean closure properties of the 3XFA classes

It is clear that 2XFA $\subset$ 3XFA $\subset$ XFA for all the XFA classes. From the considerations of Section 3.2, we obtain that the second inclusion is always proper, and a stronger separation of NFA and UFA:

Theorem 12. *The sets* 3NFA$-$UFA *and* 3UFA$-$NFA *are nonempty and neither of 3NFA and 3UFA is closed under complement.*

Proof. The 90° rotation of the billiard ball language from Definition 5 is in the first set, and the complement of this language is in the other one. The non-closures under complement follow similarly. ☐

The theorem implies that a four-way automaton cannot recognize the languages of three-way automata of a 'stronger' type. In fact, no non-trivial inclusions exist between four-way and three-way automata classes, as is implied by the following result, which is proven in [8] in a stronger form.

Theorem 13. *The set* DFA $-$ 3AFA *is nonempty.* ☐

So by combining results of [8] with the results obtained by using the billiard ball language from [11], we have obtained a complete picture of the relations between three- and four-way automata classes, on general pictures. We depict this in Figure 5, where all inclusions are proper, all inclusions between classes are given by diagram chasing, and all other pairs of classes are incomparable.

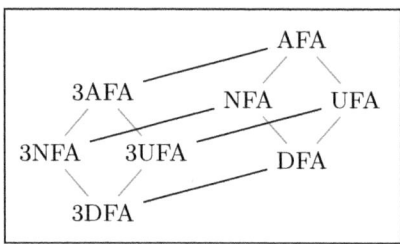

Fig. 5. Diagram of inclusions for three-way and four-way automata classes

In Section 3.3, we showed some connections between REC and the XFA classes (namely Theorems 4 and 5). It seems to be unknown whether co-AFA is actually equal to REC [3], although this seems unlikely. However, for 2-way automata, we have a natural connection between the two worlds: 2AFA is equal to a deterministic version of REC [9]. However, due to bad luck, the way the classes are usually defined we have to rotate one of them by 180° to make them coincide.

Definition 15. *DREC is the subfamily of REC obtained by using north-west deterministic tiles, that is, tilesets such that no two Wang tiles in the set may share both their north and west colors.*

With the interpretation that REC accepts a language by assigning states on the cells of a picture, this means when assigning states of a DREC grammar, knowing the north and west neighbors uniquely determines the state of the current cell. We next prove the following [9]:

Theorem 14. $2AFA^R = DREC.$

We split the proof into two inclusions. Given a picture p and a position y, we define *top left rectangle at* y to be the rectangle between $(1,1)$ and y. Given such a rectangle, we define its north child and west child as the rectangles that start from $(1,1)$ and end at $y - (1,0)$ and $y - (0,1)$, respectively. Note that $2AFA^R$ is of course just the class of languages accepted by 2-way AFA that start at the bottom right corner and can only move up and to the left.

Lemma 10. $2AFA^R \subseteq DREC.$

Proof. We construct a DREC grammar G simulating the given $2AFA^R$ A. At each tile of G, the subset of states of A is held from which there is an accepting computation of A for the top left rectangle at the position. At the top left corner, the correct subset of states of A is enforced, based on the symbol at that position. It should be obvious how to inductively find the correct state sets elsewhere, north-west deterministically, based on how the acceptance of an AFA is defined.   □

Lemma 11. $DREC \subseteq 2AFA^R$.

Proof. Given a DREC grammar G, we find a $2AFA^R$ A with the same language. The automaton A has the states $T \times S$ where T is the tileset of G, and S is a small set of helper states needed in the construction (which we leave unspecified). Let us distinquish a state $s \in S$. When A is started in state (q, s) in cell x, it accepts if and only if there is a consistent assignment of DREC states onto the cells of the top left rectangle at x such that the tile q is used as the state of cell x.

The top left corner can easily be detected by a $2AFA^R$. In the top left corner, A simply checks whether such a consistent assignment (of one tile) exists using a look-up table. In the general case, A guesses the north and west neighbors q_1 and q_2 of q with colors matching those of q. It then recursively checks these that the north and west children can be consistently colored in such a way that the corresponding q_i is used in the bottom right corner.

The algorithm works because in the recursion, in addition to consistent placements of states for the subpictures starting from the north and west neighbors existing, they must also have the same tiles in the overlapping zone due to the determinism of DREC.

Then, the initial states of A simply start such searches from all tiles $q \in T$.   □

6   Markers

In this section, we only consider deterministic automata, and work on general pictures. However, we give our automata a finite set of markers, which the automaton can carry around, drop on the cells of the picture, and later lift up again. The main results we present are from [1], although we will prove slightly stronger claims. There are many ways to formalize this idea, but most of these can be shown equivalent [1]. We choose the definition of 'physical markers', which need to be lifted before reusing, and which can be stacked on top of each other.

Of course, once k markers are given to our automata, we will obtain a hierarchy of languages based on how many markers are needed for a DFA to accept them. This hierarchy is proven infinite in [1] with a diagonalization argument. We will investigate only the beginning of this hierarchy, using concrete and natural picture languages.

The language L_{center} shows us DFA is already strengthened by the addition of one marker, since we can use the marker to implement essentially the same algorithm we used to prove $L_{center} \in$ NFA. Denoting the class of picture languages accepted by DFA with n markers by $nmDFA$, we then have the following.

Theorem 15. *DFA $\subsetneq$ 1mDFA.* □

We will devote the main part of this section to separating DFA with two markers from those with just one, illustrating interesting programming techniques for DFA with markers, and an extended pigeonhole argument for DFA with one marker.

Definition 16. *For each n, L_{nc} is the language of pictures over $\{0,1\}$ containing exactly n connected components over 1 (connected components of the graph where the vertices are the cells containing a 1, and there are edges between adjacent cells). We write $L_C = L_{1C}$. For each n, we define L_{neq} as the subset of L_{nc} containing exactly n components which are all translations of a single component.*

In [1], in order to separate 1mDFA and 2mDFA, it is proven that there exists a language $L \in$ 2mDFA whose intersection with L_{2c} is exactly L_{2eq}, but there is no such language in 1mDFA. We will instead prove the perhaps more interesting new result that $L_{2eq} \in$ 2mDFA $-$ 1mDFA.

The language L_C has been of interest to many authors in both the case of picture-walking automata and recognizability. In [20], L_C was proved to be in REC, while [8] proved it to be outside 3AFA. In [1], it was shown that a DFA can accept L_C with one marker, and it was implicitly conjectured that this is not possible with a regular DFA.

Conjecture 5. [1] The connected patterns are not in DFA.

Note that a proof through REC will certainly not work, since it is easy to write an AFA for L_C [7]. Neither does a direct pigeonhole argument seem possible.

Let us explain the construction of [1] for L_C with a one marker DFA, since it also nicely illustrates some of the techniques we will need for showing $L_{2eq} \in$ 2mDFA. First, we give some slightly informal terminology. By a local property P we mean a set of $(p, x, m_1, \ldots, m_k)$, where $x, m_i \in \text{dom}(p)$, x representing the head of the automaton and m_i the positions of markers. We say P is an n-marker property, if a DFA with n unused markers started in cell x of picture p with some fixed k markers of its at $m_1, \ldots, m_k$ can check whether $(p, x, m_1, \ldots, m_k) \in P$, then returning to x carrying again the n markers.

So, let $p \in \Sigma_*^*$ be a binary picture. We say a column j of p is left separating if it contains a 1, there is a 1 in p somewhere to the left of column j, and

$$\nexists i : p[i, j] = 1 \wedge p[i, j-1] = 1.$$

That is, left separating columns contain 1's all of which are separated from all 1's to the left of them. Similarly, we define right separating columns. It is clear that the 1's of p are connected if and only if there are no left or right separating columns, and on each column, two 1's with only 0's between them are connected.

Theorem 16. $L_{1C} \in$ 1mDFA.

Proof. Checking that no column is left or right separating is easy. Next, the automaton reads the columns top-down, from left to right. At each 1 which is not the last in its column, we check that it is in the same component as the next 1 in the column. More precisely, we note that the local property 'The current cell x contains a 1 and the next 1 in this column (at y) is in the same component.' is a 1-marker property: the automaton drops the marker and then follows the edge of the area of 1's (using the classical labyrinth algorithm) until it either returns to x, or it finds a 1 under the marker at x with only 0's in between. Note that the latter is a 0-marker property and thus checkable during the search. In the first case, y is not in the same component, and we conclude $p \notin L_C$. If the second case always occurs, we conclude $p \in L_C$.   □

We will first count that there are exactly two components, by extending the previous argument. This technique can be used to prove $L_{nC} \in nmDFA$, although, as in the case of one component, it is unknown whether this is optimal.

Theorem 17. $L_{2C} \in 2mDFA$.

Proof. We define the shoulder of p as the cell containing a 1 seen first during a columnwise top-down left to right search, and similarly we define the shoulder of a component of 1's in p. We define the top left component as the component of 1's that the shoulder of p belongs to. Note that being on the outer border of the top left component is a 1-marker property, since whether the current cell is the shoulder is a 0-marker property. For brevity, we call the top left component A.

We now start a columnwise top-down left to right search over the picture. At each column, we continue down without permanently dropping any markers, while staying in A. This is possible since we can check whether the first cell we see is in A (since it is necessarily on the outer border of its component), and we can check whether the component changes as we move down the column. If another component is not seen during the whole search, we conclude there is just one component, and thus $p \notin L_{2C}$.

If a different component is seen at any time, we drop one of the markers. Note that it is then dropped at the outer border Y of some component B. The marker will not be moved again, so in effect, whether a cell is on the outer border of either A or B has turned into a 1-marker property. Just as importantly, if B is completely *within* A, whether a cell belongs to the edge of the corresponding hole of A is now also a 1-marker property. Let X be the set of cells on this edge, or the outer border of A if B is not within A. We obtain that belonging to X and belonging to Y are both one-marker properties.

After marking X and Y, we continue the search as previously, now essentially with a one marker automaton. When we enter the first component on a column at x, we immediately conclude $p \notin L_{2C}$ if this x is not in A or B (since, again, x is at the outer border of *some* component). After either A or B has been entered, the automaton continues down the column while the component does not change. When it does change, we immediately reject if it does not go from X

to Y or from Y to X. If there are exactly two components, only such transitions happen, and if only such transitions happen, A and B are of course the only components.   □

Next, let us compare these two components.

Theorem 18. $L_{2eq} \in 2mDFA$.

Proof. By the previous theorem, we can assume there are exactly two components. Again, let A be the top left component and B the other one. Now, B cannot be inside A or we can directly reject p. This is easy to check during the previous algorithm, and even easier to do directly. Knowing this, we first locate the shoulder of B: we do a columnwise search until either the first component seen is not A or the component changes after a cell of A. We keep the other marker – the B marker – here, and move the other marker to the shoulder of A. The two markers will stay exactly this vector v away from each other during the rest of the search.

Now, consider the components as sets of vectors and p as the set of indices where it contains a 1. We will check that $A + v \subset p$ and $B - v \subset p$, which clearly implies $A + v = B$. For the first inclusion, we do a columnwise search using both markers at once, and whenever the A-marker is in a cell of A, we check that the B-marker is on top of a 1. The second inclusion is done symmetrically.

In order to keep track of when the A-marker is on a cell of A, we start with the markers at the shoulders of the corresponding components. Both markers are moved down simultaneously while the A-marker stays within A, skipping 0's in the usual way, and we check the B-marker is on top of a 1 whenever the A-marker is. When the A-marker changes component, the head must have entered B. We then continue until the component changes again: we must have returned to A, and the search continues normally.   □

Now, let us prove also the negative result – that two markers are in fact necessary for L_{2eq}.

Theorem 19. $L_{2eq} \notin 1mDFA$.

Proof. Suppose A is a 1mDFA with $\mathcal{L}(A) = L_{2eq}$ and let A have k states. We assume A directly returns to the domain of the picture if it enters its border.

We will use a similar argument as the one in the proof of Theorem 3. For each picture p we define the function f_p mapping 'incoming ID's' to 'outgoing ID's', although this time we also allow 'accept', 'reject' and 'loop' in the codomain. (We could avoid this by repeating the argument of [23] for one-marker DFA.) The function f_p completely characterizes the behavior of A on p as long as the automaton enters the picture without carrying the marker. We say that two pictures p and q are A-equivalent if $f_p = f_q$.

The basic idea for applying the pigeonhole principle here is that while the marker stays on one side of the picture, the original partition into A-equivalent pictures applies to the other side, and the marker cannot be moved across the middle of the picture more than a linear amount of times without the automaton entering a cycle. Let us make this idea more precise.

Let n be fixed, and let X be the set of all pictures of size $n \times n$ containing exactly one connected component with 0's around it. To each element of X^2 we naturally associate a picture of shape $n \times 2n$ where the two elements of X are simply concatenated (note that such a picture is in L_{2c}). The left- and right-hand sides of these pictures p are called the LHS and the RHS. We will show that for large enough n, A will accept all pictures in $Y \times Y$ for some subset Y of X with more than one element, which is a contradiction.

We note that for some n-independent D, $|X| \geq 2^{Dn^2}$ for large enough n by using a suitable skeleton of 1's to allow rows on which bits can be chosen freely while keeping the pattern connected. Let P be the partition of X into A-equivalence classes, and again note that there are at most Cn^{Cn} components in P for some C depending only on A. We write X/P for a maximal subset Z of X containing only A-equivalent elements (tie-breaking in some natural way). Of course, $|Z| \geq \frac{|X|}{|P|}$.

Let us restrict our attention to $Y_0 = X/P$. Consider the pictures Y_0^2. The automaton cannot accept any of the pictures in this set without moving the marker on the right side, since the RHS are all A-equivalent. For some subset $Y_0' \subset Y_0$ of size at least $\frac{|Y_0|}{kn}$, A exits the LHS with the marker the first time the same way in all pictures of $Y_0'^2$. Note that the content of the RHS does not influence how the marker leaves the LHS, since all pictures in Y_0 are A-equivalent. If Y_0' has more than one element, A must eventually move the marker back to the LHS. Again, it does this the same way on some subset $Y_1 \subset Y_0'$ of size $\frac{|Y_0'|}{kn}$.

If Y_1 has more than one element, the marker eventually has to be moved on the right side, but now this can be done in only $kn - 1$ ways, since otherwise the automaton enters an infinite loop on all pictures of Y_1^2, which is impossible because $Y_1^2 \cap L_{2eq} \neq \emptyset$. We thus obtain sets Y_1' and Y_2, and similarly Y_i' and Y_{i+1} for all i until the size of one of these reaches 0 or 1. Note that $|Y_{i+1}| \geq \frac{|Y_i|}{(kn-i)^2}$. Since the automaton necessarily loops when i reaches kn, we have obtained $|X/P| \leq (kn)!^2$. But we can show this is false for large enough n, using a standard argument:

$$\frac{2^{Dn^2}}{Cn^{Cn}(kn)!^2} \leq \frac{|X/P|}{(kn)!^2} \leq 1 \iff$$

$$Dn^2 - Cn \log Cn - 2\log(kn)! \leq \log \frac{|X/P|}{(kn)!^2} \leq 0$$

which is clearly not the case, using String's approximation for $\log(kn)!$. This is a contradiction, and thus $L_{2eq} \notin 1\text{mDFA}$. $\Box$

Corollary 6. $1\text{mDFA} \subsetneq 2\text{mDFA}$. $\Box$

7   Algorithmic Questions

Picture walking automata admit efficient parsing of pictures. Pictures of a language can be recognized in linear time on the size $n = |p| \times \bar{p}$ of the picture p.

In the case of DFA, the linear time recognition can be done using logarithmic space: By Theorem 1 we can assume the DFA halts on every input picture so it is enough to run the DFA until it halts. This happens within $O(n)$ steps, and during the computation one stores the current state and position of the DFA using $O(\log n)$ space.

In the cases of NFA, UFA and AFA, linear space is enough. Consider the directed graph whose vertices are the instantaneous descriptions (q, x) of the automaton, and edges are given by the successor relation $l_1 \longrightarrow l_2$. For NFA, picture recognition amounts to finding a path from an initial ID to an accepting ID, which can be solved using standard depth-first search. In AFA, the vertices are classified as universal and existential. A simple linear-time algorithm executes a DFS search from the final ID's backwards, progressively marking new vertices that lead to acceptance. A universal vertex is marked only when all its followers become marked, while for existential vertices it is enough to have one follower marked. A picture is in the language iff an initial ID gets marked.

Another family of algorithmic questions concern the languages recognized by given automata. Here the situation is quite different, and undecidability usually prevails. The basic question is the emptiness problem: does the given automaton accept any pictures ? It was mentioned already in [1] that unary emptiness for DFA is undecidable because Minsky machines can be simulated. Recall that Minsky's two-counter machine without an input tape consists of a deterministic finite automaton and two counters, each storing one non-negative integer. The machine can detect when either counter is zero. The machine changes its state according to a deterministic transition rule. The new state only depends on the old state and on the results of the tests that check whether either counter is equal to zero. The transition rule also specifies whether to increment, decrement or keep unchanged the counters.

It is well known that two-counter machines can simulate Turing machines [16]. In particular, it is undecidable whether a given machine reaches a specified halting state h when started in its initial state i with both counters initialized to value 0.

A 2-counter machine can be interpreted as a DFA that operates on the (infinite) quadrant of the plane and has the same finite states as the 2-counter machine. The position of the DFA encodes the two counter values: Position (x, y) represents counter values x and y. Increments and decrements of the counters correspond to horizontal and vertical steps of the DFA on the plane.

Any computation of a 2-counter machine can therefore be simulated by a DFA inside a sufficiently large rectangle. The dimensions of the rectangle have to be at least as large as the largest counter values used during the accepting computation. Zero values of the counters can be recognized as these correspond to the machine stepping on the top or left boundary of the rectangle. If the DFA tries to step on the right or the bottom border, it enters an error state e that indicates that the rectangle was not large enough.

It is clear that the DFA outlined above accepts exactly the rectangles of sizes $w \times h$ where w and h are greater than the largest values in the two counters, respectively, used during the accepting run of the Minsky machine. If the Minsky machine does not halt, the DFA does not accept any rectangle. Hence we have the following.

Theorem 20. *It is undecidable if a given unary DFA accepts any rectangles [1].*
□

In the restricted models the situation is more interesting, and various decidability questions have been investigated among three-way automata [5,13,19]. In [5], the unary emptiness was shown to be decidable among 3NFA, and in [19] the result was extended to arbitrary alphabets. It was also shown in [19] that unary emptiness is undecidable among 3AFA, and even among 2AFA.

Theorem 21. *The emptiness problem is*

- *decidable among 3NFA [5,19],*
- *undecidable among unary 2AFA [19].*   □

Finally, we mention another application of the Minsky machine simulation by DFA, stating that the sizes of unary squares recognized by DFA can form a very sparse set [12].

Theorem 22. *Let $\{a_1, a_2, \ldots\}$ be any recursively enumerable set of positive integers. There exists a DFA that recognizes the language of unary squares of sizes $b_i \times b_i$ for $i = 1, 2, \ldots$ where $b_i > a_i$ for all $i = 1, 2, \ldots$.*   □

The theorem is proved easily using the DFA simulation of a two-counter machine. First, an NFA is build where non-determinism is used to select the initial counter values for the Minsky machine. Such limited non-determinism can then be removed using the trick from [23]. See [12] for the details.

8   Conclusions

To conclude, we collect the open problems from the text into a single list.

Conjecture 1. *The language of unary squares with prime side length are not in DFA.*

Conjecture 2. *DFA is a proper subset of NFA when restricted to unary squares.*

Conjecture 3. *UFA is not closed under union.*

Conjecture 4. *NFA = FNFA on three-dimensional pictures.*

Conjecture 5. *The connected patterns are not in DFA.*

Acknowledgements. We would like to thank Ilkka Törmä for helpful remarks on marker automata, and for reviewing the section on them.

References

1. Blum, M., Hewitt, C.: Automata on a 2-dimensional tape. In: FOCS, pp. 155–160. IEEE, Los Alamitos (1967)
2. Giammarresi, D., Restivo, A.: Two-dimensional languages, pp. 215–267. Springer-Verlag New York, Inc., New York (1997), http://portal.acm.org/citation.cfm?id=267871.267875
3. Ibarra, O.H., Jiang, T., Wang, H.: Some results concerning 2-d on-line tessellation acceptors and 2-d alternating finite automata. In: Tarlecki, A. (ed.) MFCS 1991. LNCS, vol. 520, pp. 221–230. Springer, Heidelberg (1991)
4. Inoue, K., Takanami, I.: Three-way tape-bounded two-dimensional turing machines. Inf. Sci. 17(3), 195–220 (1979)
5. Inoue, K., Takanami, I.: A note on decision problems for three-way two-dimensional finite automata. Inf. Process. Lett. 10(4/5), 245–248 (1980)
6. Inoue, K., Takanami, I.: A survey of two-dimensional automata theory. Inf. Sci. 55(1-3), 99–121 (1991)
7. Inoue, K., Takanami, I., Taniguchi, H.: Two-dimensional alternating turing machines. Theor. Comput. Sci. 27, 61–83 (1983)
8. Ito, A., Inoue, K., Takanami, I.: A note on three-way two-dimensional alternating turing machines. Inf. Sci. 45(1), 1–22 (1988)
9. Ito, A., Inoue, K., Takanami, I.: Deterministic two-dimensional on-line tessellation acceptors are equivalent to two-way two-dimensional alternating finite automata through 180-rotation. Theor. Comput. Sci. 66(3), 273–287 (1989)
10. Ito, A., Inoue, K., Takanami, I., Taniguchi, H.: Two-dimensional alternating turing machines with only universal states. Information and Control 55(1-3), 193–221 (1982)
11. Kari, J., Moore, C.: New results on alternating and non-deterministic two-dimensional finite-state automata. In: Ferreira, A., Reichel, H. (eds.) STACS 2001. LNCS, vol. 2010, pp. 396–406. Springer, Heidelberg (2001)
12. Kari, J., Moore, C.: Rectangles and squares recognized by two-dimensional automata. In: Karhumäki, J., Maurer, H., Păun, G., Rozenberg, G. (eds.) Theory Is Forever. LNCS, vol. 3113, pp. 134–144. Springer, Heidelberg (2004)
13. Kinber, E.B.: Three-way automata on rectangular tapes over a one-letter alphabet. Inf. Sci. 35(1), 61–77 (1985)
14. Ladner, R.E., Lipton, R.J., Stockmeyer, L.J.: Alternating pushdown and stack automata. SIAM J. Comput. 13(1), 135–155 (1984)
15. Lindgren, K., Moore, C., Nordahl, M.: Complexity of two-dimensional patterns. Journal of Statistical Physics 91, 909–951 (1998), http://dx.doi.org/10.1023/A:1023027932419
16. Minsky, M.L.: Computation: finite and infinite machines. Prentice-Hall, Inc., Englewood Cliffs (1967)
17. Okazaki, T., Zhang, L., Inoue, K., Ito, A., Wang, Y.: A note on two-dimensional probabilistic finite automata. Inf. Sci. 110(3-4), 303–314 (1998)
18. Parikh, R.: On context-free languages. J. ACM 13(4), 570–581 (1966)
19. Petersen, H.: Some results concerning two-dimensional turing machines and finite automata. In: Reichel, H. (ed.) FCT 1995. LNCS, vol. 965, pp. 374–382. Springer, Heidelberg (1995)

20. Reinhardt, K.: On some recognizable picture-languages. In: Brim, L., Gruska, J., Zlatuška, J. (eds.) MFCS 1998. LNCS, vol. 1450, pp. 760–770. Springer, Heidelberg (1998)
21. Rosenfeld, A.: Picture Languages: Formal Models for Picture Recognition. Academic Press, Inc., Orlando (1979)
22. Salo, V.: Classes of picture languages defined by tiling systems, automata and closure properties. Master's thesis, University of Turku (2011)
23. Sipser, M.: Halting space-bounded computations. In: FOCS, pp. 73–74. IEEE, Los Alamitos (1978)
24. Szepietowski, A.: Some remarks on two-dimensional finite automata. Inf. Sci. 63(1-2), 183–189 (1992)

Identity Problems, Solvability of Equations and Unification in Varieties of Semigroups Related to Varieties of Groups

Ondřej Klíma and Libor Polák*

Department of Mathematics and Statistics,
Masaryk University,
Kotlářská 2, 611 37 Brno,
Czech Republic
{klima,polak}@math.muni.cz
http://www.math.muni.cz

Abstract. In this survey we consider three kinds of algorithmic questions concerning varieties of semigroups. We are interested in identity problems, in the solvability of a system of equations and in the structure of all solutions of a given system. We study them in significant varieties of semigroups, monoids, groups, completely simple semigroups, completely regular semigroups (in particular semigroups satisfying the identity $x^r \approx x$) and involutary semigroups (in particular star regular semigroups and inverse semigroups).

Keywords: identity problems, equations, unification, varieties of semigroups, varieties of unary semigroups.

1   Introduction

Everybody knows how to solve a system of linear equations over real numbers or how to solve a quadratic equation over complex numbers. In many parts of mathematics and their applications (in particular in theoretical computer science) there is a strong need to solve various kinds of equations in various domains. The sides of equations are certain terms, the domain can be specified basically in two ways.

Firstly, it could be a concrete finite algebra. It this case the solvability problem is decidable and one is interested in the complexity of the decision problem – see e.g. [18] for results in the case of finite semigroups. We do not consider such questions in this contribution. Secondly, we can specify our domain by identities it satisfies; in fact we discuss the solvability problem for the corresponding variety. Here we need to understand the free algebras, i.e., we have to solve the identity problems there.

* The authors were supported by the Ministry of Education of the Czech Republic under the project MSM 0021622409 and by the Grant no. 201/09/1313 of the Grant Agency of the Czech Republic.

W. Kuich and G. Rahonis (Eds.): Bozapalidis Festschrift, LNCS 7020, pp. 214–227, 2011.

After this introductory section, we start to formalize our concepts. We will use semigroups, monoids, unary semigroups and unary monoids. We recall what kind of terms are appropriate in those situations.

In Section 3 we consider the identity problems for particular varieties, we mention how rewriting of terms can be used.

We study successively groups, completely simple semigroups, completely regular semigroups (in particular semigroups satisfying the identity $x^r \approx x$), involutary semigroups (in particular star regular semigroups and inverse semigroups).

The last two sections of our paper are devoted to the unification. In Section 4 we consider the solvability (of systems) of equations with constants in varieties mentioned above. The final section deals with the structure of all solutions of a given system of equations. Often, like in Abelian groups, there is the most general solution. Other possibility, which can happen, is to have finite or countable set of independent solutions such that each other solution can be derived from them. In some cases such set does not exist.

Summarizing there are three kinds of questions for each particular variety:

- the identity problem,
- the solvability of systems of equations (with constants),
- the unification type.

Of course, the first question should be solved first, since it is needed for considering the second and the third question. Note that the first question is trivial in semigroups whereas the second one is extremely deep result by Makanin. It seems that there is no significant relationship between the second and the third question.

Our choice of varieties is strongly influenced by authors interests and by the significance of those structures for theoretical computer science (in particular for automata theory and automated deduction). There are no new results in our paper. We tried to collect and comment results which are interesting for us in a transparent way.

We will touch only several results concerning unification. We advise the reader to extremely valuable surveys of Baader and Siekmann [4] and Baader and Snyder [5]. Another source worth to study algorithmic problems in varieties is a survey by Kharlampovich and Sapir [13].

We considered various types of semigroups. Notice that the famous 10th Hilbert's problem is nothing else as deciding solvability of equations with constants in commutative rings with 1.

2   Preliminaries: Terms and Free Algebras

In our paper we will meet numerous classes of algebras being varieties in a sense of universal algebra. They are of the following signatures consisting of

1. binary operation · in case of semigroups,
2. binary · and nullary 1 in case of monoids,

3. binary · and unary $(\ )^{-1}$ in case of unary semigroups, and
4. binary ·, nullary 1 and unary $(\ )^{-1}$ in case of unary monoids.

We speak about varieties of algebras of type $1 - 4$. In all considered algebras the binary operation · is associative and 1 is the neutral element, we do not mention this fact any more.

Let $X = \{x_1, x_2, \dots\}$ be a fixed set of *variables* (one uses $x, y, z, \dots$ in concrete situations). Our *terms* (often called *words*) over X in cases $1 - 4$ will be from the

1. semigroup X^+,
2. monoid $X^* = X^+ \cup \{1\}$,
3. unary semigroup U (defined below), and
4. unary monoid $U^1 = U \cup \{1\}$ (the neutral element 1 adjoined).

Let U be the smallest subsemigroup of $(X \cup \{(,)^{-1}\})^+$ containing X satisfying $u \in U$ implies $(u)^{-1} \in U$.

Morphisms between members of a given variety (of a type $1 - 4$) are mappings preserving multiplication, the neutral element for types 2 and 4, and the unary operation for types 3 and 4.

An algebra F is *free* over X in a variety $\mathcal{V}$ (of a type $1 - 4$) if

- $F \in \mathcal{V}$,
- $X \subseteq F$ and X generates F,
- for each $A \in \mathcal{V}$, each mapping $\alpha : X \to A$ extends to a morphism $F \to A$.

It is well-known that the free algebras in varieties always exist and that they are determined uniquely up to isomorphism. For

1. the variety $\mathcal{S}$ of all semigroups,
2. the variety $\mathcal{M}$ of all monoids,
3. the variety of all unary semigroups,
4. the variety of all unary monoids,

the free algebras over X have very transparent models, namely X^+, X^*, U, U^1, respectively.

3   Identity Problems

For a variety $\mathcal{V}$ (of type $1 - 4$) let $\sim_{\mathcal{V}}$ be the relation on the corresponding free algebra F of all terms consisting of all identities valid in $\mathcal{V}$. It is well-known that the corresponding quotient algebra $F/\sim_{\mathcal{V}}$ is free over X in $\mathcal{V}$.

A *solution of the identity problem* for a given variety $\mathcal{V}$ consists in an effective description of the relation $\sim_{\mathcal{V}}$. Basically, our algorithms are of two kinds:

- they compare certain *invariants* (often a kind of induction is used),
- they use *rewriting*.

We recall the basics of rewriting in case of varieties of semigroups. Consider a relation $\to$ on X^+, called the *rewrite relation*. (Below we denote by ρ^* the reflexive transitive closure of $\rho \subseteq X^+ \times X^+$.) The relation $\to$ is

- *terminating* if there is no infinite sequence $u_1 \to u_2 \to \ldots$, $u_1, u_2, \ldots \in X^+$,
- *confluent* if for each $u, v, w \in X^+$ with $v \leftarrow^* u \to^* w$, there exists $t \in X^+$ such that

$$v \to^* t \leftarrow^* w ,$$

- *canonical* if it is terminating and confluent.

For a canonical relation $\to$, each $u \in X^+$ has exactly one *canonical form* $\overrightarrow{u}$, i.e., $u \to^* \overrightarrow{u}$ and there is no $v \in X^+$ such that $\overrightarrow{u} \to v$.

Given a variety $\mathcal{V}$ of semigroups, our goal is to find an effectively described canonical rewrite relation $\to$ such that the equivalence relation generated by $\to$ is exactly $\sim_{\mathcal{V}}$. In this case, for all $u, v \in X^+$, we have $u \sim_{\mathcal{V}} v$ if and only if $\overrightarrow{u} = \overrightarrow{v}$.

For varieties of type $2 - 4$ we use obvious modifications.

Note that the identity problem is absolute trivial for $\mathcal{S}$ and $\mathcal{M}$: two terms are related if and only if they are equal.

3.1   Groups and Completely Regular Semigroups

We consider the class of all groups as a variety of unary monoids; we denote it by $\mathcal{G}$. It is determined by identities

$$x \cdot (x)^{-1} \approx 1 \quad \text{and} \quad (x)^{-1} \cdot x \approx 1 .$$

A well-known solution of the identity problem in $\mathcal{G}$ follows.

Proposition 1. *The rules*

- $u(pq)^{-1}v \to u(q)^{-1}(p)^{-1}v, \; p \in U, \; u, v \in U^1,$
- $u((p)^{-1})^{-1}v \to upv, \; p \in U, \; u, v \in U^1,$
- $up(p)^{-1}v \to uv, \; p \in U, \; u, v \in U^1,$
- $u(p)^{-1}pv \to uv, \; p \in U, \; u, v \in U^1$

determine a canonical rewrite relation on U^1 such that the equivalence relation it generates is exactly $\sim_{\mathcal{G}}$.

The identity problem for *Abelian groups* ($\mathcal{AG}$, in notation) and Abelian groups with $x^s \approx 1$ are easy to solve: $u \sim_{\mathcal{AG}} v$ iff the numbers of occurrences of each variable in canonical forms of u and v in $\mathcal{G}$ are the same. In $\mathcal{AG}(x^s \approx 1)$ one counts the occurrences modulo s.

There is extensive literature dealing with famous problems and numerous deep results concerning the Burnside groups $\mathcal{G}(x^s \approx 1)$. One could start from [7] and follow the items quoted there. Effective descriptions of the identity problem for the varieties $\mathcal{G}(x^s \approx 1)$, for $s = 2, 3, 4, 6$ is presented in Hall's book [12],

Chapter 18. Moreover, it is shown there that these varieties are *locally finite*, i.e., finitely generated members are finite. The case s odd, $s \geq 665$, is positively treated in Adian's book [1]. Note that these varieties are not locally finite.

Natural generalizations of groups are completely regular semigroups $\mathcal{CR}$ and inverse semigroups $\mathcal{I}$ (see Subsection 3.3). Completely regular semigroups are, by definition, unions of groups. They are given by the identities

$$x(x)^{-1} \approx (x)^{-1}x, \quad x(x)^{-1}x \approx x \quad \text{and} \quad ((x)^{-1})^{-1} \approx x.$$

A prominent subvariety of $\mathcal{CR}$ is $\mathcal{CS}$ formed by all completely simple semigroups which are defined by the additional identity $xyx(xyx)^{-1} \approx x(x)^{-1}$. Note that a completely regular semigroup is a semilattice of completely simple semigroups. For a variety $\mathcal{V}$ of groups, one can consider the varieties $\mathcal{CR}(\mathcal{V})$ and $\mathcal{CS}(\mathcal{V})$ of all completely regular (completely simple, respectively) semigroups with subgroups in $\mathcal{V}$. For a solution of the identity problem for $\mathcal{CS}(\mathcal{V})$, see Gerhard and Petrich [8].

3.2   Semigroups Satisfying $x^r \approx x$

For a fixed $r \geq 2$, we denote the class of all semigroups satisfying the identity $x^r \approx x$, by $\mathcal{S}(x^r \approx x)$. These are exactly the completely regular semigroups with subgroups satisfying $x^{r-1} \approx 1$. Therefore the inverses of elements are exactly the $(r-2)$-th powers. Thus we can remain in the signature of semigroups.

We translate the identity problem for this variety $\mathcal{S}(x^r \approx x)$ to the identity problem for the variety $\mathcal{G}(x^{r-1} \approx 1)$ of all groups satisfying the identity $x^{r-1} \approx 1$ as follows. For $u \in X^+$, we define:

- $\mathsf{c}(u)$ – the set of all variables in u,
- $\mathsf{0}(u)$ – the longest initial segment of u in $|\mathsf{c}(u)| - 1$ variables,
- $\mathsf{1}(u)$ – the longest final segment of u in $|\mathsf{c}(u)| - 1$.

Moreover, if $n = |\mathsf{c}(u)| \geq 2$, we define its *characteristic sequence* $[u]$ as the sequence of all maximal segments of u in $n - 1$ variables.

Example 1. For $u = xyxyztyzxztx$, we have

$$[u] = (xyxyz, yztyz, yzxz, zxztx).$$

Theorem 1 (Kad'ourek, Polák [16]). *Let $p, q \in X^+$ and let $\sim\ =\ \sim_{\mathcal{S}(x^r \approx x)}$. Then*

(i) For $|\mathsf{c}(p)| = 1$, it holds $p \sim q$ if and only if

$$\mathsf{c}(p) = \mathsf{c}(q) \quad \text{and} \quad \mathcal{G}(x^{r-1} \approx 1) \models p \approx q.$$

(ii) If $|\mathsf{c}(p)| \geq 2$, then it holds $p \sim q$ if and only if

$$\mathsf{c}(p) = \mathsf{c}(q), \ \mathsf{0}(p) \sim \mathsf{0}(q), \ \mathsf{1}(p) \sim \mathsf{1}(q),$$

and $\mathcal{G}(x^{r-1} \approx 1) \models \phi(p_0)\phi(p_1) \ldots \phi(p_m) \approx \phi(q_0)\phi(q_1) \ldots \phi(q_n)$

where $[p] = (p_0, \ldots, p_m)$, $[q] = (q_0, \ldots, q_n)$ and ϕ maps words into new variables such that $\phi(s) = \phi(t)$ iff $s \sim t$.

Note that the case of idempotent semigroups, i.e., $r = 2$, was previously solved by Green and Rees [9]. Other possibility of solving the identity problem in idempotent semigroups is a usage of rewriting.

Theorem 2 (Siekmann and Szabó [31]). *Let $\mathcal{B} = \mathcal{S}(x^2 \approx x)$ be the variety of all idempotent semigroups. Then the rules*

- $up^2v \to upv$, $p \in X^+$, $u, v \in X^*$,
- $upqrv \to uprv$, $p, q, r \in X^+$, $u, v \in X^*$, $\mathsf{c}(q) \subseteq \mathsf{c}(p) = \mathsf{c}(r)$

determine a canonical rewrite relation on X^+ such that the equivalence relation it generates is exactly $\sim_\mathcal{B}$.

A recent paper [16] discusses the possibilities of using rewriting in varieties of idempotent semigroups. Numerous valuable references can be found there, in particular references to several works by Baader.

A solution of the identity problem for $\mathcal{CR}$ is a more advanced version of Theorem 1 – see [15]. Similarly for $\mathcal{CR}(\mathcal{V})$.

3.3   Involutary Semigroups

In this subsection the unary operation is denoted by $'$. An *involutary semigroup* is a unary semigroup $(S, \cdot, ')$ satisfying the identities

$$(xy)' \approx y'x' \quad \text{and} \quad (x')' \approx x.$$

There is a natural model of a free involutary semigroup. Let $X' = \{x_1', x_2', \dots\}$ be a disjoint copy of X. Let I be free semigroup over $X \cup X'$ with the unary operation $'$ given by $(y_1 \dots y_k)' = y_k' \dots y_1'$, $(x')' = x$ for $y_1, \dots, y_k \in X \cup X'$, $x \in X$. Terms for involutary semigroups can be taken from U but it is more convenient to consider the elements of I.

Star regular semigroups, in notation $\mathcal{SR}$, are involutary semigroups satisfying the identity $xx'x \approx x$.

Theorem 3 (Polák [26]). *The rule*

- $upp'pv \to upv$, $p \in I$, $u, v \in I^1$

determine a canonical rewrite relation on I such that the equivalence relation it generates is exactly $\sim_{\mathcal{SR}}$.

Completely simple star regular semigroups are given by an additional identity $xyy'x' \approx xx'$. Similarly as above one gets a canonical rewrite relation for them by orienting the identity, i.e., one uses the rule $upqq'p'v \to upp'v$, $p, q \in I$, $u, v \in I^1$. Clearly, for groups one can use only $upp'v \to uv$, $p \in I$, $u, v \in I^1$. Denote the canonical form of $u \in I$ by $\overrightarrow{u}^\mathcal{G}$.

Inverse monoids $\mathcal{IM}$ are given by the identities for star regular semigroups and by the additional identity $xx'yy' \approx yy'xx'$. They are up to isomorphism exactly unary submonoids of the monoids of all partial bijections of a set into itself. They also play very significant rule in automata theory. A solution of the identity problem follows.

Theorem 4 (Scheiblich [30] and Munn [23])
Let $y_1, \ldots, y_m, z_1, \ldots, z_n \in X \sqcup X'$. Then $y_1 \ldots y_m \sim_{\mathcal{IM}} z_1 \ldots z_n$, if and only if

$$\overrightarrow{y_1 \ldots y_m}^{\mathcal{G}} = \overrightarrow{z_1 \ldots z_n}^{\mathcal{G}} \quad \text{and} \quad \{\overrightarrow{y_1}^{\mathcal{G}}, \ldots, \overrightarrow{y_1 \ldots y_m}^{\mathcal{G}}\} = \{\overrightarrow{z_1}^{\mathcal{G}}, \ldots, \overrightarrow{z_1 \ldots z_n}^{\mathcal{G}}\}.$$

4 Equations

In this section we overview some results concerning solvability of equations with constants in significant varieties of (unary) semigroups and monoids. We assume that our alphabet X is a disjoint union of two countable sets of *constants* $C = \{c_1, c_2, \ldots\}$ and *unknowns* $W = \{w_1, w_2, \ldots\}$. Thus we work in a free semigroup/monoid of terms F_X, which is the free semigroup X^+, or the free monoid X^*, or the free unary semigroup U over X, or the free unary monoid U^1 over X. In every case, F_C is a subalgebra of F_X.

A *substitution* is a mapping $\sigma : X \to F_X$ non-identical only on a finite set of unknowns from W. Its extension to an endomorphism of F_X is again denoted by σ. An *equation* $L \stackrel{?}{=} R$ is a pair of terms $L, R \in F_X$. For a variety $\mathcal{V}$ of semigroups (of type $1-4$) we say that a substitution σ is a $\mathcal{V}$-*solution* of the equation $L \stackrel{?}{=} R$ if $\sigma(L) \sim_{\mathcal{V}} \sigma(R)$, i.e., $\mathcal{V} \models \sigma(L) \approx \sigma(R)$. Note that there exists a solution of the equation $L \stackrel{?}{=} R$ if and only if there exists a solution $\sigma : F_X \to F_X$ such that $\sigma(x) \in F_C$ for all $x \in X$.

For a finite set of equations

$$E = \{L_1 \stackrel{?}{=} R_1, \ldots, L_k \stackrel{?}{=} R_k\}, \quad \text{where } L_1, \ldots, L_k, R_1, \ldots, R_k \in F_X,$$

a substitution σ is a $\mathcal{V}$-solution of the system E if

$$\sigma(L_1) \sim_{\mathcal{V}} \sigma(R_1), \ldots, \sigma(L_k) \sim_{\mathcal{V}} \sigma(R_k).$$

Then we denote by $\mathsf{U}_{\mathcal{V}}(E)$ the set of all $\mathcal{V}$-solutions of E.

For a given variety $\mathcal{V}$ (of type $1-4$) we consider the first basic problem, namely a *solvability problem*, where an instance of the problem is a finite system of equations E and the question is whether $\mathsf{U}_{\mathcal{V}}(E) \neq \emptyset$ or not.

Apart from the solvability problem there is the so-called *elementary solvability problem*, where only equations without constants are considered. Such equations are uninteresting in the case of varieties which are studied in this paper. Namely, whenever we have some idempotent in our free algebra we can substitute this idempotent for all unknowns and therefore every equation has a solution.

A final general remark is that it is natural to assume that the system can contain also *inequalities*. Here for a pair of terms L, R a solution σ must satisfy that $\sigma(L) \not\sim_{\mathcal{V}} \sigma(R)$.

4.1 Equations in Free Semigroups

Although the identity problem for the variety $\mathcal{S}$ is trivial, the solvability problem is very difficult and the problem was opened for many years. Finally, it was solved by Makanin.

Theorem 5 (Makanin [20]). *The solvability problem for a single equation for variety $\mathcal{S}$ is decidable.*

On the first sight, one can think that the systems of equations are more complicated to be solved but there is an easy trick how one can reduce a system of equations to a single equation. Namely, a system $E = \{L_1 \overset{?}{=} R_1, \ldots, L_k \overset{?}{=} R_k\}$ has the same set of $\mathcal{S}$-solutions as a single equation

$$L_1 a \ldots L_k a L_1 b \ldots L_k b \overset{?}{=} R_1 a \ldots R_k a R_1 b \ldots R_k b \ ,$$

see Proposition 12.1.8 in [11]. Further, there is no significant difference when one solves equations in the free semigroup X^+ or in the free monoid X^*; one needs only to discuss separately several cases given by sets of unknowns mapped to 1 in a potential solution. Thus the famous Makanin result implies that the solvability problems for varieties $\mathcal{S}$ and $\mathcal{M}$ are decidable. A reader who is interested in the topic is referred to a presentation [11] of the Makanin's proof, where a certain generalization of the result is proved, namely rational constrains can be given for every unknown.

Example 2. Let us consider the equation $xauzau \overset{?}{=} yzbxaaby$. To demonstrate that $\mathcal{S}$-solutions and $\mathcal{M}$-solutions are quite complicated, we only show some examples of solutions. First $\mathcal{M}$-solution is σ_1 given by rules $\sigma_1(x) = \sigma_1(y) = 1$, $\sigma_1(z) = a$, $\sigma_1(u) = b$. Further σ_2 given by rules $\sigma_2(x) = \sigma_2(y) = \sigma_2(z) = a$, $\sigma_2(u) = ba$ is a $\mathcal{S}$-solution. But we have also, for any natural numbers k, ℓ, a $\mathcal{S}$-solution $\tau_{k,\ell}$ given by $\tau_{k,\ell}(x) = (xab)^k x$, $\tau_{k,\ell}(y) = (xab)^\ell x$, $\tau_{k,\ell}(z) = abxa$, $\tau_{k,\ell}(u) = b(xab)^\ell x$.

Note that the equation in the previous example is taken from [11]. It is an example of a quadratic equation (every unknown occurs at most twice) for which a certain algorithm is presented there.

4.2 Equations in Locally Finite Varieties

Let $\mathcal{V}$ be a locally finite variety of semigroups (of type 1) and assume that we can solve the identity problem for this variety. Let E be an arbitrary system of equations and let C_E denote the finite set of constants and W_E denote the finite set of unknowns occurring in E. The system E has a $\mathcal{V}$-solution if and only if there exists a $\mathcal{V}$-solution σ such that $\sigma(x) \in C_E^+$ for every $x \in W_E$ and $\sigma(x) = x$ for $x \notin W_E$. Since $\mathcal{V}$ is a locally finite variety, the free semigroup F over the set C_E in the variety $\mathcal{V}$ is finite. Even, we can construct a model of F in the following way. We consider a length-lexicographical ordering[1] $\leq$ on the set C_E^+. We construct, step by step, the finite subset F' of C_E^+. In every step we consider the minimal word u in $\leq$ which was not considered yet and we add u to the actual list F' whenever u is not $\sim_\mathcal{V}$-equivalent to any word in F'. This can

---

[1] Here $u \leq v$ if and only if (i) u is shorter word than v or (ii) u and v have the same length and u is smaller in lexicographical ordering.

be effectively tested because we know how to solve the identity problem for the variety $\mathcal{V}$. Since F is finite we know that once we discover a natural number n such that in the process we do not add to F' any word of a length n. But if such a situation occurs then every word of length greater then n is $\sim_{\mathcal{V}}$-equivalent to a shorter word and we know that we can stop our procedure.

Thus for every unknown $x \in W_E$ we need to test only finitely many values $\sigma(x) \in F'$. Thus there are only finitely many substitutions which must be checked whether they are solutions.

The varieties of types $2 - 4$ can be considered similarly.

Proposition 2. *Let $\mathcal{V}$ be a locally finite variety of type $1 - 4$ such that we know the solution of the identity problem for $\mathcal{V}$. Then the solvability problem for the variety $\mathcal{V}$ is decidable.*

From Section 3 it follows the decidability of the problem for the variety of idempotent semigroups $\mathcal{S}(x^2 \approx x)$, for the group varieties $\mathcal{G}(x^s \approx 1)$ for $s = 2, 3, 4, 6$ and consequently, by [9], for the varieties of semigroups $\mathcal{S}(x^r \approx x)$ for $r = 3, 4, 5, 7$.

4.3   Equations in Varieties of Groups

Solvability of equations in the variety of all groups $\mathcal{G}$ is the second very deep result of Makanin.

Theorem 6 (Makanin [21]). *The solvability problem for the variety $\mathcal{G}$ is decidable.*

We refer to final remarks in the paper [11] and to the paper by Razborov [27] for more information on equations in free groups.

In contrast to the previous case the result concerning the variety of all Abelian groups is quite easy to prove.

Proposition 3. *The solvability problem for the variety $\mathcal{AG}$ is decidable.*

Completely different results were obtained for varieties of nilpotent groups. Let $\mathcal{N}_c$ denote the variety of all nilpotent groups of class c. It was proved by Romankov [28] that the solvability problem for the varieties $\mathcal{N}_c$, $c \geq 9$, is undecidable. This result was extended to $c \geq 3$ by Truss [32]. Look there for the whole history.

Theorem 7 (Truss [32]). *The solvability problem for a single equation for the varieties $\mathcal{N}_c$, $c \geq 3$, is undecidable.*

For the case $c = 2$ there are some partial results (see [6] for details) which suggests that the solvability problem is decidable. For example in [32] it is shown that it is decidable whether a single equation with two unknowns has a $\mathcal{N}_2$-solution.

4.4 Equations in Varieties of Completely Regular Semigroups

First we state a result on varieties of completely simple semigroups.

Theorem 8 (Polák [25]). *If the solvability problem is decidable for a variety of groups $\mathcal{V}$ then the solvability problem is decidable for the variety of completely simple semigroups $\mathcal{CS}(\mathcal{V})$.*

Note that the reduction is given in such a way that for each unknown x we prescribe the first and the last letter of $\sigma(x)$. In each such case we create a new system of equations which has a $\mathcal{V}$-solution if and only if the original system has a $\mathcal{CS}(\mathcal{V})$-solution. This means that a system of equations in n unknowns with m constants in $\mathcal{CS}(\mathcal{V})$ is equivalent to considering disjunction of m^{2n} systems in n variables with m constants in $\mathcal{V}$.

In the case of varieties $\mathcal{S}(x^r \approx x)$ one can apply a result from subsection 4.2. But one can also hope that the solvability problem for $\mathcal{S}(x^r \approx x)$ can be reduce to the solvability problem for $\mathcal{G}(x^{r-1} \approx 1)$. This is true only partially.

Theorem 9 (Klíma [17]). *Let r be such that the solvability problem in the variety $G(x^{r-1} \approx 1)$ is decidable. Then it is decidable whether a system of equations with two constants has a $\mathcal{S}(x^r \approx x)$-solution.*

The idea behind the proof of this result is that we need to distinguish, for each unknown x, whether $\sigma(x)$ contains only one constant or both and which power of which constant is the initial and final segment of $\sigma(x)$. Thus a system of equations in n variables with 2 constants in $\mathcal{S}(x^r \approx x)$ is equivalent to considering disjunction of $(4(r-1)^2 + 2(r-1))^n$ systems of equations in n variables in $\mathcal{G}(x^{r-1} \approx 1)$. The interesting point is that the number of constants in a new system is $2r$ because new constants correspond to elements in a free semigroup in the variety $\mathcal{S}(x^r \approx x)$ over single element set, i.e., to different powers of the original constants.

In the paper [17] we have suggested how one can proceed by induction with respect to number of constants. The technical problem is that considering three constants one need to know which segments in two constants in different $\sigma(x)$ are equal. These segments correspond to new constants in a new system over $\mathcal{G}(x^{r-1} \approx 1)$, and, in fact, we need to know which are equal and which are different. In other words we need to know whether certain auxiliary systems of equations and inequalities (over two constants) have solutions in $\mathcal{S}(x^r \approx x)$. Thus we need to have a generalization of Theorem 9 for system of equations and inequalities which is also given in [17].

Theorem 10 (Klíma [17]). *Let r be such that the solvability problem for a system of equations and inequalities in the variety $G(x^{r-1} \approx 1)$ is decidable. Then it is decidable whether a system of equations with three constants has a $\mathcal{S}(x^r \approx x)$-solution.*

4.5   Equations in Free Inverse Monoids

Let $\mathcal{IM}$ denote variety (of type 4) of all inverse monoids.

Theorem 11 (Rozenblat [29]). *The solvability problem for the variety $\mathcal{IM}$ is undecidable.*

To prove this result, Rozenblat needs only two constants and it was also proved there that systems over one constant are decidable. Another special case of equations containing one variable is studied in the paper [10], where the following natural idea is used. A solution of the system in an arbitrary variety $\mathcal{V}$ can be found in two steps. First, one can find a $\mathcal{U}$-solution σ for some smaller variety $\mathcal{U}$ and then one can try to modify σ to a $\mathcal{V}$-solution α which is $\sim_{\mathcal{U}}$-related to σ. Surprisingly, Deis et al. [10] proved that it is decidable whether a given $\mathcal{G}$-solution of a system of equations has $\sim_{\mathcal{G}}$-related $\mathcal{IM}$-solution of the same system. This idea can not be used to prove decidability of the solvability problem for the variety $\mathcal{IM}$, because system of equations in groups can have infinitely many solutions as we will see in the next section.

5   Unification Type

For $Y \subseteq W$ there is a quasiorder (i.e., reflexive and transitive relation) $\leq_{\mathcal{V}}^{Y}$ on substitutions, namely $\sigma \leq_{\mathcal{V}}^{Y} \tau$ iff there exists a substitution λ such that

$$(\, \forall \, y \in Y \,) \, (\lambda \circ \sigma)(y) \sim_{\mathcal{V}} \tau(y) \, .$$

In this case, we say that σ is *more general* than τ.

Let $E = \{L_1 \overset{?}{=} R_1, \dots, L_k \overset{?}{=} R_k\}$ be a system of equations and let

$$Y = (\mathsf{c}(L_1) \cup \cdots \cup \mathsf{c}(L_k) \cup \mathsf{c}(R_1) \cup \cdots \cup \mathsf{c}(R_k)) \cap W \, .$$

A set $U \subseteq \mathsf{U}_{\mathcal{V}}(E)$ is *complete* if

$$(\, \forall \, \tau \in \mathsf{U}_{\mathcal{V}}(E) \,)(\, \exists \, \sigma \in U \,) \, \sigma \leq_{\mathcal{V}}^{Y} \tau .$$

It is *minimal* if
$$(\, \forall \, \sigma, \tau \in U \,)(\, \sigma \leq_{\mathcal{V}}^{Y} \tau \implies \sigma = \tau \,) \, .$$

Note that $\emptyset$ is a minimal complete set of $\mathcal{V}$-solutions for a system E without any solution. Further, any two minimal complete sets of $\mathcal{V}$-solutions of a given system are of the same cardinality. A variety $\mathcal{V}$ is of *unification type*

(1) *unitary* if any system of equations has a minimal complete set of $\mathcal{V}$-solutions of cardinality ≤ 1,

(ω) *finitary* if any system of equations has a minimal complete set of $\mathcal{V}$-solutions of a finite cardinality,

(∞) *infinitary* if $\mathcal{V}$ is not finitary and any system of equations has a minimal complete set of $\mathcal{V}$-solutions,

(0) *nullary* if there exists a system of equations without any minimal complete set of $\mathcal{V}$-solutions.

The second basic problem for a variety $\mathcal{V}$ is to determine its unification type. We can speak also about a unification type for a single system E.

Theorem 12 (Plotkin [24]). *The variety $\mathcal{S}$ has the unification type infinitary.*

In fact, Plotkin introduced a procedure generating all minimal solutions. The type is not unitary nor finitary as the following example shows.

Example 3. The equation $ax \stackrel{?}{=} xa$ in $\mathcal{S}$ has an infinite minimal complete set of $\mathcal{S}$-solution. Namely, for every natural number k, we have $\mathcal{S}$-solution σ_k given by $\sigma_k(x) = a^k$.

In the previous example, we used the well-known fact from combinatorics on words that two words commute if and only if they are powers of the same word. The same is true in free groups. Therefore the variety $\mathcal{G}$ of all groups is of type infinitary or nullary. In [2] it is shown that in the case of equations without constants we have unification type infinitary. Unfortunately we have not found an answer for the general case.

The situation becomes more simple for Abelian groups.

Theorem 13 (Lankford et al. [19]). *The variety $\mathcal{AG}$ is of unitary type.*

Example 4. The most general solution of the equation $x^5 y^7 \stackrel{?}{=} a^8 b^2 c^3$ $\mathcal{AG}$ is:

$$\sigma(x) = a^{-4}b^{-1}c^{-5}z^{-7}, \ \sigma(y) = a^4 bc^4 z^5 \,.$$

To get all other solutions, one substitutes for z arbitrarily.

On the other hand, the situation for varieties of nilpotent groups is again quite advanced.

Theorem 14 (Albert and Lawrence [2]). *Every system of equations without constants in $\mathcal{N}_c$, $c \geq 2$, is of unification type unitary or nullary.*

For varieties of completely simple semigroups, we can state the following.

Theorem 15 (Polák [25]). *The variety $\mathcal{CS}(\mathcal{V})$ is of unification type finitary (infinitary, nullary, respectively) if and only if the same holds for the variety $\mathcal{V}$.*

Example 5. The equation

$$x^2 y^2 z^2 y^2 \stackrel{?}{=} abcabc$$

in $\mathcal{CS}(\mathcal{AG})$ has a minimal complete set of solutions of cardinality 5.

Kisielewicz in [14] determines the unification types for all varieties of commutative semigroups considering equations without constants. By Baader [3], almost all varieties of idempotent semigroups and many varieties of completely regular semigroups are of type nullary. Also Mashevitzky shows in [22] that not to be of type nullary is very restrictive for a variety of completely regular semigroups.

References

1. Adian, S.I.: The Burnside problem and identities in groups. Springer, New York (1979)
2. Albert, M.H., Lawrence, J.: Unification in varieties of groups: nilpotent varieties. Canadian Journal of Mathematics 46(6), 1135–1149 (1994)
3. Baader, F.: Unification in varieties of completely regular semigroups. In: Schulz, K.U. (ed.) IWWERT 1990. LNCS, vol. 572, pp. 210–230. Springer, Heidelberg (1992)
4. Baader, F., Siekmann, J.H.: Unification Theory. In: Gabbay, D.M., Hogger, C.J., Robinson, J.A., Siekmann, J.H. (eds.) Handbook of Logic in Artificial Intelligence and Logic Programming, vol. 2, pp. 41–125. Oxford University Press, Oxford (1994)
5. Baader, F., Snyder, W.: Unification Theory. In: Robinson, J.A., Voronkov, A. (eds.) Handbook of Automated Reasoning, pp. 446–533. Elsevier, MIT Press (2001)
6. Burke, E.K., Truss, J.K.: Unification for nilpotent groups of class 2. Forum Math. 7, 435–437 (1995)
7. O'Connor, J.J., Robertson, E.F.: A history of the Burnside problem, www-history.mcs.st-and.ac.uk/HistTopics/Burnside_problem.html
8. Gerhard, J.A., Petrich, M.: Word problem for free objects in certain varieties of completely regular semigroups. Pacific J. Math. 104, 351–360 (1983)
9. Green, J.A., Rees, D.: On semigroups in which $x^r = x$. Proc. of the Cambridge Phil. Soc. 48, 35–40 (1952)
10. Deis, T., Meakin, J., Sénizergues, G.: Equations in free inverse monoids. International Journal of Algebra and Computation 17(4), 761–795 (2007)
11. Diekert, V.: Makanin's algorithm. In: Lothaire, M. (ed.) Algebraic Combinatorics on Words, pp. 342–390. Cambridge University Press, Cambridge (2002)
12. Hall Jr., M.: Theory of Groups. Macmillan, Basingstoke (1959)
13. Kharlampovich, O., Sapir, M.: Algorithmic problems in varieties. International Journal of Algebra and Computation 12, 379–602 (1995)
14. Kisielewicz, A.: Unification in commutative semigroups. J. Algebra 200, 246–257 (1998)
15. Kad'ourek, J., Polák, L.: On the word problem for free completely regular semigroups. Semigroup Forum 34, 127–138 (1986)
16. Kad'ourek, J., Polák, L.: On free semigroups satisfying $x^r = x$. Simon Stevin 64(1), 3–19 (1990)
17. Klíma, O.: On the solvability of equations in semigroups with $x^r = x$. Contributions to General Algebra 12, 237–245 (2000)
18. Klíma, O., Tesson, P., Thérien, D.: Dichotomies in the Complexity of Solving Systems of Equations over Finite Semigroups. Theory of Computing Systems 40(3), 263–297 (2007)
19. Lankford, D.S., Butler, G., Brady, B.: Abelian group unification algorithms for elementary terms. Contemporary Mathematics 29, 193–199 (1984)
20. Makanin, G.S.: The problem of solvability of equations in free semigroups. Math. Sbornik 103, 147–236 (1977)
21. Makanin, G.S.: Equations in a free group. Izv. Akad. Nauk SSSR, Ser. Mat. 46, 1199–1273 (1982)
22. Mashevitzky, G.: Unification Types of Completely Regular Semigroups. J. Autom. Reasoning 29(2), 171–182 (2002)
23. Munn, W.D.: Free inverse semigroups. Proc. London Math. Soc. 30, 385–404 (1974)
24. Plotkin, G.: Building-in Equational Theories. Machine Intelligence 7, 73–90 (1972)

25. Polák, L.: Unification in varieties of completely simple semigroups. Contributions to General Algebra 12, 337–348 (2000)
26. Polák, L.: A solution of the word problem for free ∗-regular semigroups. Journal of Pure and Applied Algebra 157, 107–114 (2001)
27. Razborov, A.A.: On systems of equations in a free group. Izv. Akad. Nauk SSSR, Ser. Mat. 48(4), 779–832 (1984)
28. Romankov, V.A.: Unsolvability of the endomorphic reducibility problem in free nilpotent groups and in free rings. Algebra and Logic 16, 310–320 (1977)
29. Rozenblat, B.V.: Diophantine theories of free inverse semigroups. Sibirskii Matematicheskii Zhurnal 26(6), 101–107 (1986)
30. Scheiblich, H.E.: Free inverse semigroups. Proc. Amer. Math. Soc. 38, 1–7 (1973)
31. Siekmann, J.H., Szabó, P.: A Noetherian and Confluent Rewrite System for Idempotent Semigroups. Semigroup Forum 25(1), 83–110 (1982)
32. Truss, J.K.: Equation-Solving in Free Nilpotent groups Class 2 and 3. Bull. London Math. Soc. 27(1), 39–45 (1995)

Algebraic Systems and Pushdown Automata

Werner Kuich

Technische Universität Wien,
Institut für Diskrete Mathematik und Geometrie
`kuich@tuwien.ac.at`

Abstract. This survey paper serves two purposes: Firstly, we consider
cycle-free algebraic systems (with respect to a given strong convergence)
as a generalization of the usually considered proper systems (with respect
to the discrete convergence). Secondly, we develop in a parallel manner
the theory of these cycle-free algebraic systems over an arbitrary semiring
and the theory of arbitrary algebraic systems over a continuous semiring.
In both cases we prove that algebraic systems and weighted pushdown
automata are mechanisms of equal power.

1 Introduction and Preliminaries

In this paper we develop the theories of algebraic systems and weighted pushdown automata by an algebraic treatment using semirings, formal power series
and matrices. We prove that algebraic systems and weighted pushdown automata
are mechanisms of equal power.

The paper consists of this and three more sections. In Section 1 we introduce the algebraic structures needed: semirings, especially continuous semirings,
formal power series and matrices over these semirings.

In Section 2 we consider the axiomatic definition for the notion of convergence of a special type as developed in Kuich, Salomaa [15]. Examples of such
a convergence are: the discrete convergence, the Cauchy convergence, the Euler
convergence, the complete convergence and the supremum convergence. Moreover, we show how to transfer a convergence over a semiring to convergences
over formal power series and (possibly infinite) matrices.

In Section 3 we consider algebraic systems and their approximation sequences.
If the approximation sequence of an algebraic system is convergent with respect
to a certain convergence, then its limit is called strong solution. It is proved that
this strong solution is really a solution to the algebraic system if the convergence
is multiplicative. This is valid e. g. for the discrete convergence, the Cauchy
convergence and the supremum convergence.

In Section 4 we introduce weighted pushdown automata and prove the main
result of this paper: a formal power series is algebraic, i. e., its proper part is a
solution of a proper algebraic system, iff this formal power series is the behavior
of a proper weighted pushdown automaton.

In Ésik, Kuich [8] we unified two variants of the Theorem of Kleene-Schützenberger by help of partial Conway semirings: one for arbitrary semirings and

W. Kuich and G. Rahonis (Eds.): Bozapalidis Festschrift, LNCS 7020, pp. 228–256, 2011.
© Springer-Verlag Berlin Heidelberg 2011

proper finite automata; the other for Conway semirings and arbitrary finite automata. In this paper, we do the same for two variants of the equivalence of algebraic systems and weighted pushdown automata: one for cycle-free algebraic systems (with respect to a given strong convergence) and proper weighted pushdown automata (see Kuich, Salomaa [15], Petre, Salomaa [17]); the other for arbitrary algebraic systems (with respect to the supremum convergence) and proper weighted pushdown automata over a continuous semiring (see Kuich [13], Ésik, Kuich [7]).

We note that parts of the main result of this paper can be generalized to tree automata: algebraic tree systems and weighted pushdown tree automata are equivalent (see Bozapalidis [3], Kuich [14], Ésik, Kuich [6]).

We make the following notational conventions valid throughout the paper. A, Σ, Q and I, possibly indexed, always denote a semiring, an alphabet, a finite index set and an arbitrary index set, respectively.

A *monoid* consists of a nonempty set M, an associative binary operation $\cdot$ on M and of a neutral element 1 such that $m \cdot 1 = 1 \cdot m = m$ for every $m \in M$. A monoid M is called *commutative* if $m_1 \cdot m_2 = m_2 \cdot m_1$ for every $m_1, m_2 \in M$. The binary operation is usually denoted by juxtaposition and often called product.

If the operation and the neutral element of M are understood then we denote the monoid simply by M. Otherwise, we use the triple notation $\langle M, \cdot, 1 \rangle$. A commutative monoid M is often denoted by $\langle M, +, 0 \rangle$.

A *morphism* h of a monoid M into a monoid M' is a mapping $h : M \to M'$ compatible with the neutral elements and operations in $\langle M, \cdot, 1 \rangle$ and $\langle M', \circ, 1' \rangle$, i.e., $h(1) = 1'$ and $h(m_1 \cdot m_2) = h(m_1) \circ h(m_2)$ for all $m_1, m_2 \in M$.

A commutative monoid $\langle A, +, 0 \rangle$ is called *ordered* if it is equipped with a partial order $\leq$ preserved by the $+$ operation such that $0 \leq a$ holds for all $a \in A$. It then follows that $a \leq a + b$, for all $a, b \in A$. In particular, a commutative monoid $\langle A, +, 0 \rangle$ is called *naturally ordered* iff the relation $\sqsubseteq$ defined by: $a \sqsubseteq b$ iff there exists a c such that $a + c = b$, is a partial order. Morphisms of ordered monoids preserve the order.

A monoid $\langle A, +, 0 \rangle$ is called *complete* if it has sums for all families $(a_i \mid i \in I)$ of elements of A, where I is an arbitrary index set, such that the following conditions are satisfied:

(i) $\sum_{i \in \emptyset} a_i = 0$, $\sum_{i \in \{j\}} a_i = a_j$, $\sum_{i \in \{j,k\}} a_i = a_j + a_k$, for $j \neq k$,
(ii) $\sum_{j \in J} (\sum_{i \in I_j} a_i) = \sum_{i \in I} a_i$, if $\bigcup_{j \in J} I_j = I$ and $I_j \cap I_{j'} = \emptyset$ for $j \neq j'$.

A morphism of complete monoids preserves all sums. Note that any complete monoid is commutative.

Recall that a non-empty subset D of a partially ordered set P is called *directed* iff each pair of elements of D has an upper bound in D. Moreover, a function $f : P \to Q$ between partially ordered sets is *continuous* iff it preserves the least upper bound of any directed set, i.e., when $f(\sup D) = \sup f(D)$, for all directed sets $D \subseteq P$ such that $\sup D$ exists. It follows that any continuous function preserves the order.

An ordered commutative monoid $\langle A, +, 0 \rangle$ is called a *continuous monoid* if each directed subset of A has a least upper bound and the $+$ operation preserves the least upper bound of directed sets, i.e., when

$$a + \sup D = \sup(a + D),$$

for all directed sets $D \subseteq A$ and for all $a \in A$. Here, $a + D$ is the set $\{a + x \mid x \in D\}$. A morphism of continuous monoids is a continuous monoid homomorphism.

It is known that an ordered commutative monoid A is continuous iff each chain in A has a least upper bound and the $+$ operation preserves least upper bounds of chains, i.e., when $a + \sup C = \sup(a + C)$ holds for all non-empty chains C in A. (See Markowsky [16].)

Proposition 1. *Any continuous monoid $\langle A, +, 0 \rangle$ is a complete monoid equipped with the following sum operation:*

$$\sum_{i \in I} a_i = \sup\{\sum_{i \in E} a_i \mid E \subseteq I, \ E \text{ finite}\},$$

for all index sets I and all families $(a_i \mid i \in I)$ in A. Any morphism between continuous monoids is a complete monoid morphism.

By a *semiring* we mean a set A together with two binary operations $+$ and $\cdot$ and two constant elements 0 and 1 such that

(i) $\langle A, +, 0 \rangle$ is a commutative monoid,
(ii) $\langle A, \cdot, 1 \rangle$ is a monoid,
(iii) the distributivity laws $s_1 \cdot (s_2 + s_3) = s_1 \cdot s_2 + s_1 \cdot s_3$ and $(s_1 + s_2) \cdot s_3 = s_1 \cdot s_3 + s_2 \cdot s_3$ hold for every $s_1, s_2, s_3 \in S$,
(iv) $0 \cdot s = s \cdot 0 = 0$ for every $s \in A$.

A semiring A is called *commutative* if $s_1 \cdot s_2 = s_2 \cdot s_1$ for every $s_1, s_2 \in A$.

If the operations and the constant elements of A are understood then we denote the semiring simply by A. Otherwise, we use the notation $\langle A, +, \cdot, 0, 1 \rangle$.

Intuitively, a semiring is a ring (with unity) without subtraction. A typical example is the semiring of nonnegative integers $\mathbb{N}$. A very important semiring in connection with language theory is the *Boolean* semiring $\mathbb{B} = \{0, 1\}$ where $1 + 1 = 1 \cdot 1 = 1$. Clearly, all rings (with unity), as well as all fields, are semirings, e. g., integers $\mathbb{Z}$, rationals $\mathbb{Q}$, reals $\mathbb{R}$, complex numbers $\mathbb{C}$ etc.

Let $\mathbb{N}^\infty = \mathbb{N} \cup \{\infty\}$ and $\overline{\mathbb{N}} = \mathbb{N} \cup \{-\infty, \infty\}$. Then $\langle \mathbb{N}^\infty, +, \cdot, 0, 1 \rangle$, $\langle \mathbb{N}^\infty, \min, +, \infty, 0 \rangle$ and $\langle \overline{\mathbb{N}}, \max, +, -\infty, 0 \rangle$, where $+$, $\cdot$, $\min$ and $\max$ are defined in the obvious fashion (observe that $0 \cdot \infty = \infty \cdot 0 = 0$ and $(-\infty) + \infty = -\infty$), are semirings.

Let $\mathbb{R}_+ = \{a \in \mathbb{R} \mid a \geq 0\}$, $\mathbb{R}_+^\infty = \mathbb{R}_+ \cup \{\infty\}$ and $\overline{\mathbb{R}}_+ = \mathbb{R}_+ \cup \{-\infty, \infty\}$. Then $\langle \mathbb{R}_+, +, \cdot, 0, 1 \rangle$, $\langle \mathbb{R}_+^\infty, +, \cdot, 0, 1 \rangle$ and $\langle \mathbb{R}_+^\infty, \min, +, \infty, 0 \rangle$ are semirings. The semirings $\langle \mathbb{N}_+^\infty, \min, +, \infty, 0 \rangle$, $\langle \mathbb{R}_+^\infty, \min, +, \infty, 0 \rangle$ are called *tropical semirings*. Similarly, the semirings $\langle \overline{\mathbb{N}}, \max, +, -\infty, 0 \rangle$ and $\langle \overline{\mathbb{R}}_+, \max, +, -\infty, 0 \rangle$ are called *max-plus semirings* or *arctic semirings*. A further example is provided by the semiring $\langle [0, 1], \max, \cdot, 0, 1 \rangle$.

Let Σ be a finite alphabet. Then each subset of Σ^* is called *formal language* over Σ. We define, for formal languages $L_1, L_2 \subseteq \Sigma^*$, the *product* of L_1 and L_2 by

$$L_1 \cdot L_2 = \{w_1 w_2 \mid w_1 \in L_1, w_2 \in L_2\}.$$

Then $\langle 2^{\Sigma^*}, \cup, \cdot, \emptyset, \{\varepsilon\}\rangle$ is a semiring, called the *semiring of formal languages over* Σ. Here 2^U denotes the power set of a set U and $\emptyset$ denotes the empty set.

If U is a set, $2^{U \times U}$ is the set of binary relations over U. Define, for two relations R_1 and R_2, the product $R_1 \cdot R_2 \subseteq U \times U$ by

$$R_1 \cdot R_2 = \{(u_1, u_2) \mid \text{there exists an } u \in U \text{ such that} \\ (u_1, u) \in R_1 \text{ and } (u, u_2) \in R_2\}$$

and, furthermore, define

$$\Delta = \{(u, u) \mid u \in U\}.$$

Then $\langle 2^{U \times U}, \cup, \cdot, \emptyset, \Delta\rangle$ is a semiring, called the *semiring of binary relations over* U.

We now consider morphisms between semirings. Let A and A' be semirings. Then a mapping $h : A \to A'$ is a *morphism* from A into A' if $h(0) = 0$, $h(1) = 1$, $h(s_1 + s_2) = h(s_1) + h(s_2)$ and $h(s_1 \cdot s_2) = h(s_1) \cdot h(s_2)$ for all $s_1, s_2 \in A$. That is, a morphism of semirings is a mapping that preserves the semiring operations and constants. A bijective morphism is called an *isomorphism*. For instance, the semirings $\langle \mathbb{R}_+^\infty, \min, +, \infty, 0\rangle$ and $\langle [0,1], \max, \cdot, 0, 1\rangle$ are isomorphic via the mapping $x \mapsto e^{-x}$, and the semiring $\langle \mathbb{R}_+^\infty, \max, \min, 0, \infty\rangle$ is isomorphic to $\langle [0,1], \max, \min, 0, 1\rangle$ via the mapping $x \mapsto 1 - e^{-x}$.

A semiring $\langle A, +, \cdot, 0, 1\rangle$ is called *ordered* if $\langle A, +, 0\rangle$ is an ordered monoid and multiplication preserves the order. When the order on A is the natural order, $\langle A, +, \cdot, 0, 1\rangle$ is automatically an ordered semiring. A morphism of ordered semirings is an order preserving semiring morphism.

A semiring $\langle A, +, \cdot, 0, 1\rangle$ is called *continuous* if $\langle A, +, 0\rangle$ is a continuous monoid and if multiplication is continuous, i.e.,

$$a \cdot (\sup_{i \in I} a_i) = \sup_{i \in I}(a \cdot a_i) \quad \text{and} \quad (\sup_{i \in I} a_i) \cdot a = \sup_{i \in I}(a_i \cdot a)$$

for all directed sets $\{a_i \mid i \in I\}$. It follows that the distribution laws hold for infinite sums:

$$a \cdot (\sum_{i \in I} a_i) = \sum_{i \in I}(a \cdot a_i) \quad \text{and} \quad (\sum_{i \in I} a_i) \cdot a = \sum_{i \in I}(a_i \cdot a)$$

for all families $(a_i \mid i \in I)$.

A morphism of continuous semirings is a semiring morphism which is a continuous function. Note that every continuous semiring is an ordered semiring and every continuous semiring morphism is an ordered semiring morphism.

Corollary 1. *Any continuous semiring is complete.*

We now define formal power series (for expositions, see Salomaa, Soittola [19], Kuich, Salomaa [15], Berstel, Reutenauer [1], Sakarovitch [18], Ésik, Kuich [7], Droste, Kuich [5]). Let Σ be an alphabet and A a semiring. Mappings r from Σ^* into A are called *(formal) power series*. The values of r are denoted by (r, w), where $w \in \Sigma^*$, and r itself is written as a formal sum

$$r = \sum_{w \in \Sigma^*} (r, w)w.$$

The values (r, w) are also referred to as the *coefficients* of the series. The collection of all power series r as defined above is denoted by $A\langle\langle\Sigma^*\rangle\rangle$.

This terminology reflects the intuitive ideas connected with power series. We call the power series "formal" to indicate that we are not interested in summing up the series but rather, for instance, in various operations defined for series.

Given $r \in A\langle\langle\Sigma^*\rangle\rangle$, the *support* of r is the set

$$\mathrm{supp}(r) = \{w \in \Sigma^* \mid (r, w) \neq 0\}.$$

A series $r \in A\langle\langle\Sigma^*\rangle\rangle$, where every coefficient equals 0 or 1, is termed the *characteristic series* of its support L, in symbols, $r = \mathrm{char}(L)$. The subset of $A\langle\langle\Sigma^*\rangle\rangle$ consisting of all series with a finite support is denoted by $A\langle\Sigma^*\rangle$. Series of $A\langle\Sigma^*\rangle$ are referred to as *polynomials*. It will be convenient to use the notations $A\langle\Sigma \cup \{\varepsilon\}\rangle$, $A\langle\Sigma\rangle$ and $A\langle\{\varepsilon\}\rangle$ for the collection of polynomials having their supports in $\Sigma \cup \{\varepsilon\}$, Σ and $\{\varepsilon\}$, respectively.

Examples of polynomials belonging to $A\langle\Sigma^*\rangle$ for every S are 0 and aw, where $a \in A$ and $w \in \Sigma^*$, defined by:

$$(0, w) = 0 \text{ for all } w,$$
$$(aw, w) = a \text{ and } (aw, w') = 0 \text{ for } w \neq w'.$$

Often $1w$ is denoted by w.

We now introduce several operations on power series. For $r_1, r_2, r \in S\langle\langle\Sigma^*\rangle\rangle$ and $s \in A$ we define the *sum* $r_1 + r_2$, the *(Cauchy) product* $r_1 \cdot r_2$, the *Hadamard product* $r_1 \odot r_2$ and *scalar products* sr, rs, each as a series belonging to $A\langle\langle\Sigma^*\rangle\rangle$, as follows:

- $(r_1 + r_2, w) = (r_1, w) + (r_2, w)$
- $(r_1 \cdot r_2, w) = \sum_{w_1 w_2 = w} (r_1, w_1)(r_2, w_2)$
- $(r_1 \odot r_2, w) = (r_1, w)(r_2, w)$
- $(sr, w) = s(r, w)$
- $(rs, w) = (r, w)s$

for all $w \in \Sigma^*$.

It can be checked that $\langle A\langle\langle\Sigma^*\rangle\rangle, +, \cdot, 0, \varepsilon\rangle$ and $\langle A\langle\Sigma^*\rangle, +, \cdot, 0, \varepsilon\rangle$ are semirings, the semirings of formal power series resp. of polynomials over Σ and A.

We just note that the structure $\langle A\langle\langle\Sigma^*\rangle\rangle, +, \odot, 0, \mathrm{char}(\Sigma^*)\rangle$ is also a semiring (the full Cartesian product of Σ^* copies of the semiring $\langle A, +, \cdot, 0, 1\rangle$).

Clearly, the formal language semiring $\langle 2^{\Sigma^*}, \cup, \cdot, \emptyset, \{\varepsilon\}\rangle$ is isomorphic to $\langle \mathbb{B}\langle\langle \Sigma^*\rangle\rangle, +, \cdot, 0, \varepsilon\rangle$. Essentially, a transition from 2^{Σ^*} to $\mathbb{B}\langle\langle \Sigma^*\rangle\rangle$ and vice versa means a transition from L to $\mathrm{char}(L)$ and from r to $\mathrm{supp}(r)$, respectively.

We now introduce (possibly infinite) matrices. Consider two non-empty index sets I and I' and a set S. Mappings M of $I \times I'$ into S are called *matrices*. The values of M are denoted by $M_{i,i'}$, where $i \in I$ and $i' \in I'$. The values $M_{i,i'}$ are also referred to as the *entries* of the matrix M. In particular, $M_{i,i'}$ is called the (i, i')-*entry* of M. The collection of all matrices as defined above is denoted by $S^{I \times I'}$.

If both I and I' are finite, then M is called a *finite matrix*. If I or I' is a singleton, M is called a *row* or *column vector*, respectively. If $M \in S^{I \times 1}$ (resp. $M \in S^{1 \times I'}$) then we often denote the i-th entry of M, $i \in I$ (resp. $i \in I'$), by M_i instead of $M_{i,1}$ (resp. $M_{1,i}$).

Let A be a semiring. For each $i \in I$ and $i' \in I'$, consider the sets of indices $R(i) = \{i' \mid M_{i,i'} \neq 0\}$ and $C(i') = \{i \mid M_{i,i'} \neq 0\}$, respectively. Then $M \in A^{I \times I'}$ is called a *row finite* (resp. *column finite*) matrix if $R(i)$ (resp. $C(i')$) is finite for all $i \in I$ (resp. $i' \in I'$). The collection of all row finite (resp. column finite) matrices defined above is denoted by $A_R^{I \times I'}$ (resp. $A_C^{I \times I'}$).

We introduce some operations and special matrices inducing a monoid or semiring structure to matrices. For $M_1, M_2 \in A^{I \times I'}$ we define the *sum* $M_1 + M_2 \in A^{I \times I'}$ by $(M_1 + M_2)_{i,i'} = (M_1)_{i,i'} + (M_2)_{i,i'}$ for all $i \in I$, $i' \in I'$. Furthermore, we introduce the *zero matrix* $0 \in A^{I \times I'}$. All entries of the zero matrix 0 are 0. By these definitions, $\langle A^{I \times I'}, +, 0\rangle$ is a commutative monoid.

If M_1 is row finite or if M_2 is column finite or if A is complete, then, for $M_1 \in A^{I_1 \times I_2}$ and $M_2 \in A^{I_2 \times I_3}$, we define the *product* $M_1 M_2 \in A^{I_1 \times I_3}$ by

$$(M_1 M_2)_{i_1, i_3} = \sum_{i_2 \in I_2} (M_1)_{i_1, i_2} (M_2)_{i_2, i_3} \quad \text{for all } i_1 \in I_1, i_3 \in I_3.$$

Furthermore, we introduce the *matrix of unity* $E \in A^{I \times I}$. The diagonal entries $E_{i,i}$ of E are equal to 1, the off-diagonal entries E_{i_1, i_2}, $i_1 \neq i_2$, of E are equal to 0, $i, i_1, i_2 \in I$.

It is easily shown that matrix multiplication is associative, the distribution laws are valid for matrix addition and multiplication, E is a multiplicative unit and 0 is a multiplicative zero. So we infer that $\langle A_R^{I \times I}, +, \cdot, 0, E\rangle$ and $\langle A_C^{I \times I}, +, \cdot, 0, E\rangle$ are semirings and that $\langle A^{I \times I}, +, \cdot, 0, E\rangle$ is a semiring if A is complete.

If A is complete, infinite sums can be extended to matrices. Consider $A^{I \times I'}$ and define, for $M_j \in A^{I \times I'}$, $j \in J$, where J is an index set, $\sum_{j \in J} M_j$ by its entries:

$$\left(\sum_{j \in J} M_j \right)_{i,i'} = \sum_{j \in J} (M_j)_{i,i'}, \quad i \in I, i' \in I'.$$

By this definition, $A^{I \times I}$ is a complete semiring.

If A is ordered, the order on A is extended pointwise to matrices M_1 and M_2 in $A^{I \times I'}$:

$$M_1 \leq M_2 \quad \text{iff} \quad (M_1)_{i,i'} \leq (M_2)_{i,i'} \text{ for all } i \in I, i' \in I'.$$

If A is continuous then so is $A^{I \times I}$.

We now introduce blocks of matrices. Consider a matrix M in $A^{I \times I}$. Assume the existence of a non-empty index set J and of non-empty index sets I_j for $j \in J$ such that $I = \bigcup_{j \in J} I_j$ and $I_{j_1} \cap I_{j_2} = \emptyset$ for $j_1 \neq j_2$. The mapping M, restricted to the domain $I_{j_1} \times I_{j_2}$, i. e., $M : I_{j_1} \times I_{j_2} \to A$ is, of course, a matrix in $A^{I_{j_1} \times I_{j_2}}$. We denote it by $M(I_{j_1}, I_{j_2})$ and call it the (I_{j_1}, I_{j_2})-*block* of M.

We can compute the blocks of the sum and the product of matrices M_1 and M_2 (if $M_1 M_2$ is defined) from the blocks of M_1 and M_2 in the usual way:

$$(M_1 + M_2)(I_{j_1}, I_{j_2}) = M_1(I_{j_1}, I_{j_2}) + M_2(I_{j_1}, I_{j_2}),$$
$$(M_1 M_2)(I_{j_1}, I_{j_2}) = \sum_{j \in J} M_1(I_{j_1}, I_j) M_2(I_j, I_{j_2}).$$

In a similar manner the matrices of $A^{I \times I'}$ can be partitioned into blocks. This yields the computational rule

$$(M_1 + M_2)(I_j, I'_{j'}) = M_1(I_j, I'_{j'}) + M_2(I_j, I'_{j'}).$$

If we consider matrices $M_1 \in A^{I \times I'}$ and $M_2 \in A^{I' \times I''}$ partitioned into compatible blocks, i. e., I' is partitioned into the same index sets for both matrices, then we obtain the computational rule

$$(M_1 M_2)(I_j, I''_{j''}) = \sum_{j' \in J'} M_1(I_j, I'_{j'}) M_2(I'_{j'}, I''_{j''}).$$

If the semiring A is complete, there exist the following isomorphisms:

(i) The semirings

$$(A^{Q \times Q})^{I \times I}, \quad A^{(I \times Q) \times (I \times Q)}, \quad A^{(Q \times I) \times (Q \times I)}, \quad (A^{I \times I})^{Q \times Q}$$

   are isomorphic by the correspondences between

$$(M_{i_1, i_2})_{q_1, q_2}, \quad M_{(i_1, q_1), (i_2, q_2)}, \quad M_{(q_1, i_1), (q_2, i_2)}, \quad (M_{q_1, q_2})_{i_1, i_2}$$

   for all $i_1, i_2 \in I$, $q_1, q_2 \in Q$.

(ii) The semirings $A^{I \times I} \langle\!\langle \Sigma^* \rangle\!\rangle$ and $(A \langle\!\langle \Sigma^* \rangle\!\rangle)^{I \times I}$ are isomorphic by the correspondence between $(M, w)_{i_1, i_2}$ and (M_{i_1, i_2}, w) for all $i_1, i_2 \in I$, $w \in \Sigma^*$.

Observe that these correspondences are isomorphisms of complete semirings, i. e., they respect infinite sums. We will use these isomorphisms without further mention. Moreover, we will use the notation M_{i_1, i_2}, $i_1 \in I_1$, $i_2 \in I_2$, where $M \in A^{I_1 \times I_2} \langle\!\langle \Sigma^* \rangle\!\rangle$: M_{i_1, i_2} is the power series in $A \langle\!\langle \Sigma^* \rangle\!\rangle$ such that the coefficient (M_{i_1, i_2}, w) of $w \in \Sigma^*$ is equal to $(M, w)_{i_1, i_2}$. Similarly, we will use the notation (M, w), $w \in \Sigma^*$, where $M \in (A \langle\!\langle \Sigma^* \rangle\!\rangle)^{I_1 \times I_2}$: (M, w) is the matrix in $A^{I_1 \times I_2}$ whose (i_1, i_2)-entry $(M, w)_{i_1, i_2}$, $i_1 \in I_1$, $i_2 \in I_2$, is equal to (M_{i_1, i_2}, w).

Furthermore, with analogous correspondences, there exist the following isomorphisms:

(i) The semirings $(A^{Q\times Q})_R^{I\times I}$, $A_R^{(I\times Q)\times(I\times Q)}$, $A_R^{(Q\times I)\times(Q\times I)}$ and $(A_R^{I\times I})^{Q\times Q}$ and the semirings $(A^{Q\times Q})_C^{I\times I}$, $A_C^{(I\times Q)\times(I\times Q)}$, $A_C^{(Q\times I)\times(Q\times I)}$ and $(A_C^{I\times I})^{Q\times Q}$ are isomorphic.

(ii) The semiring $(A\langle\!\langle\Sigma^*\rangle\!\rangle)_R^{I\times I}$ is isomorphic to a subsemiring of $A_R^{I\times I}\langle\!\langle\Sigma^*\rangle\!\rangle$ and the semiring $(A\langle\!\langle\Sigma^*\rangle\!\rangle)_C^{I\times I}$ is isomorphic to a subsemiring of $A_C^{I\times I}\langle\!\langle\Sigma^*\rangle\!\rangle$.

2   Convergence

We introduce an axiomatic definition for the notion of convergence of a special type due to Kuich, Salomaa [15]. The notion is particularly suitable for handling equations arising in automata theory. It also gives rise to some important identities needed later on.

A mapping $\alpha : \mathbb{N} \to A$ is called a *sequence* in A. By $A^{\mathbb{N}}$ we denote the set of all such sequences. If $\alpha \in A^{\mathbb{N}}$ we use the notation $\alpha = (\alpha(n))$.

We denote by o and η the sequences defined by $o(n) = 0$ and $\eta(n) = 1$, for all $n \geq 0$, respectively. For $\alpha \in A^{\mathbb{N}}$, $c \in A$, we define $c\alpha$ and αc in $A^{\mathbb{N}}$ by $(c\alpha)(n) = c\alpha(n)$ and $(\alpha c)(n) = \alpha(n)c$, for all $n \geq 0$, respectively. For $\alpha_1, \alpha_2 \in A^{\mathbb{N}}$, we define $\alpha_1 + \alpha_2$ and $\alpha_1 \cdot \alpha_2$ by $(\alpha_1 + \alpha_2)(n) = \alpha_1(n) + \alpha_2(n)$ and $(\alpha_1 \cdot \alpha_2)(n) = \alpha_1(n)\alpha_2(n)$, for all $n \geq 0$, respectively.

Observe that $\langle A^{\mathbb{N}}, +, \cdot, o, \eta\rangle$ is a semiring. We need one further operation before giving the basic definitions of convergence.

Consider $\alpha \in A^{\mathbb{N}}$ and $a \in A$. Then $\alpha_a \in A^{\mathbb{N}}$ denotes the sequence defined by

$$\alpha_a(0) = a, \quad \alpha_a(n+1) = \alpha(n), \text{ for all } n \geq 0.$$

A *convergence* is given by a pair $\langle D, \lim\rangle$, where D is a set of convergent sequences and $\lim : D \to A$ is a limit function. Here, each set $D \subseteq A^{\mathbb{N}}$ satisfying the conditions (D1)–(D3) is called a *set of convergent sequences* in A.

(D1) $\eta \in D$.

(D2)   (i) If $\alpha_1, \alpha_2 \in D$ then $\alpha_1 + \alpha_2 \in D$.
(ii) If $\alpha \in D$ and $c \in A$ then $c\alpha, \alpha c \in D$.

(D3)   If $\alpha \in D$ and $a \in A$ then $\alpha_a \in D$.

Furthermore, each mapping $\lim : D \to A$ satisfying the following conditions (lim1)–(lim3) is called a *limit function* (on D).

(lim1) $\lim\eta = 1$.

(lim2)   (i) If $\alpha_1, \alpha_2 \in D$ then $\lim(\alpha_1 + \alpha_2) = \lim\alpha_1 + \lim\alpha_2$.
(ii) If $\alpha \in D$ and $c \in A$ then $\lim c\alpha = c\lim\alpha$ and $\lim(\alpha c) = (\lim\alpha)c$.

(lim3)   If $\alpha \in D$ and $a \in A$ then $\lim\alpha_a = \lim\alpha$.

Observe that, for all $c \in A$, the sequence $c\eta = \eta c$ is convergent independently of D and converges to c.

In what follows we often use the term "convergence in A" without explicitly specifying D and $\lim$. Sets of convergent sequences will be considered only in the case that a limit function is defined. We also use the notation $\lim_{n\to\infty} \alpha(n)$ for $\lim\alpha$. Let h be an isomorphism between A_1 and A_2, extended in the natural fashion to an isomorphism between $A_1^{\mathbb{N}}$ and $A_2^{\mathbb{N}}$. If, moreover, the sets of convergent sequences D_1 and D_2 and the limit functions $\lim_1$ and $\lim_2$ are corresponding with respect to this isomorphism then isomorphic sequences have isomorphic limits.

A notion of convergence definable for every A, referred to as *discrete convergence*, will now be discussed. This notion of convergence is the classical one considered in connection with semirings. The discrete convergence is given by the pair $\langle D_d, \lim_d \rangle$, where $D_d = \{\alpha \in A^{\mathbb{N}} \mid \text{there exists an } n_\alpha \geq 0 \text{ such that, for all } k \geq 0, \alpha(n_\alpha + k) = \alpha(n_\alpha)\}$ and $\lim_d \alpha = \alpha(n_\alpha)$.

Theorem 1. (Kuich, Salomaa [15]) D_d *is the smallest subset of $A^{\mathbb{N}}$ for which (D1)–(D3) are satisfied. If $\langle D, \lim \rangle$ is a convergence then $D_d \subseteq D$ and $\lim \alpha = \lim_d \alpha$ for all $\alpha \in D_d$.*

Example 2.1. A sequence $\alpha \in \mathbb{R}^{\mathbb{N}}$ is called a *Cauchy sequence* if for all $\varepsilon > 0$, there exists an $n_\varepsilon \geq 0$ such that $|\alpha(n_1) - \alpha(n_2)| < \varepsilon$ holds for all $n_1, n_2 \geq n_\varepsilon$.

One possible choice for the set D of convergent sequences in $\mathbb{R}$ is the set of Cauchy sequences with the usual convergence in $\mathbb{R}$. This notion of convergence in $\mathbb{R}$ is called the *Cauchy convergence*.   □

Example 2.2. A sequence $\alpha \in \mathbb{R}^{\mathbb{N}}$ is called an *Euler sequence* if the sequence $\left(\sum_{0 \leq j \leq n} \binom{n}{j} \alpha(j)/2^n \right)$ is a Cauchy sequence.

One possible choice for the set D of convergent sequences in $\mathbb{R}$ is the set of Euler sequences with the following notion of convergence:

$$\lim_E \alpha = \lim_{n \to \infty} \sum_{0 \leq j \leq n} \binom{n}{j} \alpha(j)/2^n \,,$$

where the limit on the right side denotes the Cauchy convergence.

This notion of convergence in $\mathbb{R}$ is called the *Euler convergence*. It can be shown that, for all $a \in \mathbb{R}$ with $-3 < a < 1$, $\lim_E \sum_{0 \leq j \leq n} a^i = 1/(1 - a)$ and $\lim_E a^n = 0$.

Let $a = -1$. Then $1, -1, 1, -1, \ldots$ converges to 0 and $1, 0, 1, 0, \ldots$ converges to $\frac{1}{2}$. (The reader is referred to Knopp [12] for more information concerning this example.)   □

Example 2.3. Let A be a continuous semiring. D_{sup} is the set of sequences that are ultimately nondecreasing sequences, i. e., $\alpha \in D_{\text{sup}}$ iff there exists an n_α such that $\alpha(n_\alpha + k) \leq \alpha(n_\alpha + k + 1)$ for all $k \geq 0$. Since in an continuous semiring addition and multiplication are continuous, the function $\lim_{\text{sup}} : D_{\text{sup}} \to A$ defined by $\lim_{\text{sup}} \alpha = \sup\{\alpha(n_\alpha + k) \mid k \geq 0\}$ is a limit function. $\langle D_{\text{sup}}, \lim_{\text{sup}} \rangle$ is called *supremum convergence*.   □

Example 2.4. Let A be a complete semiring. Define $D_{\text{compl}} \subseteq A^{\mathbb{N}}$ by $\beta \in D_{\text{compl}}$ if there exist an $\alpha \in A^{\mathbb{N}}$ and an n_β such that $\beta(n_\beta + k) = \sum_{0 \leq i < k} \alpha(i)$. The function $\lim_{\text{compl}} : D_{\text{compl}} \to A$ defined by $\lim_{\text{compl}} \beta = \sum_{i \geq 0} \alpha(i)$ is a limit function. $\langle D_{\text{compl}}, \lim_{\text{compl}} \rangle$ is called *complete convergence*.

Each continuous semiring is complete (see Ésik, Kuich [7]) and if A is continuous then $\lim_{\sup}\beta = \lim_{\mathrm{compl}}\beta$ for all $\beta \in D_{\mathrm{compl}}$.   □

A convergence $\langle D, \lim \rangle$ in A is called *strong* if the condition

$$\left(\sum_{0 \leq j \leq n} a^j \alpha(n-j) \right) \in D$$

is satisfied for each $\alpha \in D$ and each $a \in A$ such that $(a^n) \in D$ and $\lim_{n \to \infty} a^n = 0$. If $(a^n) \in D$ with $\lim_{n \to \infty} a^n = 0$ then a^* exists.

Theorem 2. (Kuich, Salomaa [15]) *The discrete convergence, the Cauchy convergence and the Euler convergence are strong convergences.*

A convergence $\langle D, \lim \rangle$ in A is called *multiplicative* if, for all $\alpha, \beta \in D$, $\alpha \cdot \beta \in D$ and $\lim(\alpha \cdot \beta) = (\lim \alpha) \cdot (\lim \beta)$.

Theorem 3. *The discrete convergence, the Cauchy convergence, the supremum convergence and the complete convergence are multiplicative convergences.*

Proof. The proof is obvious for the discrete convergence. The Cauchy convergence is multiplicative by Knopp [12], Theorem II.8.10. Since multiplication in a continuous semiring is continuous, the supremum convergence is multiplicative. Since multiplication of sums is distributive in complete semirings, the complete convergence is multiplicative.   □

The powers a^i, $i \geq 0$, of an element a in a semiring A are defined in the natural way, whereby $a^0 = 1$.

If $(\sum_{1 \leq j \leq n} a^i) \in D$ then we write $\lim_{n \to \infty} \sum_{1 \leq j \leq n} a^i = a^+$.

If $(\sum_{0 \leq j \leq n} a^i) \in D$ then we write $\lim_{n \to \infty} \sum_{0 \leq j \leq n} a^i = a^*$ and call it the *star* of a (with respect to the given notion of convergence).

(For $n = 0$ the range of j is empty in the sum defining a^+. In this case we consider the sum to be equal to 0.)

The next theorem shows the close interconnection between a^+ and a^*.

Theorem 4. (Kuich, Salomaa [15]) *Let $a \in A$. Then a^* exists iff a^+ exists, and $1 + a^+ = a^*$, $aa^* = a^*a = a^+$.*

Corollary 2. *If a^* exists then*

$$a^* = \sum_{0 \leq j \leq n} a^j + a^{n+1}a^* \quad and \quad a^* = \sum_{0 \leq j \leq n} a^j + a^*a^{n+1}$$

for all $n \geq 0$.

If the complete convergence in a complete semiring A is considered, a^* and a^+ exist for all $a \in A$:

$$a^* = \sum_{j \geq 0} a^j, \qquad a^+ = \sum_{j \geq 1} a^j.$$

The same holds true for continuous semirings.

We now consider *equations* of the form

$$y = ay + b, \quad a, b \in A, \tag{$*$}$$

where y is a variable. An element $s \in A$ is called a *solution* of (*) iff $s = as + b$.

Theorem 5. (Kuich, Salomaa [15]) *If a^* exists then $s = a^*b$ is a solution of (*). If, moreover, $\lim_{n \to \infty} a^n = 0$ then s is the unique solution of (*).*

Example 2.5. Let $y = ay + 1$, $-3 < a < 1$. We work with the Euler convergence. Then $\lim_{n \to \infty} a^n = 0$ and $a^* = 1/(1-a)$. Hence, the unique solution of $y = ay + 1$ is $1/(1-a)$. □

Theorem 6. *Let A be a continuous semiring and consider the supremum convergence. Then $s = a^*b$ is the least solution of (*).*

Proof. By a well-known fixed point theorem, see, e. g., Bloom, Ésik [2], Guessarian [10], Ésik, Kuich [7], Theorem 2.9. □

We now turn to the discussion of some important identities. The letters a, b stand for elements of A.

Theorem 7. (Kuich, Salomaa [15]) *(i) Assume the existence of $(a+b)^*$, a^* and $(a^*b)^*$ and, furthermore, $\lim_{n \to \infty}(a + b)^n = 0$. Then the* sum-star-identity

$$(a + b)^* = (a^*b)^*a^*$$

holds for a and b.

(ii) $(ab)^*$ exists if $(ba)^*$ exists. Whenever $(ab)^*$ exists, the product-star-identity

$$(ab)^* = 1 + a(ba)^*b$$

holds for a and b.

We now will show how to transfer a notion of convergence in A into $A\langle\langle \Sigma^* \rangle\rangle$.

Observe first that $A^{\mathbb{N}}\langle\langle \Sigma^* \rangle\rangle$ and $(A\langle\langle \Sigma^* \rangle\rangle)^{\mathbb{N}}$ are isomorphic. This isomorphism will be used in the notation below without further mention. It also follows that $D\langle\langle \Sigma^* \rangle\rangle$ can be considered as a subset of the set of sequences $(A\langle\langle \Sigma^* \rangle\rangle)^{\mathbb{N}}$.

Our next theorem shows explicitly how a notion of convergence in A can be transferred to $A\langle\langle \Sigma^* \rangle\rangle$. The main idea is that a sequence of power series determines, for each $w \in \Sigma^*$, a sequence of coefficients of w. The limits of the latter sequences determine the coefficients in the limit of our sequence of power series.

Theorem 8. (Kuich, Salomaa [15]) *Assume that $\lim : D \to A$ is a limit function. Then also the mapping $\lim : D\langle\langle \Sigma^* \rangle\rangle \to A\langle\langle \Sigma^* \rangle\rangle$ defined by*

$$\lim \alpha = \sum_{w \in \Sigma^*} \lim(\alpha, w)w$$

is a limit function.

In the sequel we will use the method of Theorem 8 to transfer the notion of convergence from A to $A\langle\langle \Sigma^* \rangle\rangle$ unless stated otherwise.

The next theorem is similar to Theorem 1.

Theorem 9. (Kuich, Salomaa [15]) *Assume that* $\lim_d \alpha = \sum_{w \in \Sigma^*} \lim_d(\alpha, w)w$ *and* $\lim \alpha = \sum_{w \in \Sigma^*} \lim(\alpha, w)w$ *are limit functions on* $D_d\langle\langle \Sigma^* \rangle\rangle$ *and* $D\langle\langle \Sigma^* \rangle\rangle$, *respectively. If* $\alpha \in D_d\langle\langle \Sigma^* \rangle\rangle$ *then* $\lim \alpha = \lim_d \alpha$.

For $k \geq 0$, we consider the truncation operator R_k defined for power series $r \in A\langle\langle \Sigma^* \rangle\rangle$ by

$$R_k(r) = \sum_{|w| \leq k} (r, w)w \, .$$

Consider the discrete convergence in A transferred to $A\langle\langle \Sigma^* \rangle\rangle$. Then, for $\alpha \in (A\langle\langle \Sigma^* \rangle\rangle)^{\mathbb{N}}$, $\alpha \in D_d\langle\langle \Sigma^* \rangle\rangle$ and $\lim \alpha = r$ iff for all $k \geq 0$, there exists an $m_k \geq 0$ such that, for all $j \geq 0$, $R_k(\alpha(m_k + j)) = R_k(\alpha(m_k))$ and then $(r, w) = (R_k(\alpha(m_k)), w)$ for all $w \in \Sigma^*$.

Corollary 3. *Consider a convergence* $\langle D, \lim \rangle$ *in* A *transferred to a convergence* $\langle D\langle\langle \Sigma^* \rangle\rangle, \lim \rangle$ *in* $A\langle\langle \Sigma^* \rangle\rangle$. *Assume that, for* $\alpha \in (A\langle\langle \Sigma^* \rangle\rangle)^{\mathbb{N}}$, *there exists, for all* $k \geq 0$, *an* $m_k \geq 0$ *such that, for all* $j \geq 0$, $R_k(\alpha(m_k + j)) = R_k(\alpha(m_k))$. *Then* $\alpha \in D\langle\langle \Sigma^* \rangle\rangle$ *and* $\lim \alpha = r$, *where* $(r, w) = (R_k(\alpha(m_k)), w)$ *for all* $w \in \Sigma^*$.

Proof. By Theorem 9. □

Theorem 10. (Kuich, Salomaa [15]) *A strong convergence in* A *transferred to a convergence in* $A\langle\langle \Sigma^* \rangle\rangle$ *is again strong.*

Theorem 11. *A multiplicative convergence in* A *transferred to a convergence in* $A\langle\langle \Sigma^* \rangle\rangle$ *is again multiplicative.*

Proof. We use for the limit functions in A and in $A\langle\langle \Sigma^* \rangle\rangle$ the same notation $\lim$ and get, for convergent seqences α and β,

$$\lim\alpha\lim\beta = (\textstyle\sum_{w_1 \in \Sigma^*} \lim(\alpha, w_1)w_1)(\sum_{w_2 \in \Sigma^*} \lim(\beta, w_2)w_2) =$$
$$\textstyle\sum_{w_1, w_2 \in \Sigma^*} \lim(\alpha, w_1)\lim(\beta, w_2)w_1 w_2 =$$
$$\textstyle\sum_{w \in \Sigma^*} (\sum_{w = w_1 w_2} \lim(\alpha, w_1)\lim(\beta, w_2))w =$$
$$\textstyle\sum_{w \in \Sigma^*} \lim(\sum_{w = w_1 w_2} (\alpha, w_1)(\beta, w_2))w =$$
$$\textstyle\sum_{w \in \Sigma^*} \lim(\alpha\beta, w)w = \lim\alpha\beta \, .$$

□

A power series $r \in A\langle\langle \Sigma^* \rangle\rangle$ is called *proper* if $(r, \varepsilon) = 0$. For a power series r, the power series $\sum_{w \in \Sigma^+} (r, w)w$ is called the *proper part* of r.

Theorem 12. (Kuich, Salomaa [15]) *If* $r \in A\langle\langle \Sigma^* \rangle\rangle$ *is proper then* $\lim_{n \to \infty} r^n = 0$ *and* r^* *exists. Moreover,*

$$r^* = \sum_{w \in \Sigma^*} (\sum_{0 \leq j \leq |w|} r^j, w)w \, .$$

In case of a strong convergence, the coefficient of ε has a great influence on the convergence behavior of a power series and gives rise to the definition of a cycle-free power series: a power series is called *cycle-free* (with respect to the given strong convergence) if $\lim_{n\to\infty}(r,\varepsilon)^n = 0$.

Theorem 13. (Kuich, Salomaa [15]) *Assume a strong convergence in A. For each cycle-free power series r, $\lim_{n\to\infty} r^n = 0$ and r^* exists.*

We transfer the term cycle-free to equations. An equation

$$ y = ry + s, \quad r, s \in A\langle\!\langle \Sigma^* \rangle\!\rangle, \qquad (*) $$

is termed *cycle-free* if r is cycle-free.

Theorem 14. (Kuich, Salomaa [15]) *Assume a strong convergence in A. Every cycle-free equation (*) has the unique solution r^*s.*

Theorem 15. (Kuich, Salomaa [15]) *Assume a strong convergence in A. For each cycle-free power series r,*

$$ r^* = (r_0^* r_1)^* r_0^* = r_0^* (r_1 r_0^*)^*, $$

where $r_0 = (r\varepsilon)\varepsilon$ and r_1 is the proper part of r.

We now turn to infinite matrices.

Assume that A is provided with a notion of convergence. We now transfer this notion of convergence in A to $A_C^{I\times I}$ where I is an arbitrary index set.

We define the set $D_C \subseteq (A_C^{I\times I})^{\mathbb{N}}$ of convergent sequences in $A_C^{I\times I}$ as follows. A sequence μ is in D_C iff

(i) for all $i, j \in I$, the sequence $\mu_{i,j}$ is in D,
(ii) for all $j \in I$, there exists a finite set $I(j) \subseteq I$ such that $\mu_{i,j} = o$, for all $i \in I \setminus I(j)$.

Intuitively, a sequence $\mu \in A_C^{I\times I}$ is convergent iff $\mu_{i,j}$ is a convergent sequence in A, for all $i \in I(j)$ and $j \in I$, and $\mu_{i,j} = o$ for all $i \in I \setminus I(j)$ and $j \in I$.

Theorem 16. (Kuich, Salomaa [15]) *Assume that $\lim : D \to A$ is a limit function. Then also the mapping $\lim_C : D_C \to A_C^{I\times I}$ defined by*

$$ (\lim_C \mu)_{i,j} = \lim \mu_{i,j} $$

is a limit function on D_C. Furthermore, $\langle D_C, \lim_C \rangle$ is a strong convergence if $\langle D, \lim \rangle$ is strong.

A clearifying remark about matrices in $(A\langle\!\langle \Sigma^* \rangle\!\rangle)_C^{I\times I}$ is now in order. Let M be in $(A\langle\!\langle \Sigma^* \rangle\!\rangle)_C^{I\times I}$ and let M' be its isomorphic copy in $A_C^{I\times I}\langle\!\langle \Sigma^* \rangle\!\rangle$. By definition, M is called *C-cycle-free* if its isomorphic copy M' is cycle-free with respect to the strong convergence $\langle D_C\langle\!\langle \Sigma^* \rangle\!\rangle, \lim \rangle$. Then M'^* is in $A_C^{I\times I}\langle\!\langle \Sigma^* \rangle\!\rangle$. But in general, the copy of M'^* in $(A\langle\!\langle \Sigma^* \rangle\!\rangle)^{I\times I}$ is not in $(A\langle\!\langle \Sigma^* \rangle\!\rangle)_C^{I\times I}$. (See Kuich, Salomaa [15], Theorem 4.2 and the discussion on pages 55/56.)

3   Algebraic Systems

In this section we consider, as a generalization of context-free grammars, algebraic systems. The defining equations are algebraic in the classical sense, i. e., are polynomial equations. Throughout this section, A denotes a *commutative* semiring.

An $A\langle \Sigma \cup \{\varepsilon\}\rangle$-*algebraic system* (briefly *algebraic system*) with *variables* in $Y = \{y, \ldots, y_n\}$, $Y \cap \Sigma = \emptyset$, is a system of equations

$$y_i = p_i, \quad 1 \le i \le n,$$

where each p_i is a polynomial in $A\langle (\Sigma \cup Y)^* \rangle$.

Defining the two column vectors

$$y = \begin{pmatrix} y_1 \\ \vdots \\ y_n \end{pmatrix} \quad \text{and} \quad p = \begin{pmatrix} p_1 \\ \vdots \\ p_n \end{pmatrix},$$

we can write our algebraic system in the matrix notation

$$y = p.$$

Intuitively, a solution to the algebraic system $y = p$ is given by n power series $\sigma_1, \ldots, \sigma_n$ in $A\langle\langle \Sigma^* \rangle\rangle$ "satisfying" the algebraic system in the sense that if each variable y_i is replaced by the series σ_i then valid equations result.

More formally, consider

$$\sigma = \begin{pmatrix} \sigma_1 \\ \vdots \\ \sigma_n \end{pmatrix} \in (A\langle\langle (\Sigma \cup Y)^* \rangle\rangle)^{n \times 1}.$$

Then we can define a morphism $h_\sigma : (\Sigma \cup Y)^* \to A\langle\langle (\Sigma \cup Y)^* \rangle\rangle$ by $h_\sigma(y_i) = \sigma_i$, $1 \le i \le n$ and $h_\sigma(x) = x$, $x \in \Sigma$.

As usual, extend h_σ to a mapping $h_\sigma : A\langle (\Sigma \cup Y)^* \rangle \to A\langle\langle (\Sigma \cup Y)^* \rangle\rangle$ by the definition

$$h_\sigma(p) = \sum_{\gamma \in (\Sigma \cup Y)^*} (p, \gamma) h_\sigma(\gamma),$$

where p is in $A\langle (\Sigma \cup Y)^* \rangle$. Observe that $h_\sigma(ap) = a h_\sigma(p)$, $h_\sigma(p + p') = h_\sigma(p) + h_\sigma(p')$ and $h_\sigma(p \cdot p') = h_\sigma(p) \cdot h_\sigma(p')$ for all $p, p' \in A\langle (\Sigma \cup Y)^* \rangle$ and $a \in A$. Hence, h_σ is a morphism. Furthermore, because p is a polynomial, we will have no difficulties with infinite sums.

A *solution* to the algebraic system $y_i = p_i$, $1 \le i \le n$, is given by a column vector $\sigma \in (A\langle\langle \Sigma^* \rangle\rangle)^{n \times 1}$ such that $\sigma_i = h_\sigma(p_i)$, $1 \le i \le n$.

Matrix notation can be used by extending the mapping h_σ entrywise to vectors and matrices. In this fashion, a solution to the algebraic system $y = p$ is given by a column vector σ such that $\sigma = h_\sigma(p)$.

An algebraic system $y_i = p_i$, $1 \leq i \leq n$, or $y = p$ is also written in the form $y_i = p_i(y_1, \ldots, y_n)$, $1 \leq i < n$, or $y = p(y)$, respectively. Then $h_\sigma(p_i)$ and $h_\sigma(p)$ can be written as $p_i(\sigma_1, \ldots, \sigma_n)$ or $p(\sigma)$, respectively. Sometimes, we write $p_i[\sigma_1|y_1, \ldots, \sigma_n|y_n]$ and $p[\sigma|y]$ for $h_\sigma(p_i)$ and $h_\sigma(p)$, respectively.

The *approximation sequence*

$$\sigma^0, \sigma^1, \sigma^2, \ldots, \sigma^j, \ldots, \quad \text{where each } \sigma^j \in (A\langle\!\langle \Sigma^* \rangle\!\rangle)^{n \times 1},$$

associated to an algebraic system $y = p(y)$ is defined as follows:

$$\sigma^0 = 0, \quad \sigma^{j+1} = p(\sigma^j), \; j \geq 0.$$

If the approximation sequence converges with respect to some given convergence in A, i.e.,

$$\lim_{j \to \infty} \sigma^j = \sigma,$$

then σ is referred to as the *strong solution* (with respect to the given convergence in A). In our next theorem we show that this strong solution is really a solution to the algebraic system if the given convergence is multiplicative. A lemma is needed before the theorem.

Lemma 1. *Let $\langle D, \lim \rangle$ be a multiplicative convergence in A. Then, for $\alpha \in (D\langle\!\langle (\Sigma \cup Y)^* \rangle\!\rangle)$ and $p(y) \in A\langle (\Sigma \cup Y)^* \rangle$,*

$$\lim_{j \to \infty} p[\alpha(j)|y] = p[\lim_{j \to \infty} \alpha(j)|y].$$

Proof. Consider first a term $t(y) = aw_0 y_{i_1} w_1 \ldots w_{k-1} y_{i_k} w_k$, $a \in A$, $w_i \in \Sigma^*$, $0 \leq i \leq k$, $k \geq 0$. Then we obtain

$$\lim_{j \to \infty}(t[\alpha(j)|y]) = \lim_{j \to \infty}(aw_0 \alpha_{i_1}(j)w_1 \ldots w_{k-1}\alpha_{i_k}(j)w_k) =$$
$$aw_0(\lim_{j \to \infty} \alpha_{i_1}(j))w_1 \ldots w_{k-1}(\lim_{j \to \infty} \alpha_{i_k}(j))w_k = t[\lim_{j \to \infty} \alpha(j)|y].$$

Since each polynomial in $A\langle (\Sigma \cup Y)^* \rangle$ is a finite sum of such terms, we obtain, for $p(y) = \sum_{1 \leq i \leq m} t_i(y)$,

$$\lim_{j \to \infty}(p[\alpha(j)|y]) = \lim_{j \to \infty}\Big(\sum_{1 \leq i \leq m} t_i[\alpha(j)|y]\Big) = \sum_{1 \leq i \leq m} \lim_{j \to \infty}(t_i[\alpha(j)|y]) =$$
$$\sum_{1 \leq i \leq m} t_i[\lim_{j \to \infty} \alpha(j)|y] = p[\lim_{j \to \infty} \alpha(j)|y].$$

$\square$

Theorem 17. *Let $\langle D, \lim \rangle$ be a multiplicative convergence in A. Consider an algebraic system $y = p(y)$ and its approximation sequence $\sigma^0 = 0$, $\sigma^{j+1} = p[\sigma^j|y]$, $j \geq 0$. Assume that this approximation sequence is convergent with limit σ. Then σ is a solution to $y = p(y)$.*

Proof. We obtain

$$p[\sigma|y] = p[\lim_{j\to\infty} \sigma^j|y] = \lim_{j\to\infty} (p[\sigma^j|y]) = \lim_{j\to\infty} \sigma^{j+1} = \lim_{j\to\infty} \sigma^j = \sigma\,.$$

Here we have applied Lemma 1 in the second equality and axiom (D3) in the fourth equality.   □

An algebraic system $y_i = p_i$, $1 \le i \le n$, is called *linear algebraic system* if $\text{supp}(p_i) \subseteq \Sigma^* Y \Sigma^* \cup \Sigma^*$. Observe now that Lemma 1 and hence Theorem 17 are valid for linear algebraic systems without the assumption that $\langle D, \lim\rangle$ is multiplicative.

Theorem 18. *Consider a linear algebraic system and assume that its approximation sequence is convergent. Then its limit is a solution to the linear algebraic system.*

In a continuous semiring, the approximation sequence of an algebraic system is always convergent in the supremum convergence and is the least solution of this algebraic system (see Ésik, Kuich [7], Theorem 2.9).

Theorem 19. *Consider an algebraic system $y = p(y)$ over a continuous semiring. Then its approximation sequence*

$$0, p(0), p^2(0), \ldots, p^j(0), \ldots$$

is convergent in the supremum convergence and its limit

$$\sup\{p^j(0) \mid j \ge 0\}$$

is the unique least solution of $y = p$.

In the theorem, $p^j(0)$ denotes the j-fold application of p to the vector 0.

A remark on equations over matrices is in order. Consider equations of the form

$$y = My + P, \quad M, P \in (A\langle\!\langle\!\langle(\Sigma\cup Y)^*\rangle\!\rangle\!\rangle)^{n\times n}\,. \tag{1}$$

Such an equation can be considered in two ways. On the one hand, it is an equation of the type (*) below Corollary 2. On the other hand, y can be considered to be a matrix of variables y_{ij}, $1 \le i, j \le n$, and the equation yields, for $1 \le j \le n$, the n linear systems

$$y_{ij} = \sum_{1\le k\le n} M_{ik} y_{kj} + P_{ij}, \quad 1 \le i \le n\,. \tag{2}$$

Clearly, each solution τ of (1) yields, for $1 \le j \le n$,

$$\tau_{ij} = \sum_{1\le k\le n} M_{ik}\tau_{kj} + P_{ij}, \quad 1 \le i \le n\,,$$

i.e., $(\tau_{ij})_{1\le i\le n}$ is a solution to the j-th linear system, $1 \le j \le n$, and vice versa. This means that, when dealing with solutions, we may consider the equation from either point of view.

Moreover, we consider equations of the form

$$y = My + p, \quad M \in (A\langle(\Sigma \cup Y)^*\rangle)^{n \times n}, \quad p \in (A\langle(\Sigma \cup Y)^*\rangle)^{n \times 1}. \quad (3)$$

Such an equation can be seen as a short version of an equation

$$z = \begin{pmatrix} M & 0 \\ 0 & 0 \end{pmatrix} z + \begin{pmatrix} 0 & p \\ 0 & 0 \end{pmatrix}$$

or, written with matrix variables u, y, v, x,

$$u = Mu, \; y = My + p,$$
$$v = 0, \quad x = 0.$$

We observe that

$$\lim_{j \to \infty} \begin{pmatrix} M & 0 \\ 0 & 0 \end{pmatrix}^j = 0 \quad \text{iff} \quad \lim_{j \to \infty} M^j = 0$$

and

$$\begin{pmatrix} M & 0 \\ 0 & 0 \end{pmatrix}^* \quad \text{exists} \quad \text{iff} \quad M^* \text{ exists}.$$

Hence, we can apply all our results on equations of type (*) to equations of type (3).

We did not treat the theory of equations of the form

$$y = My + p, \quad M \in (A\langle(\Sigma \cup Y)^*\rangle)^{n \times n}, \quad p \in (A\langle(\Sigma \cup Y)^*\rangle)^{n \times j}.$$

The reason is that we did not want to introduce A-semimodules compatible with the semiring A as regards the given convergence in A (see Kuich, Salomaa [15]).

In the next theorem we consider an algebraic system of the form $y = My + p(y)$, where $M \in (A\langle\varepsilon\rangle)^{n \times n}$ and $p(y) \in (A\langle(\Sigma \cup Y)^*\rangle)^{n \times 1}$. This means that some or all linear terms of the form ay, $a \in A$, $y \in Y$, are put together to form the matrix M.

Theorem 20. *Let $M \in (A\langle\varepsilon\rangle)^{n \times n}$ and assume that M^* exists. If σ is a solution to the algebraic system $y = M^*p(y)$, $p(y) \in (A\langle(\Sigma \cup Y)^*\rangle)^{n \times 1}$, then σ is also a solution to the algebraic system $y = My + p(y)$.*

Proof. Since σ is a solution of $y = M^*p(y)$, we infer $\sigma = M^*p(\sigma)$ and obtain

$$(My + p(y))[\sigma|y] = M\sigma + p(\sigma) = MM^*p(\sigma) + p(\sigma) =$$
$$(MM^* + E)p(\sigma) = M^*p(\sigma) = \sigma$$

by Corollary 2. □

Theorem 21. *Let $M \in (A\langle\varepsilon\rangle)^{n \times n}$. Assume that $\lim_{j \to \infty} M^j = 0$ and that M^* exists. If σ is a solution to the algebraic system $y = My + p(y)$, $p(y) \in (A\langle(\Sigma \cup Y)^*\rangle)^{n \times 1}$, then σ is also a solution to the algebraic system $y = M^*p(y)$.*

Proof. Since σ is a solution to the algebraic system $y = My + p(y)$, we obtain, for all $j \geq 0$,

$$\sigma = M\sigma + p(\sigma) = \ldots = M^{j+1}\sigma + \sum_{0 \leq i \leq j} M^i p(\sigma).$$

Since $\lim_{j \to \infty} M^j = 0$, we have $M\lim_{j \to \infty} M^j = \lim_{j \to \infty} M^{j+1} = 0$. Furthermore $M^{j+1}\sigma + \sum_{0 \leq i \leq j} M^i p(\sigma) = \sigma$ implies

$$(M^{j+1})\sigma + (\sum_{0 \leq i \leq j} M^i)p(\sigma) = \eta\sigma.$$

Hence, taking limits,

$$(\lim_{j \to \infty} M^{j+1})\sigma + (\lim_{j \to \infty} \sum_{0 \leq i \leq j} M^i)p(\sigma) = M^* p(\sigma) = \sigma.$$

and σ is a solution of $y = M^* p(y)$. $\qquad\qquad\square$

Consider a strong convergence in A. Then, by Kuich, Salomaa [15], the convergence in $A^{n \times n}$ transferred from the convergence in A is again strong. Let $y = My + p(y)$, where $M \in (A\langle\varepsilon\rangle)^{n \times n}$, $p(y) \in (A\langle(\Sigma \cup Y)^*\rangle)^{n \times 1}$ and $\mathrm{supp}(p_i(y)) \subseteq (\Sigma \cup Y)^+ \setminus Y$, $1 \leq i \leq n$, be an algebraic system. This algebraic system is called *cycle-free* (with respect to the given strong convergence) if $\lim_{j \to \infty} M^j = 0$.

Corollary 4. *Consider a strong convergence in A and a cycle-free algebraic system $y = My + p(y)$. Then the solutions of the algebraic systems $y = My + p(y)$ and $y = M^* p(y)$ coincide.*

Theorem 22. (Kuich [13], Ésik, Kuich [7], Ésik, Leiß [9]) *Let A be a continuous semiring. Then the least solution of the algebraic systems $y = My + p(y)$ and $y = M^* p(y)$, $M \in (A\langle\varepsilon\rangle)^{n \times n}$, $p(y) \in (A\langle(\Sigma \cup Y)^*\rangle)^{n \times 1}$, coincide.*

Example 3.1. Consider the cycle-free probabilistic $\mathbb{R}\langle\langle\Sigma^*\rangle\rangle$-algebraic system

$$y_1 = \tfrac{1}{4}y_2 + \tfrac{1}{4}y_3 + \tfrac{1}{3}u_1 y_1 v_1 + \tfrac{1}{6}x_1,$$
$$y_2 = \tfrac{1}{4}y_1 + \tfrac{1}{4}y_3 + \tfrac{1}{3}u_2 y_2 v_2 + \tfrac{1}{6}x_2,$$
$$y_3 = \tfrac{1}{4}y_1 + \tfrac{1}{4}y_2 + \tfrac{1}{3}u_3 y_3 v_3 + \tfrac{1}{6}x_3,$$

$u_i, v_i, x_i \in \Sigma^*$, $u_i v_i \neq \varepsilon$, $x_i \neq \varepsilon$, $1 \leq i \leq 3$. It can be written in the form $y = My + p(y)$, where

$$M = \begin{pmatrix} 0 & \tfrac{1}{4} & \tfrac{1}{4} \\ \tfrac{1}{4} & 0 & \tfrac{1}{4} \\ \tfrac{1}{4} & \tfrac{1}{4} & 0 \end{pmatrix} \quad \text{and} \quad p(y) = \begin{pmatrix} \tfrac{1}{3}u_1 y_1 v_1 + \tfrac{1}{6}x_1 \\ \tfrac{1}{3}u_2 y_2 v_2 + \tfrac{1}{6}x_2 \\ \tfrac{1}{3}u_3 y_3 v_3 + \tfrac{1}{6}x_3 \end{pmatrix}.$$

An easy proof by induction yields

$$M^n = \frac{1}{3 \cdot 2^n} U + \frac{(-1)^n}{3 \cdot 4^n} V, \quad n \geq 0,$$

where

$$U = \begin{pmatrix} 1 & 1 & 1 \\ 1 & 1 & 1 \\ 1 & 1 & 1 \end{pmatrix} \quad \text{and} \quad V = \begin{pmatrix} 2 & -1 & -1 \\ -1 & 2 & -1 \\ -1 & -1 & 2 \end{pmatrix}.$$

M^* exists in the Cauchy convergence and is given by

$$\begin{pmatrix} \frac{6}{5} & \frac{2}{5} & \frac{2}{5} \\ \frac{2}{5} & \frac{6}{5} & \frac{2}{5} \\ \frac{2}{5} & \frac{2}{5} & \frac{6}{5} \end{pmatrix}.$$

We infer by Corollary 4 that $\sigma \in (\mathbb{R}\langle\langle \Sigma^* \rangle\rangle)^{3 \times 1}$ is a solution to $y = My + p(y)$ iff it is a solution to $y = M^* p(y)$. □

Theorem 23. *Assume that σ is a solution to the algebraic system $y_i = p_i(y_1, \ldots, y_n)$, $1 \leq i \leq n$. Then there exists an algebraic system $y_i = q_i$, where each q_i is proper, with solution $\tau = \sum_{w \in \Sigma^+} (\sigma, w)w$.*

If the semiring A is continuous and σ is the least solution to $y_i = p_i$, $1 \leq i \leq n$, then τ is the least solution of $y_i = q_i$, $1 \leq i \leq n$.

Proof. We define the polynomial q_i to be the proper part of the polynomial

$$p_i((\sigma_1, \varepsilon)\varepsilon + y_1, \ldots, (\sigma_n, \varepsilon)\varepsilon + y_n).$$

The comparison of the proper parts of the equalities $\sigma_i = p_i(\sigma_1, \ldots, \sigma_n)$, $1 \leq i \leq n$, yields the equalities

$$(\sigma_i, \varepsilon)\varepsilon = p_i((\sigma_1, \varepsilon)\varepsilon, \ldots, (\sigma_n, \varepsilon)\varepsilon) \quad \text{and}$$
$$\tau_i = q_i(\tau_1, \ldots, \tau_n), \quad 1 \leq i \leq n.$$

Hence, τ is a solution to $y_i = q_i$, $1 \leq i \leq n$.

Assume now that A is continuous and consider an arbitrary solution τ' of $y_i = q_i$, $1 \leq i \leq n$. Then $\sigma' = (\sigma, \varepsilon)\varepsilon + \tau'$ is a solution of $y_i = p_i$, $1 \leq i \leq n$. Since σ is the least solution of $y_i = p_i$, $1 \leq i \leq n$, we infer that $\sigma \leq \sigma'$. But this implies $\tau \leq \tau'$. Hence, τ is the least solution of $y_i = q_i$, $1 \leq i \leq n$. □

An algebraic system $y_i = p_i$, $1 \leq i \leq n$, $p_i \in A\langle\langle (\Sigma \cup Y)^* \rangle\rangle$, is termed *proper* if $\operatorname{supp}(p_i) \subseteq (\Sigma \cup Y)^+ \setminus Y$ for all $1 \leq i \leq n$.

Theorem 24. (Kuich, Salomaa [15]) *Let $y = p$ be a proper algebraic system. Then its strong solution σ exists with respect to a multiplicative convergence. Moreover, σ is proper and it is the only proper solution.*

Proof. Let σ^k, $k \geq 0$, be the elements of the approximation sequence associated to the proper algebraic system $y = p$. We claim that for all $k, j \geq 0$,

$$R_k(\sigma^{k+j+1}) = R_k(\sigma^{k+1}).$$

(Here the truncation operator R_k is applied to vectors componentwise.) The proof of the claim is by induction on k. Clearly, the elements of σ^j, $j \geq 0$, are proper. This proves our claim for $k = 0$. For $k > 0$ we obtain, for all $j \geq 0$,

$$R_k(\sigma^{k+1}) = R_k(p(\sigma^k)) = R_k(p(R_k(\sigma^k))) = R_k(p(R_{k-1}(\sigma^k))) =$$
$$R_k(p(R_{k-1}(\sigma^{k+j}))) = R_k(p(\sigma^{k+j})) = R_k(\sigma^{k+j+1}).$$

The third equation follows by the special form of the terms in the polynomials in p. The fourth equation follows by the induction hypothesis.

Hence, by Theorems 17 and 9, the strong solution exists with respect to the given multiplicative convergence.

Let τ be a proper solution to the proper algebraic system $y = p$. We claim that, for $k \geq 0$,

$$R_k(\tau) = R_k(\sigma).$$

The proof of the claim is by induction on k. Clearly, we obtain $R_0(\tau) = R_0(\sigma) = 0$. For $k \geq 0$, we obtain

$$R_k(\sigma) = R_k(p(\sigma)) = R_k(p(R_{k-1}(\sigma))) =$$
$$R_k(p(R_{k-1}(\tau))) = R_k(p(\tau)) = R_k(\tau).$$

Consequently, $\tau = \sigma$. $\qquad\square$

The next theorem is analogous to Theorem 22.

Theorem 25. *Consider a strong convergence in the semiring A. Then a cycle-free algebraic system $y = My + p(y)$ has a unique proper solution that coincides with the unique proper solution of the proper algebraic system $y = M^*p(y)$.*

Proof. By Theorem 24 and Corollary 4. $\qquad\square$

A power series in $A\langle\langle\Sigma^*\rangle\rangle$ is called *A-algebraic* if $r = (r, \varepsilon)\varepsilon + r_1$, where r_1 is some component of the strong solution of a proper $A\langle\Sigma \cup \{\varepsilon\}\rangle$-algebraic system. The collection of all A-algebraic power series is denoted by $A^{\mathrm{alg}}\langle\langle\Sigma^*\rangle\rangle$. If the semiring is clear, we call r algebraic power series.

Corollary 5. *Consider a strong convergence in the semiring A and let r_1 be some component of the proper solution of a cycle-free algebraic system. Then $r = (r, \varepsilon)\varepsilon + r_1$ is an algebraic power series.*

Theorem 26. *Let A be a continuous semiring. Then the components of the least solution of an A-algebraic system are A-algebraic power series.*

Proof. Given an algebraic system with least solution σ, we perform the construction of Theorem 23. By Theorem 23, the proper part of σ is the least solution of this constructed system. Now we write this system in the form $y = My + p(y)$, where $M \in (A\langle\varepsilon\rangle)^{n \times n}$, $p(y) \in (A\langle\langle(\Sigma \cup Y)^*\rangle\rangle)^{n \times 1}$ and $\mathrm{supp}(p_i(y)) \subseteq (\Sigma \cup Y)^+ \setminus Y$. Then by Theorem 22, the least solution of the proper algebraic system $y = M^*p(y)$ is again the proper part of σ. Hence, by the definition of algebraic power series, σ is an algebraic power series. $\qquad\square$

4   Pushdown Automata and Algebraic Systems

In this section we generalize finite automata, and the equations introduced in Section 2. Finite automata are generalized in the following direction: An infinite set of states will be allowed in the general definition. When dealing with pushdown automata this will enable us to store the contents of the pushdown tape in the states.

Then we define $A\langle \Sigma \cup \{\varepsilon\}\rangle$-pushdown automata and consider their relation to $A\langle \Sigma \cup \{\varepsilon\}\rangle$-algebraic systems. It turns out that $a \in A^{\mathrm{alg}}\langle \Sigma^* \rangle$ iff it is the behavior of an $A\langle \Sigma \cup \{\varepsilon\}\rangle$-pushdown automaton. This generalizes the language theoretic result due to Chomsky [4] that a formal language is context-free iff it is accepted by a pushdown automaton.

Our model of an automaton will be defined in terms of a (possibly infinite) transition matrix. The semiring element generated by the transition of the automaton from one state i to another state i' in exactly k computation steps equals the (i, i')-entry in the k-th power of the transition matrix. Consider now the star of the transition matrix. Then the semiring element generated by the automaton, also called the behavior of the automaton, can be expressed by the entries (multiplied by the initial and final weights of the states) of the star of the transition matrix.

An $A\langle \Sigma \cup \{\varepsilon\}\rangle$-*automaton*

$$\mathfrak{A} = (I, M, S, P)$$

is given by

(i) a non-empty set I of *states*,
(ii) a matrix $M \in A\langle \Sigma \cup \{\varepsilon\}\rangle^{I \times I}$, called the *transition matrix*,
(iii) $S \in A\langle \Sigma \cup \{\varepsilon\}\rangle^{1 \times I}$, called the *initial state vector*,
(iv) $P \in A\langle \Sigma \cup \{\varepsilon\}\rangle^{I \times 1}$, called the *final state vector*.

An $A\langle \Sigma \cup \{\varepsilon\}\rangle$-automaton $\mathfrak{A} = (I, M, S, P)$ is called C-*cycle-free* if M is in $(A\langle \Sigma \cup \{\varepsilon\}\rangle)_C^{I \times I}$ and is C-cycle-free.

If A is complete or $\mathfrak{A}$ is C-cycle-free then the *behavior* $\|\mathfrak{A}\| \in A\langle\langle \Sigma^* \rangle\rangle$ of the $A\langle \Sigma \cup \{\varepsilon\}\rangle$-automaton $\mathfrak{A}$ is defined by

$$\|\mathfrak{A}\| = \sum_{i_1, i_2 \in I} S_{i_1} (M^*)_{i_1, i_2} P_{i_2} = SM^*P.$$

We now consider equations of the form

$$y = My + P,$$

where y is a variable, M is in $A\langle \Sigma \cup \{\varepsilon\}\rangle_C^{I \times I}$ and P is in $A\langle \Sigma \cup \{\varepsilon\}\rangle^{I \times 1}$. The remark on equations of the form (3) in Section 3 is in an analogous manner also valid for these equations.

Theorem 27. *Let* $M \in (A\langle \Sigma \cup \{\varepsilon\}\rangle)_C^{I \times I}$ *and* $P \in (A\langle \Sigma \cup \{\varepsilon\}\rangle)^{I \times J}$. *Assume that* M^* *exists. Then* M^*P *is a solution of* $y = My + P$.

If $\lim_{n\to\infty} M^n = 0$ then M^*P is the unique solution of $y = My + P$.
If A is a continuous semiring then M^*P is the least solution of $y = My + P$.

Proof. By Theorems 5 and 6.   □

$A\langle\Sigma \cup \{\varepsilon\}\rangle$-pushdown automata are finite automata (with state set Q) aug-
mented by a pushdown tape. The contents of the pushdown tape is a word
over the pushdown alphabet Γ. We consider an $A\langle\Sigma \cup \{\varepsilon\}\rangle$-pushdown automaton to be an $A\langle\Sigma \cup \{\varepsilon\}\rangle$-automaton: the state set is given by $\Gamma^* \times Q$ and its
transition matrix is in $A\langle\Sigma \cup \{\varepsilon\}\rangle^{(\Gamma^* \times Q)\times(\Gamma^* \times Q)}$. This allows us to store the
contents of the pushdown tape and the states of the finite automaton in the
states of the $A\langle\Sigma\cup\{\varepsilon\}\rangle$-pushdown automaton. Because of technical reasons, we
do not work in the semiring $A\langle\langle\Sigma^*\rangle\rangle^{(\Gamma^* \times Q)\times(\Gamma^* \times Q)}$ but in the isomorphic semi-
ring $(A\langle\langle\Sigma^*\rangle\rangle^{Q\times Q})^{\Gamma^* \times \Gamma^*}$. A matrix $M \in (A\langle\Sigma \cup \{\varepsilon\}\rangle^{Q\times Q})^{\Gamma^* \times \Gamma^*}$ is termed an
$A\langle\Sigma \cup \{\varepsilon\}\rangle$-*pushdown transition matrix* if

(i) for each $p \in \Gamma$ there exist only finitely many blocks $M_{p,\pi}$, $\pi \in \Gamma^*$, that are
unequal to 0;
(ii) for all $\pi_1, \pi_2 \in \Gamma^*$,

$$M_{\pi_1,\pi_2} = \begin{cases} M_{p,\pi} & \text{if there exist } p \in \Gamma,\ \pi' \in \Gamma^* \text{ with} \\ & \quad \pi_1 = p\pi' \text{ and } \pi_2 = \pi\pi', \\ 0 & \text{otherwise.} \end{cases}$$

The above definition implies that M is row and column finite and that an $A\langle\Sigma\cup
\{\varepsilon\}\rangle$-pushdown transition matrix has a finitary specification: it is completely
specified by its non-null blocks of the form $M_{p,\pi}$, $p \in \Gamma$, $\pi \in \Gamma^*$, and only the
following transitions are possible: if the contents of the pushdown tape is given
by $p\pi'$, the contents of the pushdown tape after a transition has to be of the
form $\pi\pi'$; moreover, the transition does only depend on the leftmost (topmost)
pushdown sympol p and not on π'. In this sense the $A\langle\Sigma \cup \{\varepsilon\}\rangle$-pushdown
transition matrix represents a proper formalization of the principle "last in—
first out".
An $A\langle\Sigma \cup \{\varepsilon\}\rangle$-*pushdown automaton*

$$\mathfrak{P} = (Q, \Gamma, M, S, p_0, P)$$

is given by

(i) a finite set Q of *states*,
(ii) a finite alphabet Γ of *pushdown symbols*,
(iii) an $A\langle\Sigma \cup \{\varepsilon\}\rangle$-*pushdown transition matrix* $M \in (A\langle\Sigma \cup \{\varepsilon\}\rangle^{Q\times Q})^{\Gamma^* \times \Gamma^*}$,
(iv) $S \in A\langle\{\varepsilon\}\rangle^{1\times Q}$, called the *initial state vector*,
(v) $p_0 \in \Gamma$, called the *initial pushdown symbol*,
(vi) $P \in A\langle\{\varepsilon\}\rangle^{Q\times 1}$, called the *final state vector*.

An $A\langle\Sigma \cup \{\varepsilon\}\rangle$-pushdown automaton $\mathfrak{P} = (Q, \Gamma, M, S, p_0, P)$ is called *C-cycle-
free* if its transition matrix M is *C-cycle-free*.

If A is complete or $\mathfrak{P}$ is C-cycle-free then the behavior $\|\mathfrak{P}\|$ of the $A\langle\Sigma\cup\{\varepsilon\}\rangle$-pushdown automaton $\mathfrak{P}$ is defined by

$$\|\mathfrak{P}\| = S(M^*)_{p_0,\varepsilon}P.$$

We now describe the computations of an $A\langle\Sigma\cup\{\varepsilon\}\rangle$-pushdown automaton. Initially, the pushdown tape contains the special symbol p_0. The $A\langle\Sigma\cup\{\varepsilon\}\rangle$-pushdown automaton now performs transitions governed by the $A\langle\Sigma\cup\{\varepsilon\}\rangle$-pushdown transition matrix until the pushdown tape is emptied. The result of these computations is given by $(M^*)_{p_0,\varepsilon}$. Multiplications by the initial state vector and by the final state vector yield the behavior of the $A\langle\Sigma\cup\{\varepsilon\}\rangle$-pushdown automaton.

Assume now that $\mathfrak{P}(\Sigma^*)$, the semiring of formal languages over Σ, is our basic semiring. We connect our definition of a $\mathbb{B}\langle\Sigma\cup\{\varepsilon\}\rangle$-pushdown automaton $\mathfrak{P} = (Q, \Gamma, M, I, p_0, P)$ by the isomorphism of $\mathfrak{P}(\Sigma^*)$ and $\mathbb{B}\langle\!\langle\Sigma^*\rangle\!\rangle$ to the usual definition of a *pushdown automaton* $\mathfrak{P}' = (Q, \Sigma, \Gamma, \delta, q_0, p_0, F)$ (see e. g., Harrison [11]), where Σ is the *input alphabet*, δ, a function from $Q \times (\Sigma \cup \{\varepsilon\}) \times \Gamma$ to the set of all finite subsets of $Q \times \Gamma^*$, is the *transition function*, $q_0 \in Q$ is the *initial state* and $F \subseteq Q$ is the set of *final states*.

Assume that a pushdown automaton $\mathfrak{P}'$ is given as above. The transition function δ defines the pushdown transition matrix M of $\mathfrak{P}$ by

$$x \in (M_{p,\pi})_{q_1,q_2} \qquad \text{iff} \qquad (q_2, \pi) \in \delta(q_1, x, p)$$

for all $q_1, q_2 \in Q, p \in \Gamma, \pi \in \Gamma^*, x \in \Sigma\cup\{\varepsilon\}$. Let now $\vdash$ be the move relation over the *instantaneous descriptions* of $\mathfrak{P}'$ in $Q\times\Sigma^*\times\Gamma^*$. Then $(q_1, w, \pi_1) \vdash^k (q_2, \varepsilon, \pi_2)$ iff $w \in ((M^k)_{\pi_1,\pi_2})_{q_1,q_2}$ and $(q_1, w, \pi_1) \vdash^* (q_2, \varepsilon, \pi_2)$ iff $w \in ((M^*)_{\pi_1,\pi_2})_{q_1,q_2}$ for all $k \geq 0, q_1, q_2 \in Q, \pi_1, \pi_2 \in \Gamma^*, w \in \Sigma^*$. Hence, $(q_0, w, p_0) \vdash^* (q, \varepsilon, \varepsilon)$ iff $w \in ((M^*)_{p_0,\varepsilon})_{q_0,q}$. Define the initial state vector S and the final state vector P by $S_{q_0} = \{\varepsilon\}, S_q = \emptyset$ if $q \neq q_0, P_q = \{\varepsilon\}$ if $q \in F, P_q = \emptyset$ if $q \notin F$. Then a word w is accepted by the pushdown automaton $\mathfrak{P}'$ by both final state and empty store iff $w \in S(M^*)_{p_0,\varepsilon}P = \|\mathfrak{P}\|$.

In our first theorem we show that an $A\langle\Sigma\cup\{\varepsilon\}\rangle$-pushdown automaton can be regarded as an $A\langle\Sigma\cup\{\varepsilon\}\rangle$-automaton.

Theorem 28. *Consider an $A\langle\Sigma\cup\{\varepsilon\}\rangle$-pushdown automaton $\mathfrak{P}$. If A is complete or $\mathfrak{P}$ is C-cycle-free then there exists an $A\langle\Sigma\cup\{\varepsilon\}\rangle$-automaton $\mathfrak{A}$ such that $\|\mathfrak{A}\| = \|\mathfrak{P}\|$.*

Proof. Let $\mathfrak{P} = (Q, \Gamma, M, S, p_0, P)$. We define the $A\langle\Sigma\cup\{\varepsilon\}\rangle$-automaton $\mathfrak{A} = (\Gamma^* \times Q, M', S', P')$ by $M'_{(\pi_1,q_1),(\pi_2,q_2)} = (M_{\pi_1,\pi_2})_{q_1,q_2}, S'_{(p_0,q)} = S_q, S'_{(\pi,q)} = 0,$ if $\pi \neq p_0, P'_{(\varepsilon,q)} = P_q, P'_{(\pi,q)} = 0,$ if $\pi \neq \varepsilon$. Then

$$\|\mathfrak{A}\| = S'M'^*P'$$

$$= \sum_{(\pi_1,q_1),(\pi_2,q_2)\in\Gamma^*\times Q} S'_{(\pi_1,q_1)}(M'^*)_{(\pi_1,q_1),(\pi_2,q_2)}P'_{(\pi_2,q_2)}$$

$$= \sum_{q_1, q_2 \in Q} S'_{(p_0, q_1)} (M'^*)_{(p_0, q_1), (\varepsilon, q_2)} P'_{(\varepsilon, q_2)} =$$

$$= \sum_{q_1, q_2 \in Q} S_{q_1} ((M^*)_{p_0, \varepsilon})_{q_1, q_2} P_{q_2} = S(M^*)_{p_0, \varepsilon} P = \| \mathfrak{P} \|.$$

$\square$

An $A\langle \Sigma \cup \{\varepsilon\}\rangle$-pushdown transition matrix $M \in ((A\langle\!\langle \Sigma^* \rangle\!\rangle)^{Q \times Q})^{\Gamma^* \times \Gamma^*}$ is called *proper* if, for all $p \in \Gamma$ and $\pi \in \Gamma^*$, $(M_{p, \pi}, \varepsilon) \neq 0$ implies $|\pi| \geq 2$. An $A\langle \Sigma \cup \{\varepsilon\}\rangle$-pushdown automaton is called *proper* if its $A\langle \Sigma \cup \{\varepsilon\}\rangle$-pushdown transition matrix is proper.

Theorem 29. *Consider a strong convergence in A. Then every proper $A\langle \Sigma \cup \{\varepsilon\}\rangle$-pushdown transition matrix M is C-cycle-free. Moreover, $(M^*)_{p, \varepsilon}$ is proper for all $p \in \Gamma$.*

Proof. For the proof we show at first that $\lim_{C} {}_{n \to \infty} (M^n, \varepsilon) = 0$, where the basic convergence in A is the discrete convergence.

By the definition of $\lim_C$ this means that the following conditions (i) and (ii) are satisfied.

(i) For all $\pi_2 \in \Gamma^*$, there exists an $n(\pi_2) \geq 0$ such that

$$(M^{n(\pi_2)+k}, \varepsilon)_{\pi_1, \pi_2} = 0$$

holds for all $\pi_1 \in \Gamma^*$ and all $k \geq 0$.

(ii) For all $\pi_2 \in \Gamma^*$, there exists a finite set $I(\pi_2) \subseteq \Gamma^*$ such that

$$(M^k, \varepsilon)_{\pi_1, \pi_2} = 0$$

holds for all $\pi_1 \in \Gamma^* \setminus I(\pi_2)$ and all $k \geq 0$.

We claim that (i) and (ii) are satisfied with $n(\pi_2) = |\pi_2| + 1$ and $I(\pi_2) = \{\pi \mid |\pi| \leq |\pi_2|\}$.

We show first that, for all $k \geq 0$,

$$(M^k, \varepsilon)_{\pi_1, \pi_2} = 0 \quad \text{if} \quad |\pi_2| \leq |\pi_1| + k - 1.$$

This is obviously true for $k = 0$ or $\pi_1 = \varepsilon$.

Assume that $k > 0$ and $|\pi_2| \leq |\pi_1| + k$. Then

$$(M^k, \varepsilon)_{p\pi_1, \pi_2} = \sum_{\pi \in \Gamma^*, \, |\pi| \geq 2} (M, \varepsilon)_{p, \pi} (M^{k-1}, \varepsilon)_{\pi\pi_1, \pi_2}.$$

Since $|\pi_2| \leq |\pi_1| + k \leq |\pi\pi_1| + k - 2$ for $|\pi| \geq 2$, it follows that $(M^{k-1}, \varepsilon)_{\pi\pi_1, \pi_2} = 0$ for $|\pi| \geq 2$, implying $(M^k, \varepsilon)_{p\pi_1, \pi_2} = 0$.

Condition (i) is now a consequence of the fact that $|\pi_2| \leq |\pi_1| + n(\pi_2) + k - 1$ holds for all $k \geq 0$. Similarly, the condition (ii) is a consequence of the fact that $|\pi_2| \leq |\pi_1| + k - 1$ holds for all $\pi_1 \in \Gamma^* \setminus I(\pi_2)$ and all $k \geq 0$.

The second sentence is proved by the fact that $|\varepsilon| \leq |p| + k - 1$ for all $k \geq 0$.

For an arbitrary strong convergence in A, the theorem is now proved by Theorem 9.   □

Because of this theorem, we always use C-convergence in connection with proper $A\langle\Sigma \cup \{\varepsilon\}\rangle$-pushdown transition matrices.

Consider an $A\langle\Sigma \cup \{\varepsilon\}\rangle$-pushdown automaton with $A\langle\Sigma \cup \{\varepsilon\}\rangle$-pushdown transition matrix M and let $\pi = \pi_1\pi_2$ be a word over the pushdown alphabet Γ. Then our next proposition states that emptying the pushdown tape with contents π has the same effect (i. e., $(M^*)_{\pi,\varepsilon}$) as emptying first the pushdown tape with contents π_1 (i. e., $(M^*)_{\pi_1,\varepsilon}$) and afterwards (i. e., multiplying) the pushdown tape with contents π_2 (i. e., $(M^*)_{\pi_2,\varepsilon}$).

Proposition 2. (Kuich, Salomaa [15], Ésik, Kuich [7])
Let $M \in (A\langle\Sigma \cup \{\varepsilon\}\rangle^{Q\times Q})^{\Gamma^* \times \Gamma^*}$ be an $A\langle\Sigma \cup \{\varepsilon\}\rangle$-pushdown transition matrix. Assume that A is a continuous semiring or M is a proper $A\langle\Sigma\cup\{\varepsilon\}\rangle$-pushdown matrix. Then

$$(M^*)_{\pi_1\pi_2,\varepsilon} = (M^*)_{\pi_1,\varepsilon}(M^*)_{\pi_2,\varepsilon}$$

holds for all $\pi_1, \pi_2 \in \Gamma^*$.

Let $M \in (A\langle\Sigma\cup\{\varepsilon\}\rangle^{Q\times Q})^{\Gamma^* \times \Gamma^*}$ be an $A\langle\Sigma\cup\{\varepsilon\}\rangle$-pushdown transition matrix and let $\{y_p \mid p \in \Gamma\}$ be an alphabet of variables. We define $y_\varepsilon = \varepsilon$ and $y_{p\pi} = y_p y_\pi$ for $p \in \Gamma$, $\pi \in \Gamma^*$, and consider the $A\langle\Sigma \cup \{\varepsilon\}\rangle^{Q\times Q}$-algebraic system

$$y_p = \sum_{\pi \in \Gamma^*} M_{p,\pi} y_\pi, \quad p \in \Gamma.$$

Given matrices $T_p \in A\langle\langle\Sigma^*\rangle\rangle^{Q\times Q}$ for all $p \in \Gamma$, we define matrices $T_\pi \in A\langle\langle\Sigma^*\rangle\rangle^{Q\times Q}$ for all $\pi \in \Gamma^*$ as follows: $T_\varepsilon = E$, $T_{p\pi} = T_p T_\pi$, $p \in \Gamma$, $\pi \in \Gamma^*$. By these matrices we define a matrix $\tilde{T} \in (A\langle\langle\Sigma^*\rangle\rangle^{Q\times Q})^{\Gamma^*\times 1}$: the π-block of $\tilde{T}$ is given by T_π, $\pi \in \Gamma^*$, i. e., $\tilde{T}_\pi = T_\pi$.

In the sequel, $F \in (A\langle\Sigma \cup \{\varepsilon\}\rangle^{Q\times Q})^{\Gamma^*\times 1}$ is defined by $F_\varepsilon = E$ and $F_\pi = 0$ if $\pi \in \Gamma^+$.

Theorem 30. (Kuich, Salomaa [15], Ésik, Kuich [7])
Let $M \in ((A\langle\Sigma\cup\{\varepsilon\}\rangle)^{Q\times Q})^{\Gamma^* \times \Gamma^*}$ be an $A\langle\Sigma\cup\{\varepsilon\}\rangle$-pushdown transition matrix. If $(T_p)_{p\in\Gamma}$, $T_p \in A^{Q\times Q}$, is a solution of $y_p = \sum_{\pi\in\Gamma^*} M_{p,\pi} y_\pi$, $p \in \Gamma$, then $\tilde{T} \in (A^{Q\times Q})^{\Gamma^*\times 1}$ is a solution of $y = My + F$.

Proof. Since M is an $A\langle\Sigma \cup \{\varepsilon\}\rangle$-pushdown transition matrix, we obtain, for all $p \in \Gamma$ and $\pi \in \Gamma^*$,

$$(M\tilde{T})_{p\pi} = \sum_{\pi_1\in\Gamma^*} M_{p\pi,\pi_1}\tilde{T}_{\pi_1} = \sum_{\pi_2\in\Gamma^*} M_{p\pi,\pi_2\pi}\tilde{T}_{\pi_2\pi} =$$

$$= \sum_{\pi_2\in\Gamma^*} M_{p,\pi_2}\tilde{T}_{\pi_2}\tilde{T}_\pi = (M\tilde{T})_p\tilde{T}_\pi.$$

Since $(T_p)_{p\in\Gamma}$ is a solution of $y_p = \sum_{\pi\in\Gamma^*} M_{p,\pi}y_\pi$, $p \in \Gamma$, we infer that $\tilde{T}_p = T_p = \sum_{\pi\in\Gamma^*} M_{p,\pi}T_\pi = \sum_{\pi\in\Gamma^*} M_{p,\pi}\tilde{T}_\pi = (M\tilde{T})_p$. Hence, $(M\tilde{T} + F)_{p\pi} = (M\tilde{T})_{p\pi} = \tilde{T}_p\tilde{T}_\pi = \tilde{T}_{p\pi}$, $p \in \Gamma$, $\pi \in \Gamma^*$. Additionally, we have $\tilde{T}_\varepsilon = E$ and $(M\tilde{T} + F)_\varepsilon = F_\varepsilon = E$. This implies that $\tilde{T}$ is a solution of $y = My + F$. $\square$

Theorem 31. *Let $M \in ((A\langle\Sigma \cup \{\varepsilon\}\rangle)^{Q\times Q})^{\Gamma^*\times\Gamma^*}$ be an $A\langle\Sigma \cup \{\varepsilon\}\rangle$-pushdown transition matrix and assume that M^* exists. Then $((M^*)_{p,\varepsilon})_{p\in\Gamma}$ is a solution to the $A\langle\langle\Sigma^*\rangle\rangle$-algebraic system $y_p = \sum_{\pi\in\Gamma^*} M_{p,\pi}y_\pi$.*

If the semiring A is continuous it is the least solution. If M is a proper $A\langle\Sigma \cup \{\varepsilon\}\rangle$-pushdown matrix then it is the unique proper solution.

Proof. We first show that $((M^*)_{p,\varepsilon})_{p\in\Gamma}$ is a solution of the $A\langle\Sigma \cup \{\varepsilon\}\rangle^{Q\times Q}$-algebraic system by substituting $(M^*)_{\pi,\varepsilon}$ for y_π:

$$\sum_{\pi\in\Gamma^*} M_{p,\pi}(M^*)_{\pi,\varepsilon} = (M^+)_{p,\varepsilon} = (M^*)_{p,\varepsilon}, \quad p \in \Gamma.$$

Assume now that A is continuous and $(T_p)_{p\in\Gamma}$ is a solution of $y_p = \sum_{\pi\in\Gamma^*} M_{p,\pi}y_\pi$. Then, by Theorem 30, $\tilde{T}$ is a solution of $y = My+F$. Since M^*F is the least solution of this equation, we infer that $M^*F \leq \tilde{T}$. This implies $(M^*F)_\pi = (M^*)_{\pi,\varepsilon} \leq \tilde{T}_\pi = T_\pi$ for all $\pi \in \Gamma^*$. Hence, $(M^*)_{p,\varepsilon} \leq T_p$ for all $p \in \Gamma$, and $((M^*)_{p,\varepsilon})_{p\in\Gamma}$ is the least solution of $y_p = \sum_{\pi\in\Gamma^*} M_{p,\pi}y_\pi$, $p \in \Gamma$.

If M is proper then, by Theorem 29, $(M^*)_{p,\varepsilon}$ is proper for all $p \in \Gamma$. Moreover, by the definition of a proper pushdown matrix, the algebraic system $y_p = \sum_{\pi\in\Gamma^*} M_{p,\pi}y_\pi$ is proper. Hence, by Theorem 24, $((M^*)_{p,\varepsilon})_{p\in\Gamma}$ is the unique proper solution to the proper algebraic system. $\square$

Let $\mathfrak{P} = (Q, \Gamma, M, S, p_0, P)$ be an $A\langle\Sigma\cup\{\varepsilon\}\rangle$-pushdown automaton and consider the $A\langle\Sigma \cup \{\varepsilon\}\rangle$-algebraic system

$$y_0 = S(\sum_{\pi\in\Gamma^*} M_{p_0,\pi}y_\pi)P,$$

$$y_p = \sum_{\pi\in\Gamma^*} M_{p,\pi}y_\pi, \quad p \in \Gamma,$$

written in matrix notation: y_p is a $Q \times Q$-matrix whose (q_1, q_2)-entry is the variable $[q_1, p, q_2]$, $p \in \Gamma$, $q_1, q_2 \in Q$; if $\pi = p_1 \ldots p_r$, $r \geq 1$, then the (q_1, q_2)-entry of y_π is given by the (q_1, q_2)-entry of $y_{p_1} \ldots y_{p_r}$, $p_1, \ldots, p_r \in \Gamma$; y_0 is a variable. Hence, the variables of the above $A\langle\Sigma \cup \{\varepsilon\}\rangle$-algebraic system are y_0, $[q_1, p, q_2]$, $p \in \Gamma$, $q_1, q_2 \in Q$.

Corollary 6. *Let $\mathfrak{P} = (Q, \Gamma, M, S, p_0, P)$ be an $A\langle\Sigma\cup\{\varepsilon\}\rangle$-pushdown automaton and assume that M^* exists. Then $\|\mathfrak{P}\|$, $((M^*)_{p,\varepsilon})_{p\in\Gamma}$ is a solution of the $A\langle\Sigma \cup \{\varepsilon\}\rangle$-algebraic system*

$$y_0 = S(\sum_{\pi\in\Gamma^*} M_{p_0,\pi}y_\pi)P,$$

$$y_p = \sum_{\pi\in\Gamma^*} M_{p,\pi}y_\pi, \quad p \in \Gamma.$$

If the semiring A is continuous it is the least solution. If $\mathfrak{P}$ is proper it is the unique proper solution.

Corollary 7. *Let $\mathfrak{P}$ be an $A\langle \Sigma \cup \{\varepsilon\}\rangle$-pushdown automaton. Assume that A is a continuous semiring or that $\mathfrak{P}$ is proper. Then the behavior of $\mathfrak{P}$ is an element of $A^{\mathrm{alg}}\langle\!\langle \Sigma^* \rangle\!\rangle$.*

We now want to show the converse to Corollary 7.

Theorem 32. *Let $r \in A^{\mathrm{alg}}\langle\!\langle \Sigma^* \rangle\!\rangle$. Then there exists a proper $A\langle \Sigma \cup \{\varepsilon\}\rangle$-pushdown automaton $\mathfrak{P}$ such that $\|\mathfrak{P}\|$ equals the proper part of r.*

Proof. By definition, the proper part r_1 of r is a component of the unique proper solution of a proper algebraic system. Consider the proper algebraic system $y_i = p_i$, $1 \le i \le n$, with unique proper solution σ and assume $\sigma_1 = r_1$.

We now define the $A\langle \Sigma \cup \{\varepsilon\}\rangle$-pushdown automaton

$$\mathfrak{P} = (\{q\}, \Sigma \cup Y, M, q, y_1, \{\varepsilon\})$$

by

$$M_{y_i, y_j \gamma} = (p_i, y_j \gamma)\varepsilon + \sum_{x \in \Sigma} (p_i, x y_j \gamma)x \quad \text{for} \quad \gamma \in (\Sigma \cup Y)^*,\ 1 \le i, j \le n,$$

$$M_{y_i, x\gamma} = \sum_{x' \in \Sigma} (p_i, x' x \gamma)x' \quad \text{for} \quad \gamma \in (\Sigma \cup Y)^*,\ x \in \Sigma,\ 1 \le i \le n,$$

$$M_{y_i, \varepsilon} = \sum_{x \in \Sigma} (p_i, x)x \quad \text{for} \quad 1 \le i \le n,$$

$$M_{x, \varepsilon} = x \quad \text{for} \quad x \in \Sigma,$$

$$M_{p, \pi} = 0, \quad p \in \Sigma \cup Y,\ \pi \in (\Sigma \cup Y)^*, \text{ in all other cases.}$$

Observe that $\mathfrak{P}$ has a single state only. Hence, the entries of M defined above are in $A\langle \Sigma \cup \{\varepsilon\}\rangle$. Since $\mathfrak{P}$ is proper, by Theorem 29, M^* exists and $(M^*)_{y_i, \varepsilon}$, $1 \le i \le n$, is proper.

We write our algebraic system $y_i = p_i$, $1 \le i \le n$, in the form

$$y_i = \sum_{1 \le j \le n} \sum_{\gamma \in (\Sigma \cup Y)^+} (p_i, y_j \gamma)y_j \gamma + \sum_{1 \le j \le n} \sum_{\gamma \in (\Sigma \cup Y)^*} \sum_{x \in \Sigma} (p_i, x y_j \gamma)x y_j \gamma$$
$$+ \sum_{\gamma \in (\Sigma \cup Y)^*} \sum_{x \in \Sigma} \sum_{x' \in \Sigma} (p_i, x' x \gamma)x' x \gamma + \sum_{x \in \Sigma} (p_i, x)x.$$

Replacing now $(p_i, y_j \gamma)\varepsilon + \sum_{x \in \Sigma}(p_i, x y_j \gamma)x$, $\sum_{x' \in \Sigma}(p_i, x' x \gamma)x'$ and $\sum_{x \in \Sigma}(p_i, x)x$ according to the definitions given above, we obtain

$$y_i = \sum_{1 \le j \le n} \sum_{\gamma \in (\Sigma \cup Y)^*} M_{y_i, y_j \gamma} y_j \gamma + \sum_{x \in \Sigma} \sum_{\gamma \in (\Sigma \cup Y)^*} M_{y_i, x\gamma} x \gamma + M_{y_i, \varepsilon}.$$

We now replace the variables y_i by $(M^*)_{y_i, \varepsilon}$ and observe that $x = (M^*)_{x, \varepsilon}$. Then, by Theorems 29 and 31, we conclude that $(M^*)_{y_i, \varepsilon}$, $1 \le i \le n$, is a proper solution to our proper algebraic system. Therefore, by Theorem 31, $\sigma_i = (M^*)_{y_i,\varepsilon}$, $1 \le i \le n$.

Hence,

$$\|\mathfrak{P}\| = (M^*)_{y_1,\varepsilon} = \sigma_1 \, .$$

$\square$

Corollary 8. *The power series r is in $A^{\mathrm{alg}}\langle\!\langle \Sigma^* \rangle\!\rangle$ iff there exists a proper $A\langle \Sigma \cup \{\varepsilon\}\rangle$-pushdown automaton $\mathfrak{P}$ such that $\|\mathfrak{P}\|$ equals the proper part of r.*

Corollary 9. *Let A be a continuous semiring. Then $r \in A^{\mathrm{alg}}\langle\!\langle \Sigma^* \rangle\!\rangle$ iff there exists an $A\langle \Sigma \cup \{\varepsilon\}\rangle$-pushdown automaton $\mathfrak{P}$ such that $\|\mathfrak{P}\| = r$.*

Corollary 10. (Chomsky [4]) *A formal language is context-free iff it is accepted by a pushdown automaton.*

If our basic semiring is $\mathbb{N}^\infty\langle\!\langle \Sigma^* \rangle\!\rangle$, we can draw some even stronger conclusions. In our next result we consider, for a given pushdown automaton $\mathfrak{P}' = (Q, \Sigma, \Gamma, \delta, q_0, p_0, F)$, the number of distinct computations from the *initial instantaneous description* (q_0, w, p_0) for w to an *accepting instantaneous description* $(q, \varepsilon, \varepsilon)$, $q \in F$.

Theorem 33. (Kuich [13]) *Let L be a formal language over Σ and let $d : \Sigma^* \to \mathbb{N}^\infty$. Then the following two statements are equivalent:*

(i) There exists a context-free grammar with terminal alphabet Σ such that the number (possibly ∞) of distinct leftmost derivations of w, $w \in \Sigma^*$, from the start variable is given by $d(w)$.

(ii) There exists a pushdown automaton with input alphabet Σ such that the number (possibly ∞) of distinct computations from the initial instantaneous description for w, $w \in \Sigma^*$, to an accepting instantaneous description is given by $d(w)$.

A pushdown automaton with input alphabet Σ is termed *unambiguous* iff, for each word $w \in \Sigma^*$ that is accepted, there exists a unique computation from the initial instantaneous description for w to some accepting instantaneous description.

Corollary 11. *A formal language is generated by an unambiguous context-free grammar iff it is accepted by an unambiguous pushdown automaton.*

References

1. Berstel, J., Reutenauer, C.: Les séries rationelles et leurs langages, Masson (1984); English translation: Rational Series and Their Languages. EATCS Monographs on Theoretical Computer Science, vol. 12. Springer, Heidelberg (1988)
2. Bloom, S.L., Ésik, Z.: Iteration Theories. EATCS Monographs on Theoretical Computer Science. Springer, Heidelberg (1993)
3. Bozapalidis, S.: Context-free series on trees. Information and Computation 169, 186–229 (2001)

4. Chomsky, N.: Context-free grammars and pushdown storage. MIT Res. Lab. of Elect., Quarterly Prog. Rep. 65, 187–194 (1962)
5. Droste, M., Kuich, W.: Semirings and formal power series. In: Droste, M., Kuich, W., Vogler, H. (eds.) Handbook of Weighted Automata, pp. 3–28. Springer, Heidelberg (2009)
6. Ésik, Z., Kuich, W.: Formal tree series. J. Automata, Languages and Combinatorics 8, 219–285 (2003)
7. Ésik, Z., Kuich, W.: Modern Automata Theory, http://www.dmg.tuwien.ac.at/kuich
8. Ésik, Z., Kuich, W.: A unifying Kleene Theorem for weighted finite automata. In: Calude, C.S., Rozenberg, G., Salomaa, A. (eds.) Rainbow of Computer Science. LNCS, vol. 6570, pp. 76–89. Springer, Heidelberg (2011)
9. Ésik, Z., Leiß, H.: Greibach normal form in algebraically complete semirings. In: Bradfield, J.C. (ed.) CSL 2002 and EACSL 2002. LNCS, vol. 2471, pp. 135–150. Springer, Heidelberg (2002)
10. Guessarian, I.: Algebraic Semantics. LNCS, vol. 99. Springer, Heidelberg (1981)
11. Harrison, M.A.: Introduction to Formal Language Theory. Addison-Wesley, Reading (1978)
12. Knopp, K.: Theorie und Anwendungen der unendlichen Reihen. 3. Aufl. Springer, Heidelberg (1931)
13. Kuich, W.: Semirings and formal power series: Their relevance to formal languages and automata theory. In: Rozenberg, G., Salomaa, A. (eds.) Handbook of Formal Languages, vol. 1, ch. 9, pp. 609–677. Springer, Heidelberg (1997)
14. Kuich, W.: Pushdown tree automata, algebraic tree systems, and algebraic tree series. Information and Computation 165, 69–99 (2001)
15. Kuich, W., Salomaa, A.: Semirings, Automata, Languages. EATCS Monographs on Theoretical Computer Science, vol. 5. Springer, Heidelberg (1986)
16. Markowsky, G.: Chain-complete posets and directed sets with applications. Algebra Universalis 6, 53–68 (1976)
17. Petre, I., Salomaa, A.: Algebraic systems and pushdown automata. In: Droste, M., Kuich, W., Vogler, H. (eds.) Handbook of Weighted Automata, pp. 257–289. Springer, Heidelberg (2009)
18. Sakarovitch, J.: Éléments de théorie des automates, Vuibert (2003)
19. Salomaa, A., Soittola, M.: Automata-Theoretic Aspects of Formal Power Series. Springer, Heidelberg (1978)

Where Automatic Structures Benefit from Weighted Automata

Dietrich Kuske

Institut für Theoretische Informatik,
Technische Universität Ilmenau, Germany

Abstract. In this paper, we report on applications of weighted automata in the theory of automatic structures. All (except one) result were known before, but their proof using weighted automata is novel. More precisely, we prove that the extension of first-order logic by the infinity $\exists^\infty$, the modulo $\exists^{(p,q)}$, and the (new) boundedness quantifier $\boxminus$ is decidable. The first two quantifiers are handled using closure properties of the class of recognizable formal power series and the fact that the preimage of a value under a recognizable formal power series is regular if the semiring is finite. Our reasoning regarding the boundedness quantifier uses Weber's decidability result of finite-valued rational transductions. We also show that the isomorphism problem of automatic structures is undecidable using an undecidability result on recognizable formal power series due to Honkala.

1 Introduction

The idea of an automatic structure goes back to Büchi and Elgot who used finite automata to decide, e.g., Presburger arithmetic [12]. In essence, a structure is automatic if the elements of the universe can be represented as strings from a regular language and every relation of the structure can be recognized by a finite automaton with several heads that proceed synchronously. Automaton decidable theories [17] and automatic groups [13] are similar concepts. A systematic study was initiated by Khoussainov and Nerode [19] who also coined the name "*automatic structure*". They received increasing interest over the last years [4,5,9,21,22,28,1,20,25,3,32,26,24,6]; the surveys [29,2] give excellent overviews of the results in this area. One of the main motivations for investigating automatic structures is that their first-order theories are decidable. From the beginning, researchers were also interested in possible extensions of this result to stronger logics. The first part of this paper contributes to this search by (1) providing a new proof technique and (2) providing a further extension of first-order logic with this favorable property.

Another natural line of research dealt with the question which structures from a given class $\mathfrak{C}$ are automatic, i.e., can be represented by finite automata. There are only very few results in this direction (for instance, the characterisations are known for ordinals [10], Boolean algebras [20], and finitely generated groups [28]) and, as it turns out, the first two characterisations are accompanied by the

W. Kuich and G. Rahonis (Eds.): Bozapalidis Festschrift, LNCS 7020, pp. 257–271, 2011.

decidability of the isomorphism problem of the automatic members in the given class $\mathfrak{C}$. On the other hand, it was shown that the isomorphism problem for all automatic structures is undecidable and even Σ_1^1-complete [20]. But these undecidability proofs depend crucially on binary non-transitive relations. The second part of this paper shows that weighted automata can be used to prove this undecidability for equivalence relations, a class of structures with a particularly simple transitive relation [24]. Honkala proved that it is undecidable whether a weighted automaton over $(\mathbb{N}, +, \cdot, 0, 1)$ takes on all values from $\mathbb{N}$. We reduce this to our isomorphism problem. This proof technique differs from the original one in [24] only in that it makes the role of the weighted automata more transparent.

2   Preliminaries

Let Γ be an alphabet and $w \in \Gamma^*$ be a finite word over Γ. The length of w is denoted by $|w|$.

2.1   Structures

A *signature* is a finite set τ of relational symbols, where every symbol $R \in \tau$ has some fixed arity m_R. Then a τ-*structure* $\mathcal{A}$ consists of a non-empty universe A and, for every $R \in \tau$, an m_R-ary relation $R^{\mathcal{A}} \subseteq A^{m_R}$. Note that we only consider relational structures. Let us fix a τ-structure $\mathcal{A} = (A, (R^{\mathcal{A}})_{R \in \tau})$, where $R^{\mathcal{A}} \subseteq A^{m_R}$. To simplify notation, we will write $a \in \mathcal{A}$ for $a \in A$. In the rest of the paper, we will often identify a symbol $R \in \tau$ with its interpretation $R^{\mathcal{A}}$.

2.2   Automatic Structures

Let us fix $m \in \mathbb{N}$ and a finite alphabet Γ. Let $\# \notin \Gamma$ be an additional padding symbol and set $\Gamma_{\#} = \Gamma \cup \{\#\}$. We will write $\Gamma_{\#}^m$ for $(\Gamma_{\#})^m$. For words $w_i \in \Gamma^*$ $(1 \le i \le n)$ we define the *convolution* $w_1 \otimes w_2 \otimes \cdots \otimes w_m$, which is a word over the alphabet $\Gamma_{\#}^m$, as follows: Let $w_i = a_{i,1} a_{i,2} \cdots a_{i,k_i}$ with $a_{i,j} \in \Gamma$ and $k = \max\{k_1, \ldots, k_m\}$. For $k_i < j \le k$ define $a_{i,j} = \#$. Then

$$w_1 \otimes \cdots \otimes w_m = (a_{1,1}, \ldots, a_{m,1}) \cdots (a_{1,k}, \ldots, a_{m,k}).$$

Thus, for instance $aba \otimes bbabb = (a, b)(b, b)(a, a)(\#, b)(\#, b)$.

An m-*dimensional (synchronous) automaton* over Γ is just a finite automaton A over the alphabet $\Gamma_{\#}^m$ such that $L(A) \subseteq \{w_1 \otimes \cdots \otimes w_m \mid w_1, \ldots, w_m \in \Gamma^*\}$. Such an automaton defines an m-ary relation

$$R(A) = \{(w_1, \ldots, w_m) \mid w_1 \otimes \cdots \otimes w_m \in L(A)\}.$$

An m-ary relation $R \subseteq (\Gamma^*)^m$ is *automatic* if it is accepted by some m-dimensional automaton or, equivalently, if the language $R^{\otimes} = \{w_1 \otimes \cdots \otimes w_m \mid (w_1, \ldots, w_m) \in R\} \subseteq (\Gamma_{\#}^m)^*$ is regular.

An *automatic presentation* is a tuple $P = (\Gamma, A_0, (A_R)_{R \in \tau})$, where:

- Γ is an alphabet.
- A_0 is a finite automaton over the alphabet Γ.
- τ is a signature, as before m_R is the arity of the symbol $R \in \tau$.
- For every $R \in \tau$, A_R is an m_R-dimensional automaton over the alphabet Γ such that $R(A_R) \subseteq L(A_0)^{m_R}$.

The structure presented by P is

$$\mathcal{A}(P) = (L(A_0), (R(A_R))_{R \in \tau}).$$

A structure $\mathcal{A}$ is called *automatic* if there exists an automatic presentation P such that $\mathcal{A} \cong \mathcal{A}(P)$.

By SA, we denote the set of all automatic presentations. Similar notions of automaticity can be based on finite tree automata, on ω-string, and on ω tree automata. The corresponding sets of presentations are denoted TA, ωSA, and ωTA, resp., but this paper will only be concerned with the set SA.

Examples

- All finite structures $\mathcal{A}$ are automatic with alphabet the universe of $\mathcal{A}$. While there are many infinite automatic structures (see below), there are no infinite automatic fields [20].
- The complete binary tree with universe $\{0, 1\}^*$, together with the binary relations "first son" S_0, "second son" S_1, "prefix" $\leq$, and "equal length" is automatic.
- Presburger arithmetic $(\mathbb{N}, +)$ is automatic: the alphabet is $\{0, 1\}$, the language of A_0 is $\{0, 1\}^*1$ where the word $a_0 a_1 \ldots a_n$ represents the number $\sum_{0 \leq i \leq n} a_i 2^i$. Differently Skolem arithmetic $(\mathbb{N}, \cdot)$ is not automatic [4]. Blumensath also showed that Skolem arithmetic is tree-automatic [4].
- The linear order $(\mathbb{Q}, \leq)$ is automatic: the universe is $\{0, 1\}^*$ with $u < v$ if and only if $(u \wedge v)0$ is a prefix of u or $(u \wedge v)1$ is a prefix of v (where $u \wedge v$ is the longest common prefix of u and v). This presentation is even "automatic-homogeneous": Let $u_1, \ldots, u_n$ and $v_1, \ldots, v_n$ be increasing sequences of equal length. Then there is an automatic automorphism f of $(\{0, 1\}^*, \leq)$ mapping u_i to v_i [23]. The rational line is a particular Fraïssé-limit, other examples are the random graph and the universal and homogeneous poset [16]. It is known that many such limits are not automatic [9,20].
- The rewrite graph $(\Sigma^*, \rightarrow)$ of every semi-Thue system and therefore the configuration graph of every Turing machine are automatic.
- The extension of this configuration graph by the binary relation of reachability is in general not automatic. But for pushdown automata, the configuration graph with reachability $(Q\Gamma^*, \rightarrow, \rightarrow^*)$ is automatic: a configuration is represented by the control state followed by the stack content.
- The theory of automatic structures was preceded by that of automatic groups [13] and semigroups [7]. In terms of automatic structures, a semigroup is automatic (in the original sense) if its Cayley-graph has an automatic presentation such that $L(A_0)$ forms a rational cross-section of the (semi-)group. Many natural groups and semigroups were shown to be automatic and therefore to have automatic Cayley-graphs:

- rational monoids [30],
- virtually free finitely generated, virtually free Abelian finitely generated, and hyperbolic groups [13],
- singular Artin monoids of finite type [8], and
- graph products of such monoids [15].

In contrast, it seems that not many infinite groups are automatic in the sense of this article. For instance, a finitely generated group is automatic if and only if it is virtually Abelian [28]. Braun and Strüngmann showed that every automatic torsion-free Abelian group is the extension of $(\mathbb{Z}^k, +)$ for some $k \in \mathbb{N}$ by a direct sum of finitely many Prüfer groups [6]. This implies Tsankov's celebrated result that $(\mathbb{Q}, +)$ is not automatic [32].

- An ordinal α is automatic if and only if $\alpha < \omega^\omega$ [10]. This proof was later generalized to show that the Hausdorff rank of every automatic linear order is finite [22]. This characterization (together with Theorem 3.2 below) can be used to show that the isomorphism of automatic ordinals is decidable. But note that this does not hold for automatic linear orders [24].
- Let $\mathcal{B}$ denote the Boolean algebra of all finite and co-finite subsets of $\mathbb{N}$. Then an infinite Boolean algebra is automatic if and only if it is a finite power of $\mathcal{B}$. Again, this characterisation leads to the decidability of the isomorphism of automatic Boolean algebras [20].

3   Definable Relations

3.1   The Classical Result on First-Order Logic FO

Fix a signature τ. Then let V be a countably infinite set of variables. Formulas of FO are then built according to the following formation rules (where α and β are formulas, $x, y, y_1, \ldots, y_k \in V$ are variables, and R is a k-ary relation symbol):

(L1) $x = y$ (L4) $\neg \alpha$
(L2) $R(y_1, \ldots, y_k)$ (L5) $\exists x \colon \alpha$
(L3) $\alpha \vee \beta$

We next recall the semantics of formulas from FO. To this aim, let $\mathcal{A}$ be a τ-structure with universe A. An *interpretation in $\mathcal{A}$* is a function $f \colon V \to A$. Given such an interpretation, we set $\mathcal{A} \models_f \varphi$ (read as "φ holds in $\mathcal{A}$ under the interpretation f") if and only if one of the following hold

(S1) $\varphi = (x = y)$ and $f(x) = f(y)$.
(S2) $\varphi = (R(y_1, \ldots, y_k))$ and $(f(y_1), \ldots, f(y_k)) \in R^{\mathcal{A}}$.
(S3) $\varphi = (\alpha \vee \beta)$ and $\mathcal{A} \models_f \alpha$ or $\mathcal{A} \models_f \beta$.
(S4) $\varphi = \neg \alpha$ and not $\mathcal{A} \models_f \alpha$.
(S5) $\varphi = \exists x \colon \alpha$ and there exists $a \in \mathcal{A}$ with $\mathcal{A} \models_{f[\frac{a}{x}]} \alpha$ where $f[\frac{a}{x}]$ is the interpretation that differs from f only in that it maps x to a.

It is an easy exercise to show the following: let $\mathcal{A}$ be a τ-structure, φ a formula, and suppose $f(y) = g(y)$ for all $y \in \text{free}(\varphi)$, the set of variables occurring freely in φ. Then $\mathcal{A} \models_f \varphi$ if and only if $\mathcal{A} \models_g \varphi$. Assuming a fixed tuple of variables $(y_1, \ldots, y_n)$ with free$(\varphi) \subseteq \{y_1, \ldots, y_n\}$, we can therefore simply write $\mathcal{A} \models \varphi(f(y_1), \ldots, f(y_n))$ for $\mathcal{A} \models_f \varphi$. In particular, for *sentences* (i.e., formulas without free variables), it makes sense to write $\mathcal{A} \models \varphi$.

Let $\varphi \in$ FO be some formula with free$(\varphi) \subseteq \{x_1, \ldots, x_n\}$ and let $\mathcal{A}$ be some τ-structure. Then

$$\varphi^{\mathcal{A}} = \{(u_1, \ldots, u_n) \in \mathcal{A}^n \mid \mathcal{A} \models \varphi(u_1, \ldots, u_n)\}$$

is a relation on the universe of the structure $\mathcal{A}$ that represents the semantics of the formula φ. The study of this relation is central in model theory [16] as well as in computable model theory (for computable instead of arbitrary structures) [14]. Consequently, they have also been studied in the context of automatic structures in which case $\varphi^{\mathcal{A}}$ is a relation on the set of words Γ^* for some alphabet Γ. The most important result is that they are effectively automatic (see below). Before we prove this result, we make the following definition: for a relation $R \subseteq X^{n+1}$ with $n \geq 0$, define the relation $(\exists R) \subseteq X^n$ by

$$(\exists R) = \{(x_1, \ldots, x_n) \in X^n \mid \text{ there exists } x \in X \text{ with } (x_1, \ldots, x_n, x) \in R\}.$$

Let $X = \Gamma^*$ and $n = 0$. Then we get $(\exists R) \subseteq (\Gamma^*)^0 = \{()\}$ and consequently $(\exists R)^{\otimes} \subseteq \{\varepsilon\}$. We then have the following:

Proposition 3.1. *If $R \subseteq (\Gamma^*)^{n+1}$ is automatic with $n \geq 0$, then $(\exists R)$ is automatic and an automaton accepting $(\exists R)^{\otimes}$ can be computed from an automaton accepting $R^{\otimes}$.*

Proof. Let proj: $\Gamma_{\#}^{n+1} \to \Gamma_{\#}^n$ be the projection that deletes the last component. We naturally extend it to a monoid homomorphism proj: $(\Gamma_{\#}^{n+1})^* \to (\Gamma_{\#}^n)^*$. Now let $u_1, \ldots, u_n \in \Gamma^*$. Then we have

$$
\begin{aligned}
u_1 \otimes \cdots \otimes u_n \in (\exists R)^{\otimes} &\iff (u_1, u_2, \ldots, u_n) \in (\exists R) \\
&\iff \text{there is some } u \in \Gamma^* \text{ with } (u_1, u_2, \ldots, u_n, u) \in R \\
&\iff \text{there is some } u \in \Gamma^* \text{ with } u_1 \otimes \cdots \otimes u_n \otimes u \in R^{\otimes} \\
&\iff (u_1 \otimes \cdots \otimes u_n)(\#, \ldots, \#)^* \cap \text{proj}(R^{\otimes}) \neq \emptyset \\
&\iff (u_1 \otimes \cdots \otimes u_n) \in \text{proj}(R^{\otimes})((\#, \ldots, \#)^*)^{-1}
\end{aligned}
$$

where $KL^{-1} = \{x \mid \text{ there is some } y \in L \text{ with } xy \in K\}$ for two languages K and L. Hence $(\exists R)^{\otimes} = \text{proj}(R^{\otimes})((\#, \ldots, \#)^*)^{-1}$. Since $R^{\otimes}$ is regular and proj is a monoid morphism, it follows that $(\exists R)^{\otimes}$ is regular, i.e., that $(\exists R)$ is automatic. $\square$

Using this proposition, we easily get the following central result.

Theorem 3.2 (cf. [17,19]). *Let $P = (\Gamma, A_0, (A_R)_{R \in \tau})$ be an automatic presentation and $\varphi \in$ FO a formula with free$(\varphi) \subseteq \{y_1, \ldots, y_n\} \subseteq V$. Then the relation $\varphi^{\mathcal{A}(P)}$ is automatic. Even more, an n-dimensional automaton for this relation can be computed from P and φ.*

Proof (sketch). The automaton is constructed by induction on the structure of the formula φ (where we assume that α and β are formulas such that $\alpha^{\mathcal{A}(P)}$ and $\beta^{\mathcal{A}(P)}$ are effectively regular):

- if $\varphi = (x_i = x_j)$, then an automaton for $\varphi^{\mathcal{A}(P)}$ is obtained from A_0.
- if $\varphi = R(x_1, \ldots, x_n)$, then $\varphi^{\mathcal{A}(P)} = R(A_R)$ is effectively automatic.
- if $\varphi = (\alpha \vee \beta)$, then $\varphi^{\mathcal{A}(P)} = \alpha^{\mathcal{A}(P)} \cup \beta^{\mathcal{A}(P)}$ is effectively automatic.
- if $\varphi = \neg \alpha$, then $\varphi^{\mathcal{A}(P)} = L(A_0)^n \setminus \alpha^{\mathcal{A}(P)}$ is effectively automatic.
- if $\varphi = \exists x \colon \alpha$, then $\varphi^{\mathcal{A}(P)} = (\exists \alpha^{\mathcal{A}(P)})$ which is effectively automatic by Prop. 3.1. □

Our principal aim is to extend first-order logic by additional quantifiers and prove the corresponding extension of Theorem 3.2. To simplify terminology, we introduce the following notation.

Definition 3.3. *Let $\dashv$ be a function that maps $2^{(\Gamma^*)^{n+k}}$ to $2^{(\Gamma^*)^n}$ for all $n \in \mathbb{N}$ and some fixed $k \in \mathbb{N}$. Then $\dashv$ preserves automaticity effectively if one can compute an n-dimensional automaton accepting $\dashv(R(A))$ from the $(n+k)$-dimensional automaton A.*

For $n = 0$, we have $\dashv(R(A)) \subseteq \{()\}$, i.e., $(\dashv(R(A)))^\otimes \subseteq \{\varepsilon\}$ and therefore $\varepsilon \in (\dashv(R(A)))^\otimes$ if and only if $\dashv(R(A)) \neq \emptyset$. Thus, to show that $\dashv(R(A))$ is effectively regular (at least for $n = 0$), we have to device an algorithm that decides whether $\dashv(R(A))$ is empty or not. In other words, the property that $\dashv$ preserves automaticity effectively is a generalization of the decidability of the emptiness problem for $\dashv(R(A))$.

Using this definition, Prop. 3.1 can now be phrased more concisely: The function $\exists$ that maps $R \subseteq (\Gamma^*)^{n+1}$ to $(\exists R) \subseteq (\Gamma^*)^n$ preserves automaticity effectively.

3.2   The Infinity Quantifier

For a relation $R \subseteq X^{n+1}$ with $n \geq 0$ let

$$(\exists^\infty R) = \{(x_1, \ldots, x_n) \in X^n \mid \text{there are infinitely many } x \in X \text{ with } (x_1, \ldots, x_n, x) \in R\}.$$

We will show that $\exists^\infty$ preserves automaticity effectively. A short proof of this fact was given by Blumensath [4] using Theorem 3.2. Our admittedly longer proof uses classical results from the theory of weighted automata and therefore fits into the setting of this paper.

Consider the complete semiring $\mathbb{N}_\infty = (\mathbb{N} \cup \{\infty\}, +, \cdot, 0, 1)$ with $0 \cdot \infty = \infty \cdot 0 = 0$.

Lemma 3.4. *From an $n+1$-dimensional automaton $\mathcal{A}$, one can construct an $\mathbb{N}_\infty$-weighted automaton $\mathcal{B}$ over the alphabet $\Gamma_\#^n$ such that*

$$(\|\mathcal{B}\|, V) = \begin{cases} |\{u \mid (u_1, u_2, \ldots, u_n, u) \in R\}| & \text{if } V = u_1 \otimes u_2 \otimes \cdots \otimes u_n, \\ 0 & \text{otherwise.} \end{cases}$$

Proof. Since $R^\otimes$ is regular, the mapping

$$1_{R^\otimes} : (\Gamma_\#^{n+1})^* \to \mathbb{N}_\infty : U \mapsto \begin{cases} 1 & \text{if } U \in R^\otimes \\ 0 & \text{otherwise} \end{cases}$$

is a recognizable and therefore rational formal power series. Recall that the projection morphism proj from the proof of Prop. 3.1 is length-preserving. Hence the formal power series $1_{R^\otimes} \circ \text{proj}^{-1}$ defined by

$$S_1 = 1_{R^\otimes} \circ \text{proj}^{-1} : (\Gamma_\#^n)^* \to \mathbb{N}_\infty : V \mapsto \sum((1_{R^\otimes}, U) \mid \text{proj}(U) = V)$$

is rational by [11, Prop. 3.6(ii)]. Note that $(S_1, V) = |\text{proj}^{-1}(V) \cap R^\otimes|$. Hence, for $V = (u_1 \otimes u_2 \otimes \cdots \otimes u_n)(\#, \ldots, \#)^k$, we have

$$(S_1, V) = |\{u \in \Gamma^* \mid (u_1, \ldots, u_n, u) \in R \text{ and } |u| = \max(|u_1|, |u_2|, \ldots, |u_n|) + k\}|.$$

If V is not of this form, then $(S_1, V) = 0$.

Next let del: $(\Gamma_\#^n)^* \to (\Gamma_\#^n)^*$ be the monoid morphism with $\text{del}(\#, \ldots \#) = \varepsilon$ and $\text{del}(A) = A$ for all other letters $A \in \Gamma_\#^n$. In other words, del deletes all occurrences of $(\#, \ldots, \#)$ from a word. Since the semiring $\mathbb{N}_\infty$ is complete, the formal power series

$$S_2 = S_1 \circ \text{del}^{-1} : (\Gamma_\#^n)^* \to \mathbb{N}_\infty : V \mapsto \sum((S_1, W) \mid \text{del}(W) = V)$$

is rational and therefore recognizable by [11, Prop. 3.6(ii)]. Note that $(S_2, u_1 \otimes u_2 \otimes \cdots \otimes u_n)$ is the number of words u with $(u_1, \ldots, u_n, u) \in R$.

All the results cited have effective proofs, so from the above arguments we can extract an algorithm for the construction of a weighted automaton $\mathcal{B}$ for the series S_2. □

We now come to the central result on $\exists^\infty$:

Proposition 3.5 ([4]). *The function $\exists^\infty$ that maps $R \subseteq (\Gamma^*)^{n+1}$ to $(\exists^\infty R)$ preserves automaticity effectively.*

Proof. Let $\mathcal{B}$ be the weighted automaton from Lemma 3.4.

Now consider the semiring $(\{0, 1, \infty\}, \max, \cdot, 0, 1)$ with $0 \cdot \infty = \infty \cdot 0 = 0$. Then $h : \mathbb{N}_\infty \to \{0, 1, \infty\}$ with $h(0) = 0$, $h(\infty) = \infty$, and $h(m) = 1$ for all $0 < m < \infty$ is a semiring homomorphism. Hence, by [31, Prop. 4.5],

$$S = h \circ ||\mathcal{B}|| : (\Gamma_\#^n)^* \to \{0, 1, \infty\} : V \mapsto h((||\mathcal{B}||, V))$$

is a recognizable formal power series into a finite semiring. Therefore the set

$$\{V \in (\Gamma_\#^n)^* \mid (S, V) = \infty\}$$

is effectively regular [31, Prop. 6.3]. Since the set $((\Gamma^*)^n)^\otimes$ of convolutions of n words over Γ is regular, also the set

$$\{V \in (\Gamma_\#^n)^* \mid (S, V) = \infty\} \cap ((\Gamma^*)^n)^\otimes$$

is effectively regular.

It remains to verify that this set equals $(\exists^\infty R)^\otimes$: So let $V = (u_1 \otimes \cdots \otimes u_n)$. Then $(\|\mathcal{B}\|, V)$ is the number of words u with $(u_1, \ldots, u_n, u) \in R$. Hence $(S, V) = \infty$ if and only if $(\|\mathcal{B}\|, V) = \infty$ if and only if $V \in (\exists^\infty R)^\otimes$ which proves the claim. $\square$

3.3   The Modulo Quantifier

For a relation $R \subseteq X^{n+1}$ with $n \geq 0$ and $0 \leq p < q$ let

$$(\exists^{(p,q)} R) = \{(x_1, \ldots, x_n) \in X^n \mid |\{x \in X \mid (x_1, \ldots, x_n, x) \in R\}| \text{ is}$$
$$\text{finite and congruent } p \text{ modulo } q\}.$$

Proposition 3.6 ([21]). *The function $\exists^{(p,q)}$ that maps $R \subseteq (\Gamma^*)^{n+1}$ to the relation $(\exists^{(p,q)} R)$ preserves automaticity effectively uniformly in (p,q). In other words, there exists one algorithm that takes as input p, q, and an automaton for R and returns an automaton for $(\exists^{(p,q)} R)$.*

The result was first stated in [21] where one finds a proof for the case $q = 2$, a proof of the general case can be found in [29, Thm. 3.19]. Both these proofs construct the automaton for $(\exists^{(p,q)} R)$ directly. Differently, the new proof below is based on the theory of weighted automata.

Proof. First note that $R \setminus ((\exists^\infty R) \times \Gamma^*)$ is the set of tuples $(u_1, \ldots, u_{n+1}) \in R$ such that there are only finitely many $u \in \Gamma_\#^*$ that can replace u_{n+1}, i.e., that satisfy $(u_1, \ldots, u_n, u) \in R$. Hence

$$(\exists^{(p,q)} R) = (\exists^{(p,q)} (R \setminus ((\exists^\infty R) \times \Gamma^*))).$$

Since $R \setminus ((\exists^\infty R) \times \Gamma^*)$ is effectively automatic by Prop. 3.5 (in conjunction with the effective closure of automatic relations under Boolean operations and direct products), it suffices to consider the case $R = R \setminus ((\exists^\infty R) \times \Gamma^*)$.

Let $\mathcal{B}$ be the weighted automaton from Lemma 3.4.

If V is the convolution of n words, then $(\|\mathcal{B}\|, V) \in \mathbb{N}$ by our assumption on R. If V is not a convolution of n words, then $(\|\mathcal{B}\|, V) = 0 \in \mathbb{N}$. It follows that $\|\mathcal{B}\|$ is even a rational formal power series over the natural semiring $(\mathbb{N}, +, \cdot, 0, 1)$ (simply replace all weights ∞ in $\mathcal{B}$ by 0 or any other natural number).

We now consider the (semi)ring $\mathbb{Z}/q\mathbb{Z} = (\{0, 1, \ldots, q-1\}, +, \cdot, 0, 1)$. Applying [11, Prop. 3.5], we find that

$$S \colon (\Gamma_\#^n)^* \to \mathbb{Z}/q\mathbb{Z} \colon V \mapsto (\|\mathcal{B}\|, V) \bmod q$$

is a rational formal power series into a finite semiring. As in the proof of Prop. 3.5, it follows that the set

$$\{V \in (\Gamma_\#^n)^* \mid (S, V) = p\} \cap ((\Gamma^*)^n)^\otimes$$

is effectively regular.

It remains to verify that this set equals $(\exists^{(p,q)} R)^\otimes$: So let $V = (u_1 \otimes \cdots \otimes u_n)$. Then $(\|\mathcal{B}\|, V)$ is the number of words u with $(u_1, \ldots, u_n, u) \in R$. Hence $(S, V) = p$ if and only if $(\|\mathcal{B}\|, V) \equiv p \pmod{q}$ if and only if $V \in (\exists^{(p,q)} R)^\otimes$ which proves the claim. $\square$

3.4   The Boundedness Quantifier

For a relation $R \subseteq X^{n+2}$ with $n \geq 0$ let

$$(\exists R) = \{(x_1, \ldots, x_n) \in X^n \mid \text{there is some } m \in \mathbb{N} \text{ such that}$$
$$|\{z \mid (x_1, \ldots, x_n, y, z) \in R\}| \leq m \text{ for all } y \in X\}.$$

We will show that $(\exists R)$ is effectively automatic if R is automatic. A central notion in this proof is that of a *finite valued* function $f \colon X \to 2^Y$ by which we mean that there is some $m \in \mathbb{N}$ such that $|f(y)| \leq m$ for all $y \in X$. We first handle the case $n = 0$.

Lemma 3.7. *If $R \subseteq (\Gamma^*)^2$ is automatic, then $(\exists R)$ is automatic and an automaton accepting $(\exists R)^\otimes$ can be computed from an automaton accepting $R^\otimes$.*

Proof. As explained after Definition 3.3, we have to decide whether the function

$$\Gamma^* \to 2^{\Gamma^*} \colon y \mapsto \{z \in \Gamma^* \mid (y, z) \in R\}$$

is finite valued. Since R is an automatic relation, it is a rational transduction. Hence the result follows from [33]. □

From now on, let $n \geq 1$. It will be convenient to consider $(\exists R)$ as the intersection of the following two relations:

$$(\exists_{\leq} R) = \{(x_1, \ldots, x_n) \in (\Gamma^*)^n \mid \text{there is some } m \in \mathbb{N} \text{ such that}$$
$$|\{z \mid (x_1, \ldots, x_n, y, z) \in R\}| \leq m \text{ for all } y \in \Gamma^*$$
$$\text{with } |y| \leq |x_i| \text{ for all } 1 \leq i \leq n\}$$

$$(\exists_{>} R) = \{(x_1, \ldots, x_n) \in (\Gamma^*)^n \mid \text{there is some } m \in \mathbb{N} \text{ such that}$$
$$|\{z \mid (x_1, \ldots, x_n, y, z) \in R\}| \leq m \text{ for all } y \in \Gamma^*$$
$$\text{with } |y| > |x_i| \text{ for all } 1 \leq i \leq n\}$$

Lemma 3.8. *The function $\exists_{\leq}$ that maps $R \subseteq (\Gamma^*)^{n+2}$ to $(\exists_{\leq} R)$ preserves automaticity effectively.*

Proof. For notational convenience, we only prove this lemma for $n = 1$, the general case can easily be shown along the same lines.

Let A be a deterministic finite automaton accepting $R^\otimes$. Let Q denote its set of states, ι the initial state, and F the set of accepting states. For three words $x, y, z \in \Gamma^*$ and a state $q \in Q$, we write $q.(x, y, z)$ for the state reached from q when executing $x \otimes y \otimes z$. Finally, let $\ell_q = |\{z'' \in \Gamma^+ \mid q.(\varepsilon, \varepsilon, z'') \in F\}| \in \mathbb{N}_\infty$ for $q \in Q$.

Now let $x, y \in \Gamma^*$ with $|y| \leq |x|$. Then we have

$$|\{z \mid (x, y, z) \in R\}| = |\{z \in \Gamma^{\leq |x|} \mid (x, y, z) \in R\}| + |\{z \in \Gamma^{> |x|} \mid (x, y, z) \in R\}|$$

$$\leq |\Gamma^{\leq |x|}| + \sum_{q \in Q} \left(\begin{array}{c} |\{z' \in \Gamma^{=|x|} \mid \iota.(x, y, z') = q\}| \\ \cdot |\{z'' \in \Gamma^+ \mid q.(\varepsilon, \varepsilon, z'') \in F\}| \end{array} \right)$$

$$\leq |\Gamma^{\leq |x|}| + |\Gamma^{=|x|}| \cdot \sum_{q \in H} \ell_q$$

where H is the set of all states $\iota.(x, y, z')$ for some $z' \in \Gamma^{=|x|}$.

Hence $x \notin (\text{Ǝ}_{\leq} R)$ if and only if there exist some $y, z' \in \Gamma^*$ with $|y| \leq |x| = |z'|$ and $\ell_{\iota.(x,y,z')} = \infty$. But this is a regular property, so also $(\text{Ǝ}_{\leq} R)$ is regular.

Note that ℓ_q can be computed from the automaton A and the state q. Hence an automaton for $(\text{Ǝ}_{\leq} R)$ can be computed.   □

Lemma 3.9. *The function* $\text{Ǝ}_>$ *that maps* $R \subseteq (\Gamma^*)^{n+2}$ *to* $(\text{Ǝ}_> R) \subseteq (\Gamma^*)^n$ *preserves automaticity effectively.*

Proof. We use the notation from the first paragraph of the proof of Lemma 3.8 (except for ℓ_q). Now let $x, y \in \Gamma^*$ with $|y| > |x|$ and write $y = y'y''$ with $|y'| = |x|$. Then we have

$$|\{z \mid (x,y,z) \in R\}| = |\{z \in \Gamma^{\leq |x|} \mid (x,y,z) \in R\}| + |\{z \in \Gamma^{> |x|} \mid (x,y,z) \in R\}|$$

$$\leq |\Gamma^{\leq |x|}| + \sum_{q \in Q} \left(\begin{array}{c} |\{z' \in \Gamma^{=|x|} \mid \iota.(x,y',z') = q\}| \\ \cdot |\{z'' \in \Gamma^+ \mid q.(\varepsilon, y'', z'') \in F\}| \end{array} \right)$$

$$\leq |\Gamma^{\leq |x|}| + |\Gamma^{|x|}| \cdot \sum_{q \in H} |\{z'' \in \Gamma^+ \mid q.(\varepsilon, y'', z'') \in F\}|$$

where H is the set of all states $\iota.(x,y',z')$ for some $z' \in \Gamma^{=|x|}$.

We now define a set $F' \subseteq Q$ of states (that plays the role of the set of states q with $\ell_q \in \mathbb{N}$ from the proof of Lemma 3.8). Let $q \in F'$ if the mapping

$$\Gamma^* \to 2^{\Gamma^*} : y'' \mapsto \{z'' \in \Gamma^+ \mid q.(\varepsilon, y'', z'') \in F\} \tag{1}$$

is finite valued. Then it is easily seen that $x \notin (\text{Ǝ}_> R)$ if and only if there are words $y', z' \in \Gamma^{=|x|}$ with $\iota.(x, y', z') \notin F'$. But this is a regular property, so also $(\text{Ǝ}_> R)$ is regular.

By [33], one can decide from the automaton A and the state q whether the mapping (1) is finite valued. Hence an automaton for $(\text{Ǝ}_> R)$ can be computed.   □

Proposition 3.10. *The function* Ǝ *that maps* $R \subseteq (\Gamma^*)^{n+2}$ *to* $(\text{Ǝ}R)$ *preserves automaticity effectively.*

Proof. The proof is immediate by Lemmas 3.7, 3.8, and 3.9 since, for $n \geq 1$, we have $(\text{Ǝ}R) = (\text{Ǝ}_{\leq} R) \cap (\text{Ǝ}_> R)$ and since automatic relations are effectively closed under intersection.   □

This proposition and consequently its proof are genuinely new, the case $n = 2$ was used in [24] to show that the set of presentations of trees of finite height is decidable.

3.5   Summary and Model Checking

Recall that first-order formulas were defined by the formation rules (L1-5). The central theorem on first-order logic was proved using that the existential quantifier preserves automaticity effectively. This statement was also shown for other "quantifiers". We will now extend first-order logic by these quantifiers, state the extended version of Theorem 3.2, and infer that the model checking problem for this large logic and automatic structures is decidable.

Formulas of FO_{ext} are built according to the formation rules (L1-9) (where α is a formula):

(L6) $\exists^{\infty}x\colon \alpha$ (L8) $\exists(y,z)\colon \alpha$
(L7) $\exists^{(p,q)}x\colon \alpha$ for $0 \le p < q$ (L9) $\mathbf{0}^k(y_1,\ldots,y_k)\colon \alpha$ for $k \ge 1$

We set $\mathcal{A} \models_f \varphi$ if and only if one of (S1-9) hold:

(S6) $\varphi = \exists^{\infty}x\colon \alpha$ and there exist infinitely many $a \in \mathcal{A}$ with $\mathcal{A} \models_{f[\frac{a}{x}]} \alpha$. For instance, $\forall y \, \neg\exists^{\infty}z\colon E(y,z)$ says of a directed graph that it has finite out-degree.

(S7) $\varphi = \exists^{(p,q)}x\colon \alpha$ and the number of elements $a \in \mathcal{A}$ with $\mathcal{A} \models_{f[\frac{a}{x}]} \alpha$ is finite and congruent p modulo q. For instance, $\exists^{(0,2)}x\colon x = x$ expresses that a structure is finite and has an even number of elements.

(S8) $\varphi = \exists(y,z)\colon \alpha$ and there exists $m \in \mathbb{N}$ such that, for all $a \in \mathcal{A}$, the set

$$\{b \in \mathcal{A} \mid \mathcal{A} \models_{f[\frac{ab}{yz}]} \alpha\}$$

contains at most m elements. In other words, $\exists(y,z)\colon \alpha$ expresses the existence of some natural number m such that any element y has at most m partners z that make α true. For instance, $\exists(y,z)\colon E(y,z)$ says of a directed graph that it has bounded out-degree.

(S9) $\mathbf{0}^k(y_1,\ldots,y_k)\colon \alpha$ for some $k \ge 1$ and there exists some infinite set $X \subseteq \mathcal{A}$ such that $\mathcal{A} \models_g \alpha$ for all $a_1,\ldots,a_k \in X$ with $g = f[\frac{a_1\ldots a_k}{y_1\ldots y_k}]$. For instance, $\mathbf{0}^2(y,z)\colon (y = z \vee (E(y,z) \wedge E(z,y))$ expresses that a directed graph has an infinite clique. For this reason, $\mathbf{0}^k$ is called *Ramsey quantifier*. In [29], it is shown that the Ramsey quantifier preserves automaticity effectively (uniformly in k).

Theorem 3.11. *Let $P = (\Gamma, A_0, (A_R)_{R\in\tau})$ be an automatic presentation and $\varphi \in FO_{ext}$ a formula with $free(\varphi) \subseteq \{y_1,\ldots,y_n\} \subseteq V$. Then the relation $\varphi^{\mathcal{A}(P)}$ is automatic. Even more, an n-dimensional automaton for this relation can be computed from P and φ.*

Proof. The proof is an immediate extension of the proof of Theorem 3.2 using in addition Prop. 3.5, 3.6, 3.10 and [29, Thm. 3.20].   □

Corollary 3.12. *The set of pairs (P, φ) where P is an automatic presentation and φ a sentence from FO_{ext} with $\mathcal{A}(P) \models \varphi$ is decidable.*

Proof. From Theorem 3.11, we get an automaton A accepting $(\varphi^{\mathcal{A}(P)})^{\otimes} \subseteq \{\varepsilon\}$. Then ε is accepted by A if and only if $\mathcal{A}(P) \models \varphi$.   □

Open question. In [26], we present a generalisation of the Ramsey quantifier and show Corollary 3.12 for the extension of first-order logic by all the infinity, the modulo, and this generalized quantifier (and therefore the Ramsey quantifier). In other words, we have two extensions of the logic with infinity, modulo, and Ramsey quantifier (namely by the boundedness quantifier and by the generalized Ramsey quantifier) such that Corollary 3.12 holds. But it is not clear whether these two extensions together give decidability.

4   Isomorphism

The *isomorphism problem* for automatic structures is the set of pairs of automatic presentations P and Q such that $\mathcal{A}(P) \cong \mathcal{A}(Q)$. Using a result by Honkala on weighted automata over $(\mathbb{N}, +, \cdot, 0, 1)$, we show in this section that the isomorphism problem for automatic structures is not decidable. More precisely, we present a single structure $\mathcal{E}_{\text{good}} = (V, \sim)$ where $\sim$ is an equivalence relation on V such that the set of automatic presentations of $\mathcal{E}_{\text{good}}$ is not decidable.

An *equivalence structure* is a structure $\mathcal{E} = (V, \sim)$ such that V is at most countably infinite and $\sim$ is an equivalence relation on V. Let $\mathcal{E}_{\text{good}}$ be the equivalence structure with universe $a^*b^*c^*$ and $a^k b^\ell c^m \sim a^{k'} b^{\ell'} c^{m'}$ if and only if $k + \ell = k' + \ell'$ and $m = m'$. It has, for every $n \in \mathbb{N}$, infinitely many equivalence classes of size $n + 1$, and no infinite equivalence class.

Now let $\mathcal{B}$ be a weighted automaton over the natural semiring $(\mathbb{N}, +, \cdot, 0, 1)$ and the alphabet Σ. Without changing the behavior of $\mathcal{B}$, i.e., the formal power series $\|\mathcal{B}\| : \Sigma^* \to \mathbb{N}$, we can assume $\mathcal{B}$ to be normalized, i.e., all the initial and final weights are 0 or 1.

From $\mathcal{B}$, we now construct an equivalence structure as follows: Let $m \in \mathbb{N}$ be the maximal weight appearing in $\mathcal{B}$. Let Γ be the set of states of $\mathcal{B}$ together with all pairs (a, k) where $a \in \Sigma$ and $0 \le k < m$. Then the universe of the equivalence structure $\mathcal{E}_{\mathcal{B}}$ is the set of sequences

$$\rho = q_0 (a_1, k_1) q_1 (a_1, k_1) q_1 \ldots (a_n, k_n) q_n \$^\ell$$

such that k_i is properly smaller than the weight of the transition (q_{i-1}, a_i, q_i) in $\mathcal{B}$ (for $1 \le i \le n$), the entry weight of q_0 and the final weight of q_n are 1. The sequence ρ and the sequence

$$\rho' = q_0' (b_1', k_1') q_1' (b_1', k_1') q_1' \ldots (b_{n'}, k_{n'}) q_{n'} \$^{\ell'}$$

are equivalent ($\rho \sim \rho'$) if

$$a_1 a_2 \ldots a_n = b_1 b_2 \ldots b_{n'} \text{ and } \ell = \ell'.$$

Then the equivalence class of ρ with respect to $\sim$ has precisely $(\|\mathcal{B}\|, a_1 a_2 \ldots a_n)$ elements. If $\sim$ has one equivalence class of size k, then it has infinitely many such equivalence classes because of the suffix from $\$^*$. Hence we get

$$\mathcal{E}_{\text{good}} \cong \mathcal{E}_{\mathcal{B}} \iff \forall n \in \mathbb{N} \exists u \in \Sigma^* : (\|\mathcal{B}\|, u) = n + 1.$$

By [18], this latter question is undecidable. Hence we showed

Theorem 4.1 ([24]). *The isomorphism for automatic equivalence structures is undecidable (more precisely: Π_1^0-complete since it is in Π_1^0 by [29]).*

One can easily transform an equivalence structure $\mathcal{E} = (V, \sim)$ into a tree $T_\mathcal{E}$ of height 2: V is the root, $V/\sim$ is the set of children of the root, and V is the set of leaves. A leaf $v \in V$ is a child of $a \in V/\sim$ if and only if $v \in a$. It is even possible to construct an automatic presentation of this tree. Then

$$\mathcal{E} \cong \mathcal{E}' \iff T_\mathcal{E} \cong T_{\mathcal{E}'}$$

implies

Corollary 4.2 ([24]). *The isomorphism problem for automatic trees of height 2 is undecidable (more precisely. Π_1^0-hard).*

In [24], we showed that the isomorphism problem for automatic trees of height n is complete for Π_{2n-3}^0 and the isomorphism problem for all order trees is Σ_1^1-complete; the above result is the (idea of the) base case. Prior to these results, it was known that the isomorphism problem for automatic successor trees is Σ_1^1-complete [20,27].

Open question. In [24], we also show that the isomorphism problem for automatic linear orders is Σ_1^1-complete. Recall that a linear order is *scattered* if it does not contain a copy of $(\mathbb{Q}, \leq)$. We also showed that the isomorphism problem for scattered automatic linear orders is significantly simpler (namely, in Δ_ω^0, i.e., reducible to true first order arithmetic). But it is still not known whether this problem is decidable (we only have an undecidability proof for tree-automatic scattered linear orders).

Acknowledgement. I thank Martin Huschenbett for reading two earlier versions of this paper very carefully.

References

1. Bárány, V.: Invariants of automatic presentations and semi-synchronous transductions. In: Durand, B., Thomas, W. (eds.) STACS 2006. LNCS, vol. 3884, pp. 289–300. Springer, Heidelberg (2006)
2. Bárány, V., Grädel, E., Rubin, S.: Automata-based presentations of infinite structures. In: Finite and Algorithmic Model Theory, pp. 1–76. Cambridge University Press, Cambridge (2011)
3. Bárány, V., Kaiser, Ł., Rubin, S.: Cardinality and counting quantifiers on omega-automatic structures. In: STACS 2008, pp. 385–396. IFIB Schloss Dagstuhl (2008)
4. Blumensath, A.: Automatic structures. Tech. rep., RWTH Aachen (1999)
5. Blumensath, A., Grädel, E.: Automatic Structures. In: LICS 2000, pp. 51–62. IEEE Computer Society Press, Los Alamitos (2000)

6. Braun, G., Strüngmann, L.: Breaking up finite automata presentable torsion-free abelian groups. International Journal of Algebra and Computation (accepted, 2011)
7. Campbell, C., Robertson, E., Ruškuc, N., Thomas, R.: Automatic semigroups. Theoretical Computer Science 250, 365–391 (2001)
8. Corran, R., Hoffmann, M., Kuske, D., Thomas, R.: Singular Artin monoids of finite type are automatic. In: Dediu, A.-H., Inenaga, S., Martín-Vide, C. (eds.) LATA 2011. LNCS, vol. 6638, pp. 250–261. Springer, Heidelberg (2011)
9. Delhommé, C.: Automaticité des ordinaux et des graphes homogènes. C. R. Acad. Sci. Paris, Ser. I 339, 5–10 (2004)
10. Delhommé, C., Goranko, V., Knapik, T.: Automatic linear orderings (2003) (manuscript)
11. Droste, M., Kuich, W.: Semirings and formal power series. In: Droste, M., Kuich, W., Vogler, H. (eds.) Handbook of Weighted Automata, pp. 3–28. Springer, Heidelberg (2009)
12. Elgot, C.: Decision problems of finite automata design and related arithmetics. Trans. Am. Math. Soc. 98, 21–51 (1961)
13. Epstein, D., Cannon, J., Holt, D., Levy, S., Paterson, M., Thurston, W.: Word Processing In Groups. Jones and Bartlett Publishers, Boston (1992)
14. Ershov, Y., Goncharov, S., Nerode, A., Remmel, J. (eds.): Handbook of recursive mathematics, vol. 1, 2. Elsevier, Amsterdam (1998)
15. Fohry, E., Kuske, D.: On graph products of automatic and biautomatic monoids. Semigroup Forum 72, 337–352 (2006)
16. Hodges, W.: Model Theory. Cambridge University Press, Cambridge (1993)
17. Hodgson, B.: On direct products of automaton decidable theories. Theoretical Computer Science 19, 331–335 (1982)
18. Honkala, J.: On the problem whether the image of an ℕ-rational series equals ℕ. Fundamenta Informaticae 73(1-2), 127–132 (2006)
19. Khoussainov, B., Nerode, A.: Automatic presentations of structures. In: Leivant, D. (ed.) LCC 1994. LNCS, vol. 960, pp. 367–392. Springer, Heidelberg (1995)
20. Khoussainov, B., Nies, A., Rubin, S., Stephan, F.: Automatic structures: richness and limitations. Log. Methods in Comput. Sci. 3(2) (2007)
21. Khoussainov, B., Rubin, S., Stephan, F.: Definability and regularity in automatic structures. In: Diekert, V., Habib, M. (eds.) STACS 2004. LNCS, vol. 2996, pp. 440–451. Springer, Heidelberg (2004)
22. Khoussainov, B., Rubin, S., Stephan, F.: Automatic linear orders and trees. ACM Transactions on Computational Logic 6(4), 675–700 (2005)
23. Kuske, D.: Is Cantor's theorem automatic? In: Vardi, M.Y., Voronkov, A. (eds.) LPAR 2003. LNCS, vol. 2850, pp. 332–345. Springer, Heidelberg (2003)
24. Kuske, D., Liu, J., Lohrey, M.: The isomorphism problem on classes of automatic structures. In: LICS 2010, pp. 160–169. IEEE Computer Society Press, Los Alamitos (2010)
25. Kuske, D., Lohrey, M.: First-order and counting theories of ω-automatic structures. Journal of Symbolic Logic 73, 129–150 (2008)
26. Kuske, D., Lohrey, M.: Some natural problems in automatic graphs. Journal of Symbolic Logic 75(2), 678–710 (2010)
27. Nies, A.: Describing groups. Bulletin of Symbolic Logic 13(3), 305–339 (2007)
28. Oliver, G., Thomas, R.: Automatic presentations for finitely generated groups. In: Diekert, V., Durand, B. (eds.) STACS 2005. LNCS, vol. 3404, pp. 693–704. Springer, Heidelberg (2005)

29. Rubin, S.: Automata presenting structures: A survey of the finite string case. Bulletin of Symbolic Logic 14, 169–209 (2008)
30. Sakarovitch, J.: Easy multiplications. I. The realm of Kleene's Theorem. Information and Computation 74, 173–197 (1987)
31. Sakarovitch, J.: Rational and recognizable series. In: Droste, M., Kuich, W., Vogler, H. (eds.) Handbook of Weighted Automata, pp. 405–453. Springer, Heidelberg (2009)
32. Tsankov, T.: The additive group of the rationals does not have an automatic presentation (2009), http://arxiv.org/abs/0905.1505
33. Weber, A.: On the valuedness of finite transducers. Acta Informatica 27, 749–780 (1990)

Survey:
Weighted Extended Top-Down Tree Transducers
Part III — Composition⋆

Aurélie Lagoutte[1] and Andreas Maletti[2],⋆⋆

[1] École normale supérieure de Cachan,
Département Informatique,
61, avenue du Président Wilson, 94235 Cachan cedex, France
`aurelie.lagoutte@ens-cachan.fr`
[2] Universität Stuttgart,
Institute for Natural Language Processing,
Azenbergstraße 12, 70174 Stuttgart, Germany
`andreas.maletti@ims.uni-stuttgart.de`

Abstract. In this survey (functional) compositions of weighted tree transformations computable by weighted extended top-down tree transducers are investigated. The existing results in the literature are explained and illustrated. It is argued, why certain compositions are not possible in the general case, and 3 informed conjectures provide an insight into potentially 3 new composition results that extend and complement the existing results. In particular, if all were true, then the beautiful symmetry in the composition results for weighted top-down and bottom-up tree transducers would be recovered.

Keywords: weighted tree transducer, top-down tree transducer, composition, deletion, copying.

1   Motivation

Weighted tree transducers [32,14,19] (also called 'tree series transducers') are a joint generalization of the unweighted tree transducer (such as the top-down tree transducer [41,42] or the bottom-up tree transducer [43]) and the weighted tree automaton [6,9,1,30,17,8,7]. A good overview over both predecessors is presented in [20]. For a more detailed historic account and an in-depth introduction into weighted extended top-down tree transducers, we refer the reader to the first part [36] of this survey. A popular application area that has driven tree transducer research in the past few years is (syntax-based) machine translation [28,27]. The second part [37] of this survey attempts to present a high-level perspective on some of the essential problems and algorithms used in this application domain.

⋆ The work was carried out while the first author was an intern at *Universität Stuttgart*.
⋆⋆ Financially supported by the German Research Foundation (DFG) grant MA/4959/1-1.

W. Kuich and G. Rahonis (Eds.): Bozapalidis Festschrift, LNCS 7020, pp. 272–308, 2011.

In this part of the survey, we will investigate compositions of weighted extended (top-down) tree transducers. Such a tree transducer computes a weighted relation between input and output trees (i.e., it assigns a weight to each pair of input and output trees). Two such relations can be composed in the usual manner with the only difference that a weight (for example, a confidence or a probability) is returned each time "membership is tested". Using the real numbers as weight structure, we compose two weighted relations $\tau_1 \colon A \times B \to \mathbb{R}$ and $\tau_2 \colon B \times C \to \mathbb{R}$ by requiring that

$$(\tau_1 \,;\, \tau_2)(a, c) = \sum_{b \in B} \tau_1(a, b) \cdot \tau_2(b, c) \tag{1}$$

for all $a \in A$ and $c \in C$. For the sake of simplicity, let us assume that B is finite. Equation (1) can be imagined in an operational manner. The first process transforms the input a into an intermediate product b at a certain cost $\tau_1(a, b)$. This intermediate product is then fed into the second process, which transforms it into a final product at cost $\tau_2(b, c)$. Thus, the components are executed in a sequential manner. Traditionally, the multiplicative operation of the weight structure (typically, a semiring [25,23]) is used to combine weights of processes that are executed in series (or in sequence). Consequently, we multiply the weights $\tau_1(a, b) \cdot \tau_2(b, c)$ to obtain the cost of producing c from a via the intermediate product b. Naturally, there might be a choice of intermediate products that are all suitable to some degree to produce the output c. Thus, we sum over all the possibilities of producing c from a.

Compositions have been and are used in a number of application areas as diverse as machine translation [46] and functional program optimization [29]. The complexity of a given tree transformation problem can be tackled and broken down into smaller pieces with the help of (de-)composition in a *divide-and-conquer* approach. Once all the subproblems have been solved, the individual solutions can be recombined with the help of composition. This approach is used in [46], where a translation model is broken into 3 smaller pieces:

- a reordering component, which changes the order of subtrees but keeps the trees otherwise intact,
- an insertion component, which has the ability to spontaneously add subtrees to the output of the translation, and
- a translation component, which just translates the words (or phrases) occurring in the input tree.

These components can now be trained and optimized individually (even from different resources). However, since the evaluation of composition chains can be very inefficient [40], automatic procedures that "compose" finite representations of such weighted relations are desirable. Naturally, the obtained finite representation should compute the composition of the weighted relations computed by the input representations. As expected, the finite representation discussed in this survey is the weighted extended top-down tree transducer (xtt) together with its variants.

In contrast to the first part [36] of this survey, we do not require a complete semiring [25,23] here, which yields that we have to avoid infinite summations. This change prompts a minor change in the definition of the model because we have to disallow rules that contain no input and output symbol at all. In fact, infinite sums also restrict the potential compositions because we have to guarantee that the sum in (1) is finite. This is achieved by two simple conditions, of which each is sufficient to guarantee the finiteness of the sum in (1). Moreover, we introduce a simple variant of our main model that has rule identifiers in order to simplify the composition constructions. The additional indirection via identifiers allows us to construct the same rule several times under different names. In this way we can obtain a closer and more direct relationship between the rules of the input xtt and the composed xtt.

In the main part of this survey, we investigate compositions of xtt. In other words, given two xtt M and N, we want to construct another xtt that computes the composition of the weighted tree transformations computed by the xtt M and N. It is known that already in the unweighted setting, this cannot always be achieved, and we will consider two important cases:

- compositions of an xtt with a top-down tree transducer, and
- compositions of selected xtt with top-down tree transducers that can additionally have ε-rules.

The former case has been investigated in the unweighted case by [12,5] and these results were partially lifted to the weighted setting in [14,33,34]. We recall all the relevant results and complement them by three conjectured results that handle the missing cases. More precisely, we conjecture:

- that a constant xtt can be composed with a linear top-down tree transducer (see Conjecture 11), where the property 'constant' will be introduced here,
- that a deterministic xtt can be composed with a nondeleting top-down tree transducer (see Conjecture 13), and
- that a constant and deterministic xtt can be composed with any top-down tree transducer (see Conjecture 14).

We explain why these conjectured cases cause additional problems, which are due to the presence of weights. While we will not present a formal construction for each case, we present a generic composition construction and then indicate how to modify it to obtain a formal construction for the individual cases.

The latter case, in which we compose an xtt with a top-down tree transducer with ε-rules, was investigated in [39] in the unweighted setting. Here, we extend the results of [39] to the weighted setting and conjecture a new result (see Conjecture 23), which is again based on the new property 'constant'. Overall, our conjectured results complement the existing results nicely, and if all were true, then we would obtain the beautiful symmetry in the weighted setting that is known from compositions [12,5] for unweighted top-down and bottom-up tree transducers [43].

2 Notation

The set of all nonnegative integers is $\mathbb{N}$, and we let $[n] = \{i \in \mathbb{N} \mid 1 \leq i \leq n\}$ for every $n \in \mathbb{N}$. We fix the set $X = \{x_i \mid i \in \mathbb{N}\}$ of (formal) variables. The set of all finite words (or sequences) over a set S is S^*, where ε is the empty word. The concatenation of the words $v, w \in S^*$ is $v.w$ or simply vw. The length of a word $w \in S^*$ is denoted by $|w|$. An *alphabet* Σ is a nonempty and finite set, of which the elements are called *symbols*. For every alphabet Q, we let $Q(S) = \{q(s) \mid q \in Q, s \in S\}$. The set $T_\Sigma(S)$ of Σ-trees[1] *with leaf labels S* is the smallest set T such that $S \subseteq T$ and $\sigma(t_1, \ldots, t_k) \in T$ for every $k \in \mathbb{N}$, $\sigma \in \Sigma$, and $t_1, \ldots, t_k \in T$. We often omit qualifications like 'for all $k \in \mathbb{N}$' if it is obvious from the context that k is a nonnegative integer. Moreover, we generally assume that $\Sigma \cap S = \emptyset$, and thus we write $\sigma()$ simply as σ for every $\sigma \in \Sigma$. Given another alphabet Δ, we treat elements of $T_\Delta(T_\Sigma(S))$ and $Q(T_\Sigma(S))$ as particular trees of $T_{Q \cup \Sigma \cup \Delta}(S)$.[2] Finally, we write T_Σ for $T_\Sigma(\emptyset)$.

Next, we define a few operations on trees. The set $\mathrm{pos}(t) \subseteq \mathbb{N}^*$ of *positions* of a tree $t \in T_\Sigma(S)$ is inductively defined by $\mathrm{pos}(s) = \{\varepsilon\}$ for every $s \in S$ and

$$\mathrm{pos}(\sigma(t_1, \ldots, t_k)) = \{\varepsilon\} \cup \{i.w \mid i \in [k], w \in \mathrm{pos}(t_i)\}$$

for every $\sigma \in \Sigma$ and $t_1, \ldots, t_k \in T_\Sigma(S)$. The set $\mathrm{pos}(t)$ of positions is (totally) ordered by the lexicographic order on $\mathbb{N}^*$. Let $t, t' \in T_\Sigma(S)$ and $w \in \mathrm{pos}(t)$. The *label* of t at position w is $t(w)$, and the *w-rooted subtree* of t is $t|_w$. Formally, these notions can be defined by $s(\varepsilon) = s|_\varepsilon = s$ for every $s \in S$ and

$$\sigma(t_1, \ldots, t_k)(\varepsilon) = \sigma \qquad\qquad \sigma(t_1, \ldots, t_k)(i.v) = t_i(v)$$
$$\sigma(t_1, \ldots, t_k)|_\varepsilon = \sigma(t_1, \ldots, t_k) \qquad\qquad \sigma(t_1, \ldots, t_k)|_{i.v} = t_i|_v$$

for every $\sigma \in \Sigma$, $t_1, \ldots, t_k \in T_\Sigma(S)$, $i \in [k]$, and $v \in \mathrm{pos}(t_i)$. For every subset $L \subseteq \Sigma \cup S$ of labels and $s \in S$, we let $\mathrm{pos}_L(t) = \{w \in \mathrm{pos}(t) \mid t(w) \in L\}$ and $\mathrm{pos}_s(t) = \mathrm{pos}_{\{s\}}(t)$. The expression $t[u]_w$ denotes the tree that is obtained from $t \in T_\Sigma(S)$ by replacing the subtree $t|_w$ at position w by $u \in T_\Delta(S)$.

The following operations implicitly always use the fixed set X of variables. We let $\mathrm{var}(t) = \{x \in X \mid \mathrm{pos}_x(t) \neq \emptyset\}$. The tree t is *linear* if every $x \in X$ occurs at most once in t. A *substitution* $\theta \colon X \to T_\Sigma(S)$ can be applied to a tree $t \in T_\Sigma(S)$, and returns the tree $t\theta$ that is obtained by replacing (in parallel) all occurrences of each variable $x \in X$ by $\theta(x)$. Formally, (i) $x\theta = \theta(x)$ for every $x \in X$, (ii) $s\theta = s$ for every $s \in S \setminus X$, and (iii) $\sigma(t_1, \ldots, t_k)\theta = \sigma(t_1\theta, \ldots, t_k\theta)$ for every $\sigma \in \Sigma$ and $t_1, \ldots, t_k \in T_\Sigma(S)$.

A *(commutative) semiring* [25,23] is an algebraic structure $(A, +, \cdot, 0, 1)$ consisting of two commutative monoids $(A, +, 0)$ and $(A, \cdot, 1)$ such that $\cdot$ distributes

---

[1] These are actually unranked trees, but our operational tree transformation model will only have finitely many rules that prescribe (and limit) the ranks of symbols, so that we could have used a ranked alphabet as well.

[2] A benefit of our approach without explicit ranks for symbols is that we can always take the union of two alphabets.

over finite sums (including the empty sum, which yields $a \cdot 0 = 0$ for all $a \in A$). The semiring Λ is *idempotent* if $1 + 1 = 1$.[3] Examples of semirings include

- the BOOLEAN semiring $(\{0, 1\}, \max, \min, 0, 1)$, which is idempotent,
- the powerset[4] semiring $(\mathcal{P}(S), \cup, \cap, \emptyset, S)$ for some set S, which is idempotent,
- the tropical semiring $(\mathbb{R} \cup \{\infty\}, \min, +, \infty, 0)$, which is also idempotent,
- the nonnegative integers $(\mathbb{N}, +, \cdot, 0, 1)$, and
- the semiring of real numbers $(\mathbb{R}, +, \cdot, 0, 1)$.

Let S and T be sets, and let $(A, +, \cdot, 0, 1)$ be a semiring. A *weighted relation r from S to T* is a mapping $r \colon S \times T \to A$. Moreover, for every mapping $f \colon S \to A$, we let $\mathrm{supp}(f) = \{s \in S \mid f(s) \neq 0\}$. Thus, $\mathrm{supp}(r) \subseteq S \times T$.

From now on, let $(A, +, \cdot, 0, 1)$ be an arbitrary semiring such that $0 \neq 1$.

Next, let us recall the weighted extended (top-down) tree transducer [11,2,26,24]. We essentially follow the definitions of [35,38], in which the corresponding unweighted device is discussed in detail. An in-depth presentation of the weighted device can be found in the first part [36] of this survey. A *(weighted) extended (top-down) tree transducer* (xtt) is a tuple $(Q, \Sigma, \Delta, I, R)$, where

- Q is a finite set of *states*,
- Σ and Δ are alphabets of *input* and *output symbols* such that $Q \cap (\Sigma \cup \Delta) = \emptyset$,
- $I \subseteq Q$ is a set of *initial states*, and
- $R \colon Q(T_\Sigma(X)) \times T_\Delta(Q(X)) \to A$ assigns *rule weights* such that $\mathrm{supp}(R)$ is finite and for every $(l, r) \in \mathrm{supp}(R)$ we have that $\{l, r\} \not\subseteq Q(X)$, l is linear, and $\mathrm{var}(r) \subseteq \mathrm{var}(l)$.[5]

For the following discussion, let $M = (Q, \Sigma, \Delta, I, R)$ be an xtt. The elements of $\mathrm{supp}(R)$ are called *rules* (of M), and we often write them as $l \to r$ instead of (l, r). We call l and r of a rule $l \to r$ the *left-* and *right-hand side*, respectively. Moreover, we write $l \to r \in R$ instead of $(l, r) \in \mathrm{supp}(R)$, and we write $l \xrightarrow{a} r \in R$ instead of $R(l, r) = a$. A rule $l \to r \in R$ is (i) *linear* if r is linear, (ii) *nondeleting* if $\mathrm{var}(r) = \mathrm{var}(l)$, and (iii) *simple* if $|\mathrm{pos}_\Sigma(l)| = 1$. In addition, the rule $l \to r$ is

- *consuming* if $|\mathrm{pos}_\Sigma(l)| \geq 1$, and an *$\varepsilon$-rule* otherwise, and
- *producing* if $|\mathrm{pos}_\Delta(r)| \geq 1$, and *erasing* otherwise.[6]

The xtt M is (i) *linear*, (ii) *nondeleting*, and (iii) a *top-down tree transducer* [32,14] (tdtt) if every rule $l \to r \in R$ is (i) linear, (ii) nondeleting, and (iii) simple, respectively. Moreover, the xtt M is BOOLEAN if $R(l, r) = 1$ for every $l \to r \in R$.

The semantics of the xtt M is given by term rewriting [13,4,38]. To simplify our composition constructions later on, we immediately present a semantics that

---

[3] By distributivity, this yields $a + a = a$ for all $a \in A$.
[4] The powerset $\mathcal{P}(S)$ of a set S is the set of its subsets; i.e., $\mathcal{P}(S) = \{U \mid U \subseteq S\}$.
[5] The restriction $\{l, r\} \not\subseteq Q(X)$, which is not present in [36], disallows rules of the form $(q(x_i), p(x_i))$. This additional restriction is necessary because we do require complete semirings [25,23], which yields that we have to avoid infinite summations.
[6] The name 'erasing' is justified by the fact that each erasing rule is consuming.

can handle "foreign" symbols, which are symbols that are not in $Q \cup \Sigma \cup \Delta$.[7] Let Σ' and Δ' be such that $\Sigma \subseteq \Sigma'$ and $\Delta \subseteq \Delta'$. The elements of $T_{\Delta'}(Q(T_{\Sigma'}(X)))$ are called *sentential forms*. A position $w \in \mathrm{pos}_Q(\xi)$ in a sentential form ξ is *reducible (for M)* if there exists a rule $l \to r \in R$ and a substitution $\theta \colon X \to T_{\Sigma'}(X)$ such that $\xi|_w = l\theta$. Let $\xi, \zeta \in T_{\Delta'}(Q(T_{\Sigma'}(X)))$ be sentential forms and $l \to r \in R$ be a rule. We say that ξ *rewrites to* ζ *using* $l \to r$, denoted by $\xi \Rightarrow_M^{(l,r)} \zeta$, if there exists a substitution $\theta \colon X \to T_{\Sigma'}(X)$ such that $\xi|_w = l\theta$ and $\zeta = \xi[r\theta]_w$ where w is the least reducible position in $\mathrm{pos}_Q(\xi)$ with respect to the lexicographic total ordering on $\mathbb{N}^*$.[8] As usual, we use ';' for relation composition, thus for example,

$$(\Rightarrow_M^{\rho_1} ; \Rightarrow_M^{\rho_2}) = \{(\xi, \zeta) \mid \exists \xi' \colon \xi \Rightarrow_M^{\rho_1} \xi' \Rightarrow_M^{\rho_2} \zeta\} \ .$$

The *(extended) weighted relation* τ_M' (or weighted tree transformation) *computed by M* is given by

$$\tau_M'(\xi, \zeta) = \sum_{\substack{\rho_1, \ldots, \rho_k \in \mathrm{supp}(R) \\ \xi \Rightarrow_M^{\rho_1}; \cdots; \Rightarrow_M^{\rho_k} \zeta}} \left(\prod_{i=1}^k R(\rho_i) \right)$$

for every $\xi, \zeta \in T_{\Delta'}(Q(T_{\Sigma'}(X)))$. The *semantics τ_M* of the xtt M is the weighted relation $\tau_M \colon T_\Sigma \times T_\Delta \to A$ such that $\tau_M(t, u) = \sum_{q \in I} \tau_M'(q(t), u)$ for every $t \in T_\Sigma$ and $u \in T_\Delta$.[9] Finally, the xtt M is *deterministic*[10] (respectively, *total*) if for all $q \in Q$ and $t \in T_\Sigma$ there exists at most (respectively, at least) one $u \in T_\Delta$ such that $(q(t), u) \in \mathrm{supp}(\tau_M')$.[11]

3   An Example Composition

We start our investigation into compositions of xtt with an example to illustrate the problem and the general principle used to solve it. Roughly speaking, given two xtt M and N we want to construct a single xtt that behaves like the two xtt M and N in sequence. Before we move to the formal definition of composition, let us introduce two example xtt and demonstrate derivations (i.e., term rewrite steps).

---

[7] A definition of the semantics without this extension can be found in [36].

[8] Given a sentential form ξ and a rule $\rho \in R$, there exists at most one sentential form ζ such that $\xi \Rightarrow_M^\rho \zeta$.

[9] Since the xtt M cannot consume symbols from $\Sigma' \setminus \Sigma$ and cannot produce symbols from $\Delta' \setminus \Delta$, the semantics τ_M does not depend on the particular choice of Σ' and Δ'.

[10] This property should correctly be called 'unambiguous', but for historical reasons we use 'deterministic' in the following.

[11] For top-down tree transducers these properties are typically defined using syntactic restrictions [12,5], which imply our corresponding semantic conditions. It requires a significant technical overhead to generalize the syntactic conditions faithfully to xtt, so we chose to present only the semantic property.

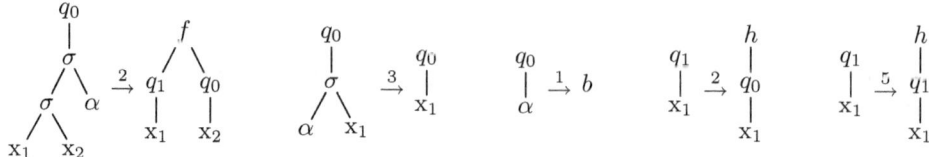

Fig. 1. The rules of the xtt M (see Example 1)

Example 1. We assume that the used semiring is the semiring $(\mathbb{R}, +, \cdot, 0, 1)$ of real numbers. Let $M = (\{q_0, q_1\}, \Sigma, \Gamma, \{q_0\}, R)$ be the xtt with input alphabet $\Sigma = \{\sigma, \alpha\}$, output alphabet $\Gamma = \{f, h, b\}$, and the following rules:

$$\rho_1 : \quad q_0(\sigma(\sigma(\sigma(x_1, x_2), \alpha)) \xrightarrow{2} f(q_1(x_1), q_0(x_2)) \qquad \rho_4 : \quad q_1(x_1) \xrightarrow{2} h(q_0(x_1))$$

$$\rho_2 : \qquad\qquad q_0(\sigma(\alpha, x_1)) \xrightarrow{3} q_0(x_1) \qquad\qquad \rho_5 : \quad q_1(x_1) \xrightarrow{5} h(q_1(x_1))$$

$$\rho_3 : \qquad\qquad\qquad q_0(\alpha) \xrightarrow{1} b \ .$$

We illustrate these rules in Fig. 1 and demonstrate their properties in the following table. Since all rules are linear and nondeleting, the xtt M is linear and nondeleting.

ε-rule	consuming	erasing	producing	linear	nondeleting
ρ_4, ρ_5	$\rho_1 - \rho_3$	ρ_2	$\rho_1, \rho_3 - \rho_5$	$\rho_1 - \rho_5$	$\rho_1 - \rho_5$

Next, let us demonstrate a derivation using M. As input and output tree we consider

$$s = \sigma(\sigma(\alpha, \sigma(\alpha, \sigma(\alpha, \alpha))), \alpha) \qquad \text{and} \qquad t = f(h(h(b)), b) \ .$$

Figure 2 shows a derivation from $q_0(s)$ to t. Its weight is

$$R(\rho_1) \cdot R(\rho_5) \cdot R(\rho_4) \cdot R(\rho_3) \cdot R(\rho_2) \cdot R(\rho_2) \cdot R(\rho_3) = 2 \cdot 5 \cdot 2 \cdot 1 \cdot 3 \cdot 3 \cdot 1 = 180$$

It is easy to verify that this is the only possible derivation from $q_0(s)$ to t. Since q_0 is the only initial state, we can conclude that $\tau_M(s, t) = 180$. □

Example 2. We keep the semiring of real numbers as our used semiring. A second xtt is given by $N = (\{p\}, \Gamma, \Delta, \{p\}, R')$, where $\Gamma = \{f, h, b\}$, $\Delta = \{\lambda, \gamma, \delta, \beta\}$, and R' contains the rules:

$$\mu_1 : \qquad p(x_1) \xrightarrow{2} \gamma(p(x_1)) \qquad\qquad \mu_4 : \quad p(h(x_1)) \xrightarrow{8} \delta(p(x_1))$$

$$\mu_2 : \quad p(f(x_1, x_2)) \xrightarrow{5} \lambda(p(x_1), p(x_2)) \qquad \mu_5 : \qquad p(b) \xrightarrow{1} \beta$$

$$\mu_3 : \quad p(f(x_1, x_2)) \xrightarrow{5} \lambda(\beta, \lambda(p(x_1), p(x_2))) \ .$$

Again, the properties of the rules are documented in the following table. We observe that the xtt N is linear and nondeleting.

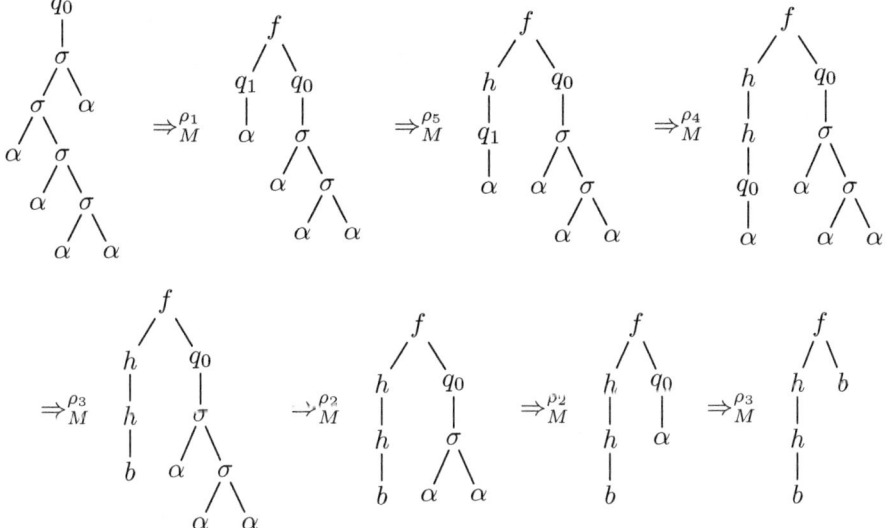

Fig. 2. A derivation from $q_0(s)$ to t in M (see Example 1)

ε-rule	consuming	erasing	producing	linear	nondeleting
μ_1	$\mu_2-\mu_5$		$\mu_1-\mu_5$	$\mu_1-\mu_5$	$\mu_1-\mu_5$

This time we illustrate a derivation for the input and output tree

$$t = f(h(h(b)), b) \qquad \text{and} \qquad u = \lambda(\beta, \lambda(\delta(\delta(\beta)), \beta)) \ .$$

Figure 3 shows the unique derivation from $p(t)$ to u. It has the weight

$$R'(\mu_3) \cdot R'(\mu_4) \cdot R'(\mu_4) \cdot R'(\mu_5) \cdot R'(\mu_5) = 5 \cdot 8 \cdot 8 \cdot 1 \cdot 1 = 320 \ . \qquad \square$$

Composition is the process of running two xtt one after the other. In this way, the output tree of the first xtt becomes the input tree of the second xtt. For

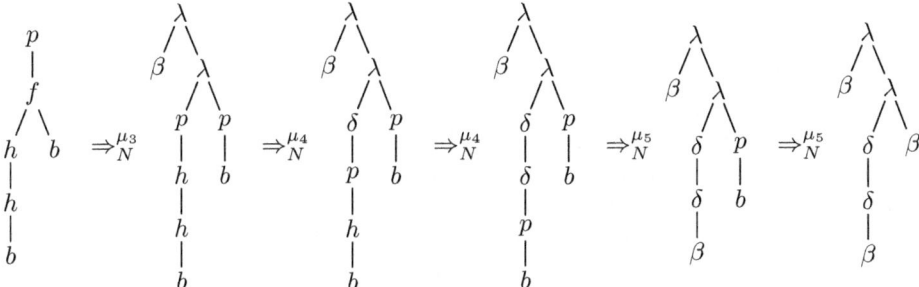

Fig. 3. A derivation from $p(t)$ to u in N (see Example 2)

example, the output tree t generated by the xtt M in Example 1 can be processed by the xtt N of Example 2, which is demonstrated in Example 2. Thus, in a composition of M and N (see Examples 1 and 2) we can transform the input tree s of Example 1 immediately to the output tree u of Example 2 by first running the xtt M on s to produce the intermediate tree t, which can then be transformed into u by N.

Formally, two weighted tree relations (of suitable type) are composed as follows. Let $\tau_1 \colon T_\Sigma \times T_\Gamma \to A$ and $\tau_2 \colon T_\Gamma \times T_\Delta \to A$ be weighted relations. Their *composition* $(\tau_1 \,;\, \tau_2) \colon T_\Sigma \times T_\Delta \to A$ is the weighted relation such that

$$(\tau_1 \,;\, \tau_2)(s, u) = \sum_{t \in T_\Gamma} \tau_1(s, t) \cdot \tau_2(t, u) \tag{2}$$

for every $s \in T_\Sigma$ and $u \in T_\Delta$. Clearly, this definition generalizes the classical definition of composition for relations. Whereas the infinite sum is not a problem in the unweighted case (where it is an infinite disjunction that becomes true once one element is true), we have to address it in the weighted case. There are essentially two options:

(i) to permit infinite sums and require that the semiring is suitably rich to handle infinite sums [25,23], which was done in [14,18,19,33,34,20] and also in the first part [36] of this survey, or
(ii) to avoid the infinite sums by restricting the weighted relations (and thus the xtt) that we allow in compositions.

In this part of the survey, we will follow the second approach by requiring that in a composition $\tau_1 \,;\, \tau_2$ we have that

$$\{t \mid (s, t) \in \mathrm{supp}(\tau_1)\} \qquad \text{or} \qquad \{t \mid (t, u) \in \mathrm{supp}(\tau_2)\} \tag{3}$$

is finite for every $s \in T_\Sigma$ and $u \in T_\Delta$. It is clear that in both cases the sum (2) in the definition of the composition $\tau_1 \,;\, \tau_2$ is finite.

Let us illustrate the general approach that is used in most composition constructions. To construct an xtt that computes the composition $\tau_M \,;\, \tau_N$ of the weighted relations computed by two xtt M and N, we need to make sure that the intermediate tree t (in Examples 1 and 2) is never constructed explicitly. To achieve this, the second xtt has to immediately consume every intermediate symbol that is produced by the first xtt M. Let us illustrate this approach by combining the two derivations of Figs. 2 and 3 such that intermediate symbols (from Γ) are consumed as soon as possible. The obtained derivation that now uses rules of both M and N is displayed in Fig. 4.

Once we have reordered the rule applications as indicated in the previous paragraph, we "glue" all rule applications that produce intermediate symbols together with the rule applications that consume these symbols. In this step, we also interpret two adjacent states (one of M and one of N) as in $p(q(s))$ as a single state $\langle p, q \rangle$. For example, based on Fig. 4 we glue the first two rule applications (of the rules ρ_1 and μ_3) together to obtain a single rule application of the rule

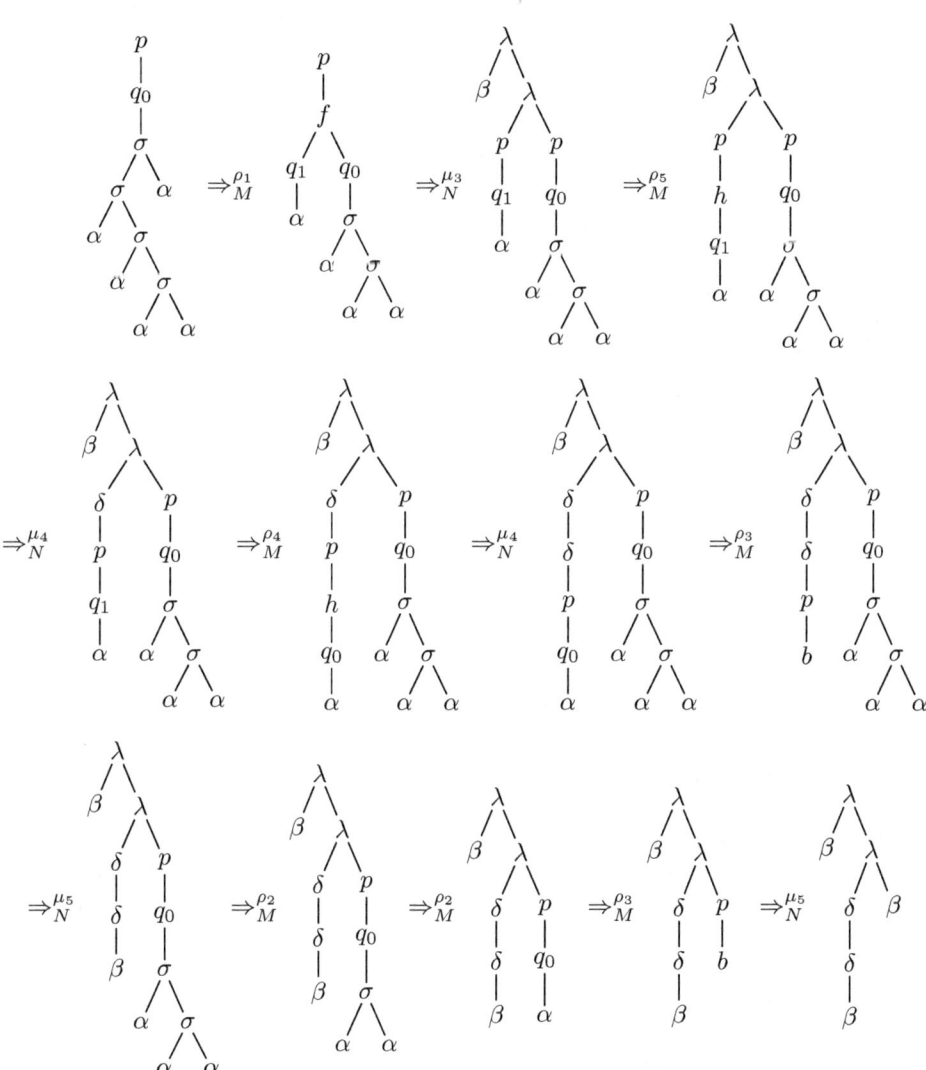

Fig. 4. Intertwined derivation from $p(q_0(s))$ to u in M and N (see Examples 1 and 2)

$$\langle p, q_0 \rangle (\sigma(\sigma(\mathrm{x}_1, \mathrm{x}_2), \alpha)) \xrightarrow{2\cdot 5} \lambda(\beta, \lambda(\langle p, q_1 \rangle(\mathrm{x}_1), \langle p, q_0 \rangle(\mathrm{x}_2))) \ ,$$

which is displayed in Fig. 5. For the rest of the discussion, we drop the distinction between rule applications and rules. In general, several rules of the first xtt need to be "glued" with several rules of the second xtt. However, it was already shown in [3] (and in the second part [37] of this survey) that this strategy does not work in general (even if both xtt M and N are linear and nondeleting). In the rest of this survey, we will thus focus on simpler cases, in which the left-hand sides of the rules of the second xtt N contain at most one input symbol. In Sect. 5 we consider compositions of an xtt M with a top-down tree transducer N. Thus, in Sect. 5 the second xtt N is such that every rule has exactly one input symbol in its left-hand side. We relax this requirement slightly in Sect. 6, where we investigate compositions of an xtt M with a top-down tree transducer N with ε-rules [39], which is an xtt in which each rule contains at most one input symbol in its left-hand side. However, before we proceed with the mentioned composition constructions we first introduce a modification of our xtt model that will prove to be useful in Sections 5 and 6.

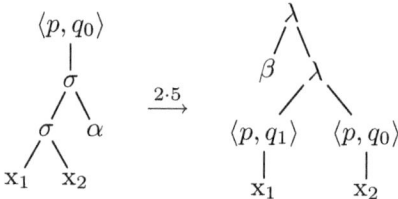

Fig. 5. Composed rule constructed from ρ_1 and μ_3 of Examples 1 and 2

4   An Equivalent Model

In this section, we introduce an alternative description for weighted extended top-down tree transducers that will be useful for our composition constructions. Essentially, we introduce explicit rule identifiers (like ρ_1–ρ_5 used in Example 1) that stand for a specific rule. A mapping that becomes part of the specification assigns weighted rules to identifiers. This indirection allows us to use multiple rules with the same left- and right-hand side and even the same weight. In our composition constructions we use this facility to establish a more concise and simpler relation between the constructed rules of the composed xtt and the original rules of the input xtt.

Definition 3. *A weighted extended (top-down) tree transducer with rule identifiers is a system* $(Q, \Sigma, \Delta, I, \mathcal{R}, \chi)$, *where*

- Q, Σ, Δ, *and* I *are the same as the corresponding elements of an xtt*,
- $\mathcal{R}$ *is a finite set of* rule identifiers, *and*
- $\chi\colon \mathcal{R} \to Q(T_\Sigma(X)) \times A \times T_\Delta(Q(X))$ *is a* rule assignment *that maps each rule identifier* $\rho \in \mathcal{R}$ *to its content* $\chi(\rho) = (l, a, r)$ *such that* $\{l, r\} \not\subseteq Q(X)$, l *is linear, and* $\mathrm{var}(r) \subseteq \mathrm{var}(l)$.

In accordance with our notation for rules, we often write $l \xrightarrow{a} r$ for elements $(l, a, r) \in Q(T_\Sigma(X)) \times A \times T_\Delta(Q(X))$. Moreover, we let $\mathrm{wt}\colon \mathcal{R} \to A$ be such that $\mathrm{wt}(\rho) = a$ for all $\rho \in \mathcal{R}$ with $\chi(\rho) = l \xrightarrow{a} r$. Intuitively, 'wt' maps a rule identifier to the weight of its identified rule.

The semantics of the xtt $M = (Q, \Sigma, \Delta, I, \mathcal{R}, \chi)$ with rule identifiers $\mathcal{R}$ is given by rewriting in essentially the same way as before. Let Σ' and Δ' be two alphabets such that $\Sigma \subseteq \Sigma'$ and $\Delta \subseteq \Delta'$ and $Q \cap (\Sigma' \cup \Delta') = \emptyset$. Again, we call a position $w \in \mathrm{pos}_Q(\xi)$ in a sentential form $\xi \in T_{\Delta'}(Q(T_{\Sigma'}(X)))$ reducible (for M) if there exists a rule $\rho \in \mathcal{R}$ with $\chi(\rho) = l \xrightarrow{a} r$ and a substitution $\theta\colon X \to T_{\Sigma'}(X)$ such that $\xi|_w = l\theta$. Now, let $\xi, \zeta \in T_{\Delta'}(Q(T_{\Sigma'}(X)))$, $\rho \in \mathcal{R}$, and $\chi(\rho) = l \xrightarrow{a} r$. We say that ξ *rewrites to* ζ *using* ρ, denoted by $\xi \Rightarrow_M^\rho \zeta$, if there exists a substitution $\theta\colon X \to T_{\Sigma'}(X)$ such that $\xi|_w = l\theta$ and $\zeta = \xi[r\theta]_w$ where w is the least reducible position in $\mathrm{pos}_Q(\xi)$ with respect to the lexicographic total order on $\mathbb{N}^*$. The *(extended) weighted relation* τ_M' computed by M is given by

$$\tau_M'(\xi, \zeta) = \sum_{\substack{\rho_1, \ldots, \rho_k \in \mathcal{R} \\ \xi \Rightarrow_M^{\rho_1}; \cdots; \Rightarrow_M^{\rho_k} \zeta}} \left(\prod_{i=1}^k \mathrm{wt}(\rho_i) \right)$$

for every $\xi, \zeta \in T_{\Delta'}(Q(T_{\Sigma'}(X)))$. As for xtt, the *semantics* τ_M of the xtt M with rule identifiers is the weighted relation $\tau_M\colon T_\Sigma \times T_\Delta \to A$ such that $\tau_M(t, u) = \sum_{q \in I} \tau_M'(q(t), u)$ for every $t \in T_\Sigma$ and $u \in T_\Delta$. The properties of rules and xtt defined in Sect. 2 generalize straightforwardly to xtt with rule identifiers.

Example 4. Let $N = (\{p\}, \Gamma, \Delta, \{p\}, \mathcal{R}, \chi)$ be the xtt with rule identifiers such that

- $\Gamma = \{f, h, b\}$ and $\Delta = \{\lambda, \gamma, \delta, \beta\}$,
- $\mathcal{R} = \{\mu_1, \ldots, \mu_7\}$, and
- the rule assignment χ is given by

$$\chi(\mu_1) = \qquad p(x_1) \xrightarrow{2} \gamma(p(x_1)) \qquad\qquad \chi(\mu_5) = p(h(x_1)) \xrightarrow{4} \delta(p(x_1))$$

$$\chi(\mu_2) = p(f(x_1, x_2)) \xrightarrow{2} \lambda(p(x_1), p(x_2)) \qquad \chi(\mu_6) = p(h(x_1)) \xrightarrow{4} \delta(p(x_1))$$

$$\chi(\mu_3) = p(f(x_1, x_2)) \xrightarrow{3} \lambda(p(x_1), p(x_2)) \qquad \chi(\mu_7) = \qquad p(b) \xrightarrow{1} \beta$$

$$\chi(\mu_4) = p(f(x_1, x_2)) \xrightarrow{5} \lambda(\beta, \lambda(p(x_1), p(x_2))) \ .$$

ε-rule	consuming	erasing	producing	linear	nondeleting
μ_1	μ_2–μ_7		μ_1–μ_7	μ_1–μ_7	μ_1–μ_7

Let $t = f(h(h(b)), b)$ and $u = \lambda(\beta, \lambda(\delta(\delta(\beta)), \beta))$ as in Example 2. Figure 6 shows a derivation from $p(t)$ to u with weight

$$\mathrm{wt}(\mu_4) \cdot \mathrm{wt}(\mu_5) \cdot \mathrm{wt}(\mu_5) \cdot \mathrm{wt}(\mu_7) \cdot \mathrm{wt}(\mu_7) = 5 \cdot 4 \cdot 4 \cdot 1 \cdot 1 = 80 \ .$$

We can construct exactly three other derivations from $p(t)$ to u using the rule sequences $\mu_4\mu_5\mu_6\mu_7\mu_7$, $\mu_4\mu_6\mu_5\mu_7\mu_7$, and $\mu_4\mu_6\mu_6\mu_7\mu_7$. All of these derivations have the same weight. Consequently, $\tau_N(t, u) = 80 + 80 + 80 + 80 = 320$.   □

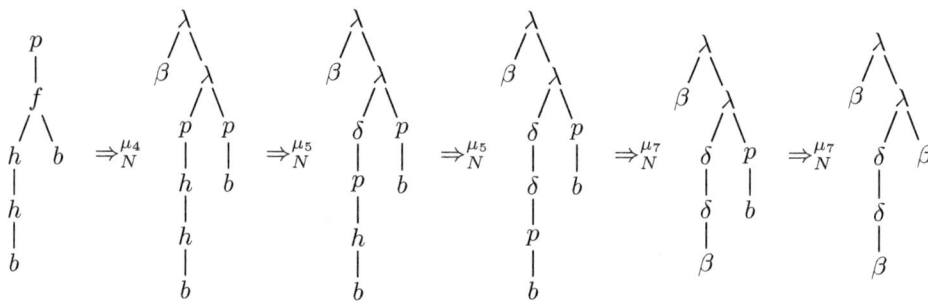

Fig. 6. One possible derivation from $p(t)$ to u in N of Example 4

We can easily see that the two models of xtt are equally expressive. For every xtt $(Q, \Sigma, \Delta, I, R)$ we can construct an equivalent xtt $(Q, \Sigma, \Delta, I, R, \chi)$ with rule identifiers by setting $\chi = \mathrm{id}_R$, where id_R is the identity on R. Conversely, given an xtt $M = (Q, \Sigma, \Delta, I, \mathcal{R}, \chi)$ with rule identifiers we can obtain an equivalent xtt $(Q, \Sigma, \Delta, I, R)$ by setting

$$R(l, r) = \sum_{\rho \in \mathcal{R} \,:\, \chi(\rho) = (l, a, r)} \mathrm{wt}(\rho) \tag{4}$$

for all $l \in Q(T_\Sigma(X))$ and $r \in T_\Delta(Q(X))$.[12] The construction is illustrated in Example 5.

Example 5. Let N be the xtt with rule identifiers of Example 4. To obtain an equivalent xtt $N' = (Q, \Gamma, \Delta, I, R)$, we

- merge the rules μ_2 and μ_3 to form the rule $p(f(x_1, x_2)) \xrightarrow{2+3} \lambda(p(x_1), p(x_2))$,
- merge the rules μ_5 and μ_6 to form the rule $p(h(x_1)) \xrightarrow{4+4} \delta(p(x_1))$, and
- keep the rules μ_1, μ_4, and μ_7 with their original weight.

In this manner, we obtain exactly the xtt of Example 2, for which we only have one derivation from $p(t)$ to u, which is shown in Fig. 3. Naturally, its weight is 320.   □

---

[12] The sum (4) returns 0, as desired, if M has no rules with left-hand side l and right-hand side r.

5   Composition with a Top-Down Tree Transducer

In this section, we will discuss compositions $\tau_M ; \tau_N$ for xtt $M = (Q, \Sigma, \Gamma, I_1, R_1)$ and $N = (P, \Gamma, \Delta, I_2, R_2)$, in which the xtt N is actually a top-down tree transducer (tdtt). Moreover, we require that the xtt M does not have any ε-rules. This restriction ensures that the set $\{t \mid (s, t) \in \mathrm{supp}(\tau_M)\}$ is finite for every $s \in T_\Sigma$ [see (3)] because each rule application consumes at least one input symbol. Hence there can only be finitely many rule applications to $q(s)$ given a state $q \in Q$ and an input tree s, which yields an upper bound on the number of potential derivations, which in turn limits the number of output symbols in each output tree. We already demonstrated in Sect. 3 that this restriction is sufficient to ensure that the sum over all intermediate trees t occurring in (2):

$$(\tau_M ; \tau_N)(s, u) = \sum_{t \in T_\Gamma} \tau_M(s, t) \cdot \tau_N(t, u)$$

is finite for all $s \in T_\Sigma$ and $u \in T_\Delta$. Thus, composition is well-defined in all cases discussed in this section.

5.1   Construction

Now we are ready to present the generic composition construction. For the sake of uniformity, we will construct more rules than strictly necessary. As already indicated in Fig. 4, the states of the composed xtt will be pairs of states with one state from each input xtt. Next, let us fix an important constant m.

- Let $c \geq |\mathrm{pos}_x(r)|$ for all $l \to r \in R_2$ and $x \in X$. Roughly speaking, c is larger than the maximal *copying degree* of N, which is the maximal number of times a variable occurs on some right-hand side of a rule of N. To keep the presentation simple, we assume that $c \geq 1$.
- Let $s \geq |\mathrm{pos}_\Gamma(r)|$ for all $l \to r \in R_1$. Consequently, s is larger than the maximal number of output symbols in a right-hand side of a rule of M.
- Finally, let $m \geq c^s$. The constant m provides an upper bound to the number of steps required by N to process a right-hand side of a rule of M.

Recall that given a sentential form $\xi \in T_\Delta(P(T_\Gamma(Q(T_\Sigma(X)))))$ and a rule $\rho \in R_1$ there exists at most one $\zeta \in T_\Delta(P(T_\Gamma(Q(T_\Sigma(X)))))$ such that $\xi \Rightarrow_M^\rho \zeta$.[13] Naturally, the same property holds for the tdtt N. To avoid an explicit conversion, we identify elements of $T_\Delta(P(Q(T_\Sigma(X))))$ with elements of $T_\Delta((P \times Q)(T_\Sigma(X)))$ in the obvious manner. Finally, we let

$$\Rightarrow_N^w = (\Rightarrow_N^{\mu_1} ; \cdots ; \Rightarrow_N^{\mu_k})$$

if $w = \mu_1 \cdots \mu_k$ with $\mu_1, \ldots, \mu_k \in R_2$.

---

[13] To match this statement to the earlier one, we have to set $\Delta' = \Delta \cup P \cup \Gamma$.

Definition 6. *The composed xtt M ; N is the xtt $(P \times Q, \Sigma, \Delta, I_2 \times I_1, \mathcal{R}, \chi)$ with rule identifiers*

$$\mathcal{R} = \{\langle \rho, p, w \rangle \mid \rho \in R_1, p \in P, w \in R_2^*, |w| \leq m\}$$

such that $\chi(\langle l \to r, p, \mu_1 \cdots \mu_k \rangle) = (p(l), a, r')$ for every $l \to r \in R_1$, $p \in P$, and rule sequence $\mu_1, \ldots, \mu_k \in R_2$ with $k \leq m$, where $r' \in T_\Delta(P(Q(X)))$ and

$$a = \begin{cases} R_1(l \to r) \cdot \prod_{i=1}^k R_2(\mu_i) & \text{if } p(l) \Rightarrow_M^{(l,r)} ; \Rightarrow_N^{\mu_1 \cdots \mu_k} r' \\ 0 & \text{otherwise.} \end{cases}$$

Clearly, the construction might return a lot of rule identifiers whose associated rules have weight 0. These rules are useless, and we typically will not report them in our examples. Moreover, we can easily see that the constructed rules never have the forbidden shape $l \to r$ with $\{l, r\} \subseteq P(Q(X))$ because the left-hand side l equals $p(l')$ for some left-hand side l' of a rule of M, which does not have ε-rules. Let us illustrate the construction on two example xtt, which we will use throughout this section.

Example 7. We again use the semiring of real numbers in this example. Moreover, let us consider the xtt M and N, which are given as follows:

$$M = (\{q\}, \Sigma, \Sigma, \{q\}, R_1) \qquad \text{and} \qquad N = (\{p_0, p\}, \Sigma, \Delta, \{p_0\}, R_2) \ ,$$

where

- $\Sigma = \{\gamma, \alpha\}$ and $\Delta = \{\sigma\} \cup \Sigma$,
- R_1 contains the rules

$$\rho_1: \quad q(\gamma(x_1)) \xrightarrow{2} \gamma(\gamma(q(x_1))) \qquad\qquad \rho_2: \quad q(\alpha) \xrightarrow{2} \alpha \ ,$$

- and R_2 contains the rules

$$\mu_1: \quad p_0(\gamma(x_1)) \xrightarrow{4} \sigma(p_0(x_1), p_0(x_1)) \qquad\qquad \mu_6: \quad p(\gamma(x_1)) \xrightarrow{1} \gamma(p(x_1))$$

$$\mu_2: \quad p_0(\gamma(x_1)) \xrightarrow{2} \sigma(p_0(x_1), p(x_1)) \qquad\qquad \mu_7: \quad p(\gamma(x_1)) \xrightarrow{3} \alpha$$

$$\mu_3: \quad p_0(\gamma(x_1)) \xrightarrow{2} \sigma(p(x_1), p_0(x_1)) \qquad\qquad \mu_8: \quad p(\alpha) \xrightarrow{1} \alpha$$

$$\mu_4: \quad p_0(\gamma(x_1)) \xrightarrow{1} \sigma(p(x_1), p(x_1))$$

$$\mu_5: \quad p_0(\alpha) \xrightarrow{1} \alpha \ .$$

ε-rule	consuming	erasing	producing	linear	nondeleting
	ρ_1, ρ_2		ρ_1, ρ_2	ρ_1, ρ_2	ρ_1, ρ_2
	μ_1–μ_8		μ_1–μ_8	μ_5–μ_8	μ_1–μ_6, μ_8

Both M and N are tdtt, M is linear and nondeleting, whereas N is neither linear nor nondeleting. Additionally, the xtt M is deterministic and total. We can set $c = 2$ and $s = 2$, and thus, we can select $m = 4$. To increase readability,

$$
\begin{array}{l}
p \\
| \\
q \\
| \\
q \\
| \\
\gamma \\
| \\
x_1
\end{array}
\Rightarrow_M^{\rho_1}
\begin{array}{l}
p \\
| \\
\gamma \\
| \\
\gamma \\
| \\
q \\
| \\
x_1
\end{array}
\Rightarrow_N^{\mu_6}
\begin{array}{l}
\gamma \\
| \\
p \\
| \\
\gamma \\
| \\
q \\
| \\
x_1
\end{array}
\Rightarrow_N^{\mu_7}
\begin{array}{l}
\gamma \\
| \\
\alpha
\end{array}
$$

Fig. 7. Derivation for rule $\langle \rho_1, p, \mu_6\mu_7 \rangle$ (see Example 7)

let $\mathcal{R}^\gamma_{p_0} = \{\mu_1, \mu_2, \mu_3, \mu_4\}$ and $\mathcal{R}^\gamma_p = \{\mu_6, \mu_7\}$. Intuitively, $\mathcal{R}^\gamma_{p_0}$ and $\mathcal{R}^\gamma_p$ are the sets of rules that consume the input symbol γ in state p_0 and p, respectively Now let us construct the composition M , N. It is the xtt

$$M \,;\, N = (Q', \Sigma, \Delta, I', \mathcal{R}, \chi)$$

with rule identifiers such that

- $Q' = \{\langle p_0, q \rangle, \langle p, q \rangle\}$ and $I' = \{\langle p_0, q \rangle\}$,
- $\mathcal{R} = \{\langle \rho_2, p_0, \mu_5 \rangle, \langle \rho_2, p, \mu_8 \rangle, \langle \rho_1, p, \mu_6\mu_6 \rangle, \langle \rho_1, p, \mu_6\mu_7 \rangle, \langle \rho_1, p, \mu_7 \rangle\} \cup \mathcal{R}'$ with

$$
\begin{aligned}
\mathcal{R}' = \; & \{\langle \rho_1, p_0, \mu_1\mu\mu' \rangle \mid \mu, \mu' \in \mathcal{R}^\gamma_{p_0}\} \,\cup \\
& \cup\, \{\langle \rho_1, p_0, \mu_2\mu\mu' \rangle \mid \mu \in \mathcal{R}^\gamma_{p_0}, \mu' \in \mathcal{R}^\gamma_p\} \,\cup \\
& \cup\, \{\langle \rho_1, p_0, \mu_3\mu\mu' \rangle \mid \mu \in \mathcal{R}^\gamma_p, \mu' \in \mathcal{R}^\gamma_{p_0}\} \,\cup \\
& \cup\, \{\langle \rho_1, p_0, \mu_4\mu\mu' \rangle \mid \mu, \mu' \in \mathcal{R}^\gamma_p\} \;.
\end{aligned}
$$

In total we have $5 + 16 + 8 + 8 + 4 = 41$ (meaningful) rule identifiers. We will not present all 41 corresponding rules, but we will show two example rules to demonstrate the construction. Let us first construct the rule for the identifier $\langle \rho_1, p, \mu_6\mu_7 \rangle$. To this end, we need to build a derivation starting at $p(q(\gamma(x_1)))$ using the rule sequence $\rho_1\mu_6\mu_7$. This derivation is illustrated in Fig. 7. We obtain the rule

$$\chi(\langle \rho_1, p, \mu_6\mu_7 \rangle) = \Big(\langle p, q \rangle(\gamma(x_1)), 2 \cdot 1 \cdot 3, \gamma(\alpha) \Big) \;.$$

Secondly, let us construct the rule for the identifier $\langle \rho_1, p_0, \mu_3\mu_7\mu_2 \rangle$. This time we need to build a derivation that starts with $p_0(q(\gamma(x_1)))$ and uses the rule sequence $\rho_1\mu_3\mu_7\mu_2$. We illustrate the derivation in Fig. 8. Consequently, we obtain the rule

$$\chi(\langle \rho_1, p_0, \mu_3\mu_7\mu_2 \rangle) = \Big(\langle p_0, q \rangle(\gamma(x_1)), 2 \cdot 2 \cdot 3 \cdot 2, \sigma(\alpha, \sigma(\langle p_0, q \rangle(x_1), \langle p, q \rangle(x_1))) \Big) \;.$$

$\square$

Our general composition construction allows us to compose an xtt with a tdtt. As we have seen, it closely follows the intuition provided in Sect. 3 and uses the xtt with rule identifiers that we introduced in Sect. 4. This has the benefit that we can obtain a direct correspondence between rule sequences of the xtt

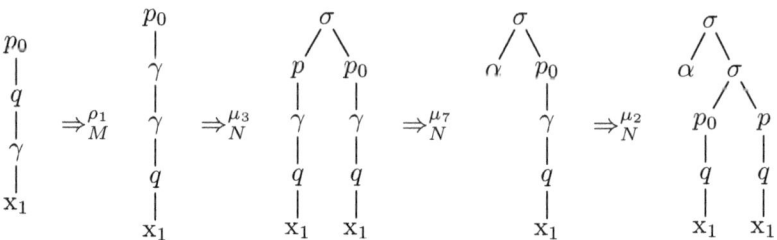

Fig. 8. Derivation for rule $\langle \rho_1, p_0, \mu_3\mu_7\mu_2 \rangle$ (see Example 7)

M and N and a rule in the composed xtt M ; N. In the standard xtt (without rule identifiers) this direct correspondence is lost since several derivations might create the same rule.[14]

Naturally, we would expect that the composed xtt M ; N computes the weighted tree transformation τ_M ; τ_N; i.e., the composition of the weighted tree transformations computed by M and N. In other words, we hope that $\tau_{M;N} = \tau_M$; τ_N. Although the rule aggregation and intertwining approach followed in the construction (and shown in Sect. 3) is reasonable, it fails to produce a correct xtt (i.e., an xtt that computes τ_M ; τ_N) in a number of cases. This is already true in the unweighted case [12,5] and the presence of weights adds a few more problematic cases, which we will discuss in the next section.

5.2   Correctness

In this section, we will investigate in which cases the composition construction (see Definition 6) actually produces an xtt that computes the composition $\tau_M;\tau_N$. In principle, the xtt M need not be a tdtt, but for the following discussion we assume that it is. The generalization to the general case is simple in almost all cases (see [38,16] for a few notable differences). Consequently, let us look at compositions of tdtt. Top-down tree transducers have been studied quite extensively in the unweighted case (see [21,22,10] for an overview). The following two slogans are known to represent properties that are unavailable in a single tdtt [12,5]:

- Nondeterminism followed by copying (non-linearity), and
- Checking (non-totality) followed by deletion.

A composition τ_M ; τ_N of two tdtt M and N can implement both properties mentioned in the slogans. Thus, these properties already restrict the potential successful compositions of tdtt. In fact, in all remaining cases shown in Table 1 the composition of the tdtt M and N is possible in the unweighted case [5, Theorem 1]. A detailed explanation of those restrictions on compositions is presented in [12,5]. Here we will focus on the particular problems that occur in the generalization of those results to the weighted case because the limitations

---

[14] The interested reader can compare our construction to [33,34].

Table 1. Cases for unweighted tdtt composition

Case	M	N
(a)		linear and nondeleting
(b)	total	linear
(c)	deterministic	nondeleting
(d)	deterministic and total	

on compositions in the unweighted case transfer immediately to our setting.[15] Thus, we will not investigate compositions that do not fulfill the requirements in Table 1.[16]

Case (a) has been partially generalized in [31, Theorem 2.4] to weighted tdtt. More precisely, it was shown that the composition succeeds if both M and N are linear and nondeleting.[17] This result was further (partially) generalized in [14, Theorem 5.18], which covers the case in which M and N are deterministic and only N is linear and nondeleting. Finally, [33, Theorem 26] presents the full generality and matches Case (a) of the unweighted setting exactly.

Theorem 8 (see [33, Theorem 26]). *If the tdtt N is linear and nondeleting, then* $\tau_{M;N} = \tau_M \; ; \tau_N$.

Case (b) is slightly problematic in the weighted setting, and the only known generalizations are actually instances of Case (d). Let us illustrate the problem. The tdtt N can delete an intermediate subtree t' that was output by M as the result of processing an input subtree s'. In the composed tdtt, the input subtree s' is deleted right away without processing it. This phenomenon is abstractly illustrated in Fig. 9. In addition, we showcase a derivation using our example xtt of Example 7 in Fig. 10 (note that only linear rules of N are used in this derivation). Thus, the actual input subtree s' and the intermediate subtree t' are not relevant in the composed tdtt. In the unweighted setting, this independence is guaranteed by the totality of M, which yields that for each input tree s there exists a translation t of it. In other words, for each input tree $s \in T_\Sigma$ and state $q \in Q$, we have

$$\sum_{t \in T_\Gamma} \Bigg(\sum_{\substack{\rho_1,\dots,\rho_k \in R_1 \\ q(s) \Rightarrow_M^{\rho_1} ; \cdots ; \Rightarrow_M^{\rho_k} t}} \Bigg(\prod_{i=1}^{k} R_1(\rho_i) \Bigg) \Bigg) = 1 \;.$$

---

[15] More precisely, the restrictions only transfer to xtt over non-rings due to a result by WANG [44,45]. A ring is a semiring that has additive inverses; i.e., there exists an element -1 such that $1 + (-1) = 0$.

[16] Although such compositions can, in principle, succeed. Mind that the counterexamples of [12,5] only generalize to non-rings. In fact, given a suitably strong ring, any composition might become possible.

[17] In fact, [31] proves closure under composition for a slightly more general class, but the mentioned result can be obtained easily by instantiating the more general construction to our weighted tdtt model.

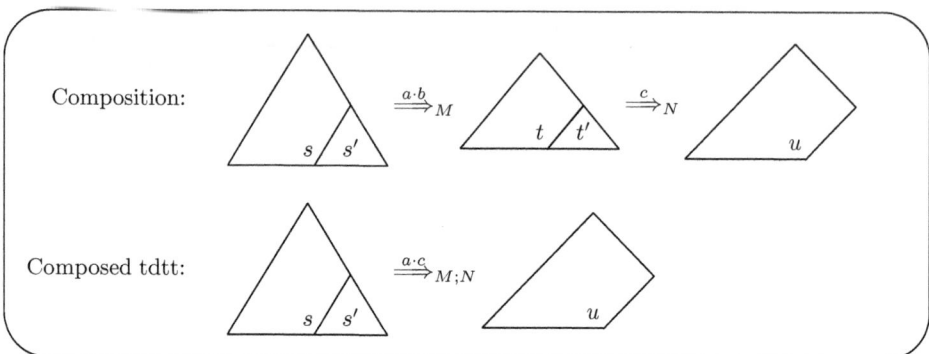

Fig. 9. Difference between composition and the composed tdtt. Atop the arrows we mark the weight and next to it the tdtt, in which the derivation happens. More precisely, weight a is charged for processing s (without s'), weight b is charged for processing s', and weight c is charged for processing t (without t').

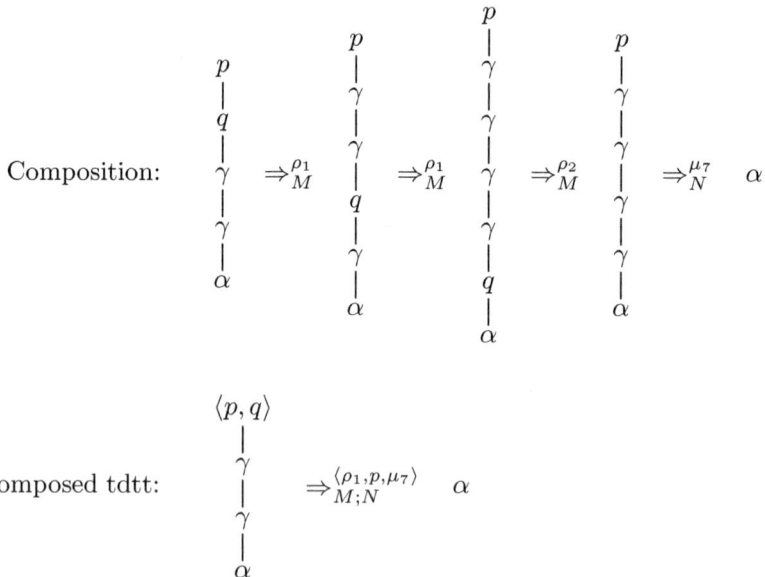

Fig. 10. Difference between composition and the composed tdtt on the tdtt M and N of Example 7. The composition charges weight $R_1(\rho_1) \cdot R_1(\rho_1) \cdot R_1(\rho_2) \cdot R_2(\mu_7) = 2^2 \cdot 2 \cdot 3$, whereas the composed tdtt only charges $2 \cdot 3$, which is the weight of the rule $\langle \rho_1, p, \mu_7 \rangle$. The charge $R_1(\rho_1) \cdot R_1(\rho_2) = 2 \cdot 2$ for processing the input subtree $\gamma(\alpha)$ is lost.

Clearly, in the BOOLEAN semiring, the previous equation is fulfilled if there is at least one derivation from $q(s)$ to some t. To obtain a generalization of the requirement for the weighted setting, we observe that the composed tdtt also ignores the input subtree s.

Definition 9. *A state $q \in Q$ is* constant *if there exists a semiring element $a \in A$ such that for every $s \in T_\Sigma$ we have*

$$\sum_{\substack{t \in T_\Gamma}} \Big(\sum_{\substack{\rho_1,\ldots,\rho_k \in R_1 \\ q(s) \Rightarrow_M^{\rho_1}; \cdots; \Rightarrow_M^{\rho_k} t}} \Big(\prod_{i=1}^{k} R_1(\rho_i) \Big) \Big) = a \ . \tag{5}$$

We also say that q is a-constant if q is constant using the semiring element a. The xtt M is constant *if all its states $q \in Q$ are constant.*

Note that the sums in (5) are always finite, which we already showed at the beginning of Sect. 5. Let us demonstrate some constant tdtt, in which all states are 1-constant. In general, different states of a constant tdtt can have different semiring elements for which they are constant.

Example 10. All of the following tdtt have only 1-constant states:

- every total tdtt over the BOOLEAN semiring,
- every BOOLEAN and total tdtt over an idempotent semiring, and
- every deterministic, total, and BOOLEAN tdtt over any semiring.

Clearly, the total tdtt M of Example 7 is not constant, which is also shows that a total tdtt is not necessarily constant. If the tdtt M is constant, then we can perfectly predict the missing weight b in the derivation of the composed tdtt in Fig. 9 and charge it for the rule that actually performs the deletion.[18] Note that our presented composition construction (see Definition 6) might fail, but the authors believe that it can be modified as indicated to obtain the following result.

Conjecture 11. If the xtt M is constant and the tdtt N is linear, then $\tau_M \, ; \, \tau_N$ can be computed by an xtt.

Note that Conjecture 11 covers all the cases (for M) mentioned in Example 10. It remains to be determined whether the indicated adjustment actually works. Moreover, depending on the semiring, it might be difficult to determine whether a state is constant, so additional syntactic requirements that lead to constant states (potentially with a weight different from 0 and 1) would be desirable.

   Case (c) is also problematic and has not been addressed in the literature. This is due to the fact that an intermediate output tree t' can be copied by N. In the composition, the weight charged for generating the tree t' from an input

---

[18] In fact, we can only predict the aggregated weight (as opposed to the weights of single deleted derivations), but that is sufficient.

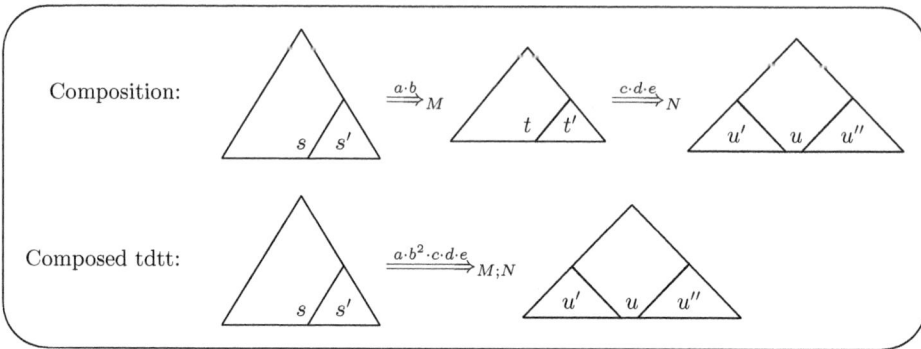

Fig. 11. Another difference between composition and the composed tdtt. Atop the arrow we mark the weight and next to it the tdtt, in which the derivation happens. More precisely, weight a is charged for processing s (without s'), weight b is charged for processing s', weight c is charged for processing t (without t'), and weights d and e are charged for processing t' (producing u' and u'', respectively).

subtree s' is charged once, but in the composed tdtt this weight is charged twice since the input subtree s' will be copied and processed twice. The process is illustrated in Fig. 11. In addition, we provide derivations using the xtt of Example 7 that demonstrate the phenomenon in Fig. 12 (note that the tdtt M of Example 7 is deterministic and we only used nondeleting rules of N in these derivations).

So again our generic composition construction (see Definition 6) might fail, but contrary to the previous case, the authors believe that this can be addressed without any further requirement. Instead of using the original state of M in all copies, the authors propose to use the corresponding state from an unweighted copy of M in all but one copies. Thus, the weight that M charges for processing the input tree would only be charged in the single copy and the other copies, which run using the unweighted copy of M, do not cause additional charges for processing the input. Since the input tdtt M is deterministic, we know that the copies will behave equally in all aspects besides the weight that they charge. Let us provide some detail.

Definition 12. *An* unweighted copy *of M is a* BOOLEAN *xtt $(Q, \Sigma, \Gamma, I_1, R'_1)$ such that*

$$l \to r \in R'_1 \quad \Longleftrightarrow \quad l \to r \in R_1 \ .$$

We will not formalize the modified construction, but we will present the essential steps. First, we take the (disjoint) union of M and an unweighted copy M' of M (by renaming all states of the copy from q to $\bar{q}$). Let us assume that a state $q \in Q$ corresponds to a state $\bar{q}$ in M', and similarly, a rule $\rho \in R_1$ corresponds to a rule $\bar{\rho}$ in M'. When processing a rule in which the tdtt N copies, we modify all but one copies to use the corresponding state from M'. Let us illustrate this

Composition:

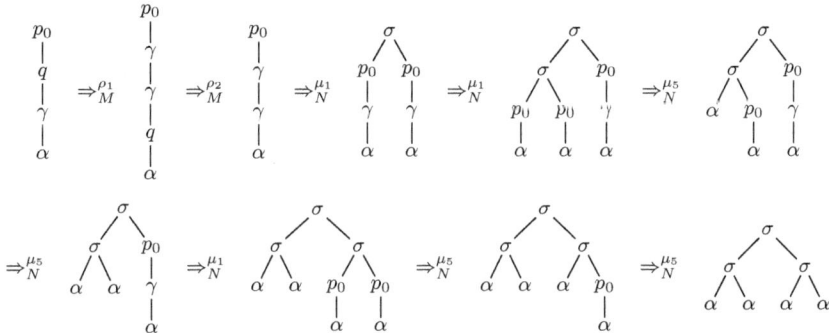

Composed tdtt:

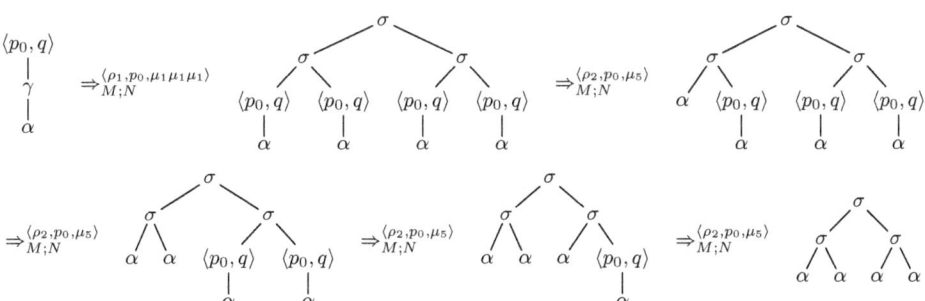

Fig. 12. Difference between composition and the composed tdtt on the tdtt M and N of Example 7. The composition (upper display) charges the weight $R_1(\rho_1) \cdot R_1(\rho_2) \cdot R_2(\mu_1)^3 \cdot R_2(\mu_5)^4 = 2 \cdot 2 \cdot 4^3 \cdot 1^4$, whereas the composed tdtt charges $2 \cdot 4^3 \cdot (2 \cdot 1)^4$. The additional weight $R_1(\rho_2)^3 = 2^3$ is charged by $M \,;N$ for processing the input subtree α (using ρ_2) three more times.

adjustment on an example rule of Example 7. Figure 13 shows the original and the modified derivation that lead to the rule $\nu = \langle \rho_1, p_0, \mu_3\mu_6\mu_1 \rangle$ of $M \; ; N$ and our new rule

$$\chi(\nu) = \Big(\langle p_0, q \rangle(\gamma(\mathbf{x}_1)), 16, \sigma(\gamma(\langle p, q \rangle(\mathbf{x}_1)), \sigma(\langle p_0, q \rangle(\mathbf{x}_1), \langle p_0, q \rangle(\mathbf{x}_1)))\Big)$$

$$\chi'(\nu) = \Big(\langle p_0, q \rangle(\gamma(\mathbf{x}_1)), 16, \sigma(\gamma(\langle p, \overline{q} \rangle(\mathbf{x}_1)), \sigma(\langle p_0, \overline{q} \rangle(\mathbf{x}_1), \langle p_0, q \rangle(\mathbf{x}_1)))\Big) \; .$$

Figure 14 shows the modified derivation corresponding to the derivation of the composed tdtt, which is displayed in Fig. 12.

Original derivation:

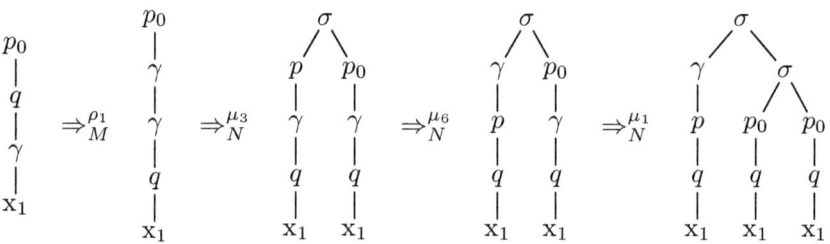

Modified derivation:

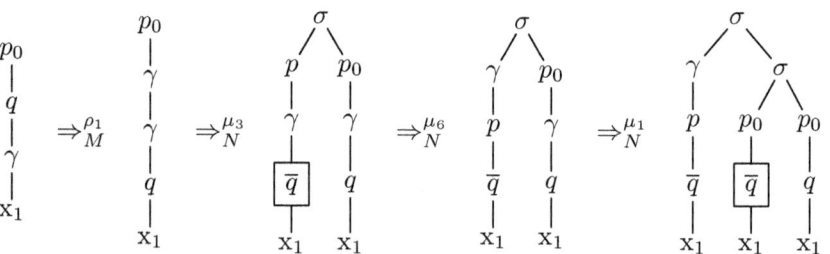

Fig. 13. Two derivations that yield rules. The upper one follows our composition construction, whereas the lower one is adjusted to address the problem of Case (c). We boxed the states that are adjusted due to copying of N. We selected to mark the left copies, but the choice is arbitrary.

Conjecture 13. If the xtt M is deterministic and the tdtt N is nondeleting, then $\tau_M \; ; \tau_N$ can be computed by an xtt.

Finally, Case (d) is essentially a combination of Cases (b) and (c). This case was first addressed by [14, Theorem 5.18], in which it was shown that a BOOLEAN, deterministic, and total tdtt M can be composed with a deterministic tdtt N. A similar statement was obtained in [33, Theorem 30], where (i) the same restrictions are placed on M and (ii) N is required to be linear. The result of [33]

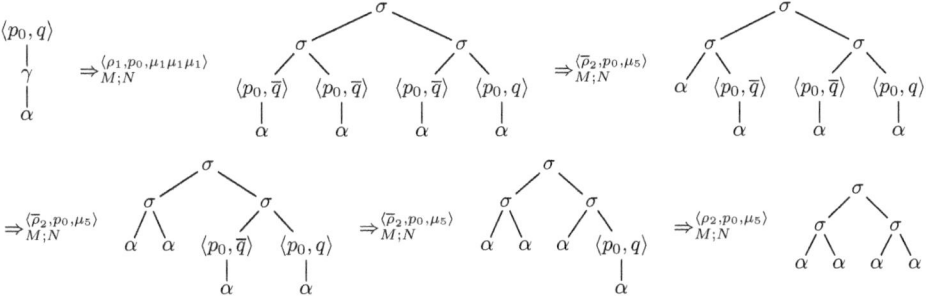

Fig. 14. Derivation using the new rules (see Fig. 12). The derivation now correctly charges the weight $R_1'(\rho_1) \cdot R_2'(\mu_1)^3 \cdot R_2(\mu_5)^4 \cdot R_1'(\rho_2) \cdot R_1'(\overline{\rho_2})^3 = 2 \cdot 4^3 \cdot 1^4 \cdot 2 \cdot 1^3$ because overlined rules charge weight 1.

clearly avoids the problematic Case (c) by requiring N to be linear. The result of [14] allows non-linear tdtt N, and it could thus be reasoned that they also had to handle the problematic Case (c). However, the requirement that the tdtt M is BOOLEAN already enforces that the additional weights (see Figs. 11 and 12) charged by the composed tdtt (constructed according to the general composition construction of Definition 6) are all 1. Thus, no modification was necessary under their assumptions. Using the indicated improvements suggested in Cases (b) and (c), the authors conjecture the following result, which covers both known results. Essentially, the authors believe that a constant and deterministic xtt M can be composed with any tdtt N.

Conjecture 14. If the xtt M is constant and deterministic and N is a tdtt, then τ_M ; τ_N can be computed by an xtt.

This concludes our investigation of compositions of tdtt. Table 2 shows the various results obtained in the weighted case. It is interesting that if Conjectures 11, 13, and 14 were true, then we would recover the beautiful symmetry that is present in the composition results [12,5] for unweighted top-down and bottom-up tree transducers [43] also in the weighted case. A summary of the composition results for weighted bottom-up tree transducers [14] can be found in Table 3, but the reader is referred to [14,33,34] for the detailed results.

6   Allowing ε-rules

This section is devoted to compositions of tree transformations computed by xtt M and N, of which the xtt N is a top-down tree transducer with ε-rules [39]. Roughly speaking, a top-down tree transducer with ε-rules is an xtt, in which simple and ε-rules are allowed. In other words, this section investigates the effect of ε-rules in N to the results of Sect. 5. In the unweighted setting, this scenario was investigated in [39], and we essentially report the results of [39], which we

Table 2. Composition results for weighted tdtt ('nondel.' abbreviates 'nondeleting' and 'det.' abbreviates 'deterministic')

Case	M	N	Reference
(a)	linear and nondel. deterministic	linear and nondel. det., linear, nondel. linear and nondel.	[31, Theorem 2.4] [14, Theorem 5.18] [33, Theorem 26]
(b)	constant	linear	Conjecture 11
(c)	deterministic	nondeleting	Conjecture 13
(d)	BOOLEAN, det., total BOOLEAN, det., total constant and det.	deterministic linear	[14, Theorem 5.18] [33, Theorem 30] Conjecture 14

adjusted to our weighted setting. Let us start with the formal definition of the requirements of this section. For the rest of this section, let $M = (Q, \Sigma, \Gamma, I_1, R_1)$ and $N = (P, \Gamma, \Delta, I_2, R_2)$ be the xtt that we want to compose.

Definition 15 (cf. [15, Definition 4] and [39, Definition 1])

- *The xtt M is* shallow *if* $|\mathrm{pos}_\Gamma(r)| \leq 1$ *for every* $l \to r \in R_1$.
- *The xtt N is a* tdtt with ε-rules *if* $|\mathrm{pos}_\Gamma(l)| \leq 1$ *for every* $l \to r \in R_2$.

Clearly, each tdtt is a tdtt with ε-rules, but a tdtt need not be shallow. Let us examine these properties for the xtt in our examples.

xtt	tdtt with ε-rules	shallow
M of Example 1	no (due to rule ρ_1)	yes
N of Example 2	yes	no (due to rule μ_3)
M of Example 7	yes (because it is a tdtt)	no (due to rule ρ_1)
N of Example 7	yes (because it is a tdtt)	yes

Now we can formally define the goal of this section. We will investigate compositions of xtt M and N such that M is shallow and N is a tdtt with ε-rules.

Table 3. Composition results for weighted bottom-up tree transducers [14] ('nondel.' abbreviates 'nondeleting' and 'det.' abbreviates 'deterministic') for comparison. Note that every weighted bottom-up tree transducer can be made total.

Case	M	N	Reference
(a)	linear, nondel. linear, nondel. linear, nondel.	linear and nondel. homomorphism	[31, Theorem 2.4] [14, Corollary 5.5] [33, Theorem 13]
(b)	linear	[total]	[33, Theorem 20]
(c)	nondeleting nondeleting	BOOLEAN, deterministic constant, deterministic	[33, Theorem 24] conjectured
(d)		BOOLEAN, homomorphism BOOLEAN, det., [total] constant, det., [total]	[14, Corollary 5.5] [33, Theorem 24] conjectured

To show that the condition that ensures well-definedness of the sum in the definition (2) of composition does not influence the results much, we additionally assume here that N does only have producing rules. In this case, there can only be finitely many rule applications generating the output tree u, which limits the size of the intermediate tree [see (3)]. Thus, all compositions are well-defined in the cases of this section.

6.1   Construction

Before we present an adaptation of the generic construction in Sect. 5.1, let us demonstrate that the generic construction fails to handle ε-rules of N in a meaningful manner.

Example 16. Let M and N be the xtt of Examples 1 and 2. Using the notions of Sect. 5, we can select $c = 1$ and $s = 1$. Consequently, we consider $m = 1$, which yields that all rule identifiers constructed in Definition 6 use at most one rule of N.[19] A derivation like the one depicted in Fig. 15, which starts with a rule of N, cannot be simulated by $M \mathbin{;} N$ because the rules constructed for $M \mathbin{;} N$ always trigger a rule of M first.   □

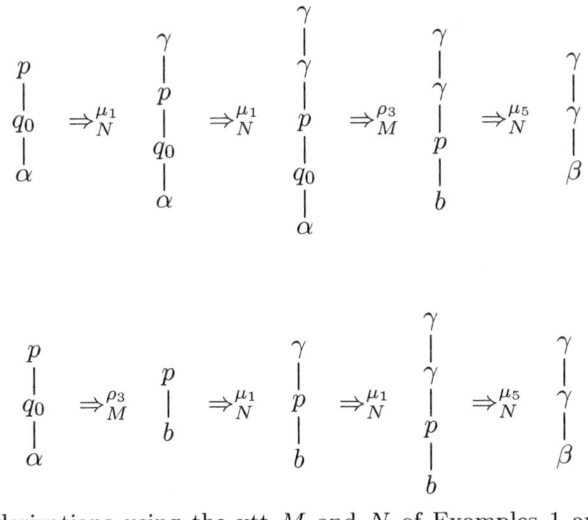

Fig. 15. Two derivations using the xtt M and N of Examples 1 and 2. The upper derivation cannot be simulated by $M;N$ since it starts with a rule of N. In principle, an unbounded number of rule applications of rule μ_1 could happen before the intermediate symbol b is consumed in the lower derivation. Thus, such derivations can, in general, also not be simulated by $M \mathbin{;} N$.

Thus, we need to adjust our construction. To avoid the problem in the lower derivation of Fig. 15, we restrict the rules of N that can be used when processing the right-hand side r of a rule $\rho \in R_1$. As in [39] we require that r is

---

[19] We could not avoid the problem, even if we would consider larger values for m.

processed only with consuming rules of N. The ε-rules of N need to fire either before ρ or after all intermediate symbols of r are fully consumed by N. This creates a problem, if the rule ρ creates 2 intermediate symbols at the same time, and the original derivation uses ε-rules after consuming one intermediate symbol but before consuming the second intermediate symbol. To avoid this problem, we already assumed in this section that M is shallow. Consequently, $m = 1$ provides an upper bound to the number of consuming rules required by N to process the right-hand side of a rule of M. This is due to the fact that there is at most one intermediate symbol in any right-hand side of a rule of M, and we can only use consuming rules of N to process it. As before, for any sentential form $\xi \in T_\Delta(P(T_\Gamma(Q(T_\Sigma(X)))))$ and rule $\rho \in R$ there exists at most one $\zeta \in T_\Delta(P(T_\Gamma(Q(T_\Sigma(X)))))$ such that $\xi \Rightarrow_M^\rho \zeta$, which also holds for the xtt N. Similarly, we recall that we identify elements of $T_\Delta(P(Q(T_\Sigma(X))))$ with elements of $T_\Delta((P \times Q)(T_\Sigma(X)))$ in the obvious manner.

Definition 17 (cf. [39, Definition 9]). *The ε-composition $M ;_\varepsilon N$ of M and N is the xtt $(P \times Q, \Sigma, \Delta, I_2 \times I_1, \mathcal{R}, \chi)$ with rule identifiers*

$$\mathcal{R} = \{\langle \rho, p, \varepsilon \rangle \mid \text{erasing } \rho \in R_1, p \in P\} \cup$$
$$\cup \{\langle \rho, p, \mu \rangle \mid \text{producing } \rho \in R_1, p \in P, \text{ consuming } \mu \in R_2\} \cup$$
$$\cup \{\langle \varepsilon, q, \mu \rangle \mid q \in Q, \ \varepsilon\text{-rule } \mu \in R_2\}$$

such that

- $\chi(\langle l \to r, p, \varepsilon \rangle) = (p(l), R_1(l \to r), p(r))$ *for every erasing rule* $l \to r \in R_1$ *and* $p \in P$,
- $\chi(\langle l \to r, p, \mu \rangle) = (p(l), a, r')$, *where*

$$a = \begin{cases} R_1(l \to r) \cdot R_2(\mu) & \text{if } p(l) \Rightarrow_M^{(l,r)} ; \Rightarrow_N^\mu r' \\ 0 & \text{otherwise} \end{cases}$$

  for every producing $l \to r \in R_1$, $p \in P$, *and consuming* $\mu \in R_2$, *and*
- $\chi(\langle \varepsilon, q, l \to r \rangle) = (l\theta, R_2(l \to r), r\theta)$, *where* $\theta(x) = q(x)$ *for every* $x \in X$, $q \in Q$, *and ε-rule* $l \to r \in R_2$.

Note that the only differences to the construction of [39] are the presence of (i) non-simple left-hand sides in rules of M and (ii) weights. Let us discuss the three sets of rule identifiers mentioned in Definition 17. Rule identifiers of the form $\langle \rho, p, \varepsilon \rangle$ refer to variants of an erasing rule ρ of R_1. For each state $p \in P$, we obtain a variant by annotating the two states (in the left- and right-hand side) by p. In other words, we perform a step using M, but since no intermediate symbol is produced, we do not perform a step using N. Second, the rule identifiers of the form $\langle \rho, p, \mu \rangle$ contain rules that are obtained in the usual way by processing the right-hand side of a producing rule of M by consuming rules of N. Since

M is shallow and N is a tdtt with ε-rules, each producing rule of M contains exactly one intermediate symbol and each consuming rule of N contains exactly one intermediate symbol. Thus, the derivation only succeeds if the producing rule of M produces exactly the symbol that the consuming rule of N consumes. These two types of rules were also present in the generic composition construction of Sect. 5.1. Finally, rule identifiers of the form $\langle \varepsilon, q, \mu \rangle$ refer to a variant of an ε-rule μ of N that is annotated with the state $q \in Q$.

Let us quickly check whether the obtained rules $l \to r$ are admissible; i.e., whether $\{l, r\} \nsubseteq P(Q(X))$. Clearly, identifiers of the form $\langle \rho, p, \varepsilon \rangle$ yield admissible rules because they contain just copies of rules of M. The same reasoning applies to rules with identifiers of the form $\langle \varepsilon, q, \mu \rangle$, which are copies of rules of N. Finally, rules with identifiers like $\langle \rho, p, \mu \rangle$ are always producing because each rule $\mu \subset R_2$ is producing. Clearly, producing rules are admissible. Next, let us illustrate the construction.

Example 18. Let $M = (\{q_0, q_1\}, \Sigma, \Gamma, \{q_0\}, R)$ and $N = (\{p\}, \Gamma, \Delta, \{p\}, R')$ be the xtt of Examples 1 and 2, respectively. The composition construction of Definition 17 yields the xtt $M ;_\varepsilon N = (P \times Q, \Sigma, \Delta, \{\langle p, q_0 \rangle\}, \mathcal{R}, \chi)$ with rule identifiers

$$\mathcal{R} = \{\langle \rho_2, p, \varepsilon \rangle, \langle \rho_1, p, \mu_2 \rangle, \langle \rho_1, p, \mu_3 \rangle, \langle \rho_3, p, \mu_5 \rangle, \langle \rho_4, p, \mu_4 \rangle, \langle \rho_5, p, \mu_4 \rangle,$$
$$\langle \varepsilon, q_0, \mu_1 \rangle, \langle \varepsilon, q_1, \mu_1 \rangle\}$$

such that

$$\chi(\langle \rho_2, p, \varepsilon \rangle) = \langle p, q_0 \rangle (\sigma(\alpha, x_1)) \xrightarrow{3} \langle p, q_0 \rangle (x_1)$$

$$\chi(\langle \rho_1, p, \mu_2 \rangle) = \langle p, q_0 \rangle (\sigma(\sigma(x_1, x_2), \alpha)) \xrightarrow{2 \cdot 5} \lambda(\langle p, q_1 \rangle (x_1), \langle p, q_0 \rangle (x_2))$$

$$\chi(\langle \rho_1, p, \mu_3 \rangle) = \langle p, q_0 \rangle (\sigma(\sigma(x_1, x_2), \alpha)) \xrightarrow{2 \cdot 5} \lambda(\beta, \lambda(\langle p, q_1 \rangle (x_1), \langle p, q_0 \rangle (x_2)))$$

$$\chi(\langle \rho_3, p, \mu_5 \rangle) = \langle p, q_0 \rangle (\alpha) \xrightarrow{1 \cdot 1} \beta$$

$$\chi(\langle \rho_4, p, \mu_4 \rangle) = \langle p, q_1 \rangle (x_1) \xrightarrow{2 \cdot 8} \delta(\langle p, q_0 \rangle (x_1))$$

$$\chi(\langle \rho_5, p, \mu_4 \rangle) = \langle p, q_1 \rangle (x_1) \xrightarrow{5 \cdot 8} \delta(\langle p, q_1 \rangle (x_1))$$

$$\chi(\langle \varepsilon, q_0, \mu_1 \rangle) = \langle p, q_0 \rangle (x_1) \xrightarrow{2} \gamma(\langle p, q_0 \rangle (x_1))$$

$$\chi(\langle \varepsilon, q_1, \mu_1 \rangle) = \langle p, q_1 \rangle (x_1) \xrightarrow{2} \gamma(\langle p, q_1 \rangle (x_1)) \ .$$

The construction of the rule corresponding to the rule identifier $\langle \rho_1, p, \mu_2 \rangle$ is illustrated in Fig. 16. □

6.2 Correctness

Let us start by recalling the two cases, in which the composition construction of [39], of which our construction in Definition 17 is an adaptation, is successful in the unweighted setting. Recall that M is shallow and N is a tdtt with ε-rules. If

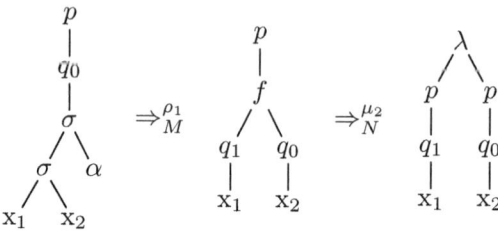

Fig. 16. Construction of the rule with identifier $\langle \rho_1, p, \mu_2 \rangle$ in Example 18

Table 4. Cases for unweighted xtt composition of M and N, where M is shallow and N is a tdtt with ε-rules

Case	M	N	Reference
(a)		linear and nondeleting	[39, Theorem 17]
(b)	total	linear	[39, Theorem 17]

- N is linear, and
- M is total or N is nondeleting,

then $\tau_M \,;\, \tau_N$ can be computed by an xtt [39, Theorem 17]. Table 4 shows these two cases, which correspond to the equally named cases in Sect. 5.

Let us start with Case (a). As in the previous section, this case does not cause further problems in the weighted setting, and we will sketch the correctness proof for our composition construction of Definition 17.

Theorem 19. *If M is shallow and N is a linear and nondeleting tdtt with ε-rules, then $\tau_{M;_\varepsilon N} = \tau_M \,;\, \tau_N$.*

Proof (sketch). Let $\xi \in P(Q(T_\Sigma))$ and $u \in T_\Delta$. We claim that there is a weight-preserving bijection between the derivations of the form

$$\xi \,(\Rightarrow_M^{\rho_1} ; \cdots ; \Rightarrow_M^{\rho_k})\,;\,(\Rightarrow_N^{\mu_1} ; \cdots ; \Rightarrow_N^{\mu_n})\,u \ ,$$

and the derivations of the form $\xi \Rightarrow_{M;_\varepsilon N}^{\nu_1} ; \cdots ; \Rightarrow_{M;_\varepsilon N}^{\nu_\ell} u$.

We construct the bijection by induction on k. Let $s \in T_\Sigma$, $p \in P$, and $q \in Q$ be such that $\xi = p(q(s))$. Next, we distinguish whether the first applied rule ρ_1 is erasing. If it is, then we start the derivation using $M \,;_\varepsilon N$ with the rule $\langle \rho_1, p, \varepsilon \rangle$, which has the same weight as ρ_1. Otherwise, the rule ρ_1 produces exactly one intermediate symbol $\gamma \in \Gamma$ that will be consumed by exactly one rule μ_i for some $i \in \mathbb{N}$. The symbol γ is consumed by exactly one rule because N is linear and nondeleting. Clearly, all rules $\mu_1, \ldots, \mu_{i-1}$ before μ_i must be ε-rules because otherwise they would consume the symbol γ. In $M \,;_\varepsilon N$ we simulate this derivation by starting with the ε-rules $\langle \varepsilon, q, \mu_1 \rangle, \ldots, \langle \varepsilon, q, \mu_{i-1} \rangle$ followed by the consuming rule $\langle \rho_1, p', \mu_i \rangle$, where p' is the (unique) state that occurs in the right-hand side of the rule μ_{i-1}.[20] Clearly, this part of the derivation has the same weight as the

---

[20] If $i = 1$, then we let $p' = p$.

product of the weight of rule ρ_1 and the weights of the rules $\mu_1, \ldots, \mu_i$. Now we covered all three cases and shortened the derivation using M. The remainder of the derivation can then be processed using the induction hypothesis. Thus, our construction relates derivations bijectively and preserves the weight. Given this bijective and weight-preserving relation, the main statement follows trivially.   □

Let us illustrate the construction used in the proof of Theorem 19.

Example 20. Let M and N be the xtt of Examples 1 and 2, and recall that $M \mathbin{;_\varepsilon} N$ is shown in Example 18. Moreover, let

$$s = \sigma(\sigma(\alpha, \sigma(\alpha, \sigma(\alpha, \alpha))), \alpha) \qquad \text{and} \qquad t = f(h(h(b)), b)$$

be an input and output tree for M as in Example 1. Figure 2 shows a derivation d_M with weight 180 from $q_0(s)$ to t using M. Moreover, let

$$u = \lambda(\beta, \lambda(\delta(\delta(\beta)), \gamma(\gamma(\beta)))) \ .$$

Figure 17 shows a derivation d_N with weight $5 \cdot 8^2 \cdot 1 \cdot 2^2 \cdot 1 = 1\,280$ from $p(t)$ to u using N. The concatenation of the two derivations gives us a derivation d from $p(q_0(s))$ to u using rules of M and N. Clearly, the weight of this derivation is $180 \cdot 1\,280 = 230\,400$.

The image of the derivations d_M and d_N by the bijection constructed in the proof of Theorem 19 is shown in Fig. 18. The first four rules in the derivation d_M are producing, and the produced symbol is immediately consumed in the corresponding step in the derivation d_N. The fifth and sixth rules in the derivation d_M are erasing rules, which are simulated by the corresponding erasing rules in the derivation d. The last rule in the derivation d_M is another producing rule, whose produced symbol b is not immediately consumed in the current step of the derivation d_N. Rather the ε-rule μ_1 is applied twice before rule μ_5 consumes the symbol b. Consequently, we have to defer the application of the rule $\langle \rho_3, p, \mu_5 \rangle$ to first allow the applications of the rule $\langle \varepsilon, q_0, \mu_1 \rangle$. The following table lists the rule applications for all three derivations and shows the correspondence.

Step	1	2	3	4	5	6	7
d_M:	ρ_1	ρ_5	ρ_4	ρ_3	ρ_2	ρ_2	ρ_3
d_N:	μ_3	μ_4	μ_4	μ_5			$\mu_1 \ \mu_1 \ \mu_5$
d:	$\langle \rho_1, p, \mu_3 \rangle$	$\langle \rho_5, p, \mu_4 \rangle$	$\langle \rho_4, p, \mu_4 \rangle$	$\langle \rho_3, p, \mu_5 \rangle$	$\langle \rho_2, p, \varepsilon \rangle$	$\langle \rho_2, p, \varepsilon \rangle$	$\langle \varepsilon, q_0, \mu_1 \rangle \ \langle \varepsilon, q_0, \mu_1 \rangle \ \langle \rho_3, p, \mu_5 \rangle$

The weight of the derivation d is

$$(2 \cdot 5) \cdot (5 \cdot 8) \cdot (2 \cdot 8) \cdot (1 \cdot 1) \cdot 3 \cdot 3 \cdot 2 \cdot 2 \cdot (1 \cdot 1) = 230\,400 \ ,$$

which coincides with the expected result.   □

Let us move on to Case (b), in which we experience the same problem with deleted subtrees as in Sect. 5. We refer the reader to the discussion of Case (b) in Sect. 5 for an illustration of the problem and examples. Here, we avoid the problem by requiring (i) that the xtt M is BOOLEAN and (ii) that the semiring A is idempotent (i.e., $1 + 1 = 1$). This yields that the xtt M is essentially

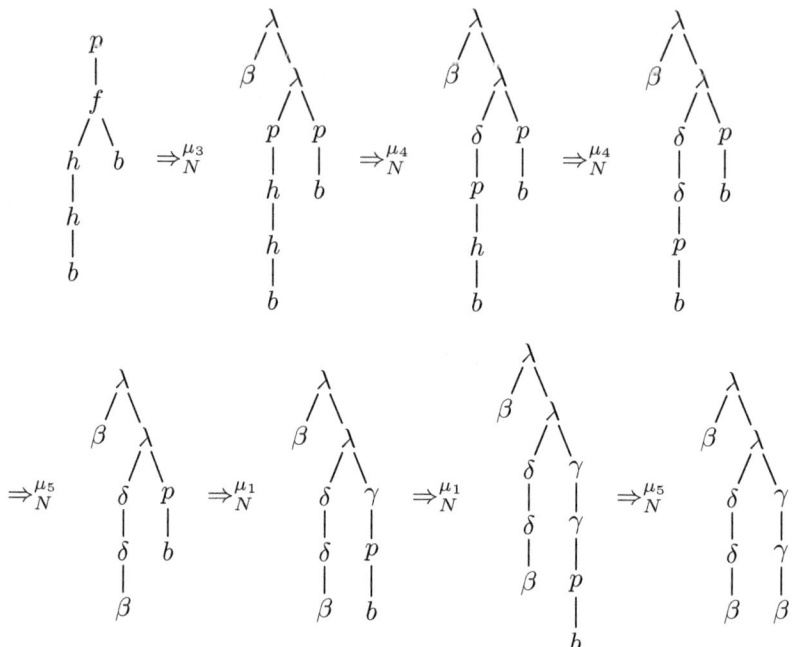

Fig. 17. A derivation from $p(t)$ to u using N (see Example 20)

unweighted and constant (with weight 1). We observe that $\tau_M(s,t) = 1$ for all $(s,t) \in \mathrm{supp}(\tau_M)$ because M is BOOLEAN and A is idempotent, and for every $s \in T_\Sigma$ there exists $t \in T_\Gamma$ such that $(s,t) \in \mathrm{supp}(\tau_M)$ due to the totality of M.

Theorem 21. *If the shallow xtt M is total and* BOOLEAN, *the tdtt N with ε-rules is linear, and the semiring A is idempotent, then $\tau_{M;_\varepsilon N} = \tau_M \, ; \tau_N$.*

Proof (sketch). Let $\xi \in P(Q(T_\Sigma))$ and $u \in T_\Delta$. We claim that there is a weight-preserving surjective mapping f from the derivations of the form

$$\xi \, (\Rightarrow_M^{\rho_1} ; \cdots ; \Rightarrow_M^{\rho_k}) \, ; (\Rightarrow_N^{\mu_1} ; \cdots ; \Rightarrow_N^{\mu_n}) \, u \ ,$$

and the derivations of the form $\xi \Rightarrow_{M;_\varepsilon N}^{\nu_1} ; \cdots ; \Rightarrow_{M;_\varepsilon N}^{\nu_\ell} u$.

Clearly, the derivation sequence $\rho_1 \cdots \rho_k \mu_1 \cdots \mu_n$ is successful. Let $\perp$ be a fresh symbol, and let $l \to r \in R_1$ be a rule of M. The *mutilated copy* of $l \to r$ is the rule $l \to \perp(r)$. We denote the mutilated copy of $\rho \in R_1$ by $\bar{\rho}$. Next, we obtain a rule sequence $\rho'_1 \cdots \rho'_k$ from $\rho_1 \cdots \rho_k$ by replacing maximally many rules ρ_i by their mutilated copy $\overline{\rho_i}$ such that

$$\xi \, (\Rightarrow_M^{\rho'_1} ; \cdots ; \Rightarrow_M^{\rho'_k}) \, ; (\Rightarrow_N^{\mu_1} ; \cdots ; \Rightarrow_N^{\mu_n}) \, u \ .$$

In order words, the new rule sequence is still a successful derivation from ξ to u. Clearly, this derivation can only be successful if N ignores (i.e., deletes) the subtrees created by mutilated rules because N cannot process the symbol $\perp$. In the

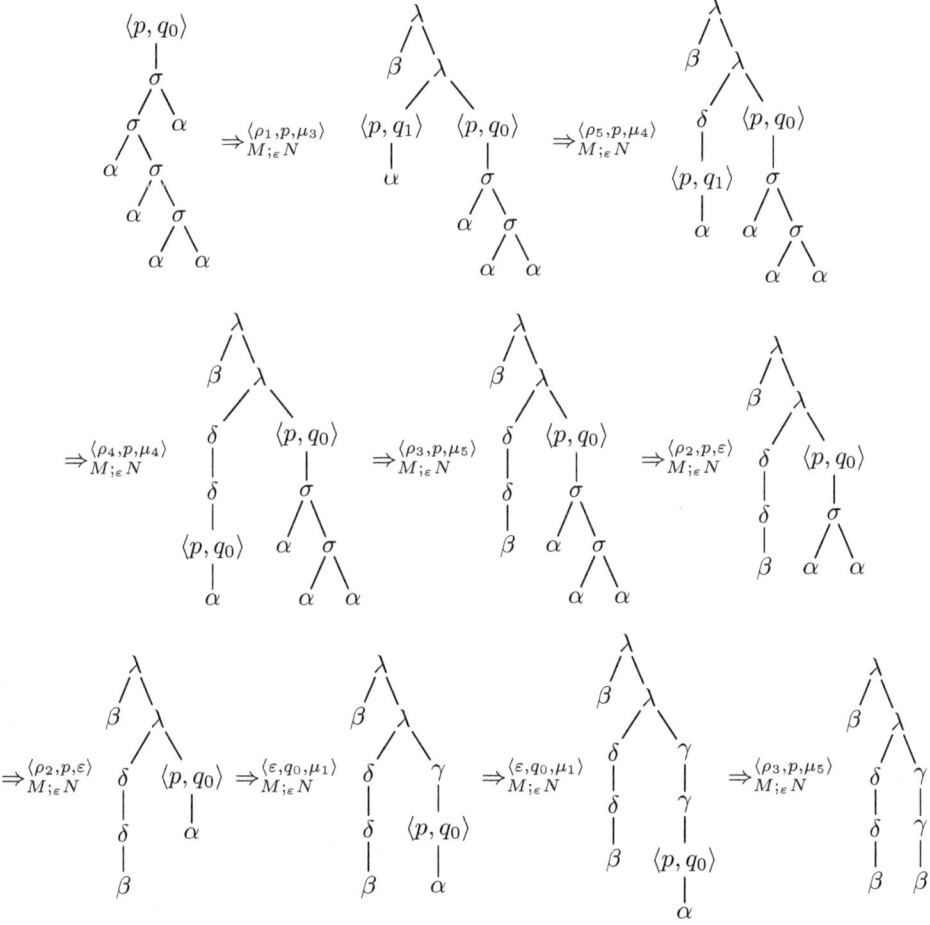

Fig. 18. The matching derivation from $\langle p, q_0 \rangle(s)$ to u using $M \mathbin{;_\varepsilon} N$ (see Example 20)

next step we drop all mutilated rules from the rule sequence $\rho'_1 \cdots \rho'_k$ and relate the obtained rule sequence in the same way as in the proof of Theorem 19 to the derivation of the composed xtt. The obtained derivation using the composed xtt has the same weight as the original derivation because we only dropped rules of R_1, which have weight 1 because M is BOOLEAN. It is not difficult to see that this mapping is surjective because we can always recover one subderivation for parts that we dropped due to the totality of M. This approach is illustrated in an example following the proof.

Now we complete the proof as follows:

$$(\tau'_M \,;\, \tau'_N)(\xi, u) = \sum_{\substack{\rho_1,\ldots,\rho_k \in R_1 \\ \mu_1,\ldots,\mu_n \in R_2 \\ \xi(\Rightarrow_M^{\rho_1};\cdots;\Rightarrow_M^{\rho_k});(\Rightarrow_N^{\mu_1};\cdots;\Rightarrow_N^{\mu_n})u}} \left(\prod_{i=1}^{k} R_1(\rho_i) \cdot \prod_{i=1}^{n} R_2(\mu_i) \right)$$

$$= \sum_{\substack{\nu_1,\ldots,\nu_\ell \in \mathcal{R} \\ d:\ \xi \Rightarrow_{M\,;_\varepsilon N}^{\nu_1};\cdots;\Rightarrow_{M\,;_\varepsilon N}^{\nu_\ell} u}} \left(\sum_{d' \in f^{-1}(d)} \left(\prod_{i=1}^{\ell} \mathrm{wt}(\nu_i) \right) \right)$$

(because the second sum is never empty due to surjectivity of f)

$$= \sum_{\substack{\nu_1,\ldots,\nu_\ell \in \mathcal{R} \\ \xi \Rightarrow_{M\,;_\varepsilon N}^{\nu_1};\cdots;\Rightarrow_{M\,;_\varepsilon N}^{\nu_\ell} u}} \left(\prod_{i=1}^{\ell} \mathrm{wt}(\nu_i) \right) = \tau'_{M\,;_\varepsilon N}(\xi, u)$$

because A is idempotent and $f^{-1}(d) \neq \emptyset$. Thus, we conclude that $M \,;_\varepsilon N$ computes $\tau_M \,;\, \tau_N$. $\square$

Let us illustrate the construction in the proof of Theorem 21 on an example.

Example 22. Let us consider Fig. 19, which is a minor variation of Fig. 10. Figure 19 displays two derivations that we want to relate. Obviously, the rule sequence of M is $\rho_1\rho_1\rho_2$. Now we need to mutilate the rules in the rule sequence. We start with the most aggressive attempt and mutilate every rule in the sequence to obtain $\overline{\rho_1\rho_1\rho_2}$. Figure 20 shows that the derivation is no longer successful for this rule sequence. Thus, we try the sequence $\rho_1\overline{\rho_1\rho_2}$, which indeed still delivers a successful derivation as depicted in Fig. 20. Next, we reduce the sequence by taking out all mutilated rules. We obtain just ρ_1. Now we combine the producing rule ρ_1 as usual with the consuming rule μ_7 of the second rule sequence and relate them to the rule $\langle \rho_1, p, \mu_7 \rangle$ of the composed tdtt. $\square$

Following the ideas of Sect. 5, the authors suspect that instead of idempotence and a total and BOOLEAN xtt M, we can simply require that M is constant. This leads to our final conjecture, which would generalize Theorem 21. We collect all obtained results of this section in Table 5.

Conjecture 23. If the shallow xtt M is constant and the tdtt N with ε-rules is linear, then $\tau_M \,;\, \tau_N$ can be computed by an xtt.

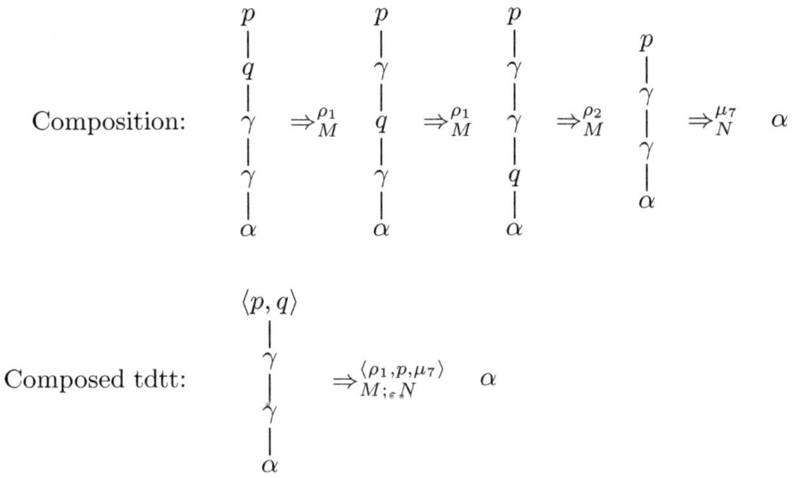

Composition:

$$p \;|\; q \;|\; \gamma \;|\; \gamma \;|\; \alpha \quad \Rightarrow_M^{\rho_1} \quad p \;|\; \gamma \;|\; q \;|\; \gamma \;|\; \alpha \quad \Rightarrow_M^{\rho_1} \quad p \;|\; \gamma \;|\; \gamma \;|\; q \;|\; \alpha \quad \Rightarrow_M^{\rho_2} \quad p \;|\; \gamma \;|\; \gamma \;|\; \alpha \quad \Rightarrow_N^{\mu_7} \quad \alpha$$

Composed tdtt:

$$\langle p, q \rangle \;|\; \gamma \;|\; \gamma \;|\; \alpha \quad \Rightarrow_{M\,;_\varepsilon\,N}^{\langle \rho_1, p, \mu_7 \rangle} \quad \alpha$$

Fig. 19. Relating rule sequences

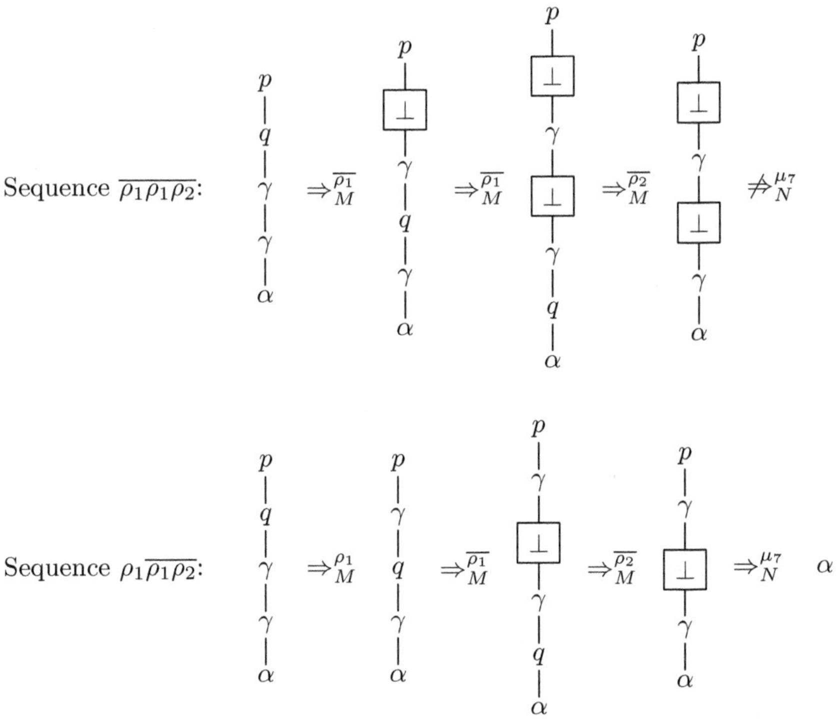

Sequence $\overline{\rho_1}\overline{\rho_1}\rho_2$:

$$p \;|\; q \;|\; \gamma \;|\; \gamma \;|\; \alpha \quad \Rightarrow_M^{\overline{\rho_1}} \quad p \;|\; \boxed{\perp} \;|\; \gamma \;|\; q \;|\; \gamma \;|\; \alpha \quad \Rightarrow_M^{\overline{\rho_1}} \quad p \;|\; \boxed{\perp} \;|\; \gamma \;|\; \boxed{\perp} \;|\; \gamma \;|\; q \;|\; \alpha \quad \Rightarrow_M^{\overline{\rho_2}} \quad p \;|\; \boxed{\perp} \;|\; \gamma \;|\; \boxed{\perp} \;|\; \gamma \;|\; \alpha \quad \not\Rightarrow_N^{\mu_7}$$

Sequence $\rho_1\overline{\rho_1}\rho_2$:

$$p \;|\; q \;|\; \gamma \;|\; \gamma \;|\; \alpha \quad \Rightarrow_M^{\rho_1} \quad p \;|\; \gamma \;|\; q \;|\; \gamma \;|\; \alpha \quad \Rightarrow_M^{\overline{\rho_1}} \quad p \;|\; \gamma \;|\; \boxed{\perp} \;|\; \gamma \;|\; q \;|\; \alpha \quad \Rightarrow_M^{\overline{\rho_2}} \quad p \;|\; \gamma \;|\; \boxed{\perp} \;|\; \gamma \;|\; \alpha \quad \Rightarrow_N^{\mu_7} \quad \alpha$$

Fig. 20. Testing two mutilated sequences. The upper one is too aggressive and the rule μ_7 of N is not applicable anymore. The lower sequence represents the sought sequence because the derivation is still successful. After the deletion of the mutilated rules, we thus obtain just ρ_1.

Table 5. Cases for weighted xtt composition, where M is a shallow xtt and N is a tdtt with ε-rules. In the second line (which uses Theorem 21), we additionally need to require that the semiring is idempotent.

Case	M	N	Reference
(a)		linear and nondeleting	Theorem 19
(b)	total and BOOLEAN	linear	Theorem 21
	constant	linear	Conjecture 23

References

1. Alexandrakis, A., Bozapalidis, S.: Weighted grammars and Kleene's theorem. Inf. Process. Lett. 24(1), 1–4 (1987)
2. Arnold, A., Dauchet, M.: Bi-transductions de forêts. In: Michaelson, S., Milner, R. (eds.) ICALP 1976, pp. 74–86. Edinburgh University Press, Edinburgh (1976)
3. Arnold, A., Dauchet, M.: Morphismes et bimorphismes d'arbres. Theoret. Comput. Sci. 20(4), 33–93 (1982)
4. Baader, F., Nipkow, T.: Term Rewriting and All That. Cambridge University Press, Cambridge (1998)
5. Baker, B.S.: Composition of top-down and bottom-up tree transductions. Inform. and Control 41(2), 186–213 (1979)
6. Berstel, J., Reutenauer, C.: Recognizable formal power series on trees. Theoret. Comput. Sci. 18(2), 115–148 (1982)
7. Borchardt, B.: The Theory of Recognizable Tree Series. Ph.D. thesis, Technische Universität Dresden (2005)
8. Borchardt, B., Vogler, H.: Determinization of finite state weighted tree automata. J. Autom. Lang. Combin. 8(3), 417–463 (2003)
9. Bozapalidis, S., Louscou-Bozapalidou, O.: The rank of a formal tree power series. Theoret. Comput. Sci. 27(1-2), 211–215 (1983)
10. Comon, H., Dauchet, M., Gilleron, R., Jacquemard, F., Lugiez, D., Löding, C., Tison, S., Tommasi, M.: Tree automata techniques and applications (2007), http://tata.gforge.inria.fr
11. Dauchet, M.: Transductions inversibles de forêts. Thèse 3ème cycle, Université de Lille (1975)
12. Engelfriet, J.: Bottom-up and top-down tree transformations: A comparison. Math. Systems Theory 9(3), 198–231 (1975)
13. Engelfriet, J.: Top-down tree transducers with regular look-ahead. Math. Systems Theory 10(1), 289–303 (1976)
14. Engelfriet, J., Fülöp, Z., Vogler, H.: Bottom-up and top-down tree series transformations. J. Autom. Lang. Combin. 7(1), 11–70 (2002)
15. Engelfriet, J., Lilin, E., Maletti, A.: Extended multi bottom-up tree transducers. In: Ito, M., Toyama, M. (eds.) DLT 2008. LNCS, vol. 5257, pp. 289–300. Springer, Heidelberg (2008)
16. Engelfriet, J., Lilin, E., Maletti, A.: Composition and decomposition of extended multi bottom-up tree transducers. Acta Inform. 46(8), 561–590 (2009)
17. Ésik, Z., Kuich, W.: Formal tree series. J. Autom. Lang. Combin. 8(2), 219–285 (2003)

18. Fülöp, Z., Gazdag, Z., Vogler, H.: Hierarchies of tree series transformations. Theoret. Comput. Sci. 314(3), 387–429 (2004)
19. Fülöp, Z., Vogler, H.: Tree series transformations that respect copying. Theory Comput. Systems 36(3), 247–293 (2003)
20. Fülöp, Z., Vogler, H.: Weighted tree automata and tree transducers. In: Droste, M., Kuich, W., Vogler, H. (eds.) Handbook of Weighted Automata, ch. 9, pp. 313–403. Springer, Heidelberg (2009)
21. Gécseg, F., Steinby, M.: Tree Automata. Akadémiai Kiadó, Budapest (1984)
22. Gécseg, F., Steinby, M.: Tree languages. In: Rozenberg, G., Salomaa, A. (eds.) Handbook of Formal Languages, vol. 3, ch. 1, pp. 1–68. Springer, Heidelberg (1997)
23. Golan, J.S.: Semirings and their Applications. Kluwer Academic, Dordrecht (1999)
24. Graehl, J., Knight, K., May, J.: Training tree transducers. Comput. Linguist. 34(3), 391–427 (2008)
25. Hebisch, U., Weinert, H.J.: Semirings — Algebraic Theory and Applications in Computer Science. Algebra, vol. 5. World Scientific, Singapore (1998)
26. Knight, K., Graehl, J.: An overview of probabilistic tree transducers for natural language processing. In: Gelbukh, A.F. (ed.) CICLing 2005. LNCS, vol. 3406, pp. 1–24. Springer, Heidelberg (2005)
27. Koehn, P.: Statistical Machine Translation. Cambridge University Press, Cambridge (2010)
28. Koehn, P., Och, F.J., Marcu, D.: Statistical phrase-based translation. In: NAACL 2003, pp. 48–54. Association for Computational Linguistics (2003)
29. Kühnemann, A.: Benefits of tree transducers for optimizing functional programs. In: Arvind, V., Sarukkai, S. (eds.) FST TCS 1998. LNCS, vol. 1530, pp. 146–158. Springer, Heidelberg (1998)
30. Kuich, W.: Formal power series over trees. In: Bozapalidis, S. (ed.) DLT 1997, pp. 61–101. Aristotle University of Thessaloniki (1998)
31. Kuich, W.: Full abstract families of tree series I. In: Karhumäki, J., Maurer, H.A., Paun, G., Rozenberg, G. (eds.) Jewels are Forever, pp. 145–156. Springer, Heidelberg (1999)
32. Kuich, W.: Tree transducers and formal tree series. Acta Cybernet. 14(1), 135–149 (1999)
33. Maletti, A.: Compositions of tree series transformations. Theoret. Comput. Sci. 366(3), 248–271 (2006)
34. Maletti, A.: The Power of Tree Series Transducers. Ph.D. thesis, Technische Universität Dresden (2006)
35. Maletti, A.: Compositions of extended top-down tree transducers. Inform. and Comput. 206(9-10), 1187–1196 (2008)
36. Maletti, A.: Survey: Weighted extended top-down tree transducers — Part I: Basics and expressive power. Acta Cybernet (2011), preprint available at: http://www.ims.uni-stuttgart.de/~maletti/pub/mal11.pdf
37. Maletti, A.: Survey: Weighted extended top-down tree transducers — Part II: Application in machine translation. Fund. Inform (2011)
38. Maletti, A., Graehl, J., Hopkins, M., Knight, K.: The power of extended top-down tree transducers. SIAM J. Comput. 39(2), 410–430 (2009)
39. Maletti, A., Vogler, H.: Compositions of top-down tree transducers with ε-rules. In: Yli-Jyrä, A., Kornai, A., Sakarovitch, J., Watson, B. (eds.) FSMNLP 2009. LNCS (LNAI), vol. 6062, pp. 69–80. Springer, Heidelberg (2010)

40. May, J., Knight, K., Vogler, H.: Efficient inference through cascades of weighted tree transducers. In: ACL 2010, pp. 1058–1066. Association for Computational Linguistics (2010)
41. Rounds, W.C.: Mappings and grammars on trees. Math. Systems Theory 4(3), 257–287 (1970)
42. Thatcher, J.W.: Generalized[2] sequential machine maps. J. Comput. System Sci. 4(4), 339–367 (1970)
43. Thatcher, J.W.: Tree automata: An informal survey. In: Aho, A.V. (ed.) Currents in the Theory of Computing, ch. 4, pp. 143–172. Prentice Hall, Englewood Cliffs (1973)
44. Wang, H.: On characters of semirings. Houston J. Math. 23(3), 391–405 (1997)
45. Wang, H.: On rational series and rational languages. Theoret. Comput. Sci. 205(1-2), 329–336 (1998)
46. Yamada, K., Knight, K.: A decoder for syntax-based statistical MT. In: ACL 2002, pp. 303–310. Association for Computational Linguistics (2002)

Valuations of Weighted Automata:
Doing It in a Rational Way

Ingmar Meinecke

Institut für Informatik, Universität Leipzig, D-04109 Leipzig, Germany
meinecke@informatik.uni-leipzig.de

Abstract. We study Kleene's theorem about the equivalence of automata and expressions in a quantitative setting both for finite and infinite words. The quantities originate from valuation monoids and ω-indexed valuation monoids which cover not only semirings but also cost models like average cost, long-run peaks of resource consumption, or discounting sums of rewards. For finite words we deduce the characterization of weighted automata by regular weighted expressions directly from Kleene's theorem. For infinite words we define three different behaviors of weighted Büchi automata depending on the way runs are evaluated. Depending on the properties of the underlying ω-indexed valuation monoid, we explore the connections between the different behaviors of weighted Büchi automata and ω-regular weighted expressions. Again, we use classical results on ω-languages to derive results in the quantitative setting.

1 Introduction

There are some results in theoretical computer science which seem to last forever. One is Kleene's theorem stating that finite automata and regular expressions define the same class of languages [31]. Certainly, regular expressions are a popular formalism to describe properties and patterns. We use them e.g. in the Emacs editor when writing this article. As far as theoretical aspects are concerned, the equivalence between automata and expressions is very useful because the two concepts are of different nature (graphs and well-structured terms) and can be applied to different problems. Thus, it is not surprising that variants and generalizations of Kleene's theorem were shown both in a qualitative and in a quantitative setting. We skip the qualitative results for numerous discrete structures and turn immediately to the quantitative ones.

A theorem by Schützenberger [39] generalized Kleene's theorem for weighted automata and weighted expressions over finite words with weights from a semiring. Semirings are a very powerful concept because they comprise many instances like the natural numbers with addition and multiplication or the tropical semiring important in optimization. In the 1980s, results analog to the one of Schützenberger were also obtained for trees by Berstel and Reutenauer [3] and by Alexandrakis and Bozapalidis [1]. Later on, the theorem was shown for a semiring weighted setting for traces [11], pictures [4,36], or series-parallel posets [33], to give just a few examples. An overview of these results and their proofs can be found in the chapters [27,28,29,38] of the Handbook of Weighted Automata [13].

W. Kuich and G. Rahonis (Eds.): Bozapalidis Festschrift, LNCS 7020, pp. 309–346, 2011.

Here, we confine ourselves to finite and infinite words. For infinite words several Kleene-Schützenberger results can be found in the literature: for ω-languages by Büchi [5] and for the weighted setting by Ésik and Kuich [25,27] for semiring-semimodule pairs, by Droste, Kuske, and Kuich [14,32] for discounting, and by Droste and Vogler [23] for bounded lattices. Especially the semiring-semimodule setting by Ésik and Kuich is a very general one relying on an algebraic and equatioal approach.

In 2008, another kind of weighted automata which does not fit into the semiring setting was suggested by Chatterjee, Doyen, and Henzinger [7]. In these automata, real numbers are attached to the transitions and then the whole finite or infinite run, i.e., a finite or infinite sequence of weights, is evaluated by a function. Such a function can compute the average or the limit superior of the values occuring along the run. Chatterjee, Doyen, and Henzinger [7,8,9,10] were mainly interested in decidability issues for concrete valuation functions as well as in comparing the expressive power of their automata models for different valuation functions. This opened another way of thinking about weighted automata. Instead of multiplying weights locally by semiring multiplication, the weight of a run is computed globally by a valuation function. Non-determinism is still resolved by a commutative monoid operation as it is done for semiring weighted automata.

In [20,21], Droste and Meinecke generalized and unified such settings by introducing *valuation monoids* (for finite words) and ω-*indexed valuation monoids* (for infinite words). These concepts comprise semirings, average, limit superior, limit average, or discounting as considered in [7] as well as the complete star-omega-semirings [25,27] or the semirings used in [15,37]. One interesting point about (ω-indexed) valuation monoids is the fact that nothing else than a commutative monoid (or a complete one for infinite words) and a valuation function is needed to define the behavior of weighted automata. But in the course of giving a characterization by weighted MSO logic [20] or by weighted regular expressions [21,22], additional operations and properties have to be amended for the underlying valuation monoid. However, these operations and properties are quite different for logic on one side and expressions on the other one. In a semiring setting, this is somehow hidden because "everything" is defined by semiring multiplication. Most comprehensive properties are needed for a characterization by expressions. The defining equations of these *Cauchy valuation monoids* as we call them have close resemblance to properties of semirings. Nevertheless, we cover structures which are not semirings.

In this paper, we deal with the characterization of weighted automata by expressions. Our contribution is twofold. First, we extend the work started in [21,22]. Especially, we explore weighted Büchi automata over ω-indexed automata in more detail. In [21], we suggested to evaluate an infinite run as follows: Use *(i)* a Büchi condition, *(ii)* a valuation function for finite sequences, and *(iii)* an ω-indexed valuation function for infinite sequences. Then evaluate the finite sequences of weights between two consecutive acceptance states by the valuation function and, finally, combine these infinitely many intermediate results by the ω-indexed valuation function. This procedure defines the *(Büchi) behavior* of a weighted Büchi automaton and guarantees the necessary link between finite and ω-automata in order to establish a Kleene-like result also for infinite words. However, in [7] the value of an infinite run was computed without intermediate results. We refer to such a kind of valuation as the *unconditional behavior*. To show that the Büchi behavior of a weighted Büchi automaton is ω-rational, i.e., definable by ω-regular weighted expressions, we were in need of a property which we called the *partition property* [21]. Now we use the defining equation of this property to define a third behavior, the so-called *cumulative behavior*.[1] This is similar to the first one but this time the run is partitioned with regard to only one single accepting state. This is done for all accepting states appearing infinitely often within the run. Now the possible occurence of different accepting states is seen as non-determinism and resolved by the sum, i.e., the monoid operation. Hence, the weight of a run is a sum of m weights if m different accepting states appear infinitely often along the run.

We explore the connections between these three behaviors of weighted Büchi automata and ω-rationality, recall the results of [21,22], and obtain several new results depending on the properties of the ω-indexed valuation monoid. One such property is *uniformity* which is very similar to an infinitary associativity law for the product of complete star-omega semirings [25,27]. It states that the valuation of an ω-sequence of weights does not depend on the chosen intermediate points. This is the case when valuating finite runs by supremum and infinite ones by limit superior. However, if we take average and limit superior average, then this property is not satisfied. Using uniformity, we can show a Kleene-Schützenberger-like result for ω-indexed valuation monoids comprising all complete star-omega semirings. In [21,22], we could do so only for structures covering idempotent complete star-omega semirings. Last but not least, we show that weighted Büchi and weighted Muller automata define the same class of (Büchi) behaviors provided the ω-indexed valuation monoid is uniform.

Another focus of this paper is the proof method which is of combinatoric nature and tries to make as much use as possible of the classical results for languages and ω-languages. Kuske [34] showed that Schützenberger's theorem can be derived directly from Kleene's theorem without repeating the proofs of Kleene's theorem in the semiring setting. Here, we apply this method to our result for finite words and give this way an alternative proof compared to the one in [21,22]. The proof is very similar to the one in [34] and shows once again that Kleene's early result somehow already contains a lot of other results in its very concepts and arguments. For infinite words we apply this approach at least for one result using unambiguous ω-regular expressions [2].

Certainly, the Kleene-like results obtained here are not the end of the line. There are settings in discounting where the discount factor depends also on the action executed [16,17]. This cannot be covered by our approach. Another line of research are locally finite structures missing distributivity [23]. A notion of local finiteness is not yet clear for valuation monoids. We guess: "Kleeneism" is here to stay.

2   Weighted Automata and Expressions on Finite Words

In this chapter, we introduce the basic concepts: valuation monoids, weighted automata, rational operations, and regular weighted expressions – for finite words. We show that weighted automata and expressions define the same class of series even in a very general quantitative setting. This result was already shown in [21,22] but here we give another

---

[1] The first idea of this notion is due to Manfred Droste.

proof following the one in [34] for semiring weighted automata. It turns out that the equivalence between automata and expressions over valuation monoids follows from Kleene's result for languages of finite words.

2.1   Weighted Automata over Valuation Monoids

Let $\mathbb{N}$ denote the positive integers and $\mathbb{N}_0$ the non-negative integers. Let Σ be an alphabet. By Σ^+ we denote the set of non-empty finite words. For $w = a_1 \ldots a_n \in \Sigma^+$ with $a_i \in \Sigma$ for $i \in \{1, \ldots, n\}$, let $|w| = n$ be the *length* of w and $\mathrm{dom}(w) = \{1, \ldots, |w|\}$ be the *domain* of w.

Weights in a weighted automaton $\mathcal{A}$ will be attached to transitions. To compute the weight of a word w in $\mathcal{A}$, the runs of $\mathcal{A}$ on w have to be evaluated and, then, the obtained values have to be summarized to a single weight. For this, we introduce quantitative structures which will cover exactly these two requests.

Definition 2.1. *A* valuation monoid $\mathbb{D} = (D, +, \mathrm{val}, \mathbb{0})$ *consists of a commutative monoid* $(D, +, \mathbb{0})$ *and a* valuation function $\mathrm{val} : D^+ \to D$ *such that*

- $\mathrm{val}(d) = d$ *for all* $d \in D$ *and*
- $\mathrm{val}(d_1, \ldots, d_n) = \mathbb{0}$ *whenever* $d_i = \mathbb{0}$ *for some* $i \in \{1, \ldots, n\}$.

Now we can define weighted automata and their behavior.

Definition 2.2. *A* weighted (finite) automaton $\mathcal{A} = (Q, I, T, F, \mu)$ *(for short: a wfa) over the alphabet* Σ *and a valuation monoid* $\mathbb{D} = (D, +, \mathrm{val}, \mathbb{0})$ *consists of a finite state set* Q, *a set* $I \subseteq Q$ *of initial states, a set* $F \subseteq Q$ *of final states, a set of transitions* $T \subseteq Q \times \Sigma \times Q$, *and a weight function* $\mu : T \to D$.

A weighted automaton is a usual finite automaton equipped with weights for the transitions. Moreover, the automaton can be assumed to be *total* as in [7], i.e., for every $q \in Q$ and every $a \in \Sigma$ there is some $q' \in Q$ with $(q, a, q') \in T$. This can be achieved by adding transitions with weight $\mathbb{0}$. *Runs* $R = (t_i)_{1 \leq i \leq n}$ with $t_i = (q_{i-1}, a_i, q_i) \in T$ are defined as finite sequences of matching transitions. We call the word $w = \ell(R) = a_1 a_2 \ldots a_n$ the *label* of the run R and R a run on w. For a run R, $\mu(R) = \left(\mu(t_i)\right)_{1 \leq i \leq n}$ is the sequence of the transition weights of R and $\mathrm{wgt}(R) = \mathrm{val}(\mu(R))$ is the *weight* of R. A run is *successful* if it starts in an initial state $q_0 \in I$ and ends in a final state $q_n \in F$. We denote the set of successful runs of $\mathcal{A}$ by $\mathrm{succ}(\mathcal{A})$. The *behavior* of $\mathcal{A}$ is the function $\|\mathcal{A}\| : \Sigma^+ \to D$ defined by

$$\|\mathcal{A}\|(w) = \sum_{\substack{R \in \mathrm{succ}(\mathcal{A}) \\ \ell(R) = w}} \mathrm{val}\left(\mu(R)\right)$$

for every $w \in \Sigma^+$. If there is no successful run on w, then $\|\mathcal{A}\|(w) = \mathbb{0}$. Every function $S : \Sigma^+ \to D$ is called a *series* (or a *quantitative language* as in [7]) over Σ^+. If S is the behavior of some weighted automaton, then S is called *recognizable*.

Remark 2.3. Classical weighted automata are defined over *semirings* [39,24,13]. $\mathbb{K} = (K, +, \cdot, 0, 1)$ is a semiring if $(K, +, \cdot)$ is a commutative monoid, $(K, \cdot, 1)$ is a monoid, multiplication $\cdot$ distributes over addition $+$, and 0 is absorbing for multiplication, i.e., $k \cdot 0 = 0 \cdot k = 0$ for all $k \in K$. Semirings can be modeled by valuation monoids: We define $\text{val}(d_1, \dots, d_n) = d_1 \cdot \dots \cdot d_n$. Then $\mathbb{D}_{\mathbb{K}} = (K, +, \text{val}, 0)$ is a valuation monoid.

A weighted automaton $\mathcal{A}_{\mathbb{K}} = (Q, \lambda, \mu, \gamma)$ over an alphabet Σ and a semiring $\mathbb{K}$ consists of a finite state set Q, the transitional weight function $\mu : Q \times \Sigma \times Q \to K$, and the initial and final weights $\lambda, \gamma : Q \to K$, respectively. Note that μ can also be considered as a mapping from Σ to $K^{Q \times Q}$ and, thus, be extended to a homomorphism $\mu : \Sigma^+ \to K^{Q \times Q}$. Then the behavior of $\mathcal{A}_{\mathbb{K}}$ is defined as $\|\mathcal{A}_{\mathbb{K}}\|(w) = \lambda \cdot \mu(w) \cdot \gamma$ where λ and γ are understood as a row and column vector of dimension $|Q|$, respectively. Thus, weights are multiplied along a run and the weights of all runs over one word are summed up.

Since we consider only non-empty words, $\mathcal{A}_{\mathbb{K}}$ can be normalized. This comprises $\lambda(Q), \gamma(Q) \subseteq \{0, 1\}$. Let $I = \{q \in Q \mid \lambda(q) = 1\}$, $F = \{q \in Q \mid \gamma(q) = 1\}$, and $T = \{(p, a, q) \in Q \times \Sigma \times Q \mid \mu(p, a, q) \neq 0\}$. Then $\mathcal{A} = (Q, I, T, F, \mu|_T)$ is a wfa over Σ and the valuation monoid $\mathbb{D}_{\mathbb{K}}$ with $\|\mathcal{A}\| = \|\mathcal{A}_{\mathbb{K}}\|$.

Remark 2.4. A *bimonoid*, cf. [18], is a structure $\mathbb{K} = (K, +, \cdot, 0, 1)$ consisting of two monoids $(K, +, 0)$ and $(K, \cdot, 1)$. If $(K, +, 0)$ is a commutative monoid and 0 is absorbing for the second operation $\cdot$, we call $\mathbb{K}$ a *strong bimonoid*. Every strong bimonoid can be seen as a valuation monoid $\mathbb{D}_{\mathbb{K}} = (K, +, \text{val}, 0)$ with $\text{val}(k_1, \dots, k_n) = k_1 \cdots k_n$. A range of examples of strong bimonoids not being semirings can be found in [18]. One class of examples for strong bimonoids are non-distributive bounded lattices.

In [18,23], weighted automata over strong bimonoids were considered. Their behavior is defined similarly to the ones over semirings. Along a run the weights are multiplied and the weights of all successful runs on a word $w \in \Sigma^+$ are summed up. Similarly to the case of semirings, we can construct for a weighted automaton $\mathcal{A}_{\mathbb{K}}$ over the strong bimonoid $\mathbb{K}$ a wfa $\mathcal{A}$ over $\mathbb{D}_{\mathbb{K}}$ such that $\|\mathcal{A}\| = \|\mathcal{A}_{\mathbb{K}}\|$.

Example 2.5. Consider $(\mathbb{R} \cup \{-\infty\}, \max, \text{last}, -\infty)$ with $\text{last}(d_1, \dots, d_n) = d_n$ if $d_i \neq -\infty$ for $i \in \{1, \dots, n\}$ and $\text{last}(d_1, \dots, d_n) = -\infty$ otherwise. This structure yields a valuation monoid where the weight of the last transition determines the weight of the whole run. In fact, this valuation monoid can be derived from a strong bimonoid (where a neutral element for last has to be added).

However, there are important examples for valuation monoids which do not fit neither into the semiring nor the bimonoid setting.

Example 2.6. $(\mathbb{R} \cup \{-\infty\}, \max, \text{avg}, -\infty)$ with $\text{avg}(d_1, \dots, d_n) = \frac{1}{n} \sum_{i=1}^{n} d_i$ is a valuation monoid. A weighted automaton over this valuation monoid takes the arithmetic mean of the weights of the transitions and resolves non-determinism by max. To take the average of the weights along a run was suggested in [7].

Example 2.7. $(\mathbb{R} \cup \{\infty\}, \min, \mathrm{maj}, \infty)$ is a valuation monoid. Here, the valuation function is a majority function. Let $\mathrm{most}(d_1, \ldots, d_n)$ be the set of values occuring most often in $(d_1, \ldots, d_n)$ (note that there may be several ones). Then $\mathrm{maj}(d_1, \ldots, d_n) = \max\big(\mathrm{most}(d_1, \ldots, d_n)\big)$ whenever all $d_i \neq \infty$, otherwise it is ∞.

Another example is *discounting*. Weighted automata with discounting were already explored extensively in the literature [7,14,16,17,32].

Example 2.8. $(\mathbb{R} \cup \{-\infty\}, \max, \mathrm{disc}_\lambda, -\infty)$ with $\mathrm{disc}_\lambda(d_0, \ldots, d_n) = \sum_{i=0}^{n} \lambda^i d_i$ for some $\lambda > 0$ is a valuation monoid. A weighted automaton over this valuation monoid evaluates a run by the discounted sum of the weights appearing along the run, i.e., the later the weight occurs the greater is its discounting. This kind of discounting is normally referred to as *exponential discounting*.

More general notions of discounting using semiring endomorphisms were considered in the literature, cf. [14,16,17]. In [16,17], the discount factors depend also from the actions executed. This kind of discounting is not covered by valuation monoids because a valuation function depends only on the weights of the transitions taken by the machine but not on the actions. However, if we choose the endomorphisms independent of the actions we can model this situation by valuation monoids.

Example 2.9. Let $\mathbb{K} = (K, +, \cdot, 0, 1)$ be a semiring and $\varphi : \mathbb{K} \to \mathbb{K}$ a semiring endomorphism, i.e., $\varphi(0) = 0$, $\varphi(1) = 1$, $\varphi(k + k') = \varphi(k) + \varphi(k')$, and $\varphi(k \cdot k') = \varphi(k) \cdot \varphi(k')$ for all $k, k' \in K$. Let φ^0 be the identity. Then $\mathbb{D}_\varphi = (K, +, \mathrm{disc}_\varphi, 0)$ with

$$\mathrm{disc}_\varphi(k_0, \ldots, k_n) = \prod_{i=0}^{n} \varphi^i(k_i)$$

is a valuation monoid. The valuation monoid of Example 2.8 is an instance of such a valuation monoid because $\varphi : d \mapsto \lambda d$ with $\lambda > 0$ is an endomorphism of the max-plus-semiring $(\mathbb{R} \cup \{-\infty\}, \max, +, -\infty, 0)$.

Another concept of discounting considered in psychologic and economic literature is *hyperbolic discounting*.

Example 2.10. Let $\lambda > 0$. Then $(\mathbb{R} \cup \{-\infty\}, \max, \mathrm{hyp}_\lambda, -\infty)$ with

$$\mathrm{hyp}_\lambda(d_0, \ldots, d_n) = \sum_{i=0}^{n} \frac{1}{1 + \lambda \cdot i} \cdot d_i$$

is a valuation monoid. Now the discounting along a run of a wfa is of hyperbolic nature.

2.2   Product and Iteration

Kleene's theorem about the equivalence of recognizable and rational languages [31] states that the recognizable languages can be build from atomic components, the letters, by non-deterministic choice, concatenation, and iteration. Over valuation monoids, non-deterministic choice is modeled by the sum of the valuation monoid as it was done for semirings. But what about the two other operations?

For semirings the analog of concatenation is the Cauchy product of two series. One considers all possible factorizations of a word and multiplies the values of the factors. So far, we do not have any multiplication for valuation monoids but only a global valuation function which suffices to define the behavior of weighted automata. One could try to define a product by means of the valuation function, i.e., to put $d \cdot d' = \text{val}(d, d')$. But if we consider Example 2.6, then the product of $\text{avg}(d_1, \ldots, d_m)$ and $\text{avg}(d'_1, \ldots, d'_n)$ would be

$$\frac{\dfrac{d_1 + \cdots + d_m}{m} + \dfrac{d'_1 + \cdots + d'_n}{n}}{2}$$

whereas the weight of the concatenation of two runs with weights $d_1, \ldots, d_m$ and $d'_1, \ldots, d'_n$ would be $\frac{d_1 + \cdots + d_m + d'_1 + \cdots + d'_n}{m+n}$ which is different from the value above. Therefore, we will introduce products which have two positive integers as parameters. These parameters represent the length of two runs to be concatenated.

Definition 2.11. *The structure* $\mathbb{D} = \big(D, +, \text{val}, (\cdot_{m,n} \mid m, n \in \mathbb{N}), \mathbb{0}\big)$ *is a* Cauchy valuation monoid *if* $(D, +, \text{val}, \mathbb{0})$ *is a valuation monoid and* $\cdot_{m,n} : D \times D \to D$ *with* $m, n \in \mathbb{N}$ *is a family of products such that for all* $d, d_i, d'_j \in D$ *and all finite subsets* $A, B \subseteq_{fin} D$:

$$\mathbb{0} \cdot_{m,n} d = d \cdot_{m,n} \mathbb{0} = \mathbb{0}, \tag{1}$$

$$\text{val}(d_1, \ldots, d_m, d'_1, \ldots, d'_n) = \text{val}(d_1, \ldots, d_m) \cdot_{m,n} \text{val}(d'_1, \ldots, d'_n), \tag{2}$$

$$\left(\sum_{d \in A} d\right) \cdot_{m,n} \left(\sum_{d' \in B} d'\right) = \sum_{d \in A, d' \in B} (d \cdot_{m,n} d'). \tag{3}$$

For the sake of notational simplicity, we will often omit the explicit notation of the parameterized products for a Cauchy valuation monoid.

Property (1) ensures that $\mathbb{0}$ is absorbing for all products. The correct concatenation of two sequences of weights is guaranteed by property (2). Finally, property (3) states distributivity of the parameterized products over sum.

Remark 2.12. Whenever the valuation monoid $\mathbb{D}_\mathbb{K} = (D, +, \text{val}, \mathbb{0})$ is derived from a semiring $\mathbb{K} = (K, +, \cdot, \mathbb{0}, \mathbb{1})$ where $\text{val}(d_1, \ldots, d_m) = d_1 \cdot \ldots \cdot d_m$, then we can choose the products just as semiring multiplication, i.e., $\cdot_{m,n} := \cdot$ for all $m, n \in \mathbb{N}$. Now (2) follows immediately and (3) is just the distributivity of the semiring. Hence, all valuation monoids derived from semirings are Cauchy.

Example 2.13. Let $(\mathbb{R} \cup \{-\infty\}, \max, \text{avg}, -\infty)$ be the valuation monoid from Example 2.6. With

$$d \cdot_{m,n} d' = \frac{m \cdot d + n \cdot d'}{m + n}$$

for all $m, n \in \mathbb{N}$ and $d, d' \in \mathbb{R} \cup \{-\infty\}$, this valuation monoid yields a Cauchy one.

Example 2.14. We consider the valuation monoid $(\mathbb{R} \cup \{-\infty\}, \max, \text{last}, -\infty)$ from Example 2.5. Recall that we defined $\text{last}(d, d') = d'$ if $d \neq -\infty$ and $\text{last}(-\infty, d') = -\infty$. Now we put $d \cdot_{m,n} d' = \text{last}(d, d')$ for all $m, n \in \mathbb{N}$ and $d, d' \in \mathbb{R} \cup \{-\infty\}$. With these products we obtain a Cauchy valuation monoid.

Example 2.15. Let $\mathbb{D}_\varphi = (K, +, \text{disc}_\varphi, \mathbb{0})$ be the valuation monoid of Example 2.9 which uses an endomorphism $\varphi : \mathbb{K} \to \mathbb{K}$ for a semiring $\mathbb{K} = (K, +, \cdot, \mathbb{0}, \mathbb{1})$. We put $k \cdot_{m,n} k' = k \cdot \varphi^m(k')$ for all $m, n \in \mathbb{N}$ and $k, k' \in K$. Together with these products $\mathbb{D}_\varphi$ is a Cauchy valuation monoid.

The discounting valuation monoid $(\mathbb{R} \cup \{-\infty\}, \max, \text{disc}_\lambda, -\infty)$ with $\lambda > 0$ from Example 2.8 with $d \cdot_{m,n} d' = d + \lambda^m d'$ for all $m, n \in \mathbb{N}$ and $d, d' \in \mathbb{R} \cup \{-\infty\}$ is an instance of such a Cauchy valuation monoid.

For the valuation monoids from Examples 2.7 (majority function) and 2.10 (hyperbolic discounting) we cannot define products such that we could turn these valuation monoids into Cauchy ones. Here, property (2) cannot be guaranteed since we would need explicit knowledge of the sequences of weights themselves and not only about their lengths and their valuations as it is demanded for the products of Cauchy valuation monoids.

By using parameterized products of a Cauchy valuation monoid, we can define Cauchy product and iteration of series.

Definition 2.16. *Let $\mathbb{D} = (D, +, \text{val}, \mathbb{0})$ be a Cauchy valuation monoid and let $S, S' : \Sigma^+ \to D$ be two series. Then we define the* sum $S + S'$ *and the* Cauchy product $S \cdot S'$ *of S and S' by $(S + S')(w) = S(w) + S'(w)$ and*

$$(S \cdot S')(w) = \sum_{\substack{w=uv \\ u,v \in \Sigma^+}} \left(S(u) \cdot_{|u|,|v|} S'(v) \right)$$

for all $w \in \Sigma^+$ where we sum up over all factorizations of w into $u, v \in \Sigma^+$. We put $S^1 = S$ and $S^{n+1} = S^n \cdot S$ for all $n \geq 1$.

Now the iteration S^+ *is defined for every $w \in \Sigma^+$ by the finite sum*

$$S^+(w) = \sum_{n=1}^{|w|} S^n(w).$$

Sum, Cauchy product, and iteration are called *rational operations* on series. They determine the semantics of the *regular weighted expressions E* which are defined by the grammar, cf. [13,39],

$$E ::= d.a \mid (E + E) \mid (E \cdot E) \mid (E)^+$$

where $d \in D$ and $a \in \Sigma$. Let $da : \Sigma^+ \to D$ denote the series $da(w) = d$ if $w = a$ and $da(w) = \mathbb{0}$ otherwise. Then we put

$$[\![d.a]\!] = da, \quad [\![(E + E')]\!] = [\![E]\!] + [\![E']\!],$$
$$[\![(E \cdot E')]\!] = [\![E]\!] \cdot [\![E']\!], \quad [\![(E)^+]\!] = [\![E]\!]^+.$$

For $n \in \mathbb{N}$ and a regular weighted expression E we denote by E^n the weighted expression $\underbrace{((\ldots((E \cdot E) \cdot E)\ldots) \cdot E)}_{n\,\text{times}}$. We will omit parentheses in regular expressions if the bracketing is obvious.

For semirings the definition of Cauchy product and iteration coincides with the classical definitions, cf. Remark 2.12. In this case, the Cauchy product is associative. But in the setting of valuation monoids, this is in general not the case. But using property (3), we can show easily the distributivity of the Cauchy product over sum:

Proposition 2.17 ([21]). *Let $\mathbb{D}$ be a Cauchy valuation monoid and $S, S_1, S_2 : \Sigma^+ \to D$. Then $S \cdot (S_1 + S_2) = S \cdot S_1 + S \cdot S_2$ and $(S_1 + S_2) \cdot S = S_1 \cdot S + S_2 \cdot S$.*

For Cauchy valuation monoids normalized automata suffice to recognize all recognizable series. Borrowing the notion of normalized finite automata, a wfa $\mathcal{A} = (Q, I, T, F, \mu)$ is *normalized* if $I = \{q_0\}$ and $F = \{q_f\}$ are singletons, and

$$(p, a, q) \in T \implies (q \neq q_0 \wedge p \neq q_f)$$

for all $(p, a, q) \in Q \times \Sigma \times Q$. Then we have

Lemma 2.18. *Let $\mathbb{D}$ be a Cauchy valuation monoid. If $S : \Sigma^+ \to D$ is recognizable, then there is a normalized wfa $\mathcal{A}$ with $\|\mathcal{A}\| = S$.*

Proof. Let $\mathcal{B} = (Q, I, T, F, \mu)$ be a wfa recognizing S. We define a normalized wfa $\mathcal{A} = (Q \,\dot\cup\, \{q_0, q_f\}, \{q_0\}, T', \{q_f\}, \mu')$ by

$$
\begin{aligned}
T' = T &\cup \{(q_0, a, p) \mid p \in Q \wedge \exists q \in I : (q, a, p) \in T\} \\
&\cup \{(p, a, q_f) \mid p \in Q \wedge \exists q \in F : (p, a, q) \in T\} \\
&\cup \{(q_0, a, q_f) \mid \exists p \in I, q \in F : (p, a, q) \in T\} \quad \text{and}
\end{aligned}
$$

$$
\mu'(p', a, q') = \begin{cases}
\mu(p', a, q') & \text{if } p', q' \in Q, \\
\sum_{p \in I} \mu(p, a, q') & \text{if } p' = q_0, q' \in Q, \\
\sum_{q \in F} \mu(p', a, q) & \text{if } p' \in Q, q' = q_f, \\
\sum_{p \in I, q \in F} \mu(p, a, q) & \text{if } p' = q_0, q' = q_f.
\end{cases}
$$

Obviously, $\mathcal{A}$ is normalized. For $w = a \in \Sigma$ we have immediately $\|\mathcal{A}\|(w) = \|\mathcal{B}\|(w)$ by the definition of $\mathcal{A}$. Now let $w \in \Sigma^+$ with $|w| > 1$. Then

$$
\|\mathcal{A}\|(w) = \sum_{\substack{R' \in \mathrm{succ}(\mathcal{A}) \\ \ell(R') = w}} \mathrm{val}(\mu'(R'))
$$

with $R' = (q_0, a_1, p_1)(p_1, a_2, p_2) \ldots (p_{n-1}, a_n, q_f)$ and $w = a_1 \ldots a_n$

$$
= \sum_{\substack{R' \in \mathrm{succ}(\mathcal{A}) \\ \ell(R') = w}} \mathrm{val}\left(\sum_{p \in I} \mu(p, a_1, p_1), \mu(p_1, a_2, p_2), \ldots, \sum_{q \in F} \mu(p_{n-1}, a_n, q)\right)
$$

by applying properties (2) and (3) several times

$$
= \sum_{\substack{R' \in \mathrm{succ}(\mathcal{A}) \\ \ell(R')=w}} \sum_{p \in I} \mu(p, a_1, p_1) \cdot_{1,n-1} \mathrm{val}\Big(\mu(p_1, a_2, p_2), \ldots, \sum_{q \in F} \mu(p_{n-1}, a_n, q)\Big)
$$

$$
= \sum_{\substack{R' \in \mathrm{succ}(\mathcal{A}) \\ \ell(R')=w}} \sum_{\substack{p \in I \\ q \in F}} \mu(p, a_1, p_1) \cdot_{1,n-1}
$$

$$
\Big(\mathrm{val}\big(\mu(p_i, a_{i+1}, p_{i+1})_{1 \le i \le n-2}\big) \cdot_{n-2,1} \mu(p_{n-1}, a_n, q)\Big)
$$

$$
= \sum_{\substack{R' \in \mathrm{succ}(\mathcal{A}) \\ \ell(R')=w}} \sum_{\substack{p \in I \\ q \in F}} \mathrm{val}\big(\mu(p, a_1, p_1), \mu(p_1, a_2, p_2), \ldots, \mu(p_{n-1}, a_n, q)\big)
$$

$$
= \sum_{\substack{R \in \mathrm{succ}(\mathcal{B}) \\ \ell(R)=w}} \mathrm{val}(\mu(R)) = \|\mathcal{B}\|(w)
$$

Hence, $\|\mathcal{A}\| = \|\mathcal{B}\|$.   □

2.3   A First Kleene-Like Result

To show that the class of recognizable series coincides with the one definable by regular
weighted expressions, we use a method which was demonstrated in [34] for semiring
weighted automata and conclude the equivalence in the weighted setting from the classical boolean result by Kleene [31].

A language $L \subseteq \Gamma^+$ is *recognizable* if there is a *finite automaton* $\mathcal{A} = (Q, I, T, F)$
over Γ recognizing L. Recall that a finite automaton $\mathcal{A}$ is defined like a wfa in Definition 2.2 but omitting the weight function. A word w is recognized by $\mathcal{A}$ if there is a
successful run of $\mathcal{A}$ on w. The language $\mathcal{L}(\mathcal{A})$ recognized by $\mathcal{A}$ is the set of all words
$w \in \Gamma^+$ recognized by $\mathcal{A}$. The finite automaton $\mathcal{A} = (Q, I, T, F)$ is *deterministic* if
$(p, a, q), (p, a, q') \in T$ imply $q = q'$ for all $(p, a, q), (p, a, q') \in Q \times \Gamma \times Q$.

Recall that a *regular (language) expression* H over the alphabet Γ is defined by the
grammar $H ::= \emptyset \mid a \mid H + H \mid H \cdot H \mid H^+$ with $a \in \Gamma$. The semantics $\mathcal{L}(H) \subseteq \Gamma^+$
of H is defined by

$$
\mathcal{L}(\emptyset) = \emptyset, \ \mathcal{L}(a) = \{a\}, \ \mathcal{L}(H + H') = \mathcal{L}(H) \cup \mathcal{L}(H'),
$$
$$
\mathcal{L}(H \cdot H') = \mathcal{L}(H) \cdot \mathcal{L}(H') = \{uv \in \Gamma^+ \mid u \in \mathcal{L}(H), v \in \mathcal{L}(H')\},
$$
$$
\mathcal{L}(H^+) = \mathcal{L}(H)^+ = \{u_1 \ldots u_n \in \Gamma^+ \mid n \in \mathbb{N}, u_i \in \mathcal{L}(H)\}.
$$

If there is a regular expression H with $\mathcal{L}(H) = L$, then the language $L \subseteq \Gamma^+$ is called
rational. However, one can restrict the use of the operators $+$, $\cdot$, and $^+$ in an unambiguous way. *Unambiguous regular (language) expressions* are defined semantically:

- $\emptyset$ and a for every $a \in \Gamma$ are unambiguous regular expressions,
- if H_1 and H_2 are unambiguous regular expressions with $\mathcal{L}(H_1) \cap \mathcal{L}(H_2) = \emptyset$, then
  $H_1 + H_2$ is an unambiguous regular expression,

- if H_1 and H_2 are unambiguous regular expressions such that, for every $u_i, v_i \in \mathcal{L}(H_i)$ ($i \in \{1, 2\}$) with $u_1 u_2 = v_1 v_2$, we have $u_1 = v_1$ (and, thus, $u_2 = v_2$), then $H_1 \cdot H_2$ is an unambiguous regular expression,
- if H is an unambiguous regular expression such that for all $u_i, v_j \in \mathcal{L}(H)$ for $i \in \{1, \ldots, m\}$ and $j \in \{1, \ldots, n\}$ with $u_1 u_2 \ldots u_m = v_1 v_2 \ldots v_n$, we have $m = n$ and $u_1 = v_1$ (and, thus, $u_i = v_i$ for every $i \in \{1, \ldots, m\}$), then H^+ is an unambiguous regular expression.

The following result by Kleene is one of the fundamental theorems about recognizable languages:

Theorem 2.19 (Kleene [31]). *Let $L \subseteq \Gamma^+$ be a language of finite words over Γ. Then the following are equivalent*

1. *L is recognizable,*
2. *L is recognized by a deterministic finite automaton $\mathcal{A}$,*
3. *L is rational,*
4. *$L = \mathcal{L}(H)$ for some unambiguous regular language expression H.*

We will lift the equivalence between recognizability and rationality to a quantitative setting over valuation monoids. First, we show that recognizability implies rationality.

Proposition 2.20. *Let $\mathbb{D}$ be a Cauchy valuation monoid and $S : \Sigma^+ \to \mathbb{D}$. If S is recognizable, then $S = [\![E]\!]$ for some regular weighted expression E.*

Proof. Let $\mathcal{A} = (Q, I, T, F, \mu)$ be a wfa recognizing S. We put $\Gamma = T$ and, thus, $\mathrm{succ}(\mathcal{A}) \subseteq \Gamma^+$. Let $T' = \{ (p, (p, a, q), q) \mid (p, a, q) \in T \}$. Then $\mathcal{A}' = (Q, I, T', F)$ is a finite automaton over the alphabet Γ recognizing $\mathrm{succ}(\mathcal{A})$. By Kleene's theorem, there is an unambiguous regular (language) expression H with $\mathcal{L}(H) = \mathrm{succ}(\mathcal{A})$. By induction on the sub-expressions of H, we define how to translate the regular expression H over Γ into a regular weighted expression $\mathbb{D}(H)$ over $\mathbb{D}$ and Σ:

$$\mathbb{D}(\emptyset) = \mathbb{0}.a, \quad \mathbb{D}((p, a, q)) = \mu(p, a, q).a, \quad \mathbb{D}(H + H') = \mathbb{D}(H) + \mathbb{D}(H'),$$

$$\mathbb{D}(H \cdot H') = \mathbb{D}(H) \cdot \mathbb{D}(H'), \quad \mathbb{D}(H^+) = \mathbb{D}(H)^+$$

where for $\mathbb{D}(\emptyset)$ the letter $a \in \Sigma$ is chosen arbitrarily. Moreover, we define the homomorphism $\pi : \Gamma^+ \to \Sigma^+$ by $\pi(p, a, q) = a$. Recall that $\mu : \Gamma^+ \to D^+$ is the homomorphic extension of $\mu : T \to D$. Then we have for every sub-expression G of H that

$$[\![\mathbb{D}(G)]\!](w) = \sum_{W \in \mathcal{L}(G) \cap \pi^{-1}(w)} \mathrm{val}(\mu(W)). \tag{4}$$

Indeed, the claim is true for $G = \emptyset$ and $G = (p, a, q)$. Let $G = G_1 + G_2$. Then

$$[\![\mathbb{D}(G_1 + G_2)]\!](w) = [\![\mathbb{D}(G_1) + \mathbb{D}(G_2)]\!](w) = [\![\mathbb{D}(G_1)]\!](w) + [\![\mathbb{D}(G_2)]\!](w)$$

$$= \sum_{W \in \mathcal{L}(G_1) \cap \pi^{-1}(w)} \mathrm{val}(\mu(W)) + \sum_{W \in \mathcal{L}(G_2) \cap \pi^{-1}(w)} \mathrm{val}(\mu(W))$$

$$= \sum_{W \in (\mathcal{L}(G_1) \cap \pi^{-1}(w)) \cup (\mathcal{L}(G_2) \cap \pi^{-1}(w))} \mathrm{val}(\mu(W)) \quad \text{(since } G \text{ is unambiguous)}$$

$$= \sum_{W \in \mathcal{L}(G) \cap \pi^{-1}(w)} \mathrm{val}(\mu(W))$$

for all $w \in \Sigma^+$. For $G = G_1 \cdot G_2$ we get

$$[\![\mathbb{D}(G_1 \cdot G_2)]\!](w) = [\![\mathbb{D}(G_1) \cdot \mathbb{D}(G_2)]\!](w)$$

$$= \sum_{w=uv} [\![\mathbb{D}(G_1)]\!](u) \cdot_{|u|,|v|} [\![\mathbb{D}(G_2)]\!](v)$$

$$= \sum_{w=uv} \left(\sum_{U \in \mathcal{L}(G_1) \cap \pi^{-1}(u)} \mathrm{val}(\mu(U)) \right) \cdot_{|u|,|v|} \left(\sum_{V \in \mathcal{L}(G_2) \cap \pi^{-1}(v)} \mathrm{val}(\mu(V)) \right)$$

$$= \sum_{w=uv} \sum_{\substack{U \in \mathcal{L}(G_1) \cap \pi^{-1}(u) \\ V \in \mathcal{L}(G_2) \cap \pi^{-1}(v)}} \left(\mathrm{val}(\mu(U)) \cdot_{|U|,|V|} \mathrm{val}(\mu(V)) \right) \quad \text{(due to Eq. (3))}$$

$$= \sum_{w=uv} \sum_{\substack{U \in \mathcal{L}(G_1) \cap \pi^{-1}(u) \\ V \in \mathcal{L}(G_2) \cap \pi^{-1}(v)}} \mathrm{val}(\mu(UV)) \quad \text{(due to Eq. (2))}$$

$$= \sum_{W \in \mathcal{L}(G_1 \cdot G_2) \cap \pi^{-1}(w)} \mathrm{val}(\mu(W)) \quad \text{(since } G \text{ is unambiguous)}$$

for all $w \in \Sigma^+$. Finally, we consider a sub-expression of the form G^+. Since G^+ is unambiguous, $\mathcal{L}(G^i) \cap \mathcal{L}(G^j) = \emptyset$ for all $i \neq j$ and Equation (4) holds true for every G^n with $n \in \mathbb{N}$ (due to the case of the Cauchy product shown above). Now we get

$$[\![\mathbb{D}(G^+)]\!](w) = [\![\mathbb{D}(G)^+]\!](w) = \sum_{n=1}^{|w|} [\![\mathbb{D}(G)]\!]^n(w) = \sum_{n=1}^{|w|} [\![\mathbb{D}(G^n)]\!](w)$$

$$= \sum_{n=1}^{|w|} \sum_{W \in \mathcal{L}(G^n) \cap \pi^{-1}(w)} \mathrm{val}(\mu(W))$$

$$= \sum_{W \in (\mathcal{L}(G^1) \cup \cdots \cup \mathcal{L}(G^{|w|})) \cap \pi^{-1}(w)} \mathrm{val}(\mu(W)) \quad \text{(since } \mathcal{L}(G^i) \cap \mathcal{L}(G^j) = \emptyset \text{ for } i \neq j\text{)}$$

$$= \sum_{W \in \mathcal{L}(G^+) \cap \pi^{-1}(w)} \mathrm{val}(\mu(W))$$

for every $w \in \Sigma^+$. This shows Equation (4). Using (4), we conclude

$$\|\mathcal{A}\|(w) = \sum_{\substack{R \in \mathrm{succ}(\mathcal{A}) \\ \ell(R)=w}} \mathrm{val}(\mu(R)) = \sum_{R \in \mathcal{L}(H) \cap \pi^{-1}(w)} \mathrm{val}(\mu(R)) = [\![\mathbb{D}(H)]\!](w)$$

for every $w \in \Sigma^+$ and, thus, $\|\mathcal{A}\| = [\![\mathbb{D}(H)]\!]$ for the regular weighted expression $\mathbb{D}(H)$. □

To show the converse, we have to translate arbitrary regular weigthed expressions into regular language expressions having in mind that addition in valuation monoids is in general not idempotent (but it is in the boolean semiring). This calls for a sophisticated treatment of sum, product, and iteration as it was already observed in [34]. To this end, let Σ and Γ be two alphabets and let $\pi : \Gamma^+ \to \Sigma^+$ and $\mu : \Gamma^+ \to D^+$ be homomorphisms. For a regular language expression H over Γ, let $\mathrm{alph}(H)$ be the letters of Γ occuring in H. We define inductively a relation $\sim_{\pi,\mu}$ between regular weighted expressions over $\mathbb{D}$ and Σ and regular language expressions over Γ as follows:

- $d.a \sim_{\pi,\mu} H$ if and only if $H = A$ for a letter $A \in \Gamma$ with $\pi(A) = a$ and $\mu(A) = d$,
- $E_1 + E_2 \sim_{\pi,\mu} H$ if and only if there are regular language expressions H_1 and H_2 over Γ with $\mathrm{alph}(H_1) \cap \mathrm{alph}(H_2) = \emptyset$, $H = H_1 + H_2$, and $E_i \sim_{\pi,\mu} H_i$ for $i \in \{1,2\}$,
- $E_1 \cdot E_2 \sim_{\pi,\mu} H$ if and only if there are regular language expressions H_1 and H_2 over Γ with $\mathrm{alph}(H_1) \cap \mathrm{alph}(H_2) = \emptyset$, $H = H_1 \cdot H_2$, and $E_i \sim_{\pi,\mu} H_i$ for $i \in \{1,2\}$,
- $E^+ \sim_{\pi,\mu} H$ if and only if there are regular language expressions H_1 and H_0 over Γ with $\mathrm{alph}(H_1) \cap \mathrm{alph}(H_0) = \emptyset$ such that $H = H_1 + (H_1 \cdot H_0)^+ + (H_1 \cdot H_0)^+ \cdot H_1$ and $E \sim_{\pi,\mu} H_i$ for $i \in \{1,0\}$.

Note that if $E \sim_{\pi,\mu} H$ for some regular weighted expression E, then the regular language expression H is unambiguous.

Proposition 2.21. *For $E \sim_{\pi,\mu} H$ we have*

$$[\![E]\!](w) = \sum_{W \in \mathcal{L}(H) \cap \pi^{-1}(w)} \mathrm{val}(\mu(W))$$

for all $w \in \Sigma^+$.

Proof. The proof is by induction on the structure of E. It is obvious for $E = d.a$. For $E_1 + E_2$ and $E_1 \cdot E_2$ the claim can be shown as Equation (4) in the proof of Proposition 2.20. Thus, we have still to consider E^+. Then there are regular language expressions H_1 and H_0 with $\mathrm{alph}(H_1) \cap \mathrm{alph}(H_0) = \emptyset$, $E \sim_{\pi,\mu} H_i$ for $i \in \{1,0\}$, and $H = H_1 + (H_1 \cdot H_0)^+ + (H_1 \cdot H_0)^+ \cdot H_1$. By induction on n, we can show that

$$[\![E]\!]^n(w) = \sum_{w = w_1 \ldots w_n} \sum_{\substack{W_i \in \mathcal{L}(H_{i \bmod 2}) \cap \pi^{-1}(w_i) \\ i \in \{1,\ldots,n\}}} \mathrm{val}(\mu(W_1 \ldots W_n))$$

for every $w \in \Sigma^+$. Indeed, for $n = 1$ the claim is true by induction hypothesis for E. Let the equation be true for $[\![E]\!]^n$. Then, for every $w \in \Sigma^+$,

$$[\![E]\!]^{n+1}(w)$$
$$= ([\![E]\!]^n \cdot [\![E]\!])(w)$$
$$= \sum_{w = u w_{n+1}} [\![E]\!]^n(u) \cdot_{|u|,|w_{n+1}|} [\![E]\!](w_{n+1})$$

$$= \sum_{w=uw_{n+1}} \Big(\sum_{u=w_1\ldots w_n} \sum_{\substack{W_i \in \mathcal{L}(H_{i \bmod 2}) \cap \pi^{-1}(w_i) \\ i \in \{1,\ldots,n\}}} \mathrm{val}(\mu(W_1 \ldots W_n)) \Big)$$

$$\cdot |u|,|w_{n+1}| \Big(\sum_{W_{n+1} \in \mathcal{L}(H_{n+1 \bmod 2}) \cap \pi^{-1}(w_{n+1})} \mathrm{val}(\mu(W_{n+1})) \Big)$$

and by applying first Eq. (3) and then Eq. (2)

$$= \sum_{w=w_1\ldots w_n w_{n+1}} \sum_{\substack{W_i \in \mathcal{L}(H_{i \bmod 2}) \cap \pi^{-1}(w_i) \\ i \in \{1,\ldots,n+1\}}} \mathrm{val}(\mu(W_1 \ldots W_n W_{n+1}))$$

which shows the claim for $[\![\,E\,]\!]^{n+1}$. Since $\mathrm{alph}(H_1) \cap \mathrm{alph}(H_0) = \emptyset$, every $W \in \mathcal{L}(H)$ has a unique factorization $W = W_1 W_2 \ldots W_n$ into alternating factors from $\mathcal{L}(H_1)$ and $\mathcal{L}(H_0)$. Now we get for every $w \in \Sigma^+$

$$[\![\,E^+\,]\!](w) = \sum_{n=1}^{|w|} [\![\,E\,]\!]^n(w)$$

$$= \sum_{n=1}^{|w|} \sum_{w=w_1\ldots w_n} \sum_{\substack{W_i \in \mathcal{L}(H_{i \bmod 2}) \cap \pi^{-1}(w_i) \\ i \in \{1,\ldots,n\}}} \mathrm{val}(\mu(W_1 \ldots W_n))$$

and due to the unique factorization of every $W \in \mathcal{L}(H)$

$$= \sum_{W \in \mathcal{L}(H) \cap \pi^{-1}(w)} \mathrm{val}(\mu(W))$$

which shows the statement for E^+. $\square$

To prove the next lemma, which was stated for a semiring setting in [34], we apply a different automaton construction than in [34]. This technique was already used in [23] to show closure of recognizable series under projections.

Lemma 2.22. *Let* $\mathbb{D} = (D, +, \mathrm{val}, \mathbb{0})$ *be a Cauchy valuation monoid,* $L \subseteq \Gamma^+$ *a recognizable language, and* $\pi : \Gamma^+ \to \Sigma^+$ *and* $\mu : \Gamma^+ \to D^+$ *homomorphisms. Then the series* $S : \Sigma^+ \to D$ *with*

$$S(w) = \sum_{W \in L \cap \pi^{-1}(w)} \mathrm{val}(\mu(W))$$

for all $w \in \Sigma^+$ *is recognizable.*

Proof. Let $\mathcal{A} = (Q, I, T, F)$ be a deterministic finite automaton over Γ with $\mathcal{L}(\mathcal{A}) = L$. Let $\mathcal{A}' = (Q \times \Gamma, I \times A_0, T', F', \mu')$ be the wfa over Σ and $\mathbb{D}$ where $A_0 \in \Gamma$ is an arbitrary but fixed letter and

$$(p, A) \in F' \iff p \in F,$$

$$((p, B), a, (q, A)) \in T' \iff \pi(A) = a \wedge (p, A, q) \in T, \text{ and}$$
$$\mu'((p, B), a, (q, A)) = \mu(A)$$

for every $(p, B), (q, A) \in Q \times \Gamma$ and $a \in \Sigma$.

Let $R' = ((p_i, A_i), a_{i+1}, (p_{i+1}, A_{i+1}))_{0 \le i \le n-1} \in \text{succ}(\mathcal{A}')$. By definition of T', $\pi(A_1 \ldots A_n) = a_1 \ldots a_n$. Moreover, $\mu'(R') = \mu(A_1 \ldots A_n)$. If we put $R = ((p_i, A_{i+1}, p_{i+1}))_{0 \le i \le n-1}$, then R is a successful run of $\mathcal{A}$ on $A_1 \ldots A_n$. Vice versa, for every successful run $R = ((p_i, A_{i+1}, p_{i+1}))_{0 \le i \le n-1}$ in $\mathcal{A}$ on $A_1 \ldots A_n$, the run $R' = ((p_i, A_i), a_{i+1}, (p_{i+1}, A_{i+1}))_{0 \le i \le n-1}$ is a successful run of $\mathcal{A}'$ on $a_1 \ldots a_n$. Thus, there is a bijection between the set of successful runs of $\mathcal{A}'$ on $w = a_1 \ldots a_n \in \Sigma^+$ and the set of successful runs of $\mathcal{A}$ on some $W = A_1 \ldots A_n \in \Gamma^+$ with $\pi(W) = w$. Hence, for every $w = a_1 \ldots a_n \in \Sigma^+$

$$
\begin{aligned}
\|\mathcal{A}'\|(w) &= \sum_{R' \in \text{succ}(\mathcal{A}'), \ell(R') = w} \text{val}(\mu'(R)) \\
&= \sum_{R \in \text{succ}(\mathcal{A}), \ell(R) = W, \pi(W) = w} \text{val}(\mu(W)) \\
&= \sum_{W \in L \cap \pi^{-1}(w)} \text{val}(\mu(W)) \quad \text{(since } \mathcal{A} \text{ is deterministic)} \\
&= S(w).
\end{aligned}
$$

Thus, the series S is recognized by the wfa $\mathcal{A}'$. $\qquad\square$

As a consequence, we get

Theorem 2.23 ([21]). *Let* $\mathbb{D} = (D, +, \text{val}, \mathbb{0})$ *be a Cauchy valuation monoid and* $S : \Sigma^+ \to D$. *Then* S *is recognizable if and only if* $S = [\![E]\!]$ *for some regular weighted expression* E.

Proof. By Proposition 2.20, recognizability of S implies rationality. If $S = [\![E]\!]$ for a regular weighted expression E, then we can construct a regular language expression H over an alphabet Γ and two homomorphisms $\pi : \Gamma^+ \to \Sigma^+$ and $\mu : \Gamma^+ \to D^+$ such that $E \sim_{\pi, \mu} H$. Due to Proposition 2.21,

$$S(w) = [\![E]\!](w) = \sum_{W \in \mathcal{L}(H) \cap \pi^{-1}(w)} \text{val}(\mu(W))$$

for every $w \in \Sigma^+$. By Kleene's Theorem, $\mathcal{L}(H)$ is a recognizable language. By Lemma 2.22, S is recognizable. $\qquad\square$

By Remark 2.12, the above result includes the one for semiring weighted automata by Schützenberger [39].

Corollary 2.24 ([39]). *Let* $\mathbb{K}$ *be a semiring and* $S : \Sigma^+ \to \mathbb{K}$. *Then* S *is recognizable if and only if* S *is rational.*

Also series with discounting (using the same endomorphism independent of the action executed) [14] are instances of our setting, cf. Example 2.15. Thus, we obtain as a consequence of Theorem 2.23 also a characterization of weighted automata by regular weighted expressions in the case of discounting [14].

3   Valuations of Infinite Words

For an alphabet Σ we denote by Σ^ω the set of infinite words. For $w \in \Sigma^\omega$ we put $\mathrm{dom}(w) = \mathbb{N}$. A monoid $(D, +, \mathbb{0})$ is *complete* [24] if it has infinitary sum operations $\sum_I : D^I \to D$ for any index set I such that

$$\sum_{i \in \varnothing} d_i = \mathbb{0}, \quad \sum_{i \in \{k\}} d_i = d_k, \quad \sum_{i \in \{j,k\}} d_i = d_j + d_k \text{ for } j \neq k,$$

$$\sum_{j \in J} \left(\sum_{i \in I_j} d_i \right) = \sum_{i \in I} d_i \text{ if } \bigcup_{j \in J} I_j = I \text{ and } I_j \cap I_k = \varnothing \text{ for } j \neq k.$$

Note that every complete monoid is commutative.

For a set D let $(\mathbb{N} \times D)^\omega = \{(n_i, d_i)_{i \in \mathbb{N}} \mid \forall i \in \mathbb{N} : n_i \in \mathbb{N}, d_i \in D\}$.

Definition 3.1. *An ω-indexed valuation monoid* $\mathbb{D} = (D, +, \mathrm{val}, \mathrm{val}^\omega, \mathbb{0})$ *is a complete valuation monoid* $(D, +, \mathrm{val}, \mathbb{0})$ *equipped with an ω-indexed valuation function* $\mathrm{val}^\omega :$ $(\mathbb{N} \times D)^\omega \to D$ *such that* $\mathrm{val}^\omega(n_k, d_k)_{k \in \mathbb{N}} = \mathbb{0}$ *whenever* $d_k = \mathbb{0}$ *for some* $k \in \mathbb{N}$.

We give a few examples of ω-indexed valuation monoids. Omega-automata over these structures were already studied in [7]. Let $\overline{\mathbb{R}} = \mathbb{R} \cup \{-\infty, \infty\}$ be the extended reals.

Example 3.2. We extend the valuation monoid of Example 2.6 to the ω-indexed valuation monoid $(\overline{\mathbb{R}}, \sup, \mathrm{avg}, \limsup \mathrm{avg}, -\infty)$ where we put

$$\limsup \mathrm{avg}(n_i, d_i)_{i \in \mathbb{N}} = \limsup_{k \in \mathbb{N}} \left(\frac{n_1 d_1 + \ldots + n_k d_k}{n_1 + \ldots + n_k} \right)$$

with the exception that $\limsup \mathrm{avg}(n_i, d_i)_{i \in \mathbb{N}} = -\infty$ if there is either a $d_i = -\infty$ or if there is a $j \in \mathbb{N}$ such that $d_i \neq \infty$ for all $i \geq j$ and $\limsup_{k \geq j} \left(\frac{n_j d_j + \ldots + n_k d_k}{n_j + \ldots + n_k} \right)$ $= -\infty$. With a similar definition of $\liminf \mathrm{avg}$ also $(\overline{\mathbb{R}}, \inf, \mathrm{avg}, \liminf \mathrm{avg}, \infty)$ yields an ω-indexed valuation monoid.

Example 3.3. For $d_1, \ldots, d_m \in \overline{\mathbb{R}}$ we put $\sup_{-\infty}(d_1, \ldots, d_m) = \sup(d_1, \ldots, d_m)$ if $d_i \neq -\infty$ for $i \in \{1, \ldots, m\}$ and $\sup_{-\infty}(d_1, \ldots, d_m) = -\infty$ otherwise. Moreover, we define

$$\limsup(n_i, d_i)_{i \in \mathbb{N}} = \begin{cases} \limsup_i (d_i)_{i \in \mathbb{N}} & \text{if } d_i \neq -\infty \text{ for all } i \in \mathbb{N}, \\ -\infty & \text{otherwise.} \end{cases}$$

Then $(\overline{\mathbb{R}}, \sup, \sup_{-\infty}, \limsup, -\infty)$ is an ω-indexed valuation monoid. Similarly, we can define the ω-indexed valuation monoid $(\overline{\mathbb{R}}, \inf, \inf_\infty, \liminf, \infty)$.

Example 3.4. Extending Example 2.5, $(\overline{\mathbb{R}}, \sup, \text{last}, \lim\sup, -\infty)$ yields an ω-indexed valuation monoid.

Example 3.5 (discounting [7,14]). Let $0 < \lambda < 1$ and $\overline{\mathbb{R}}_+ = \{r \in \mathbb{R} \mid r \geq 0\} \cup \{-\infty, \infty\}$. For $(n_i, d_i)_{i\in\mathbb{N}} \in (\mathbb{N} \times \overline{\mathbb{R}}_+)^\omega$ we put

$$\lim \text{disc}_\lambda(n_i, d_i)_{i\in\mathbb{N}} = \lim_{k\to\infty}\left(\lambda^0 d_1 + \lambda^{n_1} d_2 + \ldots + \lambda^{n_1+\ldots+n_{k-1}} d_k\right).$$

Then $(\overline{\mathbb{R}}_+, \sup, \text{disc}_\lambda, \lim\text{disc}_\lambda, -\infty)$ is an ω-indexed valuation monoid.

Weighted ω-automata with a Büchi acceptance condition are defined in the same way as wfa.

Definition 3.6. *A weighted Büchi automaton (for short: wba)* $\mathcal{A}=(Q, I, T, F, \mu)$ *over the alphabet* Σ *and an ω-indexed valuation monoid* $\mathbb{D} = (D, +, \text{val}, \text{val}^\omega, 0)$ *consists of a finite state set* Q, *a set* $I \subseteq Q$ *of initial states, a set* $F \subseteq Q$ *of accepting states, a set* $T \subseteq Q \times \Sigma \times Q$ *of transitions, and a weight function* $\mu : T \to D$.

However, the definition of the behavior of a weighted Büchi automaton over ω-indexed valuation monoids is more complex than for weighted finite automata. In the following, we will define three behaviors of wba depending on different valuations of the runs.

3.1   How to Run a Weighted Automaton on Infinite Words

Let $\mathcal{A} = (Q, I, T, F, \mu)$ be a wba over the alphabet Σ and the ω-indexed valuation monoid $\mathbb{D} = (D, +, \text{val}, \text{val}^\omega, 0)$. A *run* $R = (t_i)_{i\in\mathbb{N}}$ is an infinite sequence of matching transitions $t_i = (q_{i-1}, a_i, q_i)$ with *label* $\sigma = \ell(R) = a_1 a_2 \cdots \in \Sigma^\omega$. We say R is a run of $\mathcal{A}$ on σ.

Let $F(R) = \{j \in \mathbb{N} \mid q_j \in F\}$. Note that $F(R)$ can be finite or infinite. The run R is *successful* if $q_0 \in I$ and $F(R)$ is infinite, i.e., R starts in an initial state and satisfies a Büchi condition with regard to the acceptance set F. The set of successful runs of $\mathcal{A}$ is denoted by $\text{succ}(\mathcal{A})$. So far, the notions are the usual ones for Büchi automata.

But how to define the weight of a run R? A first possibility is quite simple. We just put $\text{wgt}_u(R) = \text{val}^\omega(1, \mu(t_i))_{i\in\mathbb{N}}$, i.e., we apply the ω-indexed valuation function to the sequence of weights appearing along R (the 'u' in $\text{wgt}_u(R)$ stands for *unconditional*). Basically, this was done both in [7], where, however, no acceptance condition was defined at all, and in [20] where a characterization of weighted ω-automata by means of weighted MSO-logics was given. However, this approach is not successful for general ω-indexed valuation monoids when we aim for a characterization by regular expressions. To do so, the weight of an infinite run has to be computed in a way that uses weights of finite sub-runs. These sub-runs will be determined by the acceptance states. Therefor, we consider two more valuations of runs.

So, a second possible valuation of a run, introduced in [21], is the following: Suppose $F(R)$ is infinite. Then we enumerate $F(R)$ by $j_1 < j_2 < j_3 < \ldots$ and put $j_0 = 0$. Let $R_k = (t_i)_{j_{k-1}<i\leq j_k}$ be the finite sub-run of R starting in $q_{j_{k-1}}$ and terminating in the k-th acceptance state q_{j_k}. We call $(R_k)_{k\in\mathbb{N}}$ the *F-partition* of R. Let

$\mu(R_k) = (\mu(t_i))_{j_{k-1} < i \leq j_k}$ be the finite sequence of weights from R_k. Now the *weight of R* is defined as

$$\text{wgt}(R) = \text{val}^\omega\left(|R_k|, \text{val}(\mu(R_k))\right)_{k \in \mathbb{N}} \tag{5}$$

if $F(R)$ is infinite, and $\text{wgt}(R) = \mathbb{0}$ otherwise. Intuitively, the automaton checks the weight at every acceptance state of the run, i.e., it computes the weight of the sub-runs R_k between two consecutive acceptance states by means of val and combines these values by the ω-indexed valuation function val^ω. For general ω-indexed valuation monoids, e.g. for the one given in Example 3.2, it is of importance for the value of $\text{wgt}(R)$ at which positions in R the acceptance states are located.

A third way to compute the weight of a run is as follows: This time, different acceptance states which appear infinitely often along the run are understood as a source of non-determinism. For every $q \in F$ let $F_q(R) = \{j \in \mathbb{N} \mid q_j = q\}$. Again, if $F_q(R)$ is infinite, we enumerate $F_q(R)$ by $j_1 < j_2 < j_3 < \ldots$ and put $j_0 = 0$. Let $R_k^q = (t_i)_{j_{k-1} < i \leq j_k}$ be the finite sub-run of R starting in $q_{j_{k-1}}$ and terminating in the k-th state q_{j_k} which equals q. $(R_k^q)_{k \in \mathbb{N}}$ is called a *q-partition* of R. Let $\mu(R_k^q) = (\mu(t_i))_{j_{k-1} < i \leq j_k}$ be the finite sequence of weights from R_k^q. We define the *cumulated weight of R* as

$$\text{wgt}_c(R) = \sum_{\substack{q \in F \\ |F_q(R)| = \omega}} \text{val}^\omega\left(|R_k^q|, \text{val}(\mu(R_k^q))\right)_{k \in \mathbb{N}} \tag{6}$$

if $F(R)$ is infinite, and $\text{wgt}_c(R) = \mathbb{0}$ otherwise. Here, we compute for every acceptance state $q \in F$ appearing infinitely often in R the weight as before but now we consider q as the only acceptance state. Finally, we sum up over all such $q \in F$, i.e., we resolve different accepting states as we resolve non-determinism. In general, $\text{wgt}(R) \neq \text{wgt}_c(R)$ even if $+$ is idempotent because the way R is splitted may influence the weight.

Now we can define three behaviors of the wba $\mathcal{A}$ which all will be functions from Σ^ω to D. $\|\mathcal{A}\|_u, \|\mathcal{A}\|, \|\mathcal{A}\|_c : \Sigma^\omega \to D$ are given by

$$\|\mathcal{A}\|_u(\sigma) = \sum_{\substack{R \in \text{succ}(\mathcal{A}) \\ \ell(R) = \sigma}} \text{wgt}_u(R), \quad \|\mathcal{A}\|(\sigma) = \sum_{\substack{R \in \text{succ}(\mathcal{A}) \\ \ell(R) = \sigma}} \text{wgt}(R), \text{ and}$$

$$\|\mathcal{A}\|_c(\sigma) = \sum_{\substack{R \in \text{succ}(\mathcal{A}) \\ \ell(R) = \sigma}} \text{wgt}_c(R)$$

for $\sigma \in \Sigma^\omega$; if σ has no successful run in $\mathcal{A}$, then $\|\mathcal{A}\|_u(\sigma) = \|\mathcal{A}\|(\sigma) = \|\mathcal{A}\|_c(\sigma) = \mathbb{0}$. We will call $\|\mathcal{A}\|_u$ the *unconditional behavior*, $\|\mathcal{A}\|$ the *(Büchi) behavior*, and $\|\mathcal{A}\|_c$ the *cumulative behavior* of $\mathcal{A}$.

A function $S : \Sigma^\omega \to D$ is called an *ω-series*. S is

- *unconditionally ω-recognizable* if $S = \|\mathcal{A}\|_u$ for some wba $\mathcal{A}$,
- *ω-recognizable* if there is a wba $\mathcal{A}$ with $\|\mathcal{A}\| = S$, and
- *cumulative ω-recognizable* if there is a wba $\mathcal{A}$ with $\|\mathcal{A}\|_c = S$.

Under certain conditions, the different behaviours coincide. First, we define a property of ω-indexed valuation functions val^ω stating that all partitions of an ω-sequence of weights result into the same value under val^ω. Let $(D_{fin})^\omega = \bigcup_{C \subseteq_{fin} D} C^\omega$ be the set of ω-sequences over D such that only finitely many values of D occur in the sequence.

Definition 3.7. *Let $\mathbb{D} = (D, +, \mathrm{val}, \mathrm{val}^\omega, \mathbb{0})$ be an ω-indexed valuation monoid. Then both $\mathbb{D}$ and val^ω are called* uniform *if the following holds:*
For every $(d_i)_{i\in\mathbb{N}} \in (D_{fin})^\omega$ and for all $0 = n_0 < n_1 < n_2 < \ldots$ we have

$$\mathrm{val}^\omega\big(n_k - n_{k-1}, \mathrm{val}(d_{n_{k-1}+1} \ldots d_{n_k})\big)_{k\in\mathbb{N}} = \mathrm{val}^\omega(1, d_i)_{i\in\mathbb{N}}. \tag{7}$$

In the sequel, we will denote $\mathrm{val}^\omega(1, d_i)_{i\in\mathbb{N}}$ also by $\mathrm{val}^\omega(d_i)_{i\in\mathbb{N}}$.

The operation $+$ is idempotent if $d + d = d$ for all $d \in D$. An ω-indexed valuation monoid $\mathbb{D} = (D, +, \mathrm{val}, \mathrm{val}^\omega, \mathbb{0})$ is *idempotent* if $+$ is idempotent.

Lemma 3.8. *Let $\mathbb{D} = (D, +, \mathrm{val}, \mathrm{val}^\omega, \mathbb{0})$ be a uniform ω-indexed valuation monoid and $\mathcal{A}$ a wba over Σ and $\mathbb{D}$. Then $\|\mathcal{A}\|_u = \|\mathcal{A}\|$.*
If $\mathbb{D}$ is uniform and idempotent, then $\|\mathcal{A}\|_u = \|\mathcal{A}\| = \|\mathcal{A}\|_c$.

Proof. Let $\mathbb{D}$ be uniform. For every run R of $\mathcal{A}$ we show that $\mathrm{wgt}(R) = \mathrm{wgt}_u(R)$. Indeed, for $R = (t_i)_{i\in\mathbb{N}}$ having the F-partition $(R_k)_{k\in\mathbb{N}}$ we get

$$\begin{aligned}
\mathrm{wgt}(R) &= \mathrm{val}^\omega\big(|R_k|, \mathrm{val}(\mu(R_k))\big)_{k\in\mathbb{N}} \\
&= \mathrm{val}^\omega\big(1, \mu(t_i)\big)_{i\in\mathbb{N}} && (\text{since } \mathbb{D} \text{ is uniform}) \\
&= \mathrm{wgt}_u(R).
\end{aligned}$$

If now $\mathbb{D}$ is uniform and idempotent, then with $(R_k^q)_{k\in\mathbb{N}}$ being the q-partition of R

$$\begin{aligned}
\mathrm{wgt}_c(R) &= \sum_{q\in F, |F_q(R)|=\omega} \mathrm{val}^\omega\big(|R_k^q|, \mathrm{val}(\mu(R_k^q))\big)_{k\in\mathbb{N}} \\
&= \sum_{q\in F, |F_q(R)|=\omega} \mathrm{val}^\omega\big(1, \mu(t_i)\big)_{i\in\mathbb{N}} && (\text{since } \mathbb{D} \text{ is uniform}) \\
&= \mathrm{val}^\omega\big(1, \mu(t_i)\big)_{i\in\mathbb{N}} && (\text{since } \mathbb{D} \text{ is idempotent}) \\
&= \mathrm{wgt}_u(R)
\end{aligned}$$

which shows $\mathrm{wgt}_c(R) = \mathrm{wgt}_u(R)$ for every run R of $\mathcal{A}$ and, thus, the assertion. $\square$

Example 3.9. The ω-indexed valuation monoids $(\overline{\mathbb{R}}, \sup, \sup_{-\infty}, \limsup, -\infty)$ of Example 3.3 and $(\overline{\mathbb{R}}_+, \sup, \mathrm{disc}_\lambda, \lim \mathrm{disc}_\lambda, -\infty)$ with $0 < \lambda < 1$ of Example 3.5 are uniform and, moreover, idempotent.

However, the ω-indexed valuation monoids $(\overline{\mathbb{R}}, \sup, \mathrm{avg}, \limsup \mathrm{avg}, -\infty)$ of Example 3.2 and $(\overline{\mathbb{R}}, \sup, \mathrm{last}, \limsup, -\infty)$ of Example 3.4 are not uniform as we have shown already in [22, Ex. 13]. For the sake of completeness, we repeat the arguments.

First, consider $(\overline{\mathbb{R}}, \sup, \mathrm{avg}, \limsup \mathrm{avg}, -\infty)$. We define infinetely many finite sequences $\overline{d}_i$ and $\overline{d}'_i$ of real numbers. Let $\overline{d}_i = \underbrace{1 \ldots 1}_{2^{2i-1}} \underbrace{-1 \ldots -1}_{2^{2i}}$ for all $i \in \mathbb{N}$, $\overline{d}'_1 = 11$, and $\overline{d_i'} = \underbrace{-1\ldots-1}_{2^{2i-2}}\underbrace{1\ldots 1}_{2^{2i-1}}$ for all $i \geq 2$. Then $\overline{d_1}\,\overline{d_2}\ldots = \overline{d_1'}\,\overline{d_2'}\ldots \in \overline{\mathbb{R}}^\omega$ with $\operatorname{avg}(\overline{d_i}) = -\frac{1}{3}$ for all $i \in \mathbb{N}$ and $\operatorname{avg}(\overline{d_i'}) = \frac{1}{3}$ for all $i \geq 2$. Thus,

$$\limsup \operatorname{avg}\big(|\overline{d_i}|, \operatorname{avg}(\overline{d_i})\big)_{i \in \mathbb{N}} = -\frac{1}{3} \text{ and } \limsup \operatorname{avg}\big(|\overline{d_i'}|, \operatorname{avg}(\overline{d_i'})\big)_{i \in \mathbb{N}} = \frac{1}{3}.$$

Hence, $(\overline{\mathbb{R}}, \sup, \operatorname{avg}, \limsup \operatorname{avg}, -\infty)$ is not uniform.

For $(\overline{\mathbb{R}}, \sup, \operatorname{last}, \limsup, -\infty)$ consider the finite sequences $\overline{d_i} = 10$ for all $i \in \mathbb{N}$, $\overline{d_0'} = 1$, and $\overline{d_i'} = 01$ for all $i \geq 1$. Then again $\overline{d_0}\,\overline{d_1}\ldots = \overline{d_0'}\,\overline{d_1'}\ldots \in \overline{\mathbb{R}}^\omega$, but $\limsup\big(|\overline{d_i}|, \operatorname{last}(\overline{d_i})\big)_{i \in \mathbb{N}} = 0$ whereas $\limsup\big(|\overline{d_i'}|, \operatorname{last}(\overline{d_i'})\big)_{i \in \mathbb{N}} = 1$.

3.2 Towards an Omega-Iteration

To characterize recognizable ω-series by expressions, our next task is to define suitable operations generating ω-series. Especially, a Cauchy product of two series, the first series over finite words, the second one over infinite words, and an ω-iteration of series over finite words have to be specified. As for valuation monoids, such operations require additional operations and properties of the ω-indexed valuation monoid which we will define next.

Definition 3.10. $\mathbb{D} = \big(D, +, \operatorname{val}, \operatorname{val}^\omega, (\cdot_{m,n} \mid m \in \mathbb{N}, n \in \mathbb{N} \cup \{\omega\}), \mathbb{0}\big)$ *is a* Cauchy ω-indexed valuation monoid *if* $\big(D, +, \operatorname{val}, (\cdot_{m,n} \mid m, n \in \mathbb{N}), \mathbb{0}\big)$ *is a Cauchy valuation monoid,* $\big(D, +, \operatorname{val}, \operatorname{val}^\omega, \mathbb{0}\big)$ *is an ω-indexed valuation monoid, and* $\cdot_{m,\omega} : D \times D \to D$ *for every* $m \in \mathbb{N}$ *such that for all* $d, d', d_i \in D$, *all finite subsets* $A \subseteq_{fin} D$, *and all subsets* $B \subseteq D$

$$\mathbb{0} \cdot_{m,\omega} d = d \cdot_{m,\omega} \mathbb{0} = \mathbb{0}, \tag{8}$$

$$\operatorname{val}^\omega (n_i, d_i)_{i \geq 1} = d_1 \cdot_{n_1,\omega} \operatorname{val}^\omega (n_i, d_i)_{i \geq 2}, \tag{9}$$

$$\left(\sum_{d \in A} d\right) \cdot_{m,\omega} \left(\sum_{d' \in B} d'\right) = \sum_{d \in A, d' \in B} (d \cdot_{m,\omega} d'), \tag{10}$$

and for every $C \subseteq_{fin} D$, $n_k \in \mathbb{N}$, *finite index sets* I_k, *and all* $d_{i_k} \in C$ $(i_k \in I_k)$

$$\operatorname{val}^\omega \left(n_k, \sum_{i_k \in I_k} d_{i_k}\right)_{k \in \mathbb{N}} = \sum_{(i_k)_k \in I_1 \times I_2 \times \ldots} \operatorname{val}^\omega (n_k, d_{i_k})_{k \in \mathbb{N}}. \tag{11}$$

Moreover, we will call $\mathbb{D}$ *associative if*

$$\operatorname{val}^\omega \big((n_1 + n_2, d_1 \cdot_{n_1,n_2} d_2), (n_3, d_3), (n_4, d_4), \ldots\big) = \operatorname{val}^\omega (n_i, d_i)_{i \in \mathbb{N}} \tag{12}$$

for all $(n_i, d_i)_{i \in \mathbb{N}} \in (\mathbb{N} \times D)^\omega$.

For an interpretation of these conditions it is useful to understand $\operatorname{val}^\omega$ as a parameterized infinitary product on D where the parameters are ω-sequences over $\mathbb{N}$. Property (9) is a kind of finitary associativity for $\operatorname{val}^\omega$ and the products $\cdot_{n,\omega}$. Distributivity of the parameterized products $\cdot_{n,\omega}$ over sum is given by property (10) whereas property (11) states distributivity of val^ω over finite sums. Equation (12) states a kind of associativity within the arguments of the function val^ω. By (9) and (12), we conclude

$$d_1 \cdot_{n_1,\omega} \left(d_2 \cdot_{n_2,\omega} \mathrm{val}^\omega(n_i, d_i)_{i\geq 3} \right) = (d_1 \cdot_{n_1,n_2} d_2) \cdot_{n_1+n_2,\omega} \mathrm{val}^\omega(n_i, d_i)_{i\geq 3} \quad (13)$$

for all $(n_i, d_i)_{i\in\mathbb{N}} \in (\mathbb{N} \times D)^\omega$ which states a kind of associativity between different parameterized products. This justifies the name *associativity* for property (12).

Now we can define the ω-*rational operations sum, Cauchy product,* and ω-*iteration.*

Definition 3.11. *Let $\mathbb{D}$ be a Cauchy ω-indexed valuation monoid, $S : \Sigma^+ \to D$, and $S', S'' : \Sigma^\omega \to D$. The sum $S' + S''$ and the Cauchy product $S \cdot S'$ are defined for all $\sigma \in \Sigma^\omega$ by $(S' + S'')(\sigma) = S'(\sigma) + S''(\sigma)$ and*

$$(S \cdot S')(\sigma) = \sum_{\substack{\sigma=u\sigma' \\ u\in\Sigma^+, \sigma'\in\Sigma^\omega}} S(u) \cdot_{|u|,\omega} S'(\sigma') .$$

The ω-iteration S^ω of $S : \Sigma^+ \to D$ is defined for every $\sigma \in \Sigma^\omega$ by

$$S^\omega(\sigma) = \sum_{\substack{\sigma=u_1u_2\ldots \\ u_k\in\Sigma^+, k\in\mathbb{N}}} \mathrm{val}^\omega\big(|u_k|, S(u_k)\big)_{k\in\mathbb{N}}$$

where the sum is taken over all infinite factorizations $u_1u_2\ldots$ of σ.

By property (10), the Cauchy product distributes from the left over sums.

Proposition 3.12 ([21]). *Let $\mathbb{D}$ be a Cauchy ω-indexed valuation monoid, $S : \Sigma^+ \to D$, and $S_1, S_2 : \Sigma^\omega \to D$. Then $S \cdot (S_1 + S_2) = S \cdot S_1 + S \cdot S_2$.*

The class of ω-*regular weighted expressions* over Σ and a Cauchy ω-indexed valuation monoid $\mathbb{D}$ is given by the grammar

$$E ::= E + E \mid G \cdot E \mid G^\omega$$

where G is any regular weighted expression. The semantics of E is an ω-series $[\![E]\!]$: $\Sigma^\omega \to D$ defined inductively by

$$[\![E_1 + E_2]\!] = [\![E_1]\!] + [\![E_2]\!], \ [\![G \cdot E]\!] = [\![G]\!] \cdot [\![E]\!], \ [\![G^\omega]\!] = [\![G]\!]^\omega .$$

We call an ω-series $S : \Sigma^\omega \to D$ ω-*rational* if there is an ω-regular weighted expression E with $[\![E]\!] = S$.

Next, we give examples of Cauchy ω-valuation monoids. Here, we do not verify the single properties. Those interested in more details are referred to [22].

Example 3.13. $(\overline{\mathbb{R}}, \sup, \mathrm{avg}, \limsup \mathrm{avg}, -\infty)$ from Example 3.2 with the products

$$d \cdot_{m,n} d' = \frac{m \cdot d + n \cdot d'}{m + n}, \quad d \cdot_{m,\omega} d' = \begin{cases} d' & \text{if } d \notin \{-\infty, \infty\} \text{ or } d' = -\infty, \\ d & \text{otherwise,} \end{cases}$$

is an associative Cauchy ω-indexed valuation monoid.

Recall that $\mathrm{last}(d, d') = d'$ if $d \neq \mathbb{0}$ and $\mathrm{last}(\mathbb{0}, d') = \mathbb{0}$.

Example 3.14. Let $(\overline{\mathbb{R}}, \sup, \sup_{-\infty}, \limsup, -\infty)$ be the ω-indexed valuation monoid from Example 3.3. We define $d \cdot_{m,n} d' = \sup_{-\infty}(d, d')$ and $d \cdot_{m,\omega} d' = \mathrm{last}(d, d')$. Together with these products the structure is associative and Cauchy.

Example 3.15. Consider $(\overline{\mathbb{R}}, \sup, \mathrm{last}, \limsup, -\infty)$ from Example 3.4. We define the parameterized products as follows: $d \cdot_{m,n} d' = d \cdot_{m,\omega} d' = \mathrm{last}(d, d')$. This way, we obtain again an associative Cauchy ω-indexed valuation monoid.

Example 3.16. Let $(\overline{\mathbb{R}}_+, \sup, \mathrm{disc}_\lambda, \mathrm{lim\,disc}_\lambda, -\infty)$ with $0 < \lambda < 1$ be the discounting ω-indexed valuation monoid from Example 3.5. If we put $d \cdot_{m,n} d' = d \cdot_{m,\omega} d' = d + \lambda^m d'$, we get an associative Cauchy ω-indexed valuation monoid.

Remark 3.17. Ésik and Kuich [25,26] define complete semiring-semimodule pairs and complete star-omega-semirings. Those semirings are equipped with infinite sums $\sum$ and products $\prod$ satisfying conditions similar to our properties (7) (uniformity), (9), and (11), see [26,27] for a formal definition. These semirings fit into the frame of Cauchy ω-indexed valuation monoids. We associate an ω-indexed valuation monoid as follows: The finite and infinite sums together with the zero element of the semiring are those of the ω-indexed valuation monoid. The valuation function val and the parameterized products are just semiring multiplication. The ω-indexed valuation function is given by the infinite product: $\mathrm{val}^\omega(n_i, d_i)_{i \in \mathbb{N}} = \prod_{i \in \mathbb{N}} d_i$. The ω-indexed valuation monoid defined this way is Cauchy, associative, and uniform.

A concrete instance of complete star-omega-semirings is the semiring $\mathbb{N}_0^\infty = (\mathbb{N}_0 \cup \{\infty\}, +, \cdot, 0, 1)$. Here, an infinite sum equals ∞ if and only if either a summand is ∞ or there is an infinite number of non-zero summands. The infinite product is defined as

$$\prod_{i \in \mathbb{N}} n_i = \begin{cases} 0 & \text{if } \exists j \in \mathbb{N} : n_j = 0, \\ n_1 \cdot \ldots \cdot n_k & \text{if } \forall j > k : n_j = 1, \\ \infty & \text{otherwise.} \end{cases}$$

Note that this semiring is not idempotent. By abuse of notation, we will denote the associated ω-indexed valuation monoid also by $\mathbb{N}_0^\infty$.

Let us have a look at an ω-regular weighted expession and its semantics for different ω-indexed valuation monoids.

Example 3.18. Let $\Sigma = \{a, b\}$ and $E = (1.a + 0.b)^\omega$. For the ω-indexed valuation monoid

– $(\overline{\mathbb{R}}, \sup, \sup_{-\infty}, \limsup, -\infty)$ from Example 3.14 we have

$$[\![E]\!](\sigma) = \begin{cases} 1 & \text{if there are infinitely many } a \text{ in } \sigma, \\ 0 & \text{otherwise} \end{cases}$$

for all $\sigma \in \Sigma^\omega$, i.e., E indicates if there are infinitely or only finitely many a in σ,

- for $(\overline{\mathbb{R}}, \sup, \mathrm{avg}, \limsup \mathrm{avg}, -\infty)$ from Example 3.13, E describes the long-run ratio of occurences of a in some $\sigma \in \Sigma^\omega$, e.g., $[\![E]\!](aaaabababa\ldots) = \frac{1}{2}$ and $[\![E]\!](abbabbabb\ldots) = \frac{1}{3}$,
- for $(\overline{\mathbb{R}}_+, \sup, \mathrm{disc}_{\frac{1}{2}}, \lim\mathrm{disc}_{\frac{1}{2}}, -\infty)$ from Example 3.16 we count with a discount factor $\frac{1}{2}$ the occurences of a in $\sigma \in \Sigma^\omega$, e.g., $[\![E]\!](ababab\ldots) = \sum_{i=0}^\infty \frac{1}{4^i} = \frac{4}{3}$,
- for $\mathbb{N}_0^\infty$ from Remark 3.17, $[\![E]\!](a^\omega) = 1$ and for all $\sigma \in \Sigma^\omega$ containing at least one b we have $[\![E]\!](\sigma) = 0$.

As for weighted finite automata, we can normalize a weighted Büchi automaton. A wba $\mathcal{A} = (Q, I, T, F, \mu)$ is *normalized* if $I = \{q_0\}$ and whenever $(p, a, q) \in T$ then $q \neq q_0$.

Lemma 3.19. *Let $\mathcal{A}$ be a wba over Σ and a Cauchy ω-indexed valuation monoid $\mathbb{D}$. Then there is a normalized wba $\mathcal{A}'$ over Σ and $\mathbb{D}$ such that $\|\mathcal{A}\|_u = \|\mathcal{A}'\|_u$.*

If $\mathbb{D}$ is moreover associative, there is also a normalized wba $\mathcal{A}'$ with $\|\mathcal{A}\| = \|\mathcal{A}'\|$ and $\|\mathcal{A}\|_c = \|\mathcal{A}'\|_c$.

Proof. Let $\mathcal{A} = (Q, I, T, F, \mu)$. Then we put $\mathcal{A}' = (Q \,\dot\cup\, \{q_0\}, \{q_0\}, T', F, \mu')$ where $T' = T \cup \{(q_0, a, q) \mid q \in Q \wedge \exists p \in I : (p, a, q) \in T\}$, $\mu'(p, a, q) = \mu(p, a, q)$ if $(p, a, q) \in T$ and $\mu'(q_0, a, q) = \sum_{p \in I} \mu(p, a, q)$ otherwise. Let $R' = (t'_i)_{i \in \mathbb{N}} \in \mathrm{succ}(\mathcal{A}')$ with $t'_i = (q_{i-1}, a_i, q_i)$. Then we put

$$I(R') = \{R = (t_i)_{i \in \mathbb{N}} \mid t_1 = (p, a_1, q_1) \text{ for some } p \in I \text{ and } t_i = t'_i \text{ for all } i \geq 2\}.$$

Note that $I(R'_1) \cap I(R'_2) = \emptyset$ for $R'_1 \neq R'_2$ and $\bigcup_{R' \in \mathrm{succ}(\mathcal{A}')} I(R') = \mathrm{succ}(\mathcal{A})$.

For every Cauchy ω-indexed valuation monoid $\mathbb{D}$ we can show, by using properties (9) and (10), that

$$\mathrm{wgt}_u(R') = \sum_{R \in I(R')} \mathrm{wgt}_u(R)$$

for every $R' \in \mathrm{succ}(\mathcal{A}')$ which implies immediately $\|\mathcal{A}'\|_u = \|\mathcal{A}\|_u$. If $\mathbb{D}$ is, moreover, associative, then we have also

$$\mathrm{wgt}(R') = \sum_{R \in I(R')} \mathrm{wgt}(R) \quad \text{and} \quad \mathrm{wgt}_c(R') = \sum_{R \in I(R')} \mathrm{wgt}_c(R)$$

for every $R' \in \mathrm{succ}(\mathcal{A}')$ which shows the other assertions. We give the details for $\mathrm{wgt}(R')$. Let $(R'_k)_{k \in \mathbb{N}}$ be the F-partition of R. Then

$$\mathrm{wgt}(R') = \mathrm{val}^\omega\big(|R'_k|, \mathrm{val}(\mu'(R'_k))\big)_{k \in \mathbb{N}}$$

with $n_1 = |R'_1|$ and by property (9)

$$= \mathrm{val}(\mu'(R'_1)) \cdot_{n_1,\omega} \mathrm{val}^\omega\big(|R'_k|, \mathrm{val}(\mu'(R'_k))\big)_{k \geq 2}$$

with $R'_1 = (q_0, a_1, q_1)\hat{R}'_1$ and, thus, $\mu'(R'_1) = \big(\sum_{p \in I} \mu(p, a_1, q_1)\big)\mu(\hat{R}'_1)$, and by (2)

$$= \left[\left(\sum_{p \in I} \mu(p, a_1, q)\right) \cdot_{1, n_1 - 1} \mathrm{val}(\mu(\hat{R}'_1))\right] \cdot_{n_1, \omega} \mathrm{val}^\omega\big(|R'_k|, \mathrm{val}(\mu'(R'_k))\big)_{k \geq 2}$$

and now by associativity, and, thus, (13)

$$= \left(\sum_{p\in I} \mu(p, a_1, q)\right) \cdot_{1,\omega} \left[\operatorname{val}(\mu(\hat{R}'_1)) \cdot_{n_1-1,\omega} \operatorname{val}^\omega \left(|R'_k|, \operatorname{val}(\mu'(R'_k))\right)_{k\geq 2}\right]$$

next applying property (10)

$$= \sum_{p\in I} \left(\mu(p, a_1, q) \cdot_{1,\omega} \left[\operatorname{val}(\mu(\hat{R}'_1)) \cdot_{n_1-1,\omega} \operatorname{val}^\omega \left(|R'_k|, \operatorname{val}(\mu'(R'_k))\right)_{k\geq 2}\right]\right)$$

and then by (13), (2), and (9), and with $R_1^p = (p, a_1, q_1)\hat{R}'_1$ and $R_k = R'_k$ for $k \geq 2$

$$= \sum_{p\in I} \operatorname{val}^\omega \left((|R_1^p|, \operatorname{val}(\mu(R_1^p))), (|R_2|, \operatorname{val}(\mu(R_2))), (|R_3|, \operatorname{val}(\mu(R_3))), \dots\right)$$

$$= \sum_{R\in I(R')} \operatorname{wgt}(R)$$

which shows the claim.   □

3.3   Omega-Rational Series Are Recognizable

Now we explore if the ω-rational series are behaviors of weighted Büchi automata or, to put it another way, whether the different behaviors of wba are closed under the ω-rational operations. This question was already solved in [21,22] for Büchi behaviors.

Theorem 3.20 ([21,22]). *Let $\mathbb{D}$ be a Cauchy ω-indexed valuation monoid and $S : \Sigma^\omega \to D$. If S is ω-rational, then there is a wba $\mathcal{A}$ with $S = \|\mathcal{A}\|$.*

The proof of this theorem is by induction on the structure of an ω-regular weighted expression E defining S and can be found in [22]. It makes use of Theorem 2.23 and of properties (9), (10), and (11).

Together with Lemma 3.8 we get

Corollary 3.21. *Let $\mathbb{D}$ be a uniform Cauchy ω-indexed valuation monoid and $S : \Sigma^\omega \to D$. If S is ω-rational, then there is a wba $\mathcal{A}$ with $S = \|\mathcal{A}\|_u$.*

Let $\mathbb{D}$ be a uniform and idempotent Cauchy ω-indexed valuation monoid. If $S : \Sigma^\omega \to D$ is ω-rational, then there is a wba $\mathcal{A}$ with $S = \|\mathcal{A}\|_c$.

But what about the cumulative behavior if $\mathbb{D}$ is not uniform and idempotent? We will show that for *associative* Cauchy ω-indexed valuation monoids ω-rationality implies also cumulative recognizability by a wba.

Theorem 3.22. *Let $\mathbb{D}$ be an associative Cauchy ω-indexed valuation monoid and $S : \Sigma^\omega \to D$. If S is ω-rational, then there is a wba $\mathcal{A}$ with $S = \|\mathcal{A}\|_c$.*

Proof. Let E be an ω-regular weighted expression over Σ and $\mathbb{D}$ with $[\![E]\!] = S$. We show that $[\![E]\!]$ is cumulative ω-recognizable by induction on the structure of E.

First, let $E = G^\omega$ for a regular weighted expression G. Due to Theorem 2.23 and Lemma 2.18, there is for every regular weighted expression G a *normalized* wfa $\mathcal{A} = (Q, \{q_0\}, T, \{q_f\}, \mu)$ with $\|\mathcal{A}\| = [\![G]\!]$. Let $\mathcal{A}' = (Q', \{q_0\}, T', \{q_0\}, \mu')$ be a wba with $Q' = Q \setminus \{q_f\}$, the unique initial state q_0, $T' = \{(p, a, q) \in T \mid q \neq q_f\} \cup \{(p, a, q_0) \mid (p, a, q_f) \in T\}$, the acceptance set $\{q_0\}$, and

$$\mu'(p, a, q) = \begin{cases} \mu(p, a, q) & \text{if } (p, a, q) \in T, \\ \mu(p, a, q_f) & \text{if } q = q_0. \end{cases}$$

Let $\sigma \in \Sigma^\omega$. For $R \in \mathrm{succ}(\mathcal{A}')$ let $(R_k^{q_0})_{k \in \mathbb{N}}$ be the q_0-partition of R. Then

$$\|\mathcal{A}'\|_c(\sigma) = \sum_{\substack{R \in \mathrm{succ}(\mathcal{A}') \\ \ell(R)=\sigma}} \mathrm{wgt}_c(R) = \sum_{\substack{R \in \mathrm{succ}(\mathcal{A}') \\ \ell(R)=\sigma}} \sum_{q \in \{q_0\}} \mathrm{val}^\omega\big(|R_k^q|, \mathrm{val}(\mu'(R_k^q))\big)_{k \in \mathbb{N}}$$

$$= \sum_{\substack{R \in \mathrm{succ}(\mathcal{A}') \\ \ell(R)=\sigma}} \mathrm{val}^\omega\big(|R_k^{q_0}|, \mathrm{val}(\mu'(R_k^{q_0}))\big)_{k \in \mathbb{N}}$$

$$= \sum_{\substack{\sigma = u_1 u_2 \cdots \\ u_k \in \Sigma^+}} \sum_{\substack{(R_k)_{k \in \mathbb{N}} \in (\mathrm{succ}(\mathcal{A}))^{\mathbb{N}} \\ \ell(R_k)=u_k}} \mathrm{val}^\omega\big(|R_k|, \mathrm{val}(\mu(R_k))\big)_{k \in \mathbb{N}}$$

and due to property (11)

$$= \sum_{\substack{\sigma = u_1 u_2 \cdots \\ u_k \in \Sigma^+}} \mathrm{val}^\omega\Big(|u_k|, \sum_{\substack{R_k \in \mathrm{succ}(\mathcal{A}) \\ \ell(R_k)=u_k}} \mathrm{val}(\mu(R_k))\Big)_{k \in \mathbb{N}}$$

$$= \sum_{\substack{\sigma = u_1 u_2 \cdots \\ u_k \in \Sigma^+}} \mathrm{val}^\omega\big(|u_k|, [\![G]\!](u_k)\big)_{k \in \mathbb{N}} = [\![G]\!]^\omega(\sigma)$$

and, thus, $[\![G]\!]^\omega$ is cumulative ω-recognizable.

Next, we consider $E \cdot E'$ where E is a regular weighted expression and E' an ω-regular weighted expression such that $[\![E']\!]$ is cumulative ω-recognizable. Due to Theorem 2.23, Lemma 2.18, the induction hypothesis, and Lemma 3.19, there are a normalized wfa $\mathcal{A} = (Q, \{q_0\}, T, \{q_f\}, \mu)$ with $\|\mathcal{A}\| = [\![E]\!]$ and a normalized wba $\mathcal{B} = (P, \{p_0\}, T_B, F, \mu_B)$ with $\|\mathcal{B}\|_c = [\![E']\!]$. We define a wba $\mathcal{C} = (Q \setminus \{q_f\} \cup P, \{q_0\}, T', F, \mu')$ by

$$T' = \{(q, a, q') \in T \mid q' \neq q_f\} \cup \{(q, a, p_0) \mid (q, a, q_f) \in T\} \cup T_B,$$

$$\mu'(q, a, q') = \begin{cases} \mu(q, a, q') & \text{if } (q, a, q') \in T, \\ \mu(q, a, q_f) & \text{if } q \in Q, q' = p_0, \\ \mu_B(q, a, q') & \text{if } (q, a, q') \in T_B. \end{cases}$$

With $(R_k^q)_{k \in \mathbb{N}}$ being the q-partition of a run R, we get for every $\sigma \in \Sigma^\omega$

$$\|\mathcal{C}\|_c(\sigma) = \sum_{\substack{R \in \mathrm{succ}(\mathcal{C}) \\ \ell(R)=\sigma}} \mathrm{wgt}_c(R) = \sum_{\substack{R \in \mathrm{succ}(\mathcal{C}) \\ \ell(R)=\sigma}} \sum_{q \in F} \mathrm{val}^\omega\big(|R_k^q|, \mathrm{val}(\mu'(R_k^q))\big)_{k \in \mathbb{N}}$$

with $n_1 = |R_1^q|$ and by (9)

$$= \sum_{\substack{R \in \mathrm{succ}(\mathcal{C}) \\ \ell(R)=\sigma}} \sum_{q \in F} \mathrm{val}(\mu'(R_1^q)) \cdot_{n_1,\omega} \mathrm{val}^\omega\left(|R_k^q|, \mathrm{val}(\mu'(R_k^q))\right)_{k \geq 2}$$

with $R_1^q = R_f \hat{R}_1^q$ where R_f is the sub-run of R_1^q from q_0 to p_0 and $\hat{R}_1^q$ the one from p_0 to q, we get by (2)

$$= \sum_{\substack{R \in \mathrm{succ}(\mathcal{C}) \\ \ell(R)=\sigma}} \sum_{q \in F} \left(\mathrm{val}(\mu'(R_f)) \cdot_{|R_f|,|\hat{R}_1^q|} \mathrm{val}(\mu'(\hat{R}_1^q))\right) \cdot_{n_1,\omega} \mathrm{val}^\omega\left(|R_k^q|, \mathrm{val}(\mu'(R_k^q))\right)_{k \geq 2}$$

and now, due to associativity, applying (13)

$$= \sum_{\substack{R \in \mathrm{succ}(\mathcal{C}) \\ \ell(R)=\sigma}} \sum_{q \in F} \mathrm{val}(\mu'(R_f)) \cdot_{|R_f|,\omega} \left(\mathrm{val}(\mu'(\hat{R}_1^q)) \cdot_{|\hat{R}_1^q|,\omega} \mathrm{val}^\omega\left(|R_k^q|, \mathrm{val}(\mu'(R_k^q))\right)_{k \geq 2}\right)$$

replacing R_f from q_0 to p_0 by the corresponding run $\tilde{R} \in \mathrm{succ}(\mathcal{A})$ from q_0 to q_f, putting $\hat{R}_k^q = R_k^q$ for $k \geq 2$, and applying (10) and (9)

$$= \sum_{\substack{u \in \Sigma^+, \hat{\sigma} \in \Sigma^\omega \\ \sigma = u\hat{\sigma}}} \sum_{\substack{\tilde{R} \in \mathrm{succ}(\mathcal{A}) \\ \ell(\tilde{R})=u}} \sum_{\substack{\hat{R} \in \mathrm{succ}(\mathcal{B}) \\ \ell(\hat{R})=\hat{\sigma}}} \mathrm{val}(\mu(\tilde{R})) \cdot_{|\tilde{R}|,\omega} \left(\sum_{q \in F} \mathrm{val}^\omega\left(|\hat{R}_k^q|, \mathrm{val}(\mu_B(\hat{R}_k^q))\right)_{k \in \mathbb{N}}\right)$$

and applying (10) once again

$$= \sum_{\substack{u \in \Sigma^+, \hat{\sigma} \in \Sigma^\omega \\ \sigma = u\hat{\sigma}}} \sum_{\substack{\tilde{R} \in \mathrm{succ}(\mathcal{A}) \\ \ell(\tilde{R})=u}} \mathrm{val}(\mu(\tilde{R})) \cdot_{|u|,\omega} \sum_{\substack{\hat{R} \in \mathrm{succ}(\mathcal{B}) \\ \ell(\hat{R})=\hat{\sigma}}} \left(\sum_{q \in F} \mathrm{val}^\omega\left(|\hat{R}_k^q|, \mathrm{val}(\mu_B(\hat{R}_k^q))\right)_{k \in \mathbb{N}}\right)$$

$$= \sum_{\substack{u \in \Sigma^+, \hat{\sigma} \in \Sigma^\omega \\ \sigma = u\hat{\sigma}}} \|\mathcal{A}\|(u) \cdot_{|u|,\omega} \|\mathcal{B}\|_c(\hat{\sigma}) = \left(\|\mathcal{A}\| \cdot \|\mathcal{B}\|_c\right)(\sigma) = \left(\llbracket E \rrbracket \cdot \llbracket E' \rrbracket\right)(\sigma)$$

which shows $\llbracket E \rrbracket \cdot \llbracket E' \rrbracket = \|\mathcal{C}\|_c$.

Finally, if $\llbracket E \rrbracket$ and $\llbracket E' \rrbracket$ for ω-regular weighted expressions E and E' are cumulative behaviors of wba $\mathcal{A}$ and $\mathcal{A}'$, then $\llbracket E + E' \rrbracket$ is the cumulative behavior of the disjoint union of $\mathcal{A}$ and $\mathcal{A}'$.

Thus, for every ω-regular weighted expression E there is a weighted Büchi automaton $\mathcal{A}$ with $\|\mathcal{A}\|_c = \llbracket E \rrbracket$. $\square$

Theorem 3.22 applies to all ω-indexed valuation monoids from Examples 3.13, 3.14, 3.15, and 3.16 as well as for all ω-indexed valuation monoids derived from complete star-omega-semirings like $\mathbb{N}_0^\infty$.

Now we turn to the converse: Describing behaviors of wba by expressions.

3.4   From Weighted Büchi-Automata to Omega-Expressions

The characterization of behaviors of weighted Büchi automata by means of ω-regular weighted expressions is more difficult than for finite words. However, for the cumulative behavior of a wba, the proof is rather straightforward.

Theorem 3.23. *Let $\mathbb{D}$ be a Cauchy ω-indexed valuation monoid and $\mathcal{A}$ a wba over Σ and $\mathbb{D}$. Then there is an ω-regular weighted expression E with $[\![\,E\,]\!] = \|\mathcal{A}\|_c$.*

Proof. Let $\mathcal{A} = (Q, I, T, F, \mu)$ be a weighted Büchi automaton over Σ and $\mathbb{D}$. We define for $p, q \in Q$ the following wba $\mathcal{A}^{pq} = (Q, \{p\}, T, \{q\}, \mu)$, i.e., $\mathcal{A}^{pq}$ can be entered in p only and the only acceptance state is q. Then we have for every $\sigma \in \Sigma^\omega$

$$
\begin{aligned}
\|\mathcal{A}\|_c(\sigma) &= \sum_{\substack{R \in \mathrm{succ}(\mathcal{A}) \\ \ell(R) = \sigma}} \mathrm{wgt}_c(R) = \sum_{\substack{R \in \mathrm{succ}(\mathcal{A}) \\ \ell(R) = \sigma}} \sum_{\substack{q \in F \\ |F_q(R)| = \omega}} \mathrm{val}^\omega \big(|R_k^q|, \mathrm{val}(\mu(R_k^q)) \big)_{k \in \mathbb{N}} \\
&= \sum_{p \in I, q \in F} \|\mathcal{A}^{pq}\|_c(\sigma) = \sum_{p \in I, q \in F} \|\mathcal{A}^{pq}\|(\sigma) .
\end{aligned}
$$

Now we show that $\|\mathcal{A}^{pq}\|$ can be described by an ω-regular weighted expression. Let $\mathcal{B}^{pq} = (Q, \{p\}, T', \{q\}, \mu')$ and $\mathcal{C}^q = (Q \cup \{q_I\}, \{q_I\}, T'', \{q\}, \mu'')$ be weighted finite automata with $T' = \{(p', a, q') \mid (p', a, q') \in T \wedge p' \neq q\}$, $\mu'(p', a, q') = \mu(p', a, q')$, $T'' = \{(p', a, q') \mid (p', a, q') \in T \wedge p' \neq q\} \cup \{(q_I, a, q') \mid (q, a, q') \in T\}$, and $\mu''(p', a, q') = \mu(p', a, q')$ if $p', q' \in Q$, and $\mu''(p', a, q') = \mu(q, a, q')$ if $p' = q_I$.

The successful runs in $\mathcal{C}^q$ simulate the runs in $\mathcal{A}^{pq}$ which go from q to q without passing q in between. Using properties (9) and (11), it is easy to show that $\|\mathcal{A}^{pq}\| = \|\mathcal{B}^{pq}\| \cdot \|\mathcal{C}^q\|^\omega$ whenever $p \neq q$ and $\|\mathcal{A}^{pq}\| = \|\mathcal{C}^q\|^\omega$ if $p = q$. Due to Theorem 2.23 there are regular weighted expressions G^{pq} and H^q with $[\![\,G^{pq}\,]\!] = \|\mathcal{B}^{pq}\|$ and $[\![\,H^q\,]\!] = \|\mathcal{C}^q\|$ and, thus, $\|\mathcal{A}^{pq}\|$ can be described by $G^{pq} \cdot (H^q)^\omega$ for $p \neq q$ and by $(H^q)^\omega$ for $p = q$. Finally, we have $\|\mathcal{A}\|_c = [\![\,E\,]\!]$ for the ω-regular weighted expression

$$
E = \sum_{q \in I \cap F} (H^q)^\omega + \sum_{p \in I, q \in F, p \neq q} G^{pq} \cdot (H^q)^\omega .
$$

which shows that $\|\mathcal{A}\|_c$ is ω-rational. $\qquad\square$

Together with Lemma 3.8 we get

Corollary 3.24. *Let $\mathbb{D}$ be a uniform and idempotent Cauchy ω-indexed valuation monoid and let $\mathcal{A}$ be a wba over Σ and $\mathbb{D}$. Then $\|\mathcal{A}\|_u$ and $\|\mathcal{A}\|$ are ω-rational.*

Idempotency and uniformity are strong conditions. They hold for instance for $(\overline{\mathbb{R}}, \sup, \sup_{-\infty}, \limsup, -\infty)$ from Example 3.14. But the Cauchy ω-index valuation monoid from Example 3.2 with limit average as ω-valuation function is idempotent but not uniform, cf. Example 3.9. Nevertheless, we can succeed in showing that $\|\mathcal{A}\|$ is ω-rational for every wba over this ω-index valuation monoid. However, for this, we need an additional property called the *partition property* which we will define next.

Recall that $(D_{\mathit{fin}})^\omega = \bigcup_{C \subseteq_{\mathit{fin}} D} C^\omega$. Let $\alpha = n_1 n_2 \ldots \in \mathbb{N}^\mathbb{N}$ with $n_1 < n_2 < \ldots$ be an ω-sequence of strictly increasing positive integers and $\mathrm{Im}(\alpha)$ be the image of α.

Let $\alpha^j = m_1^j m_2^j \cdots \in \mathrm{Im}(\alpha)^{\mathbb{N}}$ be infinite sub-sequences of α for every $j \in J$ where J is an arbitrary index set. We say that $(\alpha^j)_{j \in J}$ is a *finite partition of α* if J is a finite set, $\mathrm{Im}(\alpha^{j_1}) \cap \mathrm{Im}(\alpha^{j_2}) = \varnothing$ for $j_1 \neq j_2$, and $\bigcup_{j \in J} \mathrm{Im}(\alpha^j) = \mathrm{Im}(\alpha)$. This means that the sequence α is partitioned into finitely many infinite sub-sequences α^j.

In [21], we introduced the following property.

Definition 3.25. *An ω-indexed valuation monoid $\mathbb{D}$ has the* partition property *if the following holds:*

- *for every $(d_i)_{i \in \mathbb{N}} \in (D_{fin})^\omega$,*
- *for every $\alpha = n_1 n_2 \ldots \in \mathbb{N}^{\mathbb{N}}$ with $n_1 < n_2 < \ldots$,*
- *for every finite partition $(\alpha^j)_{j \in J}$ of α with $\alpha^j = m_1^j m_2^j \ldots$*

we have with $n_0 = m_0^j = 0$ ($j \in J$) that

$$
\mathrm{val}^\omega \big(n_k - n_{k-1}, \mathrm{val}(d_{n_{k-1}+1} \cdots d_{n_k}) \big)_{k \in \mathbb{N}}
$$
$$
= \sum_{j \in J} \mathrm{val}^\omega \big(m_k^j - m_{k-1}^j, \mathrm{val}(d_{m_{k-1}^j+1} \cdots d_{m_k^j}) \big)_{k \in \mathbb{N}} . \tag{14}
$$

In automata-theoretic terms, the partition property (14) guarantees that the weight $\mathrm{wgt}(R)$ of a run R is the same as the cumulated weight $\mathrm{wgt}_c(R)$. The sequence α collects the positions of the run where an accepting state $q \in F$ is passed. Let J be such that $\{q^j \mid j \in J\} \subseteq F$ is the set of accepting states which appear infinitely often in R. For every $j \in J$ the sequence α^j compasses the positions where the single accepting state q^j is traversed. Now the left hand side of (14) gives $\mathrm{wgt}(R)$ whereas the right hand side equals $\mathrm{wgt}_c(R)$.

Remark 3.26. Note that every uniform and idempotent ω-indexed valuation monoid satisfies the partition property [22, Prop. 12].

Example 3.27. The ω-indexed valuation monoids $(\overline{\mathbb{R}}, \sup, \mathrm{avg}, \limsup \mathrm{avg}, -\infty)$ from Example 3.13 and $(\overline{\mathbb{R}}, \sup, \mathrm{last}, \limsup, -\infty)$ from Example 3.15 are not uniform but have the partition property [22].

On the other side, $\mathbb{N}_0^\infty$ from Remark 3.17 is uniform but not idempotent and does not have the partition property.

Satisfaction of the partition property implies that the behavior of weighted Büchi automata is ω-rational.

Theorem 3.28 ([21,22]). *Let $\mathbb{D}$ be a Cauchy ω-indexed valuation monoid having the partition property (14). Let $\mathcal{A}$ be a wba over Σ and $\mathbb{D}$. Then $\|\mathcal{A}\|$ is ω-rational.*

Proof sketch. The proof follows exactly the lines of the one for Theorem 3.23. If $\mathcal{A} = (Q, I, T, F, \mu)$, then we put again $\mathcal{A}^{pq} = (Q, \{p\}, T, \{q\}, \mu)$ for $p, q \in Q$. Now we can show

$$
\|\mathcal{A}\| = \sum_{p \in I, q \in F} \|\mathcal{A}^{pq}\| .
$$

To do so, we have to use the partition property (14) of $\mathbb{D}$, see [21,22] for details. Now we carry on as before showing that $\|\mathcal{A}^{pq}\|$ is ω-rational. $\square$

Remark 3.29. Theorems 3.20 and 3.28 generalize previous Kleene-like results for discounting [14], for idempotent, o-complete, infinitely distributive semirings [15], and for idempotent complete star-omega-semirings [25,27].

Ésik and Kuich give in [25,27] a general Kleene-like result even for non-idempotent semiring-semimodule pairs. The direction from ω-regular weighted expressions to weighted Büchi automata is covered by Theorem 3.20. However, the proof of the other direction needs an infinitary associativity for the product (as it is satisfied by e.g. $\mathbb{N}_0^\infty$) similar to our notion of uniformity. Next, we will show that behaviors of wba are ω-rational even if the underlying ω-indexed valuation monoid is neither idempotent nor satisfies the partition property. However, we have to assume uniformity. This will cover the result for arbitrary (also non-idempotent) complete star-omega semirings [25,27] like $\mathbb{N}_0^\infty$.

We will prove the result in a similar manner like we have shown Proposition 2.20. Thus, we need unambiguous ω-regular language expressions. Recall that an ω-*regular (language) expression* H over the alphabet Γ is defined by the grammar

$$H ::= H + H \mid G \cdot H \mid G^\omega$$

where G is an arbitrary regular expression over Γ. The semantics $\mathcal{L}(H) \subseteq \Sigma^\omega$ of an ω-regular language expression H is given by

$$\mathcal{L}(H_1 + H_2) = \mathcal{L}(H_1) \cup \mathcal{L}(H_2),$$
$$\mathcal{L}(G \cdot H) = \mathcal{L}(G) \cdot \mathcal{L}(H) = \{u\sigma \in \Sigma^\omega \mid u \in \mathcal{L}(G), \sigma \in \mathcal{L}(H)\},$$
$$\mathcal{L}(G^\omega) = \mathcal{L}(G)^\omega = \{u_1 u_2 \ldots \in \Sigma^\omega \mid u_i \in \mathcal{L}(G) \text{ for all } i \in \mathbb{N}\}.$$

A language $L \subseteq \Sigma^\omega$ is called ω-*rational* if there is an ω-regular expression H with $\mathcal{L}(H) = L$. As Büchi showed [5], $L \subseteq \Sigma^\omega$ is ω-rational if and only if there is a Büchi automaton $\mathcal{A}$ with behaviour $\mathcal{L}(\mathcal{A}) = L$.

Next, we define *unambiguous ω-regular (language) expressions* semantically:

- if H_1 and H_2 are unambiguous ω-regular expressions and $\mathcal{L}(H_1) \cap \mathcal{L}(H_2) = \emptyset$, then $H_1 + H_2$ is unambiguous,
- if G is an unambiguous regular expression, H an unambiguous ω-regular expression, and $u_1\sigma_1 = u_2\sigma_2$ for $u_i \in \mathcal{L}(G)$, $\sigma_i \in \mathcal{L}(H)$ for $i \in \{1,2\}$ implies $u_1 = u_2$ (and, thus, also $\sigma_1 = \sigma_2$), then $G \cdot H$ is unambiguous,
- if G is an unambiguous regular expression and $u_1 u_2 \ldots = v_1 v_2 \ldots$ with $u_i, v_j \in \mathcal{L}(G)$ for all $i, j \in \mathbb{N}$ implies $u_1 = v_1$ (and, thus, $u_i = v_i$ for all $i \in \mathbb{N}$), then G^ω is unambiguous.

The next result is stated in [2] as a consequence of the fact that every ω-rational language is recognized by an non-ambiguous Büchi automaton which can be shown using the famous Büchi-MacNaughton theorem about the equivalence of Büchi and deterministic Muller automata (see also [6] for a sophisticated study of unambiguous Büchi automata). Since the description of ω-regular languages by unambiguous ω-regular expressions may be not so common as the one for finite words and in [2] the arguments are not given in detail, we state the result here together with its proof.

Lemma 3.30 ([2]). *Let $L \subseteq \Gamma^\omega$. Then L is recognizable by a Büchi automaton if and only if there is an unambiguous ω-regular language expression H with $\mathcal{L}(H) = L$.*

Proof. It follows from Büchi's result [5] that if H is an unambiguous ω-regular language expression, then $\mathcal{L}(H)$ is recognizable by a Büchi automaton.

Vice versa, let $L \subseteq \Gamma^\omega$ be recognizable by a Büchi automaton. Then there is an non-ambiguous Büchi automaton $\mathcal{A}$ recognizing L [2]. Here, a Büchi automaton is *non-ambiguous* if for each $\sigma \in \mathcal{L}(\mathcal{A})$ there is only one accepting run of $\mathcal{A}$ on σ. Let $\mathcal{A} = (Q, I, T, F, \mu)$ with $F = \{q_1, \ldots, q_m\}$. Moreover, we can assume that $I \cap F = \emptyset$ (otherwise introduce copies for initial states) and that every $q_j \in F$ is reachable from I. Now we partition the set of successful states as suggested in [25]. Let $\mathrm{succ}_j(\mathcal{A}) \subseteq \mathrm{succ}(\mathcal{A})$ for $j \in \{1, \ldots, m\}$ be the set of those successful runs

- which pass $q_j \in F$ infinitely often and
- which traverse $\{q_1, \ldots, q_{j-1}\} \subset F$ only finitely often.

Then we have

$$\mathrm{succ}(\mathcal{A}) = \mathrm{succ}_1(\mathcal{A}) \,\dot\cup\, \mathrm{succ}_2(\mathcal{A}) \,\dot\cup\, \ldots \,\dot\cup\, \mathrm{succ}_m(\mathcal{A}). \tag{15}$$

Let $L_j = \{\sigma \in \Gamma^\omega \mid \exists R \in \mathrm{succ}_j(\mathcal{A}) : \ell(R) = \sigma\}$. Note that every run $R \in \mathrm{succ}_j(\mathcal{A})$ can be decomposed into $R = R_1 R_2$ where

- R_1 is a finite run from some $q \in I$ to $q_j \in F$ such that
  - either q_j is not passed in between $q \in I$ to $q_j \in F$
  - or between the last two consecutive occurences of q_j at least one state from $\{q_1, \ldots, q_{j-1}\}$ is passed,
- R_2 is an infinite run starting in q_j, passing q_j infinitely often, and which is not running through $\{q_1, \ldots, q_{j-1}\}$ at all.

So, the decomposition $R = R_1 R_2$ splits R at the first occurence of q_j such that afterwards no state from $\{q_1, \ldots, q_{j-1}\}$ is traversed anymore. Note that R_1 is always non-empty since $I \cap F = \emptyset$. It is not difficult to show that the set L_{Ij} of labels $u \in \Gamma^+$ of finite runs of the form of R_1 is a rational language. Due to Kleene's Theorem 2.19, there is an unambiguous regular expression H_{Ij} with $\mathcal{L}(H_{Ij}) = L_{Ij}$. Let L_{jj} be the set of finite words which are labels of finite runs from q_j to q_j without passing $\{q_1, \ldots, q_j\}$ in between. Again, L_{jj} is rational and can be denoted by an unambiguous rational expression H_{jj}. Moreover, $L_j = L_{Ij} \cdot L_{jj}^\omega$ and, hence, $L_j = \mathcal{L}(H_{Ij} \cdot H_{jj}^\omega)$. We will show that $H_j := H_{Ij} \cdot H_{jj}^\omega$ is unambiguous.

Recall that H_{jj} is unambiguous. Now assume $u_i, v_i \in \mathcal{L}(H_{jj})$ for $i \in \mathbb{N}$ such that $u_1 u_2 \ldots = v_1 v_2 \ldots$. Since q_j is reachable from I, there is a $w \in \Gamma^+$ and a finite run on w from I to q_j. Hence, $\sigma = w u_1 u_2 \ldots = w v_1 v_2 \ldots \in \mathcal{L}(\mathcal{A})$. Assume $u_1 \neq v_1$. Then $|u_1| \neq |v_1|$. But both u_1 and v_1 label runs from q_j to q_j without passing q_j in between. Hence, there have to be two different accepting runs of $\mathcal{A}$ on σ. This contradicts $\mathcal{A}$ being a non-ambiguous Büchi automaton. Thus, $u_1 = v_1$ and H_{jj}^ω is unambiguous. Now consider $H_{Iq} \cdot H_{qq}^\omega$ where both H_{Iq} and H_{qq}^ω are unambiguous. Let $u_1 \sigma_1 = u_2 \sigma_2$ with $u_i \in \mathcal{L}(H_{Iq})$ and $\sigma_i \in \mathcal{L}(H_{qq}^\omega)$ for $i \in \{1, 2\}$. If $u_1 \neq u_2$ and so $|u_1| \neq |u_2|$, then we obtain two different accepting runs of $\mathcal{A}$ on $u_1 \sigma_1 = u_2 \sigma_2$ because u_1 and u_2 label two different finite runs from I to the first occurence of q_j such that afterwards no state from $\{q_1, \ldots, q_{j-1}\}$ is passed anymore. Again, this contradicts the non-ambiguity of $\mathcal{A}$. Hence, $u_1 = u_2$ and $H_j = H_{Ij} \cdot H_{jj}^{\omega}$ is unambiguous.

Now let $H := H_1 + \ldots + H_m$. Since $\operatorname{succ}_i(\mathcal{A}) \cap \operatorname{succ}_j(\mathcal{A}) = \emptyset$ for $i \neq j$ and $\mathcal{A}$ is non-ambiguous, the ω-expression H is unambiguous. Moreover, $L = \mathcal{L}(\mathcal{A}) = L_1 \cup \ldots \cup L_m = \mathcal{L}(H_1) \cup \ldots \cup \mathcal{L}(H_m) = \mathcal{L}(H)$. Thus, L can be defined by the ω-regular language expression H. $\qquad\square$

Using unambiguous ω-regular language expressions, we will show

Theorem 3.31. *Let $\mathbb{D}$ be a uniform Cauchy ω-indexed valuation monoid and $\mathcal{A}$ a wba over Σ and $\mathbb{D}$. Then $\|\mathcal{A}\|$ and $\|\mathcal{A}\|_u$ are ω-rational.*

Proof. Let $\mathcal{A} = (Q, I, T, F, \mu)$. By Lemma 3.8, $\|\mathcal{A}\| = \|\mathcal{A}\|_u$. Let $\mathcal{A}' = (Q, I, T', F)$ be the Büchi automaton over the alphabet $\Gamma = T$ with $T' = \{(p, (p, a, q), q) \mid (p, a, q) \in T\}$. We have $\mathcal{L}(\mathcal{A}') = \operatorname{succ}(\mathcal{A})$. Due to Lemma 3.30, there is an unambiguous ω-regular language expression H with $\mathcal{L}(H) = \operatorname{succ}(\mathcal{A})$. In the proof of Proposition 2.20, we have already defined for every unambiguous regular language expression G an associated weighted regular expression $\mathbb{D}(G)$ such that with the homomorphism $\pi : \Gamma^+ \to \Sigma^+$ where $\pi(p, a, q) = a$ and the homomorphic extension $\mu : \Gamma^+ \to D^+$ of $\mu : T \to D$

$$\mathbb{D}(G)(w) = \sum_{W \in \mathcal{L}(G) \cap \pi^{-1}(w)} \operatorname{val}(\mu(W)) \tag{16}$$

for every $w \in \Sigma^+$. First, we extend by induction the translation of sub-expressions of the ω-regular expression H to ω-regular weighted expressions over Σ as follows:

$$\mathbb{D}(H' + H'') = \mathbb{D}(H') + \mathbb{D}(H''), \ \mathbb{D}(G \cdot H') = \mathbb{D}(G) \cdot \mathbb{D}(H'), \ \mathbb{D}(G^{\omega}) = \mathbb{D}(G)^{\omega}$$

for regular sub-expression G and ω-regular sub-expression H' and H'' of H. Moreover, let $\pi : \Gamma^{\omega} \to \Sigma^{\omega}$ and $\mu : \Gamma^{\omega} \to D^{\omega}$ be the extensions of the homomorphisms above.

Recall that for $W = W_1 W_2 \cdots \in \Gamma^{\omega}$ the term $\operatorname{val}^{\omega}(\mu(W))$ is an abbreviation for $\operatorname{val}^{\omega}(1, \mu(W_i))_{i \in \mathbb{N}}$. Now we show that for any ω-regular sub-expression H' of H and any $\sigma \in \Sigma^{\omega}$

$$[\![\mathbb{D}(H')]\!](\sigma) = \sum_{W \in \mathcal{L}(H') \cap \pi^{-1}(\sigma)} \operatorname{val}^{\omega}(\mu(W)). \tag{17}$$

The case $H' = H_1 + H_2$ is handled exactly the same way as for regular sub-expression, cf. the proof of Proposition 2.20. Next let $H' = G \cdot \hat{H}$. For every $\sigma \in \Sigma^{\omega}$ we have

$$[\![\mathbb{D}(G \cdot \hat{H})]\!](\sigma) = [\![\mathbb{D}(G) \cdot \mathbb{D}(\hat{H})]\!](\sigma)$$

$$= \sum_{\sigma = u\hat{\sigma}} [\![\mathbb{D}(G)]\!](u) \cdot_{|u|, \omega} [\![\mathbb{D}(\hat{H})]\!](\hat{\sigma})$$

$$= \sum_{\sigma = u\hat{\sigma}} \left(\sum_{U \in \mathcal{L}(G) \cap \pi^{-1}(u)} \operatorname{val}(\mu(U)) \right) \cdot_{|u|, \omega} \left(\sum_{V \in \mathcal{L}(\hat{H}) \cap \pi^{-1}(\hat{\sigma})} \operatorname{val}^{\omega}(\mu(V)) \right)$$

$$= \sum_{\substack{\sigma=u\hat{\vartheta} \\ U \in \mathcal{L}(G') \cap \pi^{-1}(u) \\ V \in \mathcal{L}(\hat{H}) \cap \pi^{-1}(\hat{\sigma})}} \left(\mathrm{val}(\mu(U)) \cdot_{|U|,\omega} \mathrm{val}^\omega(\mu(V)) \right) \quad \text{(due to Eq. (10))}$$

and now applying Eq. (9) and uniformity

$$= \sum_{\substack{\sigma=u\hat{\sigma} \\ U \in \mathcal{L}(G) \cap \pi^{-1}(u) \\ V \in \mathcal{L}(\hat{H}) \cap \pi^{-1}(\hat{\sigma})}} \mathrm{val}^\omega(\mu(UV))$$

$$= \sum_{W \in \mathcal{L}(G \cdot \hat{H}) \cap \pi^{-1}(w)} \mathrm{val}^\omega(\mu(W)) \quad \text{(since } G \cdot \hat{H} \text{ is unambiguous)}$$

which shows (17) for $H' = G \cdot \hat{H}$.

Finally, let $H' = G^\omega$. Then we get

$$\llbracket \mathbb{D}(G^\omega) \rrbracket(\sigma) = \llbracket \mathbb{D}(G)^\omega \rrbracket(\sigma)$$

$$= \sum_{\sigma=u_1 u_2 \ldots} \mathrm{val}^\omega \left(|u_k|, \llbracket \mathbb{D}(G) \rrbracket(u_k) \right)_{k \in \mathbb{N}}$$

$$= \sum_{\sigma=u_1 u_2 \ldots} \mathrm{val}^\omega \left(|u_k|, \sum_{U_k \in \mathcal{L}(G) \cap \pi^{-1}(u_k)} \mathrm{val}(\mu(U_k)) \right)_{k \in \mathbb{N}}$$

$$= \sum_{\sigma=u_1 u_2 \ldots} \sum_{\substack{(U_k)_{k \in \mathbb{N}} \in \\ (\mathcal{L}(G) \cap \pi^{-1}(u_k))_{k \in \mathbb{N}}}} \mathrm{val}^\omega \left(|U_k|, \mathrm{val}(\mu(U_k)) \right)_{k \in \mathbb{N}} \quad \text{(by Eq. (11))}$$

and since $H' = G^\omega$ is unambiguous and $\mathbb{D}$ is uniform

$$= \sum_{W \in \mathcal{L}(G^\omega) \cap \pi^{-1}(\sigma)} \mathrm{val}^\omega \left(\mu(W) \right)$$

for every $\sigma \in \Sigma^\omega$ and, thus, (17) is shown. Now we conclude that for any $\sigma \in \Sigma^\omega$

$$\|\mathcal{A}\|_u(\sigma) = \sum_{\substack{R \in \mathrm{succ}(\mathcal{A}) \\ \ell(R)=\sigma}} \mathrm{wgt}_u(R) = \sum_{R \in \mathcal{L}(H) \cap \pi^{-1}(\sigma)} \mathrm{val}^\omega \left(\mu(R) \right) = \llbracket \mathbb{D}(H) \rrbracket(\sigma)$$

where the last equality is due to (17). Hence $\|\mathcal{A}\|_u = \|\mathcal{A}\|$ is ω-rational. $\qquad \square$

In the last section, we draw our attention to weighted Muller automata.

3.5 Weighted Muller Automata

Automata on infinite words can run with different acceptance conditions. One prominent condition which was used in [20] for a logical characterization of weighted automata is Muller acceptance.

Definition 3.32. A weighted Muller automaton $\mathcal{M} = (Q, I, T, \mathcal{F}, \mu)$ *(for short: a wma) over an alphabet Σ and an ω-indexed valuation monoid $\mathbb{D}$ consists of a finite state set Q, a set $I \subseteq Q$ of initial states, a set $T \subseteq Q \times \Sigma \times Q$ of transitions, a weight function $\mu : T \to D$, and a set $\mathcal{F} \subseteq 2^Q$ of accepting sets.*

The behavior of weighted Muller automata is defined similarly as the one of weighted Büchi automata but the set of successful runs is a different one. Let $R = (t_i)_{i \in \mathbb{N}}$ be a run of $\mathcal{M}$ with $t_i = (q_{i-1}, a_i, q_i)$. Then we put

$$\mathrm{Inf}(R) = \{q \in Q \mid q_i = q \text{ for infinitely many } i \in \mathbb{N}\},$$

i.e., $\mathrm{Inf}(R)$ is the set of states appearing infinitely often in R. Now a run R is *successful* if R starts in a state $q_0 \in I$ and $\mathrm{Inf}(R) \in \mathcal{F}$. We denote the set of successful runs of a wma $\mathcal{M}$ by $\mathrm{succ}_M(\mathcal{M})$. Let $R = (t_i)_{i \in \mathbb{N}} \in \mathrm{succ}_M(\mathcal{M})$. Then $\mathrm{Inf}(R) \in \mathcal{F}$. We put $\mathcal{F}(R) = \{j \subset \mathbb{N} \mid q_j \in \mathrm{Inf}(R)\}$ and $\mathcal{F}_q(R) = \{j \in \mathbb{N} \mid q_j = q\}$ for every $q \in \mathrm{Inf}(R)$. The $\mathcal{F}$-partition $R = (R_k)_{k \in \mathbb{N}}$ and the q-partition $R = (R_k^q)_{k \in \mathbb{N}}$ into finite sub-runs going from one acceptance state to the next one are defined as for wba but now with respect to $\mathcal{F}(R)$ and $\mathcal{F}_q(R)$, respectively. Then we put for every $\sigma \in \Sigma^\omega$

$$\|\mathcal{M}\|_u(\sigma) = \sum_{\substack{R \in \mathrm{succ}_M(\mathcal{M}) \\ \ell(R) = \sigma}} \mathrm{wgt}_u(R) = \sum_{\substack{R \in \mathrm{succ}_M(\mathcal{M}) \\ \ell(R) = \sigma}} \mathrm{val}^\omega(1, \mu(t_i))_{i \in \mathbb{N}},$$

$$\|\mathcal{M}\|(\sigma) = \sum_{\substack{R \in \mathrm{succ}_M(\mathcal{M}) \\ \ell(R) = \sigma}} \mathrm{wgt}(R) = \sum_{\substack{R \in \mathrm{succ}_M(\mathcal{M}) \\ \ell(R) = \sigma}} \mathrm{val}^\omega\big(|R_k|, \mathrm{val}(\mu(R_k))\big)_{k \in \mathbb{N}},$$

$$\|\mathcal{M}\|_c(\sigma) = \sum_{\substack{R \in \mathrm{succ}_M(\mathcal{M}) \\ \ell(R) = \sigma}} \mathrm{wgt}_c(R) = \sum_{\substack{R \in \mathrm{succ}_M(\mathcal{M}) \\ \ell(R) = \sigma}} \bigg(\sum_{q \in \mathrm{Inf}(R)} \mathrm{val}^\omega\big(|R_k^q|, \mathrm{val}(\mu(R_k^q))\big)_{k \in \mathbb{N}} \bigg).$$

As for weighted Büchi automata, cf. Lemma 3.8, we can show

Lemma 3.33. *Let $\mathbb{D}$ be an ω-indexed valuation monoid and $\mathcal{M}$ a wma over $\mathbb{D}$.*
If $\mathbb{D}$ is uniform, then $\|\mathcal{M}\|_u = \|\mathcal{M}\|$.
If $\mathbb{D}$ is idempotent and uniform, then $\|\mathcal{M}\|_u = \|\mathcal{M}\| = \|\mathcal{M}\|_c$.

For the unconditional behavior wba and wma recognize the same class of ω-series.

Theorem 3.34. *Let $\mathbb{D}$ be an ω-indexed valuation monoid.*

(a) *For every wba $\mathcal{A}$ over $\mathbb{D}$ there is a wma $\mathcal{M}$ over $\mathbb{D}$ with $\|\mathcal{A}\|_u = \|\mathcal{M}\|_u$.*
(b) *For every wma $\mathcal{M}$ over $\mathbb{D}$ there is a wba $\mathcal{A}$ over $\mathbb{D}$ with $\|\mathcal{A}\|_u = \|\mathcal{M}\|_u$.*

Proof. **(a)** If $\mathcal{A} = (Q, I, T, F, \mu)$ is a wba over Σ and $\mathbb{D}$, then let $\mathcal{M} = (Q, I, T, \mathcal{F}, \mu)$ with $\mathcal{F} = \{F' \subseteq Q \mid F' \cap F \neq \emptyset\}$. Then we have $\mathrm{succ}(\mathcal{A}) = \mathrm{succ}_M(\mathcal{M})$ and, thus, $\|\mathcal{A}\|_u = \|\mathcal{M}\|_u$.

(b) Let $\mathcal{M} = (Q, I, T, \mathcal{F}, \mu)$ be a wma over $\mathbb{D}$ with $\mathcal{F} = \{F_1, \ldots, F_m\}$. We construct an equivalent wba as it is done in the qualitative setting of ω-languages, cf. [30, Satz 7.17]. Let $\mathcal{A} = (Q', I', T', F', \mu')$ with

$$Q' = Q \,\dot\cup\, \bigcup_{i=1}^m \big(\{i\} \times F_i \times 2^{F_i}\big), \quad I' = I,$$

$$T' = T \cup \{(q, a, (i, \hat{q}, \emptyset)) \mid q \in Q, i \in \{1, \ldots, m\}, (q, a, \hat{q}) \in T, \hat{q} \in F_i\}$$
$$\cup \{((i, q, S), a, (i, \hat{q}, S \cup \{q\})) \mid (q, a, \hat{q}) \in T, q, \hat{q} \in F_i, S \neq F_i\}$$
$$\cup \{((i, q, S), a, (i, \hat{q}, \emptyset)) \mid (q, a, \hat{q}) \in T, q, \hat{q} \in F_i, S = F_i\}$$

$$\mu'(p', a, q') = \begin{cases} \mu(p', a, q') & \text{if } p', q' \in Q, \\ \mu(p', a, q) & \text{if } p \in Q, q' = (i, q, \emptyset), \\ \mu(p, a, q) & \text{if } p' = (i, p, S), q' = (i, q, \hat{S}), \end{cases}$$

$$F = \{(i, q, F_i) \mid i \in \{1, \ldots, m\}, q \in F_i\} \, .$$

There is a one-to-one correspondence between $\mathrm{succ}_M(\mathcal{M})$ and $\mathrm{succ}(\mathcal{A})$. For this, let $R = (t_j)_{j \in \mathbb{N}} \in \mathrm{succ}_M(\mathcal{M})$ with $t_j = (q_{j-1}, a_j, q_j)$ and $\mathrm{Inf}(R) = F_i$. Then let $k_R \in \mathbb{N}$ be the unique position for which

$$(q_{k_R - 1} \notin F_i \vee k_R - 1 = 0) \wedge (\forall k \geq k_R : q_k \in F_i) \, .$$

We define the run $R' = (t'_j)_{j \in \mathbb{N}}$ of $\mathcal{A}$ with $t'_j = (q'_{j-1}, a_j, q'_j)$ where $q'_j = q_j$ for all $j < k_R$ and $q'_j = (i, q_j, S_j)$ for $j \geq k_R$ (note that for a run R' the sets S_j are uniquely determined). Since $\mathrm{Inf}(R) = F_i$, R' passes the set $\{(i, q, F_i) \mid q \in Q\}$ infinitely often. Hence, $R' \in \mathrm{succ}(\mathcal{A})$. Moreover, $\mu'(R') = \mu(R)$. Vice versa, for every $R' \in \mathrm{succ}(\mathcal{A})$ the projection of R' to the Q-component yields a successful run in $\mathcal{M}$. Thus, we have

$$\|\mathcal{A}\|(\sigma) = \sum_{\substack{R' \in \mathrm{succ}(\mathcal{A}) \\ \ell(R') = \sigma}} \mathrm{wgt}_u(R') = \sum_{\substack{R' = (t'_j)_{j \in \mathbb{N}} \in \mathrm{succ}(\mathcal{A}) \\ \ell(R') = \sigma}} \mathrm{val}^\omega(1, \mu'(t'_j))_{j \in \mathbb{N}}$$

$$= \sum_{\substack{R \in \mathrm{succ}_M(\mathcal{M}) \\ \ell(R) = \sigma}} \mathrm{val}^\omega(1, \mu(t_j))_{j \in \mathbb{N}} = \sum_{\substack{R \in \mathrm{succ}_M(\mathcal{M}) \\ \ell(R) = \sigma}} \mathrm{wgt}_u(R) = \|\mathcal{M}\|(\sigma)$$

for every $\sigma \in \Sigma^\omega$. $\qquad\square$

Note that the constructions of the last proof do not yield equivalent automata without $\mathbb{D}$ being uniform. This is due to the fact that the acceptance states in the wba $\mathcal{A}$ are distributed differently along a run than in the wma $\mathcal{M}$.

Lemma 3.8, Lemma 3.33, and Theorem 3.34 imply two more results.

Corollary 3.35. *Let $\mathbb{D}$ be a uniform ω-indexed valuation monoid.*

(a) *For every wba $\mathcal{A}$ over $\mathbb{D}$ there is a wma $\mathcal{M}$ over $\mathbb{D}$ with $\|\mathcal{A}\| = \|\mathcal{M}\|$.*
(b) *For every wma $\mathcal{M}$ over $\mathbb{D}$ there is a wba $\mathcal{A}$ over $\mathbb{D}$ with $\|\mathcal{A}\| = \|\mathcal{M}\|$.*

Corollary 3.36. *Let $\mathbb{D}$ be an idempotent uniform ω-indexed valuation monoid.*

(a) *For every wba $\mathcal{A}$ over $\mathbb{D}$ there is a wma $\mathcal{M}$ over $\mathbb{D}$ with $\|\mathcal{A}\|_c = \|\mathcal{M}\|_c$.*
(b) *For every wma $\mathcal{M}$ over $\mathbb{D}$ there is a wba $\mathcal{A}$ over $\mathbb{D}$ with $\|\mathcal{A}\|_c = \|\mathcal{M}\|_c$.*

For non-uniform ω-indexed valuation monoids like $(\overline{\mathbb{R}}, \sup, \mathrm{avg}, \limsup \mathrm{avg}, -\infty)$ from Example 3.13 the expressive power of wma compared to the one of wba has still to be clarified.

4 Conclusion

We have explored weighted automata and weighted expressions for valuation monoids and ω-indexed valuation monoids. We have shown several Kleene-like results in this very general setting both for finite and infinite words. An overview of our results

Table 1. An overview on weighted ω-recognizability and ω-rationality

Properties of $\mathbb{D}$	Results & Examples
– uniform – idempotent – Cauchy	The following holds: – ω-recognizability (in any running mode) and ω-rationality coincide, – $\|\mathcal{A}\|_u = \|\mathcal{A}\| = \|\mathcal{A}\|_c$ for every wba $\mathcal{A}$, – Büchi and Muller automata define the same class of series in every running mode. Examples: – $(\overline{\mathbb{R}}, \sup, \sup_{-\infty}, \limsup, -\infty)$ – $(\overline{\mathbb{R}}_+, \sup, \mathrm{disc}_\lambda, \lim \mathrm{disc}_\lambda, -\infty)$
– uniform – associative – Cauchy	The following holds: – ω-recognizability (in any running mode) and ω-rationality coincide, – $\|\mathcal{A}\|_u = \|\mathcal{A}\|$ for every wba $\mathcal{A}$, – Büchi and Muller automata define the same class of series for the unconditional and the Büchi running mode. Example: – $\mathbb{N}_0^\infty = (\mathbb{N} \cup \{\infty\}, +, \cdot, \prod, 0)$
– associative – Cauchy – partition property	The following holds: – ω-recognizability (in the Büchi and the cumulated running mode) and ω-rationality coincide, – Büchi and Muller automata define the same class of series for the unconditional running mode. Example: – $(\overline{\mathbb{R}}, \sup, \mathrm{avg}, \limsup \mathrm{avg}, -\infty)$ – $(\overline{\mathbb{R}}, \sup, \mathrm{last}, \limsup, -\infty)$

for infinite words and some classes of ω-indexed valuation monoids is given in Table 1 on page 343.

Several questions remain open. First, we would be interested in more example structures, especially for the class of uniform non-idempotent Cauchy ω-indexed valuation monoids where the example $\mathbb{N}_0^{\infty}$ is derived from a semiring. Is there a "nice" structure in this class which is not derived from a semiring? Other examples would be welcomed to show sharpness of results.

There are a few concepts not yet considered for valuation monoids. One such notion is *local finiteness* used e.g. for bimonoids [23]. Certainly, it is not complicated to define something like a locally finite valuation function. But in order to show a Kleene-like result as in [23] one has to consider also the parameterized products. For them, it is not obvious how to define local finiteness because of the parameters representing the length of words.

An important class of expressions in the Boolean setting are *star-free* expressions. Such expressions were considered in a semiring setting [19] and a similar approach could be tried in our setting.

With regard to expressions, the construction of small automata for a given expression is a vital topic. In this context, the method of partial derivatives was also transferred to the weighted setting [35]. It is not clear if such an approach can be successful for Cauchy valuation monoids. If not, is there another way to construct small automata?

There are notions of discounting where the discount factors depend on the action executed. Can our concept of a valuation function be generalized in a way, e.g. by putting val : $(\Sigma \times D)^+ \to D$, such that those scenarios are also covered?

Finally, there are trees. A first step to consider weighted tree automata over tree valuation monoids was done in [12]. There, a characterization by weighted MSO logic was given for finite trees. What about expressions?

But first and foremost, the work done by Chatterjee, Doyen, and Henzinger [7,8,9,10] should be combined with our results. Such a combination could open the way to quantitative specification languages (using logics and expressions) having good properties concerning decidability and complexity of problems like satisfiability or equivalence.

References

1. Alexandrakis, A., Bozapalidis, S.: Weighted grammars and Kleene's theorem. Inform. Process. Lett. 24(1), 1–4 (1987)
2. Arnold, A.: Rational ω-languages are non-ambiguous. Theor. Comput. Sci. 26, 221–223 (1983)
3. Berstel, J., Reutenauer, C.: Recognizable formal power series on trees. Theor. Comput. Sci. 18, 115–148 (1982)
4. Bozapalidis, S., Grammatikopoulou, A.: Recognizable picture series. J. Autom. Lang. Comb. 10(2/3), 159–183 (2005)
5. Büchi, J.R.: On a decision method in restricted second order arithmetics. In: Nagel, E., et al. (eds.) Proc. Intern. Congress on Logic, Methodology and Philosophy of Science, pp. 1–11. Stanford University Press, Stanford (1962)
6. Carton, O., Michel, M.: Unambiguous Büchi automata. Theor. Comput. Sci. 297, 37–81 (2003)

7. Chatterjee, K., Doyen, L., Henzinger, T.A.: Quantitative languages. In: Kaminski, M., Martini, S. (eds.) CSL 2008. LNCS, vol. 5213, pp. 385–400. Springer, Heidelberg (2008)
8. Chatterjee, K., Doyen, L., Henzinger, T.A.: Alternating weighted automata. In: Kutyłowski, M., Charatonik, W., Gębala, M. (eds.) FCT 2009. LNCS, vol. 5699, pp. 3–13. Springer, Heidelberg (2009)
9. Chatterjee, K., Doyen, L., Henzinger, T.A.: Expressiveness and closure properties for quantitative languages. In: LICS 2009, pp. 199–208. IEEE Comp. Soc. Press, Los Alamitos (2009)
10. Chatterjee, K., Doyen, L., Henzinger, T.A.: Probabilistic weighted automata. In: Bravetti, M., Zavattaro, G. (eds.) CONCUR 2009. LNCS, vol. 5710, pp. 244–258. Springer, Heidelberg (2009)
11. Droste, M., Gastin, P.: The Kleene-Schützenberger theorem for formal power series in partially commuting variables. Inform. and Comput. 153, 47–80 (1999)
12. Droste, M., Götze, D., Märcker, S., Meinecke, I.: Weighted tree automata over valuation monoids and their characterizations by weighted logics. In: Kuich, W., Rahonis, G. (eds.) Bozapalidis Festschrift. LNCS, vol. 7020, pp. 309–346. Springer, Heidelberg (2011)
13. Droste, M., Kuich, W., Vogler, H. (eds.): Handbook of Weighted Automata. EATCS Monographs in Theoretical Computer Science. Springer, Heidelberg (2009)
14. Droste, M., Kuske, D.: Skew and infinitary formal power series. Theor. Comput. Sci. 366, 199–227 (2006)
15. Droste, M., Püschmann, U.: Weighted Büchi automata with order-complete weights. Internat. J. Algebra Comput. 17(2), 235–260 (2007)
16. Droste, M., Rahonis, G.: Weighted automata and weighted logics with discounting. Theor. Comput. Sci. 410, 3481–3494 (2009)
17. Droste, M., Sakarovitch, J., Vogler, H.: Weighted automata with discounting. Inform. Process. Lett. 108, 23–28 (2008)
18. Droste, M., Stüber, T., Vogler, H.: Weighted finite automata over strong bimonoids. Inform. Sci. 180, 156–166 (2010)
19. Droste, M., Gastin, P.: On aperiodic and star-free formal power series in partially commuting variables. Theory Comput. Syst. 42, 608–631 (2008)
20. Droste, M., Meinecke, I.: Describing average- and longtime-behavior by weighted MSO logics. In: Hliněný, P., Kučera, A. (eds.) MFCS 2010. LNCS, vol. 6281, pp. 537–548. Springer, Heidelberg (2010)
21. Droste, M., Meinecke, I.: Regular expressions on average and in the long run. In: Domaratzki, M., Salomaa, K. (eds.) CIAA 2010. LNCS, vol. 6482, pp. 211–221. Springer, Heidelberg (2011)
22. Droste, M., Meinecke, I.: Weighted automata and regular expressions over valuation monoids. Int. J. Foundat. Comp. Sci. (in print, 2011)
23. Droste, M., Vogler, H.: Kleene and Büchi theorems for weighted automata and multi-valued logics over arbitrary bounded lattices. In: Gao, Y., Lu, H., Seki, S., Yu, S. (eds.) DLT 2010. LNCS, vol. 6224, pp. 160–172. Springer, Heidelberg (2010)
24. Eilenberg, S.: Automata, Languages, and Machines, vol. A. Academic Press, London (1974)
25. Ésik, Z., Kuich, W.: A semiring-semimodule generalization of ω-regular languages I+II. J. Autom. Lang. Comb. 10, 203–264 (2005)
26. Ésik, Z., Kuich, W.: On iteration semiring-semimodule pairs. Semigroup Forum 75, 129–159 (2007)
27. Ésik, Z., Kuich, W.: Finite Automata. In: Droste, et al. (eds.) [13], ch. 3 (2009)
28. Fichtner, I., Kuske, D., Meinecke, I.: Traces, Series-Parallel Posets, and Pictures: A Weighted Study. In: Droste, et al. (eds.) [13], ch. 10 (2009)
29. Fülöp, Z., Vogler, H.: Weighted Tree Automata and Tree Transducers. In: Droste, et al. (eds.) [13], vol. ch. 9 (2009)

30. Hofmann, M., Lange, M.: Automatentheorie und Logik. Springer, Heidelberg (2011)
31. Kleene, S.: Representations of events in nerve nets and finite automata. In: Shannon, C., McCarthy, J. (eds.) Automata Studies, pp. 3–42. Princeton University Press, Princeton (1956)
32. Kuich, W.: On skew formal power series. In: Bozapalidis, S., Kalampakas, A., Rahonis, G. (eds.) CAI 2005, pp. 7–30. Aristotle University of Thessaloniki Press (2005)
33. Kuske, D., Meinecke, I.: Branching automata with costs – a way of reflecting parallelism with costs. Theor. Comput. Sci. 328, 53–75 (2004)
34. Kuske, D.: Note: Schützenberger's theorem on formal power series follows from Kleene's theorem. Theor. Comput. Sci. 401, 243–248 (2008)
35. Lombardy, S., Sakarovitch, J.: Derivatives of rational expressions with multiplicity. Theor. Comput. Sci. 332, 141–177 (2005)
36. Mäurer, I.: Characterizations of recognizable picture series. Theor. Comput. Sci. 374, 214–228 (2007)
37. Rahonis, G.: Infinite fuzzy computations. Fuzzy Sets and Systems 153(2), 275–288 (2005)
38. Sakarovitch, J.: Rational and Recognisable Power Series. In: Droste, et al. (eds.) [13], ch. 4 (2009)
39. Schützenberger, M.: On the definition of a family of automata. Inform. and Control 4, 245–270 (1961)

Selected Combinatorial Properties of Random Intersection Graphs

Sotiris Nikoletseas, Christoforos Raptopoulos, and Paul G. Spirakis

Computer Technology Institute, P.O. Box 1122, 26110 Patras, Greece
University of Patras, 26500 Patras, Greece
{nikole,spirakis}@cti.gr, raptopox@ceid.upatras.gr

Abstract. Consider a universal set $\mathcal{M}$ and a vertex set V and suppose that to each vertex in V we assign independently a subset of $\mathcal{M}$ chosen at random according to some probability distribution over subsets of $\mathcal{M}$. By connecting two vertices if their assigned subsets have elements in common, we get a random instance of a random intersection graphs model. In this work, we overview some results concerning the existence and efficient construction of Hamilton cycles in random intersection graph models. In particular, we present and discuss results concerning two special cases where the assigned subsets to the vertices are formed by (a) choosing each element of $\mathcal{M}$ independently with probability p and (b) selecting uniformly at random a subset of fixed cardinality.

1 Introduction

Random graphs, introduced by P. Erdös and A. Rényi, still continue to attract a large amount of research and interest in the communities of Theoretical Computer Science, Graph Theory and Discrete Mathematics.

There exist various models of random graphs. The most famous is the $G_{n,p}$ random graph, a sample space whose points are graphs produced by randomly sampling the edges of a graph on n vertices independently, with the same probability p. Other models have also been quite a lot investigated: $G_{n,r}$ (the "random regular graphs", produced by randomly and equiprobably sampling a graph from all regular graphs of n vertices and vertex degree r) and $G_{n,k}$ (produced by randomly and equiprobably selecting an element of the class of graphs on n vertices having k edges). For an excellent survey of these models, see [1,4].

In this work we overview some results concerning the existence and efficient construction of Hamilton cycles in random intersection graphs models. In general, a random instance of these models is constructed by first assigning independently to each vertex a random subset of a predefined universal set and then connecting two vertices if their assigned subsets have elements in common.

1.1 Motivation

Random intersection graphs may be used to model several real-life applications characterized by local interactions quite accurately (compared to the $G_{n,\hat{p}}$ model

W. Kuich and G. Rahonis (Eds.): Bozapalidis Festschrift, LNCS 7020, pp. 347–362, 2011.

where edges appear independently with probability $\hat{p}$). In particular, the $G_{n,\hat{p}}$ model seems inappropriate for describing some real world networks (like sensor and social networks) because it lacks certain features of those networks, such as a scale free degree distribution and the emergence of local clusters. One of the underlying reasons for this mismatch is its independence of the edges, in other words the missing transitivity that characterizes such networks: if vertices x and y exhibit a relationship of some kind in a real world network and so do vertices y and z, then this suggests a connection between vertices x and z, too.

For example, we consider the following scenario concerning efficient and secure communication in sensor networks: The vertices in our model correspond to sensor devices that blindly choose a limited number of resources among a globally available set of shared resources (such as communication channels, encryption keys etc). Whenever two sensors select at least one resource in common (e.g. a common communication channel, a common encryption key), a communication link is implicitly established (represented by a graph edge); this gives rise to communication graphs that look like random intersection graphs. Particularly for security purposes, the random selection of elements in our graphs can be seen as a way to establish local common keys on-line, without any global scheme for predistribution of keys. In such a case, the set of labels can be a global set of large primes (known to all) but each node selects uniformly at random only a few. Two nodes that have selected a common prime can communicate securely. Notice that no other node can know what numbers a different node has selected. Thus, the local communication is guaranteed to be secure. In the case when the shared resource is the wireless spectrum, then two nodes choosing the same label (frequency) may interfere, and the corresponding link in the intersection graph abstracts a conflict, while an independent set (vertices with no edges between them) abstracts a set of sensors that can simultaneously access the wireless medium.

Random intersection graphs in general and in particular the uniform random intersection graphs model are also relevant to and capture quite nicely social networking. Indeed, a social network is a structure made of nodes (individuals or organizations) tied by one or more specific types of interdependency, such as values, visions, financial exchange, friends, conflicts, web links etc. Social network analysis views social relationships in terms of nodes and ties. Nodes are the individual actors within the networks and ties are the relationships between the actors.

Other applications may include oblivious resource sharing in a (general) distributed setting, interactions of mobile agents traversing the web etc. Even epidemiological phenomena (like spread of disease) tend to be more accurately captured by these "interaction-sensitive" random graph models.

1.2   Background

Random intersection graphs $G_{n,m,p}$ were introduced by M. Karoński, E.R. Sheinerman and K.B. Singer-Cohen [11] and K.B. Singer-Cohen [18]. In such graphs, each one of m labels is chosen independently with probability p by each one of

n vertices, and there are edges between any vertices with overlaps in the labels chosen. Fill, Sheinerman and Singer-Cohen in [9] proved that the $G_{n,m,p}$ becomes statistically equivalent to a Bernoulli random graph (in which every edge appears independently with some probability $\hat{p}$), when the number of labels m is quite large (in fact for $m \geq n^6$, but it was conjectured that the same holds for smaller m). However, the two models seem to behave quite differently when the number of labels is less than the number of vertices. The authors of [8] find thresholds (that are optimal up to a constant factor) for the appearance of Hamilton cycles in random intersection graphs. Their approach is non-constructive, which is the essential difference between their result and the results presented here. The efficient construction of large independent sets in $G_{n,m,p}$ was studied in [13]. Also, by using a sieve method, Stark [19] gives exact formulas for the degree distribution of an arbitrary fixed vertex of $G_{n,m,p}$ for a quite wide range of the parameters of the model. In [14] the authors show that a random instance of the random intersection graphs model is a vertex expander with high probability (whp) and also provide bounds on the second largest eigenvalue of random walks on such graphs. Furthermore, using a technique due to [6], the authors provide tight bounds on the cover time of random walks on random intersection graphs. The component evolution in general random intersection graphs was studied quite recently in [5], where the authors used branching process techniques to give conditions on the existence and uniqueness of a giant component.

Godehardt and Jaworski [10] defined a different model called *uniform random intersection graphs model* $G_{n,m,\lambda}$. In this model, to each of the n vertices of the graph, a random subset of λ elements of a universal set of m elements in total is independently assigned. Two vertices u, v are then adjacent in the $G_{n,m,\lambda}$ graph if and only if their assigned sets of elements have at least one element in common. The $G_{n,m,\lambda}$ seems to behave similarly to $G_{n,m,p}$ when one can show concentration on the number of labels chosen by a vertex in the latter (which can happen for quite large λ). However, notice that for small values of λ such concentration results do not hold, and the statistical behavior of the two models is quite different. In their paper [10], the authors focused on the distribution of the number of isolated vertices in $G_{n,m,\lambda}$, as well as the distribution of vertex degrees. The vertex degree distribution of general random intersection graphs (where the choice of the label sets S_v is made according to a general distribution) was studied independently by Blonzelis [3] and Deijfen and Kets [7]. Connectivity and communication security aspects of $G_{n,m,\lambda}$ in various important settings was studied in [16,2]. A tight characterization of the independence number of $G_{n,m,\lambda}$ in the case $m < n$ was given in [12].

1.3   Roadmap

In this work, we overview some results concerning the efficient construction of Hamilton cycles in random intersection graphs and uniform random intersection graphs. These results first appeared in [12] and [17] (for additional non-constructive results see [8]).

In particular, in Section 2, we give formal definitions of the the random intersection graphs model $\mathcal{G}_{n,m,p}$ and the uniform random intersection graphs model $\mathcal{G}_{n,m,\lambda}$. In Section 3 we present a reduction of the $G_{n,m,p}$ model to the model $G_{n,k}$ (i.e. the model where we randomly and equiprobably select an element of the class of graphs on n vertices and k edges) in the case where $m = n^\alpha, \alpha > 1$. In particular, for this range of m and p this reduction enables the application of any result (algorithmic and combinatoric) concerning increasing properties of $G_{n,k}$ graphs to the $G_{n,m,p}$ graphs.

In Section 4 we present a polynomial expected time algorithm for finding Hamilton cycles in a $G_{n,m,p}$ random graph, in the case where p is constant and the number of labels is at most $\alpha\sqrt{\frac{n}{\log n}}$ for some constant α. The algorithm uses first a randomized greedy algorithm and if it fails it then uses an exhaustive algorithm to solve the problem. Furthermore, in Section 5 we present a result showing that the greedy approach gives an algorithm that succeeds with high probability in the case $m = o(\frac{n}{\log n})$ and p is not constant. We also present a polynomial time randomized algorithm that finds with high probability a Hamilton cycle in the case where p is just above the connectivity threshold for these graphs. This algorithm also serves as a way to find a quite tight bound on the probability p that ensures that with high probability the uniform random intersection graph has a Hamilton cycle.

Section 6 is concerned with the hamiltonicity of uniform random intersection graphs. We present a result showing that when the number of vertices n is at least $(1 + \epsilon)\binom{m}{\lambda}\ln\binom{m}{\lambda}$, for some constant $\epsilon > 0$ as small as possible, then $G_{n,m,\lambda}$, with $\lambda \geq 2$, has a Hamiltonian cycle with high probability (whp), i.e. a very small constant number of labels suffices to yield hamiltonicity. The proof uses the coupon collector's problem together with an interesting combinatorial construction. It also leads to a polynomial time randomized algorithm for constructing Hamiltonian cycles whp for this range of values for the parameters of the model.

Finally, we give some concluding remarks in Section 7.

2   Notation and Definition of the Models

We now formally define the two models that we study in this work. The following definition was given in [11,18].

Definition 1 (Random intersection graph). *Let $\mathcal{M}$ be a universe of m distinct elements which will be called* labels *and let V be a set of vertices. A random instance of the random intersection graph model $\mathcal{G}_{n,m,p}$ is constructed as follows: assign independently to each vertex $v \in V$ a subset S_v of the universe $\mathcal{M}$ by choosing each label $l \in \mathcal{M}$ independently with probability p. Then, connect two vertices v, u if and only if $S_v \cap S_u \neq \emptyset$, i.e. their assigned label sets have at least one label in common. In this model we also denote by L_l the set of vertices that have chosen label $l \in M$. The degree of $v \in V(G)$ will be denoted by $d_G(v)$. Also, the set of edges of $G_{n,m,p}$ will be denoted by $e(G)$.*

Consider now the bipartite graph with vertex set $V(G) \cup \mathcal{M}$ and edge set $\{(v_j, i) : i \in S_{v_j}\} = \{(v_j, i) : v_j \in L_i\}$. *We will refer to this graph as the* bipartite random graph $B_{n,m,p}$ associated to $G_{n,m,p}$.

It is easy to see that a specific graph may be produced by more than one associated bipartite graphs. Furthermore, if the number of labels is too small, then there can be some graphs that are impossible to appear in this model.

A slightly more general definition was given in [13], namely the *general random intersection graphs model* $\mathcal{G}_{n,m,\mathbf{p}}$. In this model, the probability of label selection $\mathbf{p}_l$ is label dependent.

When we assume that every vertex is allowed to choose exactly λ labels, then a completely different kind of randomization to intersection graphs is introduced. The following definition was given in [10].

Definition 2 (Uniform Random Intersection Graph). *Let $\mathcal{M}$ be a universe of m distinct elements which will be called* labels *and let V be a set of vertices. A random instance of the uniform random intersection graph model* $\mathcal{G}_{n,m,\lambda}$ *is constructed as follows: assign independently to each vertex $v \in V$ a subset S_v of the universe $\mathcal{M}$, chosen uniformly at random among all subsets of $\mathcal{M}$ of size λ. Then, connect two vertices v, u if and only if $S_v \cap S_u \neq \emptyset$, i.e. their assigned label sets have at least one label in common.*

It is worth noting that the distributions $\mathcal{G}_{n,m,p}$ and $\mathcal{G}_{n,m,\lambda}$ are quite different, especially in the case where the number of labels chosen by a vertex in the first is not concentrated around its mean value.

3   A Reduction to the $G_{n,k}$ Model

We now present a translation result between the $G_{n,m,p}$ model and the $G_{n,k}$ model of random graphs with exactly k edges, which first appeared in [17].

Let $G_{n,1,p}$ be a random intersection graph with only one label. It is obvious from the definition of the random intersection graphs model that this graph will either contain a single clique, or it will be the empty graph. We note that given that the $G_{n,1,p}$ graph has a clique of size k, then this clique is equiprobably any of the $\binom{n}{k}$ cliques of size k. Let now $\hat{p}$ denote the probability that the $G_{n,1,p}$ graph has at least one edge. Then, for $np \to 0$ we get

$$\hat{p} \overset{\text{def}}{=} P\{G_{n,1,p} \text{ is non-empty}\} = 1 - (1-p)^n - np(1-p)^{n-1} \sim \frac{1}{2}n^2p^2.$$

Now, let us return to the case of m labels. We construct a graph in the following way:

1. Initially is the empty graph with n vertices.
2. In each step $i = 1, 2, \ldots, m$ we independently add a single edge to with probability $\frac{1}{2}n^2(1+\epsilon)^2p^2$. This edge can be any of the $\binom{n}{2}$ possible edges of H with equal probability. Also, with probability $1 - \frac{1}{2}n^2(1+\epsilon)^2p^2$ we do nothing.

We note that the graph H constructed above is a multigraph, since a single edge may appear more than once. Furthermore, given that has exactly k edges, then the graph is with equal probability any of the multigraphs with k edges. The mean number of all edges of H is $m\frac{1}{2}n^2(1+\epsilon)^2p^2$. By using Chernoff bounds we get for any constant β in $(0,1)$ that

$$P\{\# \text{ of all edges of } H \leq (1-\beta)m\tfrac{1}{2}n^2(1+\epsilon)^2p^2 \} \leq e^{-\beta^2 mn^2(1+\epsilon)^2p^2/4}.$$

So, if we choose $mn^2(1+\epsilon)^2p^2 \to \infty$, then with high probability H will have at least $(1-\beta)m\frac{1}{2}n^2(1+\epsilon)^2p^2$ edges (counting multiple eges as many times as their multiplicity).

Let now $e(H)$ be the number of different edges of H (i.e. we count multiple edges as one). Then, by a slightly modified coupon collector's problem, we can see that $e(H)$ must satisfy the inequality $e(H) \geq \frac{1-\beta}{1+\gamma}mn^2(1+\epsilon)^2p^2$, with probability at least $1 - e^{-\beta^2 mn^2(1+\epsilon)^2p^2/4} - \frac{1}{\gamma^2\binom{n}{2}} \to 1$, where γ is any positive constant.

We note here that given that $e(H) = k$, the graph H is with equal probability any graph with n vertices and k edges, hence it is distributed like $G_{n,k}$. We can therefore prove the following:

Theorem 1 ([17]). *Let $\mathcal{Q}$ be an increasing property on the number of edges and suppose that for some integer k we have*

$$P\{G_{n,k} \text{ has property } \mathcal{Q}\} = 1 - g(n)$$

where $g(n) = o(1)$. Let also m, p, ϵ be such that $m = n^\alpha$, $\alpha > 1$, $np \to 0$, ϵ is a (small) positive constant and

$$k \leq \frac{1-\beta}{1+\gamma}mn^2(1+\epsilon)^2p^2 \tag{1}$$

where β, γ are any two small positive constants. Then

$$P\{H \text{ has property } \mathcal{Q}\} \geq 1 - g(n) - e^{-\beta^2 mn^2(1+\epsilon)^2p^2/4} - \frac{1}{\gamma^2\binom{n}{2}} \to 1$$

We now use Theorem 1 for the case where $\mathcal{Q}$ is the property "existence of a Hamilton cycle". We can prove the following

Corollary 1. *If $m = n^\alpha$, $\alpha > 1$, $p = \sqrt{\frac{\log n}{nm}}$ and ϵ is a small constant [1], then*

$$P\{G_{n,m,(1+2\epsilon)p} \text{ has a Hamilton cycle}\} \geq 1 - g(n) - e^{-\beta^2 mn^2(1+\epsilon)^2p^2/4} - \frac{1}{\gamma^2\binom{n}{2}} \to 1.$$

---

[1] In fact the constant ϵ can be as small as to ensure that the constant quantity $\frac{1-\beta}{1+\gamma}(1+\epsilon)^2$ is as close to 1 as possible (but greater than 1). Since β, γ are small (and can be controlled to become as small as possible), we can get a quite small value for ϵ.

Proof. It is obvious that $np \to 0$. Moreover, for $k = \frac{1}{2}n \log n + o(n \log n)$ (which ensures that $G_{n,k}$ has a Hamilton cycle with high probability) inequality (1) is satisfied. Since all the conditions of Theorem 1 are satisfied, we then have

$$P\{H \text{ has a Hamilton cycle}\} \geq 1 - g(n) - e^{-\beta^2 mn^2(1+\epsilon)^2 p^2/4} - \frac{1}{\gamma^2 \binom{n}{2}} \to 1.$$

Let us now consider a $G_{n,m,(1+2\epsilon)p}$ graph. Because of independence, we can construct such a graph by "putting together" m independent $G_{n,1,(1+2\epsilon)p}$ graphs. As in the beginning of this section, the probability that each such graph has at most one edge is

$$P\{G_{n,1,(1+2\epsilon)p} \text{ is non-empty}\} = 1 - (1 - (1 + 2\epsilon)p)^n \quad n(1 + 2\epsilon)p(1 - (1 + 2\epsilon)p)^{n-1}$$
$$\sim \frac{1}{2}n^2(1 + 2\epsilon)^2 p^2 + o(n^2 p^2) \geq \frac{1}{2}n^2(1 + \epsilon)^2 p^2).$$

Notice now that if we associate each of the m independent $G_{n,1,(1+2\epsilon)p}$ graphs with the m steps in the construction of the graph H, we see that not only do the first ones add an edge to the $G_{n,m,(1+2\epsilon)p}$ graph with higher probability than their corresponding steps do to H, but they are also likely to add more than one edge (i.e. a clique). Put more simply, the graph H can be simply derived from a sparser random intersection graph than $G_{n,m,(1+2\epsilon)p}$ by simply removing some edges (i.e. when there is a set L_l with more than two vertices in it, we simply delete some vertices from it so that it finally has exactly two). Hence the result follows. $\qquad \square$

We finally note that, as shown in [18], the value $p = \sqrt{\frac{\log n}{nm}}$ is exactly the connectivity threshold of a random intersection graph with $m = n^\alpha$, $\alpha > 1$.

4   Hamilton Cycles in $G_{n,m,p}$ with Constant p

In this section we present an expected polynomial time algorithm for constructing Hamilton Cycles in random intersection graphs with constant p and $m \leq \alpha\sqrt{\frac{n}{\log n}}$, where $\alpha < \frac{\beta p}{\sqrt{2}}$ and $\beta \in (0, 1)$ are positive constants. This algorithm was first described in [17]. It can be shown that even in this restricted range of values of p, m the problem of finding a Hamilton cycle in $G_{n,m,p}$ is non-trivial. Before presenting the algorithm we give some preliminary results.

It is easy to see that if the $G_{n,m,p}$ has a Hamilton cycle, then there exists a sequence

$$HC := l_1 \to v_1 \to l_2 \to \cdots \to l_k \to v_k \to l_{k+1}(= l_1) \tag{2}$$

that satisfies the following four conditions:

1. $l_i \in M$ and $v_i \in V$,
2. $\bigcup_{i=1}^{k} L_{l_i} = V$,
3. $l_i, l_{i+1} \in S_{v_i}$, $i = 1, 2, \ldots, k$ and
4. $v_i \neq v_j$, for any $i \neq j$.

We note here that given an HC sequence of the form (2) that satisfies the four constraints above we can construct a Hamilton cycle in time $\Theta(n \cdot m)$ (we give such a construction in the algorithm that follows this section).

Suppose now that our graph has a Hamilton cycle hence there is a sequence HC of the form (2) satisfying the four constraints above. We will refer to the integer k as the *size* of the sequence HC. Let x denote the number of different labels used by HC and let HC_{min} be a sequence of the form (2) that satisfies the four conditions, uses exactly the labels used by HC and has minimum length. If we denote by k_{min} the size of HC_{min} then it is obvious that $x \leq k_{min}$. We can also prove the following:

Lemma 1. *If k_{min} and x are defined as above, then $k_{min} \leq 1 + \frac{x(x-1)}{2}$.*

Proof. Suppose that we start "following" the sequence HC_{min} and that at some point t we have (just) met $y \geq 1$ different labels. In order to meet the next label that is different than the y seen so far we will have to make at most y steps. This is true because if we make $y+1$ or more steps then we will have met some of the y labels at least twice (from point t and on). But this would mean that the sequence HC_{min} can be shortened, which is a contradiction because HC_{min} is of minimum size. So, the size of HC_{min} can be no more than $1+1+2+\cdots+(x-1) = 1+\frac{x(x-1)}{2}$. □

Suppose now that we are interested in the number of different sequences HC_{min} that use x specific labels. Because of Lemma 1 we know that their size must be at most $1 + \frac{x(x-1)}{2}$. We now suppose that we have to fill in $1 + \frac{x(x-1)}{2}$ label-cells by choosing for each one any of x different labels and $1 + \frac{x(x-1)}{2}$ vertex-cells by choosing for each any of n different vertices. In that way, sequences with size less than the maximum will correspond to some sequence of size $1 + \frac{x(x-1)}{2}$ that has at least one label that repeats itself in the following label-cell. We easily see then that the total number of ways to fill in these cells is an upper bound to the number of different HC_{min} sequences.

In view of the above, the number of sequences of the form (2) that an exhaustive algorithm that searches for a Hamilton cycle needs to check is at most

$$\sum_{x=1}^{m} \binom{m}{x} (x \cdot n)^{1+\frac{x(x-1)}{2}} \leq \sum_{x=1}^{m} \frac{m^x}{x!} (x \cdot n)^{\frac{x^2}{2}} \leq \exp\{m^2 \log n\} = n^{m^2}. \quad (3)$$

where in the last inequality we used the upper bound on m.

4.1   Expected Polynomial Time Algorithm for Constant p

In this section we use the previous results in order to present an algorithm for finding Hamilton Cycles in a $G_{n,m,p}$ graph with constant p and $m \leq \alpha \sqrt{\frac{n}{\log n}}$, where $\alpha < \frac{\beta p}{\sqrt{2}}$ and $\beta \in (0, 1)$ are positve constants.

The algorithm first uses a greedy algorithm (steps 1-10) to form a sequence HC of the form (2) that uses all the labels and if this fails (which happens with some small probability of failure) it runs the exhaustive algorithm (step 12) implied in the previous section. Both the randomized algorithm and the exhaustive algorithm try to find a sequence HC of the form (2) that satisfies all four constraints. If either of them succeeds in finding one, then they run the following procedure that constructs a Hamilton cycle.

procedure CONSTRUCT_HAM(HC)

1. let $HC := l_1 \rightarrow v_1 \rightarrow l_2 \rightarrow \cdots \rightarrow l_k \rightarrow v_k \rightarrow l_{k+1}(= l_1)$;
2. $i = 1$; $A = V \backslash \bigcup_{i=1}^{k} \{v_i\}$;
3. while $i \leq k$ do
4.   let D_i be any ordered list of $L_{l_i} \cap A$;
5. $A = A \backslash \{D_i\}$; $i = i + 1$;
6. output $D_1 \rightarrow v_1 \rightarrow D_2 \rightarrow \cdots D_k \rightarrow v_k$;

We now show that if the HC sequence satisfies all four constraints, then the output of procedure CONSTRUCT_HAM(HC) is a Hamilton cycle. First, we note that since HC satisfies the 2nd constraint, all vertices are contained at most once in the output. Second, because of the definition of the sets D_i, step 5 of the algorithm and the 4th condition, each vertex is contained exactly once in the output. Third, because of the definition of the sets D_i and the 3rd condition, every two consecutive vertices of the output have at least one label in common, hence they are connected. Finally, we close the Hamilton cycle by noting that v_k has at least one label in common with any vertex in D_1 (i.e. label l_1).

The expected polynomial time algorithm is shown below. We denote by L_y the set of vertices having chosen label $y \in M$.

Algorithm I:
Input: The representation matrix $R_{n,m,p}$ of a graph $G_{n,m,p}$.
Output: A Hamilton cycle of the graph corresponding to $R_{n,m,p}$.

1. let $l_1, l_2, \ldots, l_m$ be a *random* ordering of the labels;
2. consider the sequence $l_1, l_2, \ldots, l_m, l_{m+1} = l_1$;
3. if $\bigcup_{i=1}^{m} L_{l_i} \neq V$ then
4.   output "The graph has no Hamilton Cycle"; exit;
5. $i = 1$; $A = V$; $HC =$ empty list;
6. while $i < m + 1$ do
7.   if $L_{l_i} \cap L_{l_{i+1}} \cap A = \emptyset$ then goto L1;
8.   select a *random* vertex $v_i \in L_{l_i} \cap L_{l_{i+1}} \cap A$;

9.   set $HC = HC \rightarrow \{l_i\} \rightarrow \{v_i\}$;
10.   set $A = A \backslash \{v_i\}$; $i = i + 1$;
11. goto L2
12. L1: if there is a sequence of the form (2) and size $k \leq 1 + \frac{m(m-1)}{2}$ that satisfies all four costraints then
13.   let HC be such a sequence; goto L2;
14. else
15.   output "The graph has no Hamilton Cycle"; exit;
16. L2: CONSTRUCT_HAM(HC);

4.2   Analysis of Algorithm I

We will first find an upper bound for the probability of failure of the greedy part of the algorithm (lines 1-10). Notice that the greedy algorithm can fail to output a solution only because of step 7. That is, the algorithm fails only in the case where two consecutive labels in the sequence $l_1, l_2, \ldots, l_m, l_{m+1} = l_1$ do not have a common vertex that is also not used to "connect" two previous consecutive labels.

Let us now denote by X_{ij} the number of vertices having chosen both labels $i, j \in M$. Obviously, X_{ij} is a random variable that is binomially distributed with parameters n and p^2. So, $E[X_{ij}] = np^2$. But from Chernoff bounds we have that for any constant β in $(0, 1)$

$$P\{X_{ij} \leq (1 - \beta)np^2\} \leq e^{-\beta^2 p^2 n/2}$$

and by applying Boole's inequality

$$P\{\exists i, j \in M : X_{ij} \leq (1 - \beta)np^2\} \leq \binom{m}{2} e^{-\beta^2 p^2 n/2} \stackrel{def}{=} \phi.$$

We have thus proven that with probability at least $1 - \binom{m}{2} e^{-\beta^2 p^2 n/2}$ every pair of labels has $\Theta(n)$ vertices in common. Bearing in mind that the vertices used by the greedy algorithm are exactly $m = o(n)$, every pair of labels is left with at least $\Theta(n) - o(n) = \Theta(n)$ vertices in common. So, the probability that the greedy algorithm fails is at most $\binom{m}{2} e^{-\beta^2 p^2 n/2}$. Moreover, it is easy to see that the running time of the greedy algorithm is $O(n \cdot m)$ in the worst case.

If the greedy algorithm fails (which means that it does not output a Hamilton cycle nor does it output "The graph has no Hamilton Cycle"), then we run the exhaustive part of the algorithm (included in step 12). In order to check if a particular sequence of the form (2) and of size at most $1 + \frac{x(x-1)}{2}$ satisfies the four constraints we need $O(n \cdot m)$ time in the worst case.

Finally, the running time of the procedure CONSTRUCT_HAM is $O(n \cdot k)$ in the worst case, where k is the size of the HC sequence. If the HC sequence is the one produced by the greedy algorithm, then $k = m$, otherwise $k \leq 1 + \frac{m(m-1)}{2}$. Therefore, by using inequality (3), it is easy to see that

$$E[\text{running time of Algorithm I}] \leq (1 - \phi)O(n \cdot m) + \phi n^{m^2} O(n \cdot m)$$

which becomes $O(n \cdot m)$ for $m \leq \alpha\sqrt{\frac{n}{\log n}}$, where $\alpha < \frac{\beta p}{\sqrt{2}}$ is a constant. It follows then that Algorithm I runs in polynomial expected time.

5   A Second Greedy Algorithm for Smaller p

In this section we will see that the greedy approach still works quite well for smaller p, in the case where m is at most $o(\frac{n}{\log n})$. This was pointed out in [17]. More specifically, we will present an algorithm that with high probability finds a Hamilton cycle in a $G_{n,m,p}$ graph, in the case where $m - o(\frac{n}{\log n})$ and p is just above the connectivity threshold of $G_{n,m,p}$, i.e. $p = \frac{\log n + h(n)}{m}$, for any function $h(n)$ that goes to ∞ as slowly as possible (see [18]).

Suppose that we partition the set of vertices V into two sets V_1, V_2 of $\frac{n}{2}$ vertices each. We then use only vertices in V_1 to make an HL sequence of the form

$$HL := l_1 \rightarrow v_1 \rightarrow l_2 \rightarrow \cdots \rightarrow l_k \tag{4}$$

that is as long as possible and satisfies the following conditions:

1. $l_i \in M$ and $v_i \in V_1$,
2. $l_i, l_{i+1} \in S_{v_i}$, $i = 1, 2, \ldots, k - 1$,
3. $v_i \neq v_j$, for any $i \neq j$ and
4. $l_x \neq l_y$, for any $x \neq y$.

We use the following procedure in order to construct a sufficiently large HL sequence:

procedure MAKE_HL(V_1, M)

1. set $F_V = V_1$; set $F_M = M$;
2. select a *random* label $l \in M$; set $F_M = F_M \backslash \{l\}$; set $HL = l$;
3. L1: if $|L_l \cap F_V| < \log n$ then goto Lo;
4. else
5.   let $v_1, \ldots, v_{\log n} \in L_l \cap F_V$;
6.   if $\exists l', v : l' \in F_M, v \in \{v_1, \ldots, v_{\log n}\}$ and $l' \in S_v$ then
7.   set $F_M = F_M \backslash \{l'\}$; $F_V = F_V \backslash \{v_1, \ldots, v_{\log n}\}$;
8.   set $HL = HL \rightarrow v \rightarrow l'$; set $l = l'$; goto L1;
9.   else goto Lo;
10. Lo: output HL

If we assume that we can perform a random selection of an element in a set of n items in constant time, the above procedure runs obviously in polynomial time.

Indeed, the loop between steps 3 to 9 is executed at most $\frac{n}{2\log n}$ times because in step 7 we reduce the set F_V by $\log n$ vertices. Furthermore, the most expensive steps of the loop are the if-condition of step 3 which can take at most $n^2/4$ time steps (because F_V and F_M have at most $n/2$ vertices each) and the if-condition of step 6 which can take at most $m^2 \log n$ time steps (because F_M has at most m labels and each set S_v has at most m labels). Hence, taking into consideration that $m = o(\frac{n}{\log n})$, we have proved the following:

Lemma 2. *Procedure MAKE_HL terminates in at most $O(\frac{n^3}{\log n})$ time steps.*

As we will now see, procedure MAKE_HL produces with high probability an HL sequence of the form (4) that has the following additional properties:

1. The size of HL is at least $m - \frac{m}{\log n}$ and
2. $\bigcup_{i=1}^{k} l_i = V$.

We will prove these in that order. First we note that steps 6-9 of the procedure can be executed at most m times, since each time step 7 is executed, the set F_M loses one label. Hence, whenever step 3 is executed, the set F_V will always consist of at least $|V_1| - m \log n = \frac{n}{2} - o(n)$ vertices. Also, it is easy to see that the label l selected in step 6 of the procedure chooses each vertex in F_V independently with probability p. So, $|L_l|$ is stochastically greater than a random variable $X \sim \mathcal{B}(|V_1| - m \log n, p)$. Clearly, $E[X] = \mu = (|V_1| - m \log n)p \geq (\frac{n}{2} - m \log n)\frac{\log n}{m} = \Theta(\frac{n \log n}{2m})$. By Chernoff bounds and Boole's inequality we then have, for any constant β,

$$P\{\text{an execution of step 3 finds } |L_l \cap F_V| \leq (1 - \beta)\mu\} \leq m e^{-\beta^2 \mu/2} \to 0.$$

We have thus proved that with probability at least $1 - \exp\left\{-\frac{\beta^2 n \log n}{6m}\right\} \to 1$, the procedure never stops due to the if-condition of step 3.

The only other way to end the procedure is by step 6 (and as we said before the number of executions of step 7 is at most m). Suppose then that at some point of the execution of the procedure the HL sequence has been extended so as to contain $m - k$ labels (i.e. the test of step 6 has been passed $m - k - 1$ times), leaving exactly k "free" labels in F_M. It is true that these labels continue to select each member of F_V independently with probability p. This means that the probability that none of the free labels contains some of the vertices $v_1, \ldots, v_{\log n}$ of step 5 is exactly $(1 - p)^{k \log n}$. Now, by Boole's inequality, for $k = \frac{m}{\log n}$, the probability that some of the $m - k - 1$ first tests of step 6 fails is at most

$$m(1 - p)^{k \log n} \leq \exp\left\{\log m - pm\right\} \leq \exp\left\{\log m - \log n\right\} \to 0.$$

So, with probability $1 - \exp\left\{\log m - \log n\right\} \to 1$ the HL sequence of procedure MAKE_HL(V_1, M) has size at least $m - \frac{m}{\log n}$.

Given now that the size of the HL sequence is at least $m - \frac{m}{\log n}$ we can see that the mean number of vertices not covered by (i.e. not contained in) any label of HL is at most

$$E[\# \text{ vertices not covered by } HL] \leq n(1-p)^{m - \frac{m}{\log n}}$$
$$\leq \exp\left\{\log n - mp + \frac{mp}{\log n}\right\} \to 0$$

for the values of m, p we have assumed. So, by Markov's inequality, the HL sequence of procedure MAKE_HL(V_1, M) covers all the vertices in V with probability at least $1 - \exp\left\{-h(n) + 1 + \frac{h(n)}{\log n}\right\} \to 1$. We have thus proved the following:

Theorem 2 ([17]). *With probability at least*

$$1 - \exp\left\{-\frac{\beta^2 n \log n}{6m}\right\} - \exp\left\{\log m - \log n\right\} - \exp\left\{-h(n) + 1 + \frac{h(n)}{\log n}\right\} \to 1$$

procedure MAKE_HL(V_1, M) constructs an HL sequence of the form (4) that covers all the vertices of the graph.

We now notice that if we manage to "close" the sequence HL in order to get a sequence HC of the form (2), then we can run CONSTRUCT_HAM(HC) to get a Hamilton cycle. This is where we use the set of vertices V_2. Suppose that we run MAKE_HL(V_2, M) but instead of choosing a random label to begin with at step 2, we select label l_1 of the HL sequence constructed by MAKE_HL(V_1, M). By symmetry, a theorem similar to Theorem 2 is also true in this case. This means that all the vertices of V will be covered by the HL' sequence constructed by MAKE_HL(V_2, M). Also, by Chernoff bounds we can see that if l_f is the last label of HL, then

$$P\{|L_l \cap V_2| \leq (1 - \beta)\frac{np}{2}\} \leq e^{-\beta^2 np/4} \to 0$$

for any constant β. Hence, label l has almost definitely some vertices in V_2.

Suppose then that at some point of the execution of MAKE_HL(V_2, M) HL' has the form

$$HL' := l_1 \to v'_1 \to \cdots \to l'_k$$

and that $L_{l_f} \cap L_{l'_k} \cap F_V \neq \emptyset$ (because of Theorem 2 such a point always exists with high probability). Then, if $v \in L_{l_f} \cap L_{l'_k} \cap F_V$, we can stop the procedure and set

$$HC := HL \to v \to l'_k \to \cdots \to v'_1 \to l_1.$$

Then it is easy to see that the sequence HC is of the form (2) and satisfies all four conditions specified in Section 4 so that we may run CONSTRUCT_HAM(HC) to get a Hamilton cycle.

Consequently, we have shown a polynomial (due to Lemma 2) greedy way of finding a Hamilton cycle in a $G_{n,m,p}$ graph with probability at least $1 - 2\exp\left\{-h(n) + 1 + \frac{h(n)}{\log n}\right\} \to 1$.

6   Hamiltonicity in $\mathcal{G}_{n,m,\lambda}$

We now turn to the hamiltonicity of uniform random intersection graphs. In this section, we present a result in [12], which states that $G_{n,m,\lambda}$ has a Hamiltonian cycle (i.e. a cycle including all n vertices) with high probability (whp) when the number of vertices is large enough, even for a very small constant $\lambda \geq 2$.

Theorem 3 ([12]). *Let* $e_0 = \binom{m}{\lambda}$. *If* $n \geq (1+\epsilon)e_0 \ln e_0$, *for some constant* $\epsilon > 0$ *as small as possible, then* $G_{n,m,\lambda}$, *with* $\lambda \geq 2$, *is Hamiltonian whp as* $n \to \infty$.

Proof. For simplicity, we will refer to a set of λ labels as an *Element*. The total number of possible Elements in $G_{n,m,\lambda}$ is then obviously e_0. Let $\mathcal{E}_e$ be the event that no vertex chooses e. By independence,

$$\Pr(\mathcal{E}_e) = \left(1 - \frac{1}{e_0}\right)^n.$$

Let X denote the mean number of Elements not chosen by any vertex. Then

$$E[X] = e_0 \Pr(\mathcal{E}_e) \leq e^{\ln e_0 - \frac{n}{e_0}} \to 0$$

for any $n \geq (1 + \epsilon)e_0 \ln e_0$. Hence, the vertices of $G_{n,m,\lambda}$ will have chosen *all* available Elements (choosing exactly 1 Element each) whp.

The proof is completed by showing in the following how this implies the existence of a Hamiltonian cycle in the case $\lambda \geq 2$. Consider an arbitrary ordering of the labels of the graph $\{l_1, l_2, \ldots, l_m\}$ and construct the sets $D_1, D_2, \ldots, D_m$, where $D_i = \{v \in V : l_i \in S_v \text{ and } l_j \notin S_v, \text{ for all } j \leq i - 1\}$, i.e. D_i is the set of vertices that have chosen label l_i and none of the labels $l_1, \ldots, l_{i-1}$. We now establish two properties that these sets have.

1. First of all, note that since the vertices of $G_{n,m,\lambda}$ have chosen every available Element, we will have that the only empty sets will be all D_i with $i = m-\lambda+2, \ldots, m$. Indeed, for all $i \leq m-\lambda+1$, there will be at least one vertex u that has $i \in S_u$ and $S_u \subseteq \{l_i, \ldots l_m\}$. Also, since every vertex chooses exactly λ distinct labels, every vertex that has chosen l_i, for $i = m - \lambda + 2, \ldots, m$, must belong to *exactly one* of $D_1, \ldots, D_{m-\lambda+1}$.

2. Second, note that by construction (of the D_is), and because of the fact that the vertices of $G_{n,m,\lambda}$ have chosen every available Element, there will be at least one edge between D_i and D_j, for all $i = 1, \ldots, m - \lambda$ and all $j = i + 1, \ldots, m - \lambda + 1$. Also, for *every* edge $\{x_i, y_{i+1}\}$ between D_i and D_{i+1}, $i = 1, \ldots, m - \lambda - 1$, there is an edge $\{x_{i+1}, y_{i+2}\}$ between D_{i+1} and D_{i+2} that satisfies $\{x_i, y_{i+1}\} \cap \{x_{i+1}, y_{i+2}\} = \emptyset$, unless $|D_{i+1}| = 1$, where

$y_{i+1} \equiv x_{j+1}$. Indeed, this is a consequence of the fact that the vertices of $G_{n,m,\lambda}$ have chosen every available Element (i.e. every combination of λ labels) whp. Finally, *all* edges $\{x_j, y_{j+1}\}$ between D_j and D_{j+1}, for *every* $j = i+2, \ldots, m - \lambda$, satisfy $\{x_i, y_{i+1}\} \cap \{x_j, y_{j+1}\} = \emptyset$, by the construction of the sets D_i.

These two properties allow us to fix a sequence of pairs $\{x_i, y_{i+1}\}$, for all $i = 1, 2, \ldots, m - \lambda$, that are disjoint, except for the case where some $|D_i| = 1$, which does not change our proof. As a final step, let y_1 be a vertex that satisfies $\{l_1, l_{m-\lambda+1}\} \subseteq S_{y_1}$, and $S_{y_1} \setminus \{l_1, l_{m-\lambda+1}\} \subseteq \{l_{m-\lambda+2}, \ldots, l_m\}$. Such a vertex exists whp and it is easy to see that it is connected to all vertices in $D_{m-\lambda+1}$.

Let now σ_i, $i = 1, \ldots, m - \lambda$ be an arbitrary ordering or the set D_i, that begins with y_i and ends with x_i. Also, let $\sigma_{m-\lambda+1}$ be an arbitrary ordering or the set $D_{m-\lambda+1}$, that begins with $y_{m-\lambda+1}$. Since every D_i is a clique, it is easy to verify that the sequence $\sigma_1 \sigma_2 \cdots \sigma_{m-\lambda} \sigma_{m-\lambda+1}$ is indeed a Hamiltonian cycle.

□

Note here that $\lambda = 2$ is in fact as small as one can have in order to achieve Hamiltonicity. Indeed, for $\lambda = 1$ the graph is disconnected (see also [2]). In this sense, the above result is optimal. Finally, note that the proof of Theorem 3 leads naturally to a randomized polynomial time (in terms of n and m) algorithm for constructing Hamiltonian cycles whp in this case.

7   Conclusions

In this work, we presented some existing results concerning the existence and efficient construction of Hamilton cycles in random intersection graphs and uniform random intersection graphs. There are several other open algorithmic and combinatorial problems concerning these two models. To name just a few, very little is known about the size of the minimum dominating set of $G_{n,m,p}$, let alone the existence of efficient algorithms that can approximate it. Furthermore, the problems of approximating the clique number or that of finding a large planted clique are also not yet investigated in random intersection graphs.

References

1. Alon, N., Spencer, J.H.: The probabilistic method. John Wiley & Sons, Inc., New York (2000)
2. Blackburn, S.R., Gerke, S.: Connectivity of the uniform random intersection graph, arXiv:0805.2814v2 [math.CO]
3. Blonzelis, M.: Degree Distribution of a Typical Vertex in a General Random Intersection Graph. Lithuanian Math. J. 48(1), 38–45 (2008)
4. Bollobás, B.: Random Graphs, 2nd edn. Cambridge University Press, Cambridge (2001)
5. Bradonjić, M., Hagberg, A., Hengartner, N.W., Percus, A.G.: Component evolution in general random intersection graphs. In: Kumar, R., Sivakumar, D. (eds.) WAW 2010. LNCS, vol. 6516, pp. 36–49. Springer, Heidelberg (2010)

6. Cooper, C., Frieze, A.: The cover time of sparse random graphs. Random Struct. Algorithms 30, 1–16 (2007)
7. Deijfen, M., Kets, W.: Random intersection graphs with tunable degree distribution and clustering,
   http://citeseerx.ist.psu.edu/viewdoc/summary?doi=10.1.1.159.3535
8. Efthymiou, C., Spirakis, P.G.: On the existence of Hamilton cycles in random intersection graphs. In: Caires, L., Italiano, G.F., Monteiro, L., Palamidessi, C., Yung, M. (eds.) ICALP 2005. LNCS, vol. 3580, pp. 690–701. Springer, Heidelberg (2005)
9. Fill, J.A., Sheinerman, E.R., Singer-Cohen, K.B.: Random intersection graphs when $m = \omega(n)$: an equivalence theorem relating the evolution of the $G(n, m, p)$ and $G(n, p)$ models. Random Struct. Algorithms 16(2), 156–176 (2000)
10. Godehardt, E., Jaworski, J.: Two models of random intersection graphs for classification. In: Opitz, O., Schwaiger, M. (eds.) Exploratory Data Analysis in Empirical Research. LNCS, pp. 67–82 (2002)
11. Karoński, M., Scheinerman, E.R., Singer-Cohen, K.B.: On random intersection graphs: the subgraph problem. Combinatorics, Probability & Computing 8, 131–159 (1999)
12. Nikoletseas, S., Raptopoulos, C., Spirakis, P.G.: Combinatorial properties for efficient communication in distributed networks with local interactions. In: 23rd IEEE International Symposium on Parallel and Distributed Processing, pp. 1–11. IEEE Press, New York (2009)
13. Nikoletseas, S., Raptopoulos, C., Spirakis, P.G.: Large independent sets in general random intersection graphs. Theor. Comput. Sci. 406(3), 215–224 (2008)
14. Nikoletseas, S., Raptopoulos, C., Spirakis, P.G.: Expander properties and the cover time of symmetric random intersection graphs. In: Kučera, L., Kučera, A. (eds.) MFCS 2007. LNCS, vol. 4708, pp. 44–55. Springer, Heidelberg (2007)
15. Penrose, M.: Random geometric graphs. Oxford Studies in Probability (2003)
16. Di Pietro, R., Mancini, L.V., Mei, A., Panconesi, A., Radhakrishnan, J.: Sensor networks that are provably resilient. In: 2nd International Conference on Security and Privacy in Communication Networks, pp. 1–10. IEEE Press, New York (2009)
17. Raptopoulos, C., Spirakis, P.G.: Simple and efficient greedy algorithms for Hamilton cycles in random intersection graphs. In: Deng, X., Du, D.-Z. (eds.) ISAAC 2005. LNCS, vol. 3827, pp. 493–504. Springer, Heidelberg (2005)
18. Singer-Cohen, K.B.: Random intersection graphs. PhD thesis, John Hopkins University (1995)
19. Stark, D.: The vertex degree distribution of random intersection graphs. Random Struct. Algorithms 24(3), 249–258 (2004)

Author Index